SPACE SCIENCE, EXPLORATION AND POLICIES SERIES

BLACK HOLES AND GALAXY FORMATION

Space Science, Exploration and Policies Series

Progress in Dark Matter Research
J. Val Blain (Editor)
2005. ISBN 1-59454-248-1

Space Science: New Research
Nick S. Maravell (Editor)
2006. ISBN 1-60021-005-8

Space Policy and Exploration
William N. Callmers (Editor)
2008. ISBN 978-1-60456-448-8

Space Commercialization and the Development of Space Law from a Chinese Legal Perspective
Yun Zhao
2009. ISBN 978-1-60692-244-6

Next Generation of Human Space Flight Systems
Alfred T. Chesley (Editor)
2009. ISBN 978-1-60692-726-7

Smaller Satellites Operations near Geostationary Orbit
Matthew T. Erdner
2009. ISBN 978-1-60741-181-9

Environmental Satellites: Weather and Environmental Information Systems
Vincent L. Webber (Editor)
2009. ISBN 978-1-60692-984-1

Nutritional Biochemistry of Space Flight
Scott M. Smith, Sara R. Zwart, Vickie Kloeris and Martina Heer
2009. ISBN 978-1-60741-641-8

Space Exploration Research
John H. Denis and Paul D. Aldridge (Editors)
2009. ISBN: 978-1-60692-264-4

Black Holes and Galaxy Formation
Adonis D. Wachter and Raphael J. Propst (Editors)
2010. ISBN: 978-1-60741-703-3

SPACE SCIENCE, EXPLORATION AND POLICIES SERIES

BLACK HOLES AND GALAXY FORMATION

ADONIS D. WACHTER
AND
RAPHAEL J. PROPST
EDITORS

Nova Science Publishers, Inc.
New York

For permission to use material from this book please contact us:
Telephone 631-231-7269; Fax 631-231-8175
Web Site: http://www.novapublishers.com

LIBRARY OF CONGRESS CATALOGING-IN-PUBLICATION DATA

Black holes and galaxy formation / editors, Adonis D. Wachter and Raphael J. Propst.
p. cm.
Includes index.
ISBN 978-1-60741-703-3 (hardcover)
1. Black holes (Astronomy) 2. Galaxies--Formation. I. Wachter, Adonis D. II. Propst, Raphael J.
QB843.B55B584 2009
523.8'875--dc22

2009025205

Published by Nova Science Publishers, Inc. ✦ New York

CONTENTS

PREFACE

Galaxies are the basic unit of cosmology. The study of galaxy formation is concerned with the processes that formed a heterogeneous universe from a homogeneous beginning. The physics of galaxy formation is complicated because it deals with the dynamics of stars, thermodynamics of gas and energy production of stars. A black hole is a massive object whose gravitational field is so intense that it prevents any form of matter or radiation to escape. It is hypothesized that the most massive galaxies in the universe—"elliptical galaxies"—grow simultaneously with the supermassive black holes at their centers, giving us much stronger evidence that black holes control galaxy formation. This book reviews new evidence in the field.

In Chapter 1, the authors examine cosmological time series from the background gravitational waves presumably generated within one-trillionth of a second after the Big Bang up to the present, and develop a causal theory of cosmological evolution. The temporal distribution of galaxies and of quasars (Las Campanas, 2dF, and Sloan surveys), as well as background gravitational waves (recordings from LIGO), show a pattern (Bios) of aperiodic oscillations with features that demonstrate causality and diversification (increase in variance with time or dimension), novelty (less recurrence than randomized copies), and temporal complexity (transformation of pattern), indicating creativity. Biotic patterns are also generated by quantum processes as described by the Schrödinger wave function three different distance scales (atomic, nuclear and Planck, time and distance scale obtaining shortly after the Big Bang) and its relativistic version, the Klein-Gordon-Fock equation but only when applied to entities with sufficient mass to slow down their velocity below the speed of light. Bios is also found in more complex biological and economic processes. Biotic patterns can be generated by recursions of sinusoidal functions such as those embodied by electromagnetic energy, which include a conservation term, such as the process equation $A(t+1) = A(t) + \sin(A(t))$, but not by random processes. Sinusoidal functions are embodied by electromagnetic energy and matter is conserved and slows down velocity. Novelty (but not Bios) is demonstrated in the Cosmic Background Radiation constituted by massless photons. The authors thus propose that cosmic evolution is a creative process initiated by causal action, bipolar interactions, and conservation of material structure at the quantum level that create diversification, novelty, and complexity (Bios), thereby initiating cosmological evolution and generating fractal biotic patterns of organization at multiple levels of organization. Instead of random change, the simplest original processes are asymmetry (Lattice Theory), symmetry of opposites (Group Theory) and continuous transformation (Topology) of tridimensional matter which are embodied in action, bipolarity (electromagnetic waves), and tripolar nuclear forces that form matter. Lattice, Group and Topology are the foundations of mathematics (Bourbaki), intellectual development (Piaget) and, we propose, cosmological evolution.

As explained in Chapter 2, the standard model of gravitational structure formation is based on the Jeans 1902 acoustic theory, neglecting nonlinear instabilities controlled by viscosity, turbulence and diffusion. A linear instability length scale equal to the sound speed times the gravitational freefall time emerges. Because the Jeans scale L_J for the hot primordial plasma is always slightly larger than the scale of causal connection *ct*, where *c* is the speed of light and *t* is the time after the big bang, it has been assumed that no plasma structures could form without guidance from a cold (so $L_{J\ CDM}$ is small) collisionless cold-dark-matter CDM fluid to give condensations that gravitationally collect the plasma. Galaxies by this CDM model are therefore produced by gradual hierarchical-clustering of CDM halos to galaxy mass over billions of years, contrary to observations showing that well formed galaxies existed much earlier. No observations exist of CDM halos. However, gravitational instability is non-linear and absolute, controlled by viscous and turbulent forces or by diffusivity at Schwarz length scales smaller than *ct*. Because the universe during the plasma epoch is rapidly expanding, the first structures formed were at density minima by fragmentation when the viscous-gravitional scale L_{SV} first matched *ct* at 30,000 years to produce protosupercluster voids and protosuperclusters. Weak turbulence produced at expanding void boundaries guide the morphology of smaller fragments down to protogalaxy size just before transition to gas at 300,000 years. The size of the protogalaxies reflect the plasma Kolmogorov scale with a linear and spiral morphology predicted by the Nomura direct numerical simulations and confirmed by deep field space telescope observations. On transition to gas the kinematic viscosity decreases by ten trillion so the protogalaxies fragment at Jeans scale clouds, each with a trillion earth-mass planets, as predicted by Gibson 1996 and observed by Schild 1996. The planets promptly form stars near the cores of the protogalaxies, but as the universe cools the planets gradually freeze to form the galaxy dark matter. Space telescope observations of the most distant galaxies confirm the Kolmogorov scale size and Nomura geometry of the protogalaxies. High resolution images of planetary nebula and supernova remnants reveal thousands of frozen hydrogen-helium dark matter planets with large atmospheres evaporated by such events. Galaxy mergers show frictional trails of young globular clusters formed in place, proving that dark matter halos of galaxies consist of dark matter planets in metastable clumps. Galaxy evolution is guided by friction from this galaxy dark matter.

In Chapter 3, a dynamical model is presented for the study of the behavior of orbits in a binary quasar. The two quasars are hosted in a pair of interacting disk galaxies. Two cases are investigated: the time independent model and the evolving model. The author's numerical experiments show that a large fraction of orbits are chaotic in the time independent case. A careful analysis suggests that several Lindblad resonances are responsible for the chaotic motion of stars in both host galaxies. In the time dependent case, the author follows the evolution of the model as one of the two interacting galaxies develops a quasar, by mass transport from the disk to the nucleus. In this interesting case, there are orbits that change character from regular to chaotic. This result strongly indicates that the regular or chaotic nature of orbits depends not only in the galactic interaction but also in the presence of the quasar. The outcomes derived from the author's dynamical model are compared with data available from observations.

As discussed in Chapter 4, black holes and other exotica might be produced at the forthcoming Large Hadron Collider if the fundamental scale of gravity is near the TeV scale . Simulations of black holes and other proposed physics beyond the Standard Model, supersymmetry and string theory, are presented here. Black hole events are simulated using the CATFISH Monte Carlo generator, simulations of string resonances use PYTHIA and supersymmetric simulations use a combination of ISAJET and PYTHIA. Our analysis shows that black holes can be discriminated from supersymmetry and string resonances can be distinguished from Standard Model processes and black hole events. In particular, isolated high transverse momentum leptons allow discrimination between black holes and supersymmetry. Black holes and string resonances can be distinguished using Z^0 bosons and photons with high transverse momentum. Additional discrimination is provided by the analysis of energy, momenta and event-shape variables.

Research work conducted in the early 1970s has shown that density fluctuation in the early universe leads inevitably, under certain circumstances, to the formation of primordial black holes (PBH). The key idea is that a black hole is formed in overdense regions when the primordial fluctuations cause the gravitational energy to considerably exceed the kinetic energy of the expanding region. Alternative theories of the formation of primordial black holes assume that this process occurs during the inflation epoch, when the universe experienced a very short period of very rapid expansion.

A review of primordial black hole formation, evolution, detection, and identification from gamma ray burst signature is discussed in Chapter 5.

In Chapter 6, the author uses a simple heuristic argument based on the conception that Planck length measures the upper resolution limit for all objects and use the quantum interest conjecture to show that the information paradox cannot be solved by the Susskind principle of black hole complementarity, and the simplest presuppositions of multiversal cosmology and the conservation laws for dark energy accretion onto black and worm-holes to see that the recent Hawking conjecture to recover information conservation can no longer work. The author's conclusion therefore is that the information paradox is not solved and that, much like a vacuum, is able to create energy pairs satisfying the Heisenberg principle, quantum gravity may create a quantum vacuum which can be able to apparently annihilate information, asking who is afraid of that possibly expectable loss. Some comments are also added to the connection between the information paradox and the origin of life in the context of the quantum multiverse.

In Chapter 7, the authors present a detailed study about the black hole evolution. They focus on the dynamical time-dependent process which requires the use of numerical techniques. For the sake of clarity they restrict their analysis to a spherically symmetric space-time. The authors set the construction of the solutions as an initial value problem consisting of a 3+1 decomposition of Einstein's equations under the BSSN formulation. The authors also show the case of a growing black hole by accretion. The type of matter accreted by the black hole is a massless scalar field minimally coupled to gravity. Among the techniques needed to achieve a dynamical Black Hole evolution, the authors present the gauge election, initial data and boundary conditions. The analysis is restricted to the geometrical aspects of the process including the following diagnostics: apparent horizon and event horizon tracking, causal structure and gauge dependence. The content of this chapter allows the understanding of the tools currently applied to more general space-times in numerical relativity like the binary black hole case.

Chapter 8 reviews the problem of the spontaneous loss of electric charge by non-rotating black holes with a cosmological constant. The pair-creation process which is allowed in the presence of a charged black hole and causes its discharge is a long-standing topic in the framework of quantum effects on a black-hole background, at least as old as the Hawking effect but still different in its origin. In order to study this process, a very effective method is represented by the Damour-Deruelle-Ruffini approach, which identifies the existence of level-crossing between positive and negative continuous energy states for charged matter fields on the given background as a signal of the quantum instability leading to the discharge of the black hole. After a review of this approach, the authors apply it to the case of charged non-rotating black holes in the presence of a non-vanishing cosmological constant Λ, and they take into account both the case of Λ negative (Anti-de-Sitter background) and the case of Λ positive (de-Sitter background). The analysis for $\Lambda < 0$ is more difficult, because of the boundary-like behavior of infinity in a AdS background. Level-crossing is shown to be allowed but not ensured, as happens also when $\Lambda = 0$. When Λ is positive, a novel feature appears, indeed level-crossing always occurs, due to the relevant and peculiar presence of two event horizons; nevertheless a significant effect of pair creation is expected only when the potential barrier between the positive and negative energy states is as narrow as possible. For $\Lambda > 0$ also the Nariai geometry and the ultracold cases are considered.

Gauss-Bonnet gravity provides one of the most promising frameworks to study curvature corrections to the Einstein action, for example, in supersymmetric string theories, because of avoiding ghosts and keeping second order field equations. The original properties of Gauss-Bonnet black holes—i.e., the presence of dilatonic hair, down limit on minimal black hole mass, and new types of space-time singularities—are reviewed in N = 4 dimensions. As the endpoint of black-hole evaporation is a very intriguing problem of modem physics, the author shows (based on the Einstein-dilaton-Gauss-Bonnet four-dimensional string gravity model) that black holes do not disappear completely and should become relics at the end of the evaporation process, so that the standard Bekenstein-Hawking evaporation formula must be modified near the last stages. This solution could represent a model for relic remnants of primordial black holes. Higher order curvature corrections also play an important role in braneworld gravity models; therefore, the author discusses the forms of black hole solutions in N > 4 dimension gravity models. The possibility of direct and indirect experimental detection of outlined black holes in astrophysics, cosmology and a new generation of colliders are discussed in Chapter 9 as well.

In Chapter 10, the authors discuss the X-ray radiation from relativistic accretion disks around supermassive black holes, supposed to exist in the centers of Active Galactic Nuclei (AGN). The focus is on the X-ray radiation, especially in the Fe K α line which originates in the innermost parts of an accretion disk. Moreover, here the authors discuss some effects which can disturb the Fe K α profile and cause its rapid and irregular variability, observed in the X-ray spectra of some AGN. The authors give attention to three such effects: perturbations in the disk emissivity, absorbtion by warm absorbers and gravitational microlensing. The X-ray emission from accretion disks around non-rotating (Schwarzschild metric), as well as rotating (Kerr metric) supermassive black holes, is discussed. The X-ray radiation of AGN is probably produced in a compact region near their central supermassive black holes, and can provide some essential information about the plasma conditions and the space-time geometry in these regions. The goal of this chapter is mainly to present a short overview of some important and recent investigations in this field.

A common consequence of accretion onto black holes is the formation of powerful, relativistic jets that escape the system. In the case of supermassive black holes at the centres of galaxies, this has been known for decades, but for stellar-mass black holes residing within galaxies like our own, it has taken recent advances to arrive at this conclusion. In Chapter 11, a review is given of the evidence that supports the existence of jets from accreting stellar-mass black holes, from observations made at optical and infrared wavelengths. In particular, it is found that, on occasion, jets can dominate the emission of these systems at these wavelengths. In addition, the interactions between the jets and the surrounding matter produce optical and infrared emission on large scales via thermal and non-thermal processes. The evidence, implications and applications in the context of jet physics are discussed. It is shown that many properties of the jets can be constrained from these studies, including the total kinetic power they contain. The main conclusion is that, like the supermassive black holes, the jet kinetic power of accreting stellar-mass black holes is sometimes comparable to their bolometric radiative luminosity. Future studies can test ubiquities in jet properties between objects, and attempt to unify the properties of jets from all observable accreting black holes, i.e., of all masses.

In recent years, Hawking radiation of black holes has been reinterpreted by Robinson and Wilczek et al. through the method of anomaly cancellation of quantum fields, and such a method has been used in the derivation of the Hawking radiation of many different types of black holes and some other black-hole-like objects with event horizons. In Chapter 12, the author studies the Hawking radiation of rotating D3-brane which appeared in superstring theories using the method of anomaly cancellation. For the sake of briefness, the author studies only the rotating D3-brane with one nonzero rotating parameter. The author first studies the effective field theory of quantum fields near the horizon of the rotating D3-brane and finds that it is equal to a two-dimensional field theory in a curved background. Then, the author calculates the energy-momentum flux of the rotating D3-brane via the approach of anomaly cancellation of gravitational fields, and finds that the energy-momentum flux of the rotating D3-brane derived from the method of anomaly cancellation forms a thermal radiation, and the thermal temperature of the radiation matches the Hawking temperature of the rotating D3-brane obtained from the black brane thermodynamics. Thus, the author can derive the Hawking radiation of the rotating D3-brane from the method of anomaly cancellation.

In Chapter 13, the author examines the effective four-dimensional world that emanates from a general class of static spherical Ricci-flat solutions in Kaluza-Klein gravity in Ddimensions. By means of dimensional reduction we obtain a family of asymptotically flat Schwarzschild-like metrics for which all the components of the Ricci tensor, except for R_{11}, are zero. Although the reduced spacetime is not empty, it is similar to a vacuum in the sense that the effective matter satisfies an equation of state which is the generalization of $(\rho + 3p) = 0$ for "nongravitating matter" in 4D. In Kaluza-Klein gravity these Schwarzschild-like metrics describe the exterior of a spherical star without rotation. In this framework, the author generalizes the well-known Buchdahl's theorem for perfect fluid spheres whose mass density does not increase outward. Without any additional assumptions, the author develops the most general expression for the compactness limit of a star and provides some numerical values for it, which in principle are observationally testable and allow comparison and contrast with different theories and exteriors. The author finds that in Kaluza-Klein gravity the compactness limit of a star can be larger than 1/2, without being a black hole: the general-relativistic upper limit $M/R < 4/9$ is increased as one goes away from the Schwarzschild vacuum exterior. The author shows how this limit depends on the number of dimensions of spacetime, and demonstrates that the effects of gravity are stronger in $4D$ than in any other number of dimensions.

As discussed in Chapter 14, models of observations from our solar system scale to the galaxy cluster scale are only moderately successful. Such observations are the pioneer anomaly; mid-range, flat rotation curves (RCs); asymmetric RCs; the spiral galaxy central mass and velocity dispersion correlation with outer galaxy parameters; and galaxy structure. Each model describes only one type of observation. The Scalar Potential Model (SPM) posits the existence of a plenum that influences matter and that is influenced by matter. The plenum obeys the heat equation, is sourced at the center of spiral galaxies, and is sunk in elliptical galaxies. The SPM is consistent with all the types of observation listed above, with rising and falling RCs in addition to flat RCs, and with inner-range and outer-range RCs. The SPM suggests a shift in fundamental principles and suggests a new model for the structure, formation and evolution of galaxies. Qualitative speculations about models of QSOs; of black holes; of clouds of shocked, outflowing hydrogen near the center of spiral galaxies; of the relationship of galaxy luminosity to many galaxy parameters; of the cause of the X-rays emanating from the center of spiral galaxies; of why the mid-range HI and H α RCs differ; of why inner-range RCs of different trace materials differ; of the cause of disks in spiral galaxies; of the cause of warps in the disks; of cooling flows; and of why radio galaxies are preferentially near elliptical galaxies follow from the SPM.

In: Black Holes and Galaxy Formation
Editors: A.D. Wachter and R.J. Propst, pp. 1-86
ISBN: 978-1-60741-703-3

Chapter 1

BIOTIC DYNAMICS OF GALACTIC DISTRIBUTION, GRAVITATIONAL WAVES, AND QUANTUM PROCESSES. A CAUSAL THEORY OF COSMOLOGICAL EVOLUTION

H. Sabelli, G. Thomas, L. Kovacevic, A. Lawandow and D. Horan
Chicago Center for Creative Development, Milwaukee School of Engineering, and University of Chicago

Abstract

Here we examine cosmological time series from the background gravitational waves presumably generated within one trillionths of a second after the Big Bang up to the present, and develop a causal theory of cosmological evolution. The temporal distribution of galaxies and of quasars (Las Campanas, 2dF, and Sloan surveys), as well as background gravitational waves (recordings from LIGO), show a pattern (Bios) of aperiodic oscillations with features that demonstrate causality and diversification (increase in variance with time or dimension), novelty (less recurrence than randomized copies), and temporal complexity (transformation of pattern) , indicating creativity. Biotic patterns are also generated by quantum processes as described by the Schrödinger wave function three different distance scales (atomic, nuclear and Planck, time and distance scale obtaining shortly after the Big Bang) and its relativistic version, the Klein-Gordon-Fock equation but only when applied to entities with sufficient mass to slow down their velocity below the speed of light. Bios is also found in more complex biological and economic processes. Biotic patterns can be generated by recursions of sinusoidal functions such as those embodied by electromagnetic energy, which include a conservation term, such as the process equation $A(t+1) = A(t) + \sin(A(t))$, but not by random processes. Sinusoidal functions are embodied by electromagnetic energy and matter is conserved and slows down velocity. Novelty (but not Bios) is demonstrated in the Cosmic Background Radiation constituted by massless photons. We thus propose that cosmic evolution is a creative process initiated by causal action, bipolar interactions, and conservation of material structure at the quantum level that create diversification, novelty, and complexity (Bios), thereby initiating cosmological evolution and generating fractal biotic patterns of organization at multiple levels of organization. Instead of random change, the simplest original processes are asymmetry (Lattice Theory), symmetry of opposites (Group Theory) and continuous transformation (Topology) of tridimensional matter which are embodied in action, bipolarity (electromagnetic waves), and tripolar nuclear forces that form matter. Lattice, Group and Topology are the foundations of mathematics (Bourbaki), intellectual development (Piaget) and, we propose, cosmological evolution.

Key Words: Bios, cosmic background radiation, cosmological evolution, galaxies, gravitational waves, quasars, quantum equations.

Introduction

The stable models of the universe that dominated discourse from Plato to the early twentieth century have been empirically refuted by the demonstration of the expansion of the universe by American physicist Hubble [1937]. Galaxies within the same cluster do not recede from each other, and they may even approach each other. Expansion is global; gravitation predominates locally.

Both gravitation and expansion are clearly directed, yet variations in the distribution of galaxies are regarded as fluctuations within a stationary random process. Stationary randomness does not explain why the universe is globally expanding, possibly at an increasing rate.

Equilibrium and randomness are often regarded as natural. It is often assumed that the universe was uniform [Springel et al., 2006] and in thermal equilibrium at its origin. The standard cosmological model assumes that the universe is homogenous and isotropic. The overall distribution of matter is regarded as a random stationary process [Peebles, 1993; Mandelbrot, 1975], and the observed hierarchy of galactic clustering is attributed to magnification of initial deviations from uniformity.

The twentieth century established cosmological evolution (Friedmann, Lemaitre, Hubble, Hoyle). There is, however, no theory that explains these processes of creative evolution. Hence change and evolution are attributed to random variation. Here we study cosmological evolution by analyzing time series of observable phenomena (Figure 1) with novel methods designed to measure creativity and non-random complexity, and to distinguish causal from random changes.

The finite velocity of light convolutes space and time in the direction away from our point of observation (the z-axis) when considering galactic scales. As the distances between galaxies and quasars are enormous, their distribution along the z-axis that represents space-time allows one to observe directly cosmological evolution by analyzing the distribution of galaxies and quasars. We have thus examined the distribution of galaxies (recorded by three recent surveys) as a function of the age of the universe.

The observation of the Cosmic Background Radiation (CBR) allows one to consider earlier cosmological times, 380,000 years after the 'Big Bang, ' when the first atoms formed. The study of gravitational waves, as recorded by the Laser Interferometer Gravitational-Wave Observatory (LIGO) may provide a way to study even earlier cosmological processes; the gravitational wave background (GWB) dates from the trillionth-of-a-second after the 'Big Bang.'

The earlier processes in what Lemaître called 'the primeval atom' are assumed to be quantum phenomena that presumably started the ongoing evolution of the universe. We thus study quantum processes with the same time series analyses. At the other end of evolution, the present, we can observe a planetary physical process, such as temperature.

The study of these very different processes reveals the existence of common features of pattern, namely increasing diversity, novelty, temporal complexity, and evidence for causal generation. This pattern has been previously discovered in biological processes and named Bios [Kauffman and Sabelli, 1998; Sabelli, 2005].

Change is universal. Spontaneous, apparently erratic fluctuations are observed at all levels of organization. The evolving universe displays structure at multiple scales and levels of organization. Empirically, the distribution of galaxies is highly heterogeneous and dynamic for at least 1000 Mpc, approximately 3000 million light years. Empirical observations show non-homogeneous and hierarchical distribution of galaxies (clusters, superclusters, filaments). Stars, galaxies, clusters of galaxies, and superclusters demonstrate an irregular hierarchical structure, which in a way continue the sequence of quarks, nuclei, atoms and molecules at the small end of the scale. The universe appears to be a hierarchical system rather than a random distribution. Galaxies and clusters of galaxies are clustered together in a rich topological structure that has come to be called the "cosmic web" [Springel et al. 2006], which at multiple points displays spatial hierarchical lattice-like organization. Tomozawa (2007) showed that the distribution of apparently circular galaxies is not uniform in the sky and that there exists a special direction in the universe in our neighborhood. A tidal force due to the mass distribution around the universe center causes the deformation of galactic shapes depending on its orientation and location relative to the center and may account for these observations. Moreover, the CBR dipole data can also be associated with the center of the universe expansion.

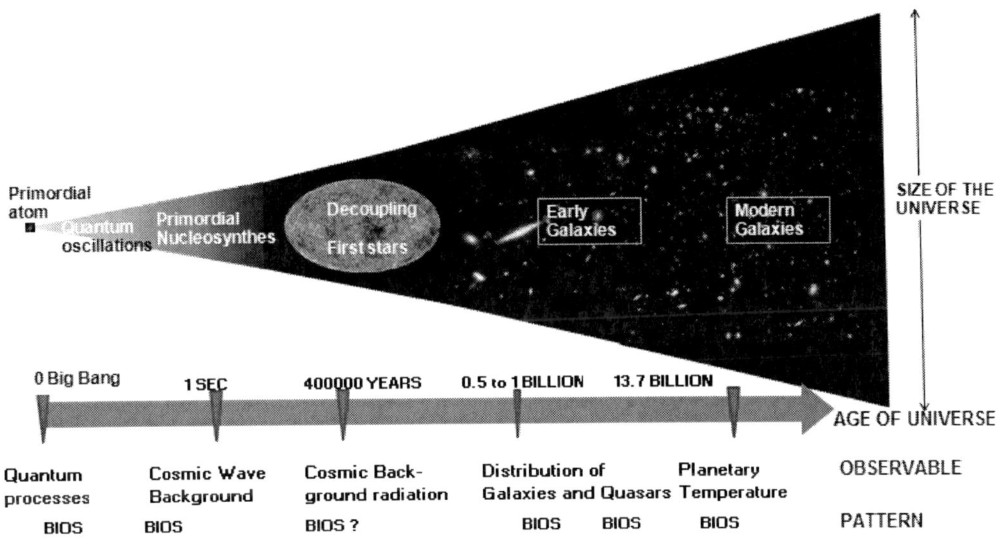

Figure 1. The evolution of the universe and the patterns observed in the available data.

Notwithstanding, it is often assumed that at its origin, the universe was uniform [Springel et al. 2006] and in thermal equilibrium, and yet, contradictorily, also in a state of extremely low entropy [Carroll, 2006] and expanding in a "Big Bang." The standard cosmological model assumes homogenous and isotropic mass distribution (Einstein's cosmological principle), and a random stationary origin for the distribution of matter [see Peebles, 1993]. The observation of large macroscopic heterogeneity in the distribution of galaxies and microscopic heterogeneity in the CBR have led to the weaker notion that homogeneity is only statistical and at large scales. According to the current view the entire universe originated in a small causally connected region.

To account for large scale homogeneity, several inflation models have been advanced [Starobinsky 1980; Guth, 1981]. Inflation models postulate that the large-scale structures derive from Gaussian random-phase distribution of quantum noise. It is proposed that the heterogeneous distribution of galaxies has been determined by random initial deviations from uniformity such as those detected in studies of the CBR. However WMAP temperature maps show anomalies that are not expected from Gaussian fluctuations and show large-scale asymmetries unexpected in a homogeneous and isotropic space [Springel et al, 2006]. How does this heterogeneity come about? The origin of the initial quantum non-homogeneity and the mechanism for inflation are unknown. The quantum fluctuation in the primeval atom are believed to be the seeds of the structure of the universe when they were magnified to cosmic size by inflation.

In contrast to random stationary models, we have proposed, based on the study of galactic distribution, to regard the evolution of the universe as a causal and creative process [Sabelli and Kovacevic, 2003]. The development of a biological organism creates a unique individual starting from a set of structures (the genes) that determine its overall features largely independently from external inputs. In a similar manner, the evolution of the universe could be a creative development from a "cosmic gene" [Sabelli and Carlson-Sabelli, 1996] rather than as the product of chance and selection. Cosmological structure may have thus originated with quantum processes that generate an expansive pattern characterized by novelty, diversification, temporal complexity (non-uniform wavelet and recurrence plots), and asymmetric statistical distribution. These features characterize Bios, and demonstrate that the processes that generate it are creative: Biotic patterns are generated mathematically with bipolar feedback models such as the process equation $A_{t+1} = A_t + g * \sin(A_t)$ that combine action (recursion), bipolar opposition (trigonometric function) and connection with previous state (A_t term), that is to say conservation. Here we demonstrate the importance of conservation through the examination of the process and the Klein-Gordon equations.

Biotic patterns are ubiquitous in natural and human processes (heartbeat intervals, respiration, sequences of bases in DNA, meteorological data, shape of rivers and shorelines, economic series) [Kauffman and Sabelli, 1998; Sabelli et al., 1997; Sabelli, 2005] but they have often been described as noise or Chaos because one can only identify Bios, and distinguish it from other patterns, with time series analyses that measure features of creativity and causality. Creativity is demonstrated by diversification (increase in variance with the length of the observation), novelty (less recurrence than randomized copies of the data) and temporal complexity (changes in pattern with time as detected by recurrence and wavelet plots); these analyses rule out random, periodic, and chaotic patterns. Causality is demonstrated by statistical correlations, recurrence measures, and most convincingly, patterns in the time series of differences between consecutive terms that excludes random walk.

These measures identify biotic patterns in the temporal distribution of galaxies [Sabelli and Kovacevic, 2003] and in the Schrödinger equation [Sabelli and Kovacevic 2006; Thomas et al, 2006]. We thus proposed that biotic processes contribute to cosmological evolution.

Our goal is to study cosmological processes in order to test this view and compare creative evolution (Bios model) with worldviews that postulate stationary randomness or entropic decay to disorder. Theories thus describe the universe in distinctly different ways:

1. The **Standard Cosmological Model** postulates a static, homogenous and isotropic universe. The observable structure represents local fluctuations. The overall distribution of matter represents a **random stationary process.**
2. The self-similar **fractal geometry** model [Mandelbrot, 1975] assumes a static and non-homogenous universe in which structure is generated by a **random stationary process** unrelated to temporal evolution.
3. Classic **thermodynamics** hypothesizes **decay from order to disorder** (maximal entropy).
4. The **Bios Theory of Creation** [Sabelli, 2005] postulates a non-static universe in which causal interactions generate a non-homogenous distribution of galaxies, the expansion of intergalactic space, and evolution from simple to complex, all of which are observed empirically.

This chapter is organized in two parts. In the first, empirical section of this chapter, we explain in detail the methods used for the analysis of the data, and we present these analyses in detail so they may become useful for specialists in cosmology, physics, and other scientific fields. The empirical studies of galaxies, quasars, cosmic background radiation, gravitational waves, quantum processes, and planetary temperature, demonstrate two fundamental features common to all these processes: (1) all these time series demonstrate creativity; the universe is not determined and static, but instead it is creative; (2) these creative processes are causal, not random. These two properties, creativity and causality, define the pattern that we call Bios [Sabelli, 2005]. The second part of the chapter is theoretical. What does the demonstration of Bios imply regarding the evolution of the universe? How is Bios produced? What are the implications regarding the origin of the evolution of the universe? Quantum mechanical equations as bipolar feedback generate Bios. Quantum processes generate creative processes denoted by biotic patterns and propagated to larger and more complex processes. Non-specialists may skim over part I, while the second part is of general interest, regardless of the discipline, because the methods, the models, and the theory apply at all levels of organization.

Empirical Studies

Methods

Data are analyzed with the Bios Data Analyzer (Sugerman, 2005) and the Bios Analyzer (http://www.inverudio.com). Since the methods that we have used are relatively new and not widely employed in physics, here we shall describe them and also the results obtained with four model time series, namely random numbers, random walks, Chaos and Bios. Chaotic and biotic series are generated by the process equation A(t) +1 = A(t) + g* sin(A(t))) with low and high values for g. Chaotic series are also generated with the logistic equation A(t) + 1 = g* A(t) * 1- A(t)).

We measure global **diversification** as the increase in standard deviation (S.D.) with the length of the series, and local diversification as the increase in S.D. with embeddings. The S.D. is computed for sets of 1, 2, 3..., N consecutive terms of the time series, starting with each term in the series. The values obtained for each embedding are averaged for the entire series, and these averages are plotted as a function of the number of embeddings. Biotic series, random walks and many creative natural processes show diversification [Sabelli et al, 2005]. Random distributions and chaotic attractors maintain a stable variance after the first few embeddings.

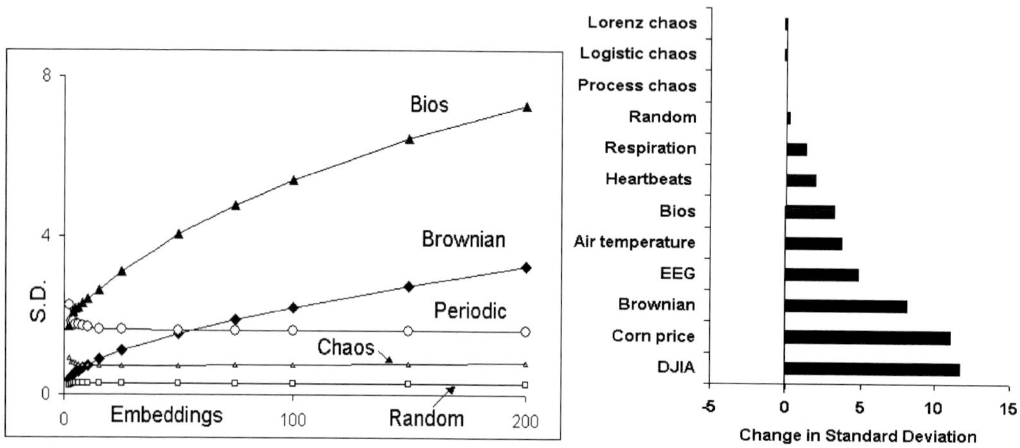

Figure 2. Left: Changes in S.D. with embedding in mathematical series. Right: Change in S.D. from 2 to 100 embeddings for empirical and mathematical series.

Recurrence plots (Figure 3) provides measures of complexity and novelty. To construct a recurrence plot, the Euclidean norms of vectors of 1, 2, 3..., N consecutive terms are calculated and compared. When their difference is less than a given margin of tolerance (within 1 % in our study), a recurrence **isometry** is counted and plotted. Distinct clusters of recurrences with organized patterns (complexes) separated by recurrence-free intervals are evident in biotic series and in random walks. Shuffling erases these complexes. Random and chaotic series show uniform recurrence plots, without clustering. We also show plots of similarity recurrences (recurrence meaning that the vectors are isometric and with the same direction) using Webber and Zbilut's [1994] program. The two types of plots are similar, except for the density of recurrence at a given embedding.

Temporal pattern and changes in pattern is detected in both the time domain (recurrence plots) and the frequency domain (wavelet plots); these are complementary opposite views, as frequency is the inverse of time (frequency = 1/t). In the frequency domain, uniformity ('white noise') is observed with random and chaotic series, and non-uniform patterns (such 1/f "noise") is observed with many natural processes, Bios and random walks. Shuffling the data erases this structure. Random data shows no structure. The pattern of the biotic series and of random walks changes in time, as observed with many natural processes. Distinct periods with different pattern can be detected in recurrence plots (Figure 4) and wavelet plots [Sabelli et al, 1997]; they indicate the episodic character of the process. In contrast random and chaotic series show temporal uniformity.

Complexity is defined by the transformation from one pattern to another (temporal complexity) rather than by the compressibility of computer programs that equates maximal complexity with randomness [Chaitin, 1987]. To study both simple and complex components of variation, we use 1, 2, ... 100 embeddings (Contrary to much literature on "embedology", there is no one appropriate or ideal embedding at which a time series must be measured).

Recurrence quantification is performed by counting isometries. *Isometry* is the number of recurrences as a percentage of the total number of possible recurrences in the sample (N x N / 2). *Consecutive isometry is* the number of recurrences that follow each other; e.g. if vector x_t is recurrent with vector x_{t+m}, then vector x_{t+1} is recurrent with vector x_{t+m+1}; consecutive isometry is calculated as percentage of isometry. Webber and Zbilut [1994] regard consecutive recurrence as a measure of determinism. We have found necessary to plot the number of consecutive isometries as a function of the radius of tolerance (*radial isometry* and *radial consecutive isometry*) to distinguish deterministic from stochastic series. *Arrangement* is a measure of complexity calculated as the ratio of consecutive isometry to the number of isometries.

Each series is compared with copies randomized by shuffling. Shuffling the data decreases the number of isometries in periodic and chaotic series, and increases it in biotic series (Figures 5 and 6). Since a recurrence is a repetition of pattern, a lower than random recurrence rate indicates that the process under consideration innovates more than chance events. **Novelty** is thus measured as the increase in recurrence isometry produced by shuffling the data [Sabelli, 2001]. Novelty is a defining feature of Bios that differentiates it from periodic and chaotic series that have larger numbers of isometry recurrences than their randomized copies. In random data, recurrence can increase or decrease with shuffling, as expected. Note that the definition of isometry differs from the definition of recurrence used by other researchers. Only the calculation of isometry allows one to compute novelty as here defined; other measures of recurrence do not manifest this feature.

Diversification, temporal complexity, and novelty are observed in biotic series generated deterministically as well as in stochastic processes. To distinguish between these two cases, we study the series of the differences between consecutive terms in these series. The difference between consecutive terms in a random walk series is of course random. In the simple biotic series generated by the process equation, the time series of the difference is chaotic, but in more complex computer models of Bios, the time series of the differences is biotic.

In summary, biotic patterns show: (A) Creative features (Diversification, Novelty, and Temporal Complexity) and (B) Evidence for causation (Partial autocorrelation, Pattern in series of differences, and Consecutive recurrence).

Frequency analysis involves **wavelet plots** constructed with the Chaos Data Analyzer [Sprott and Rowland, 2000] and **power spectrum** using the program developed by Patel [Patel and Sabelli, 2003]. In many natural processes (astronomical, physical, biological and psychological), the power spectrum often shows an inverse relation between frequency and energy [Gilden, 2001; Handel and Chung, 1993; Press, 1978; Schröeder, 1991]. The inverse power law applies to multiple levels of organization indicating their isomorphism or self-similarity. It must reflect a fundamental feature of natural processes. While 1/f pattern is often attributed to the sum of random changes, Patel and Sabelli [2003] have shown that a 1/f pattern may also be generated by deterministic actions that generate Bios.

To explore causality we also measure **autocorrelation** and partial autocorrelation using the methods of Kendal [1973].

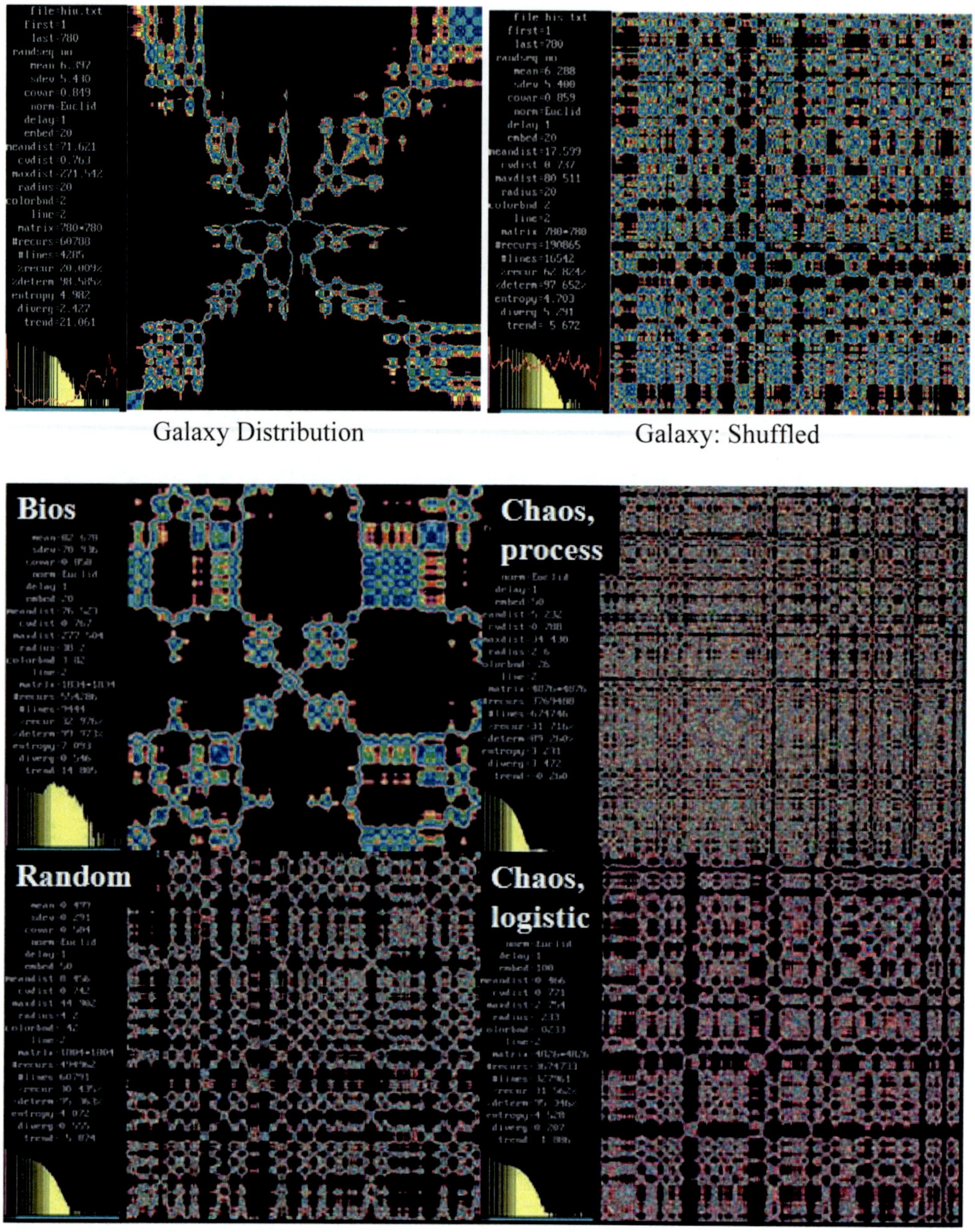

Figure 3. Recurrence plots. Top: 20 embeddings; galaxy distances (left) and shuffled copy (right). The original data has few recurrences, which are clustered (complexes). The shuffled copy: shows more recurrences that are distributed rather than separated into complexes. Middle and bottom rows show plots of series generated with mathematical models for Chaos and Bios, and random numbers.

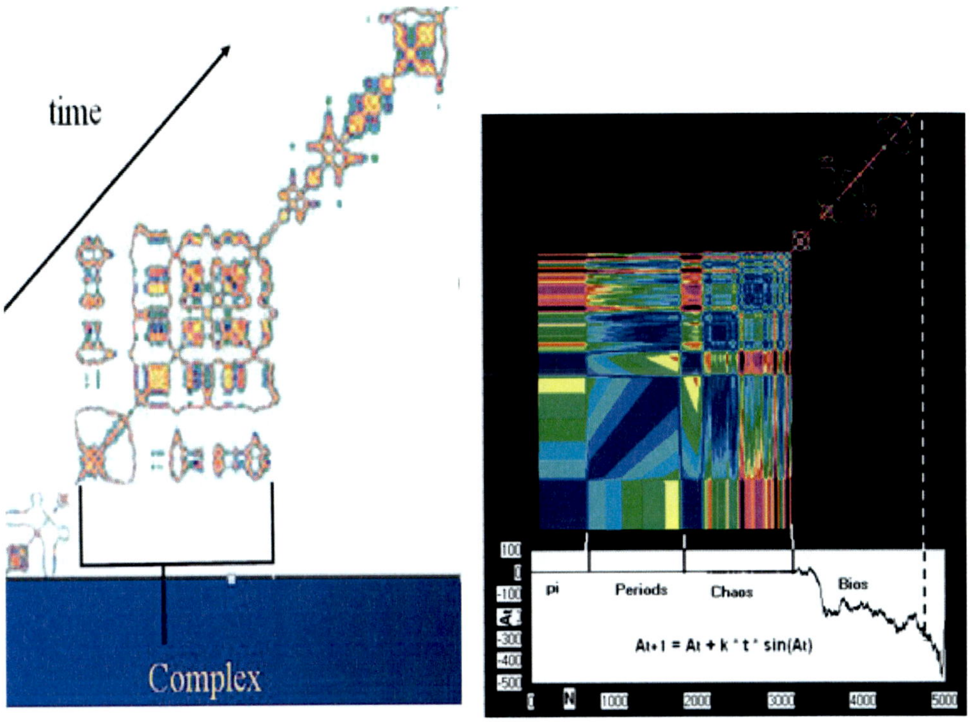

Figure 4. Recurrence plot of series generated by the relation between a time series generated with the process equation. Left: Complex during Biotic phase. Right: Changes in recurrence pattern with increased gain. Recurrences are universal and close during the initial convergence to π, universal but with a varying similarity during the periodic and chaotic phases, and scarce and nucleated in complexes during the biotic phase.

Entropy is measured as defined by de Moivre's equation $H = - \Sigma P_i \log_e (P_i)$. In simple terms, the statistical entropy of a set of numbers is measured by distributing the observed values in a histogram, and calculating the relative frequency (P_i) of values observed in each bin (class interval). Entropy is minimal if all observations fall within one bin ($H = 0$); entropy is maximal when the relative frequency of observations is the same in all bins (for N bins, this maximum value is $\log_e N$). We measure statistical entropy with a range of bins (2 to 100) and plots the results as a function of the logarithm of the number of bins. The intercept measures the **symmetry** of the data, and the slope measures the **diversity** or the uniformity of the data [Sabelli, 2005].

<table>
<tr><th></th><th>Random Walk</th><th>Bios</th><th>Chaos</th></tr>
<tr><td>Time Series:
Black
Randomized
Copy: Blue</td><td>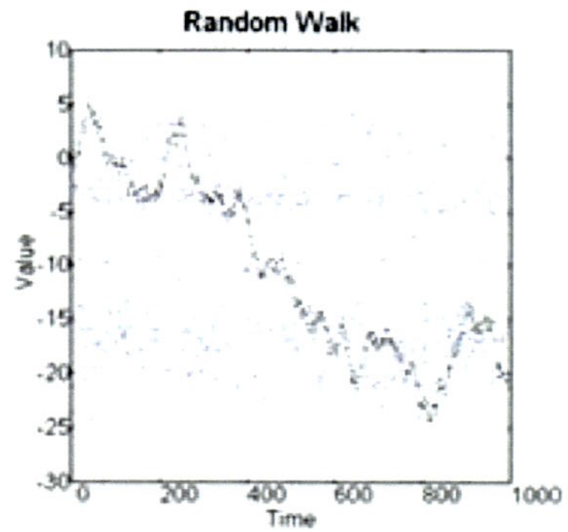
</td><td>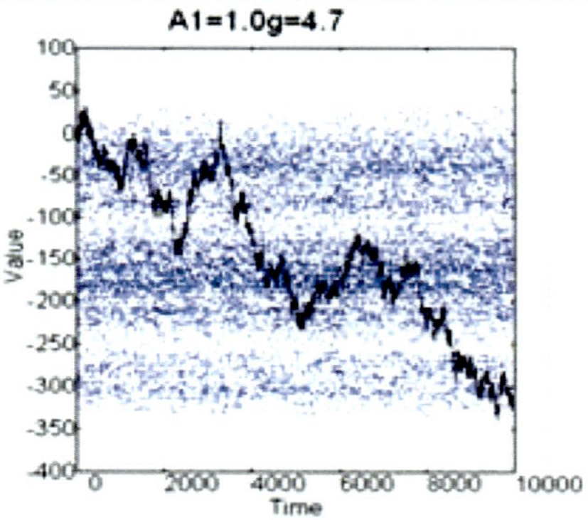
</td><td>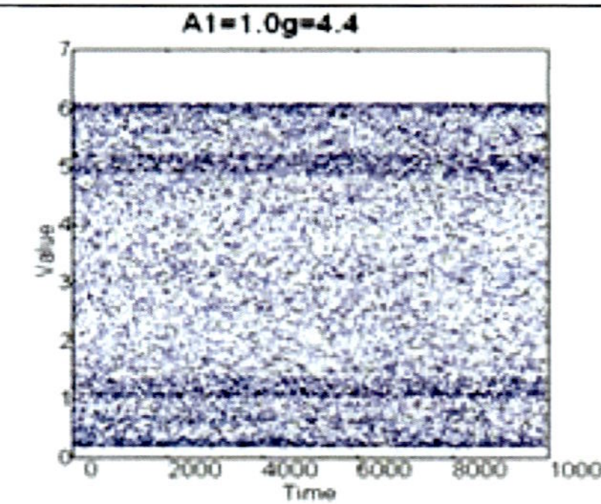
</td></tr>
<tr><td>Recurrence
Plot</td><td>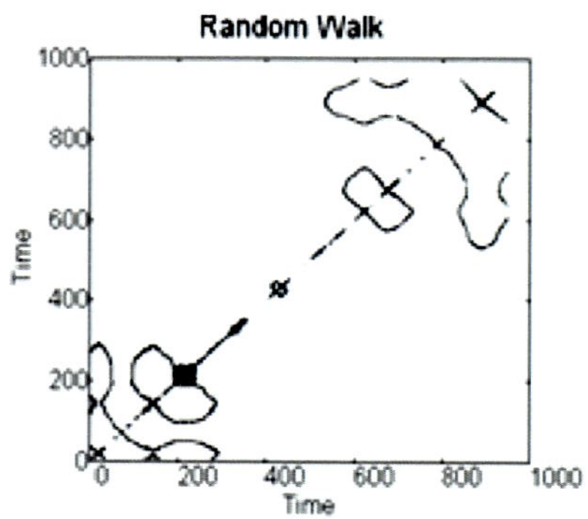
</td><td>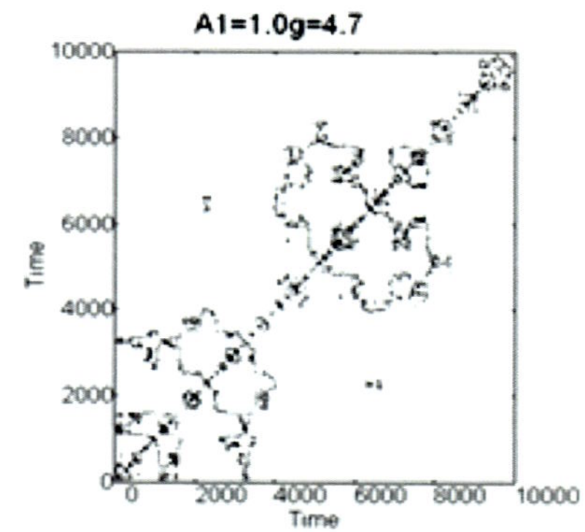
</td><td>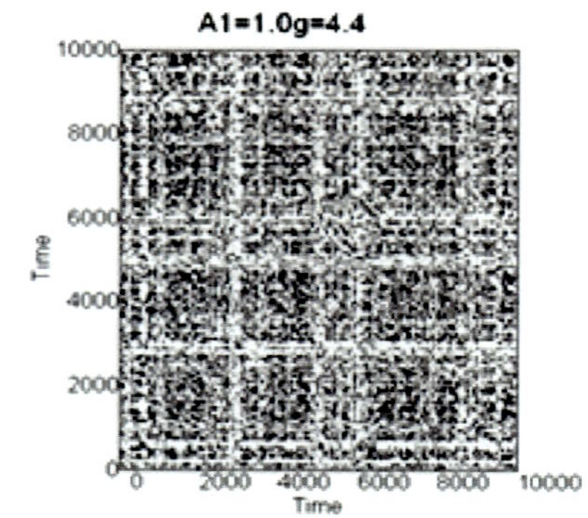
</td></tr>
</table>

	Random Walk	Bios	Chaos
Isometry	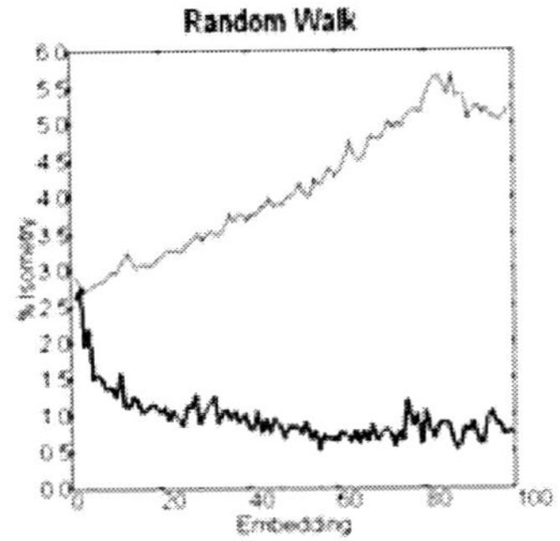	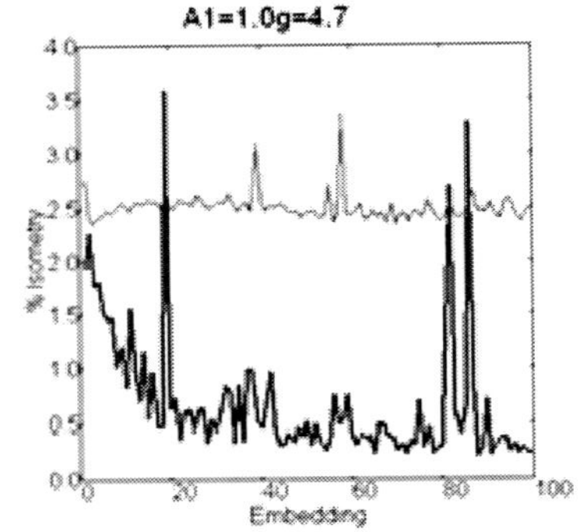	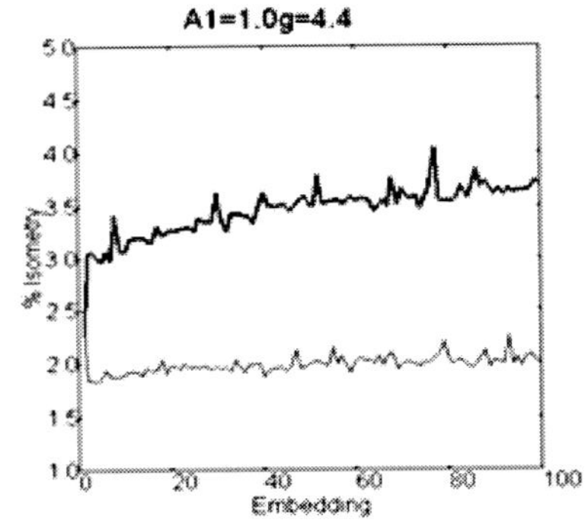
Consecutive Isometry	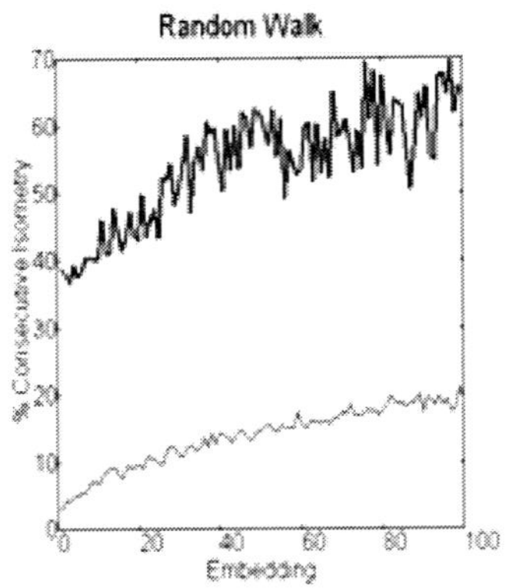	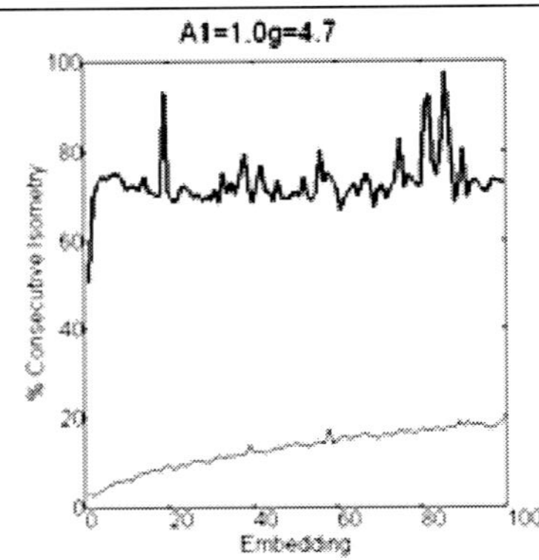	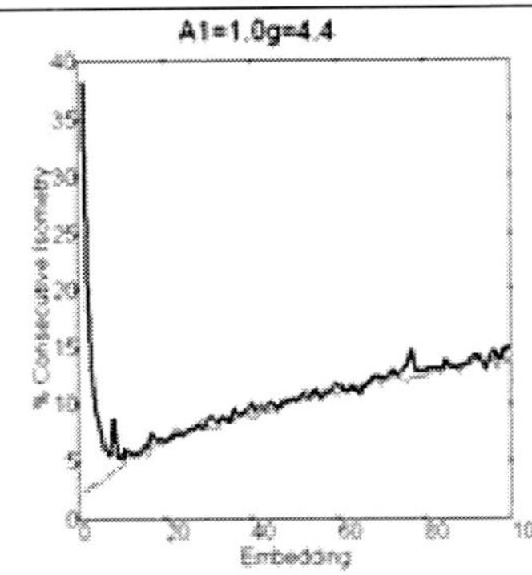

Figure 5. Continued on next page.

	Random Walk	Bios	Chaos
Arrangement	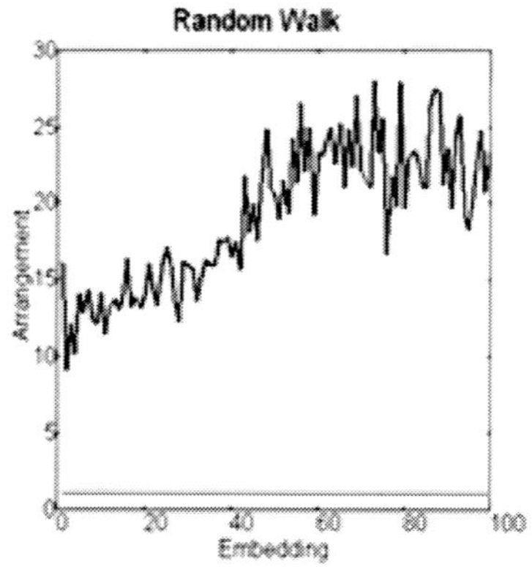	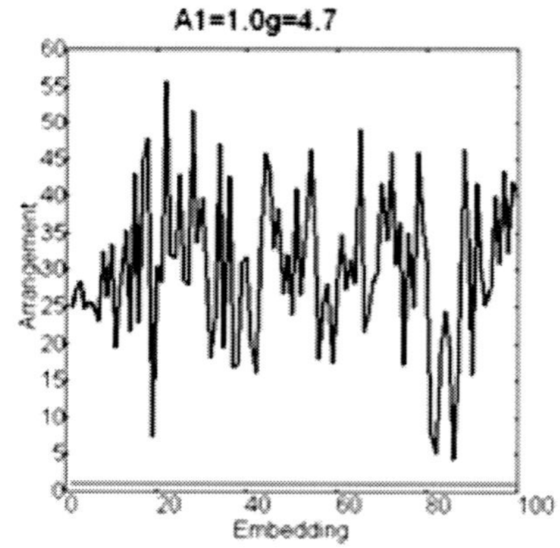	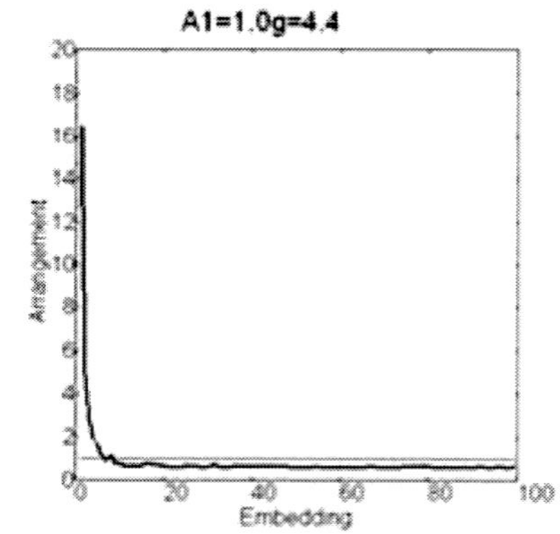
Local Diversification or Convergence to Attractor	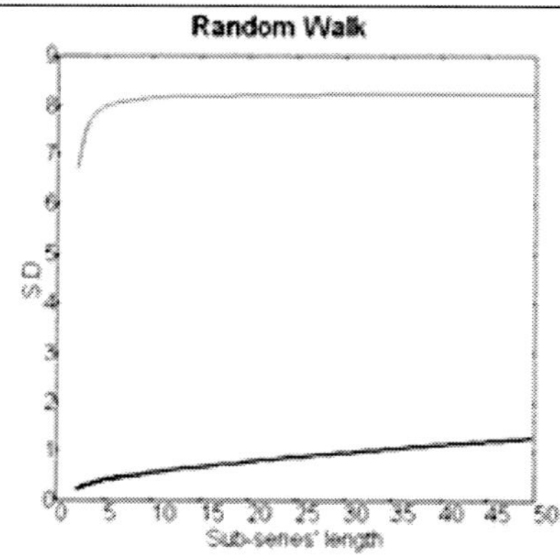	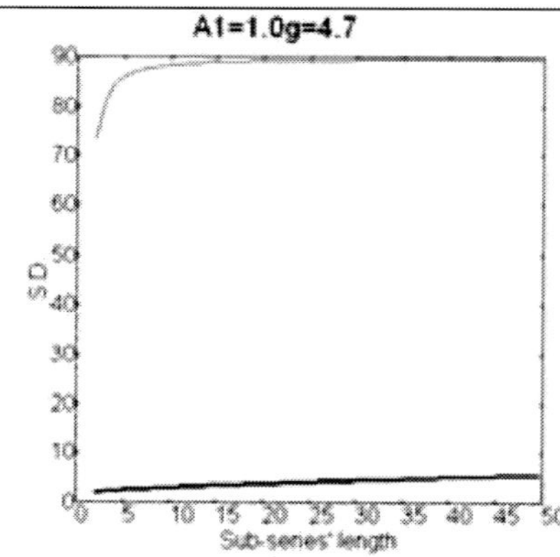	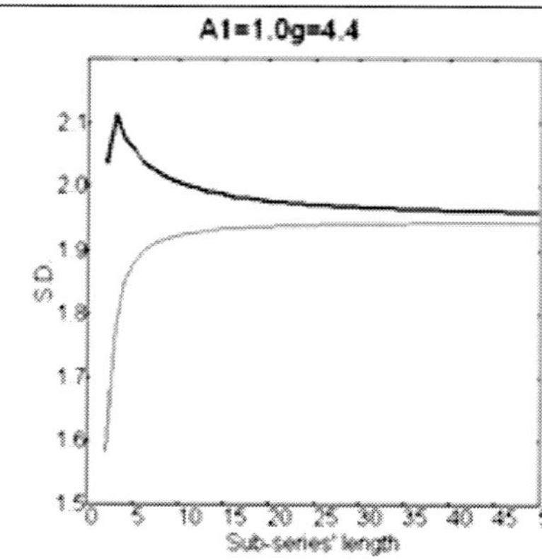

Figure 5. Continued on next page.

	Random Walk	Bios	Chaos
Process Entropy	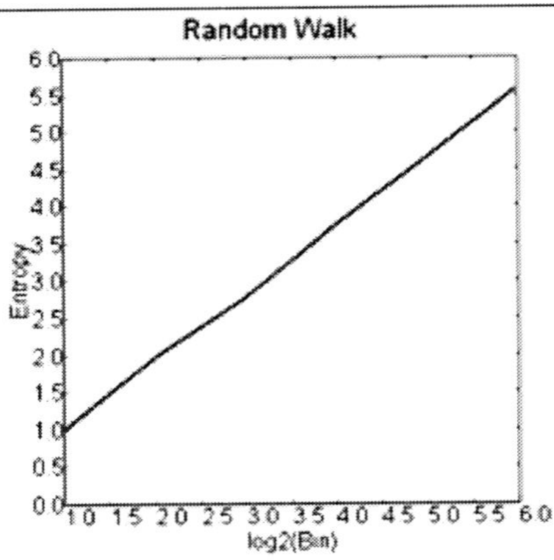	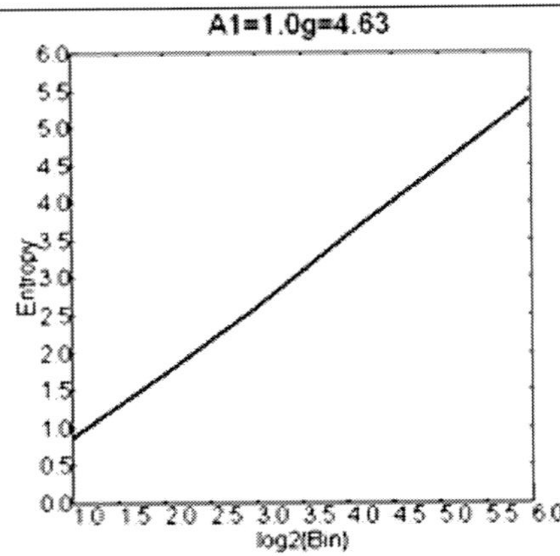	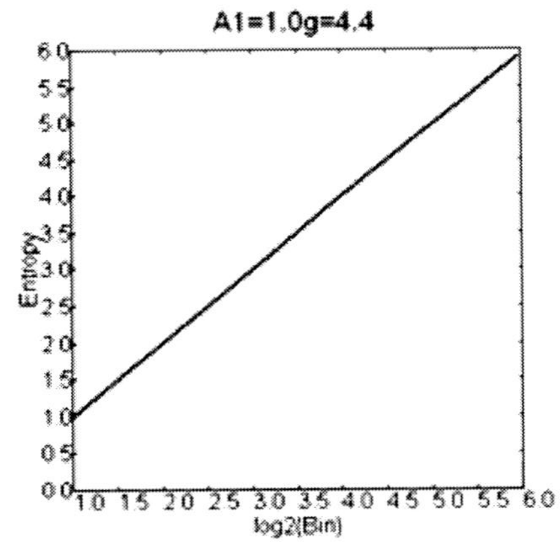

Figure 5. Comparison between three aperiodic patterns. Chaos and Bios are generated with the process equation A(t) +1 = A(t) + g* sin(A(t))) with low and high values for g. Black lines: series. Blue: shuffled copies. Top row shows the time series, which are similar for Random Walk and Bios, and very ordered and with sharp limits for Chaos. The second row shows the recurrence plot with organic forms for Random Walk and Bios, and denser, and with simple order for Chaos. The third row shows the number of isometries from 1 to 100 embeddings which is lower than the number of isometries in shuffled copies for Random Walk and Bios (novelty) and higher for Chaos. The fourth row shows the number of consecutive isometries, which is higher than the number observed in shuffled copies at all embeddings for Random Walk and Bios and only at small embeddings for Chaos. Consecutive isometry thus does not differentiate between random and causal origin. The fifth row shows arrangement, a portrait of complexity which is evident for Bios and Random Walk at all embeddings and demonstrable for Chaos only at low embeddings. The sixth row shows the changes in the S.D. as measured from 1 to 50 embeddings. The S.D. is higher in chaos than in shuffled copies, and decreases with embedding, showing convergence to an attractor. It is much lower than in shuffled copies for Bios and Random Walk, but increases with embedding demonstrating diversification. The bottom row shows the changes in measures of entropy with increased number of bins (logarithmic scale), the straight lines generated show a 45 degree slope for Chaos and Bios and the same or slightly smaller slope for Random Walk; the slope measures the diversity of the series. The intercept at 2 embeddings is 1 for Chaos and Random Walk and slightly less for Bios. This asymmetry is much more marked in natural biotic series than in the mathematical Bios.

	Random Walk: Differenced	Bios: Differenced	Chaos: Differenced
Time Series: Black Randomized Copy: Blue	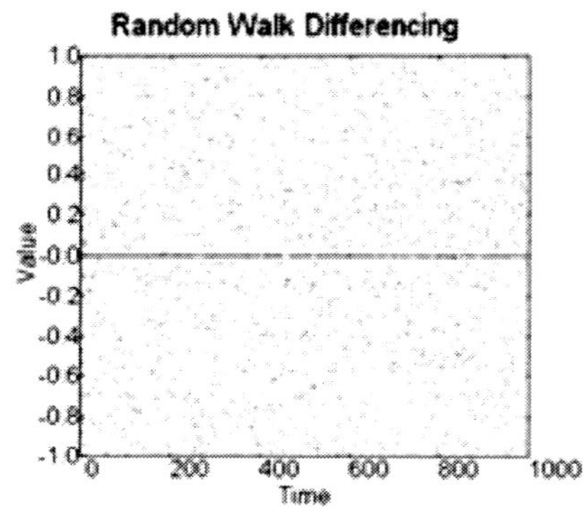	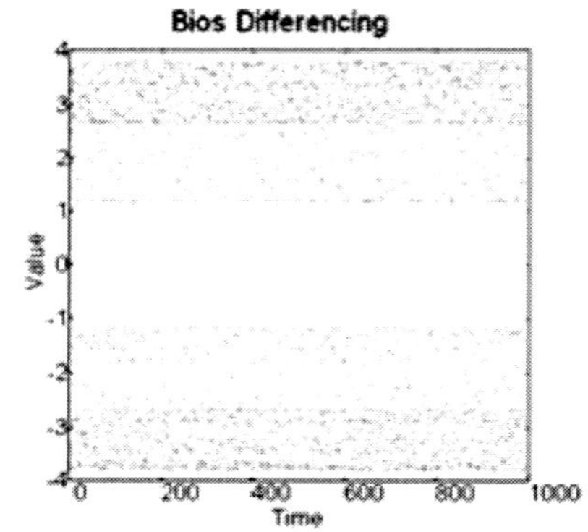	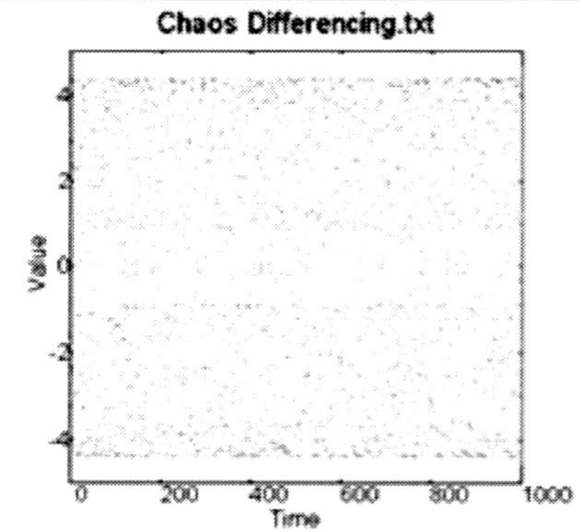
Recurrence	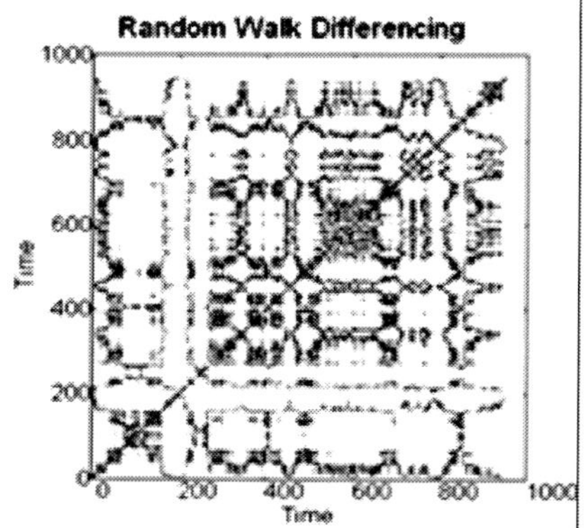	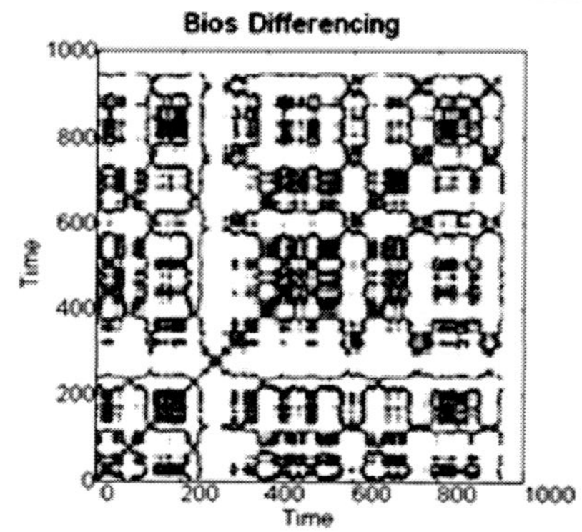	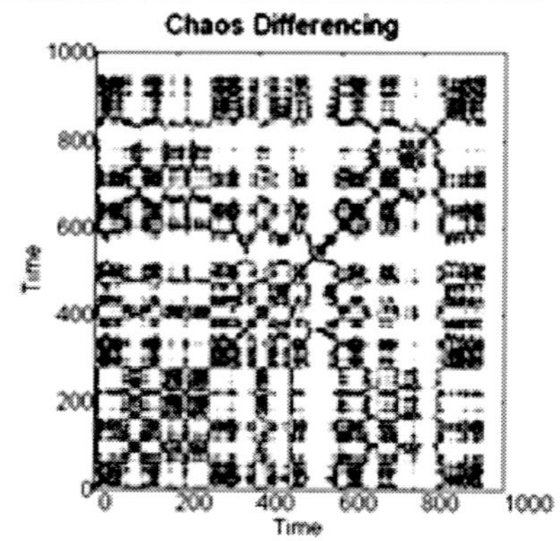

Figure 6. Continued on next page.

	Random Walk: Differenced	Bios: Differenced	Chaos: Differenced
Isometry	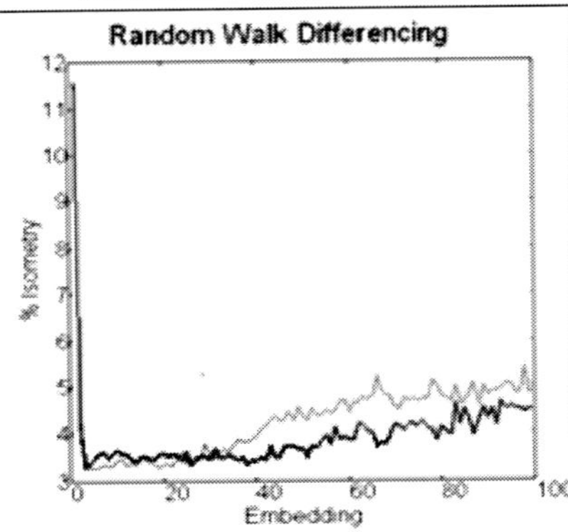	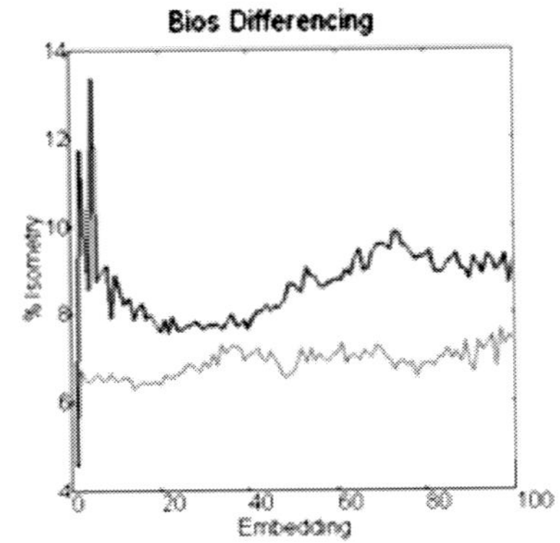	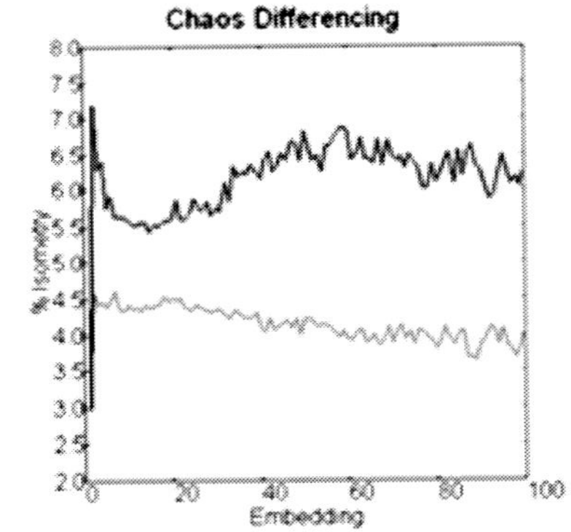
Consecutive Isometry	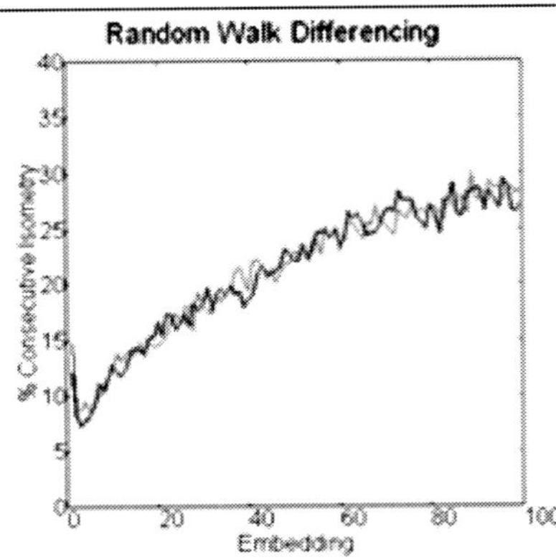	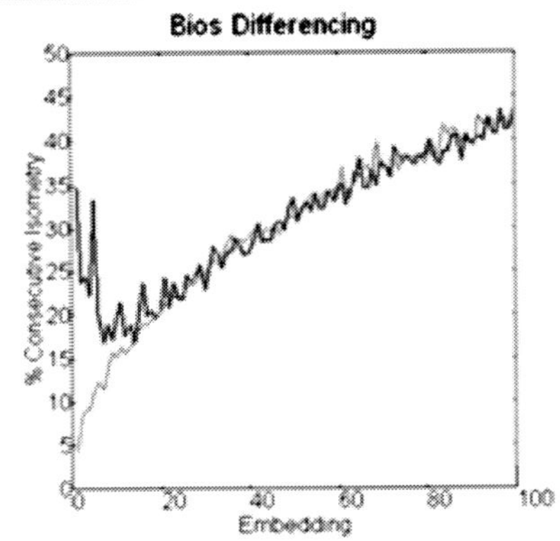	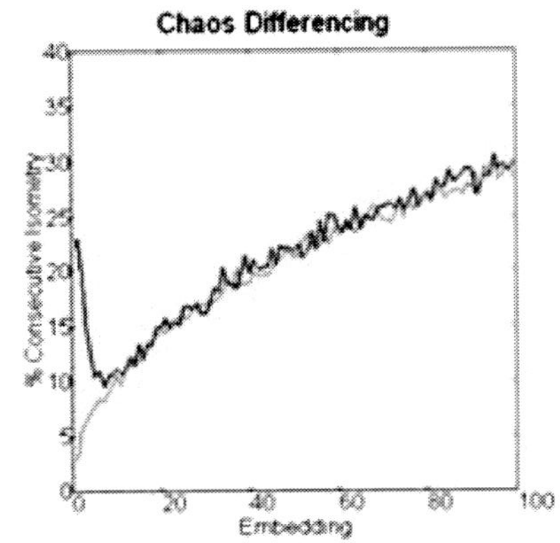

Figure 6. Continued on next page.

	Random Walk: Differenced	Bios: Differenced	Chaos: Differenced
Arrangement	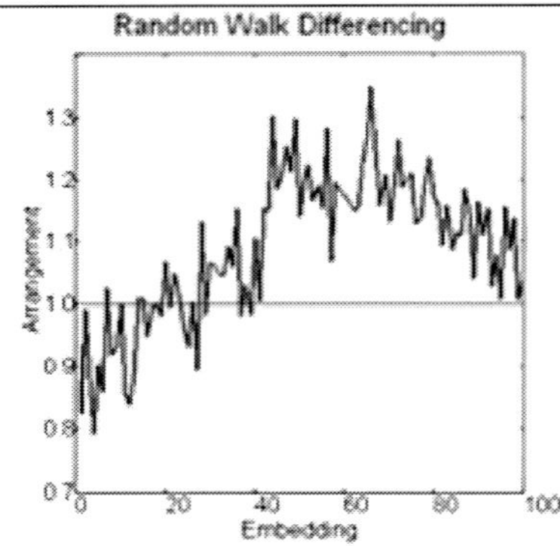	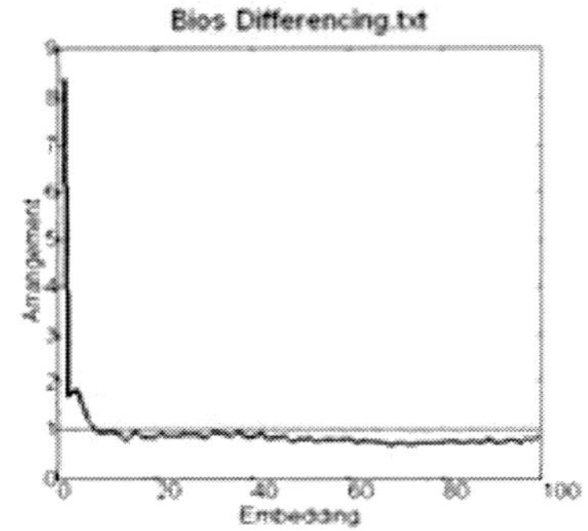	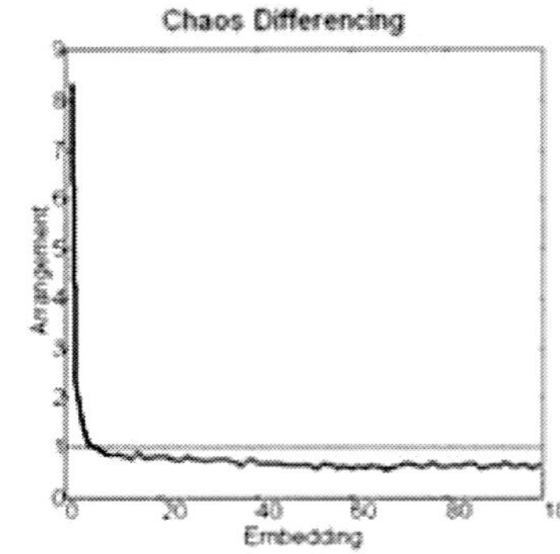
Convergence	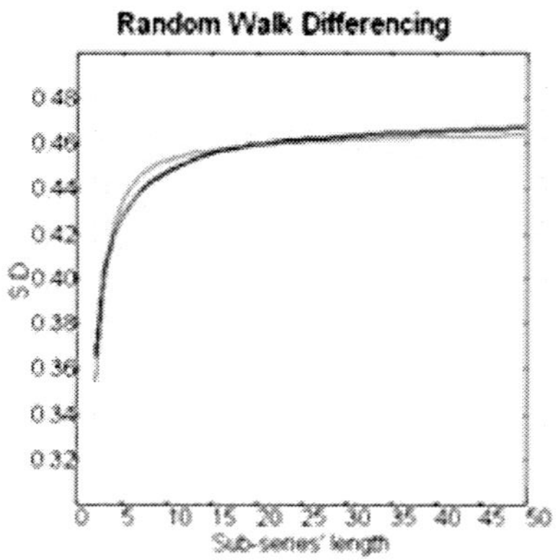	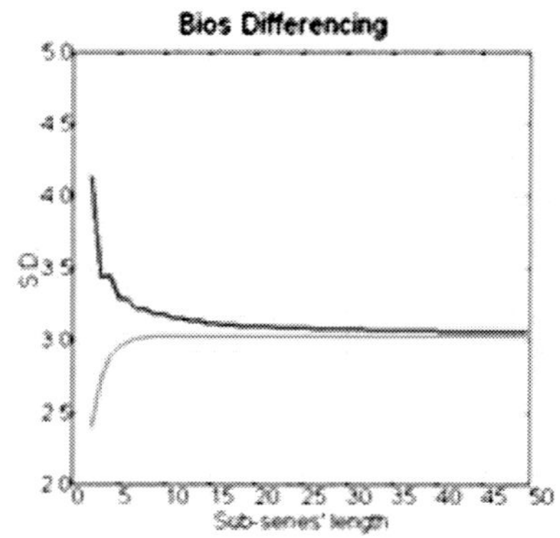	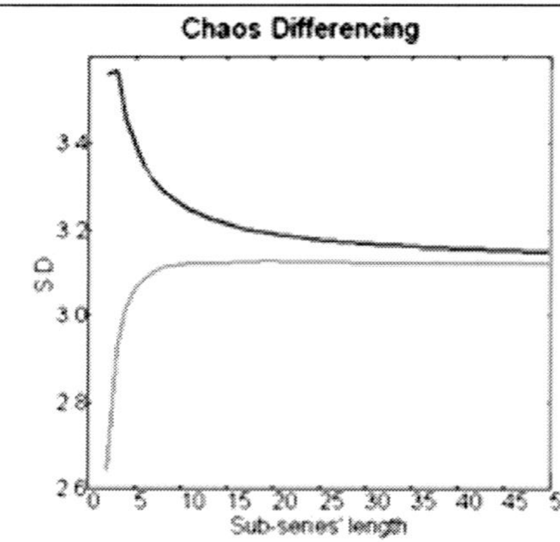

Figure 6. Comparison between three aperiodic patterns by examining the time series of differences between consecutive terms. The top row shows pattern for Bios and Chaos, but not for Random Walk. However, one sees pattern even for Random Walk in the recurrence plot (second row). The causal generation of Bios and Chaos is evident in the higher number of isometries that are in shuffled copy (third row), as well as in the higher number of consecutive isometries at low embeddings (fourth row). Arrangement is much higher in Chaos and Bios at low embeddings than in Random Walk. The S.D. is higher in Chaos and Bios than in random copies, and tends to converge rather than diversify.

Table 1 summarizes the difference between aperiodic patterns. The comparison of Biotic series with other aperiodic ones shows similarity with Random Walk for the series itself, differentiated from Chaos, and similarity with Chaos for the difference between consecutive terms, differentiated from Random Walk. Interpreting these findings we can differentiate three types of patterns according to convergence to an attractor, conservation of pattern, or creative features, and all these three types may be generated in a deterministic, or probabilistic, manner (Figure 7). The table in Figure 7 also notes the observations to be reported in this chapter.

Table 1. Time Series Analyses Comparison for Aperiodic Patterns

Time series analyses	Random	Random walk	Chaos	Bios
S.D.	Stable	Diversification	Convergence	Diversification
Recurrence plot	Uniform, stable patterns	Continual transformation, complexes	Uniform, stable patterns	Continual trans-formation, complexes
Isometry	Recurrence	Novelty	Recurrence greater than random	Novelty
Difference between consecutive terms	Random	Random	Patterned	Patterned

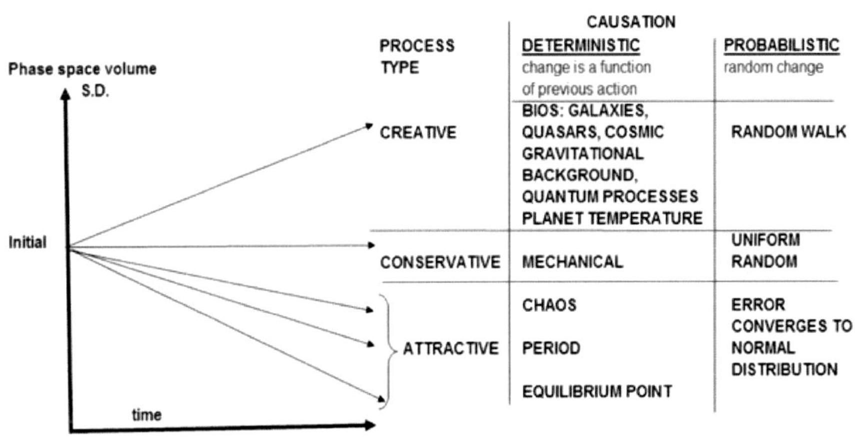

Figure 7. Comparison between three types of aperiodic patterns.

Time Series Analyses of the Temporal Distribution of Galaxies and Quasars

Analyses of the temporal distribution of galaxies as recorded in two earlier surveys ("Las Campanas Redshift Survey," and the "2-degree Field Galaxy Redshift Survey"), show life-like (biotic) patterns [Sabelli and Kovacevic, 2003] characterized by diversification, novelty,

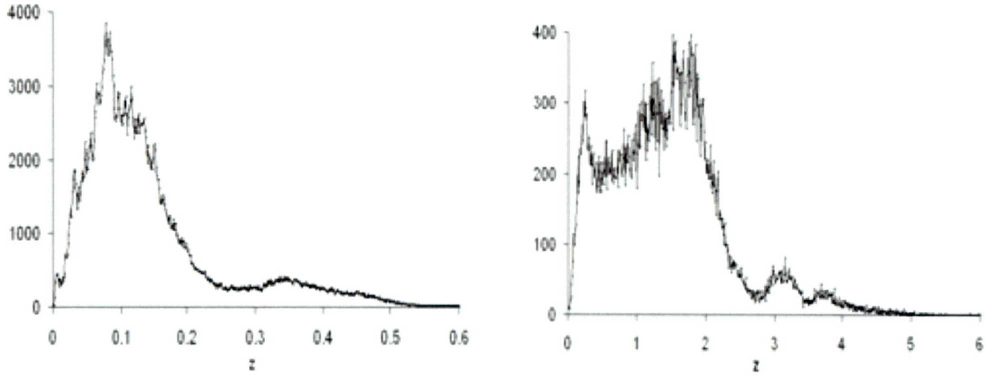

Figure 8. Distribution of galaxies (left) and of quasars (right) along the z space-time axis.

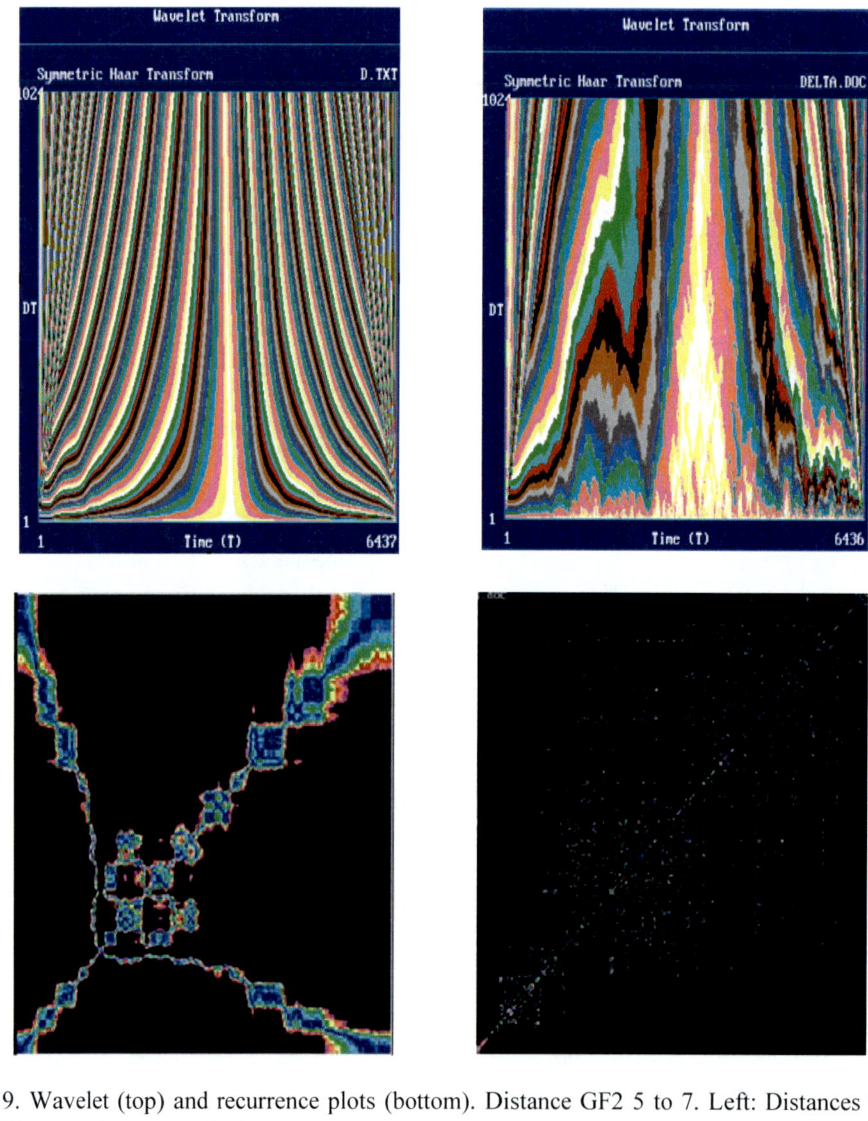

Figure 9. Wavelet (top) and recurrence plots (bottom). Distance GF2 5 to 7. Left: Distances from us. Right: Distances between galaxies.

temporal complexity (i.e. transformation from one pattern to another), and evidence for non-random change between consecutive states. A much larger survey that includes more than 675,000 galaxies and 90,000 quasars was completed by Sloan Digital Sky Galaxy Survey (SDSS) in June, 2005 [Springel et al, 2006]. This provides an opportunity to confirm or refute previous observations. We analyzed the temporal distribution of 462,879 galaxies and 61,966 quasars from the SDSS. The series analyzed are the number of galaxies in 10^{-3} red shift bins or quasars in 10^{-2} red shift bins (Figure 2). These series were compared with shuffled copies.

The temporal pattern of the number of galaxies per Mpc as a function of the age of the universe was portrayed with wavelet plots (Figures 9, 10 and 11) and recurrence plots (Figure 9 and 12).

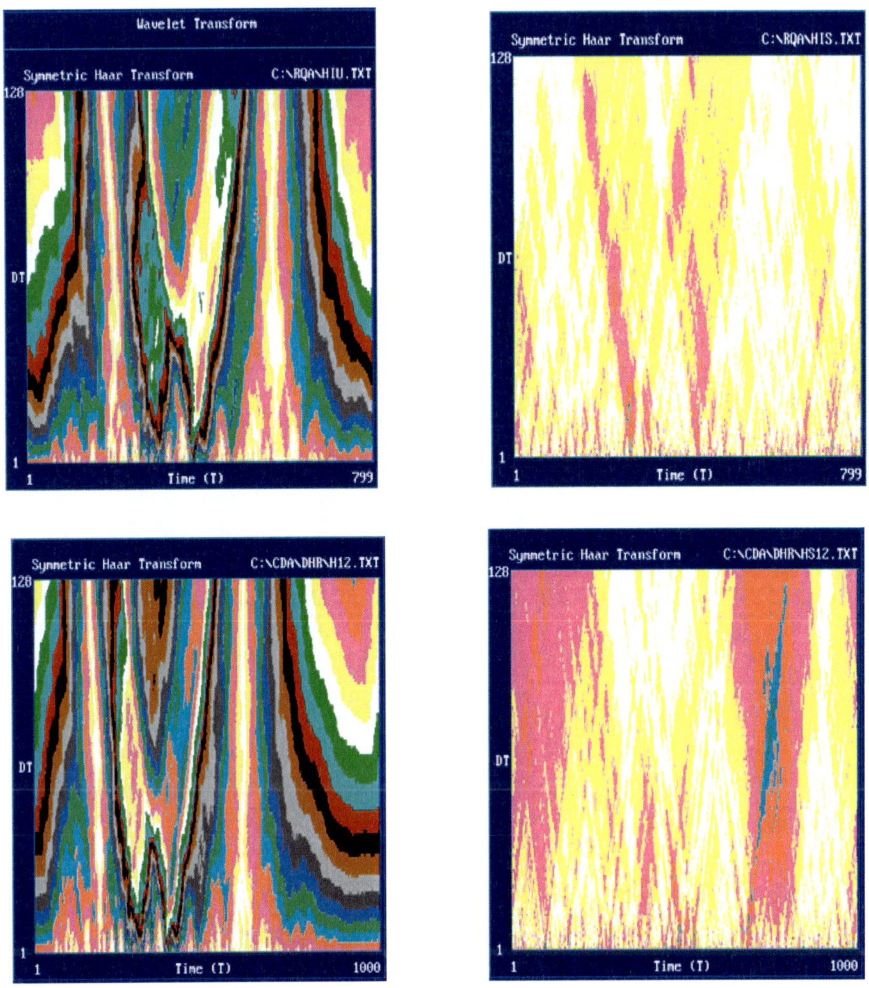

Figure 10. Wavelet plots of galactic distances in two different directions (left) and randomized (shuffled) copies (right). The pattern in the original series indicates changes in time, as observed in Bios, while pattern is erased in the shuffled copy.

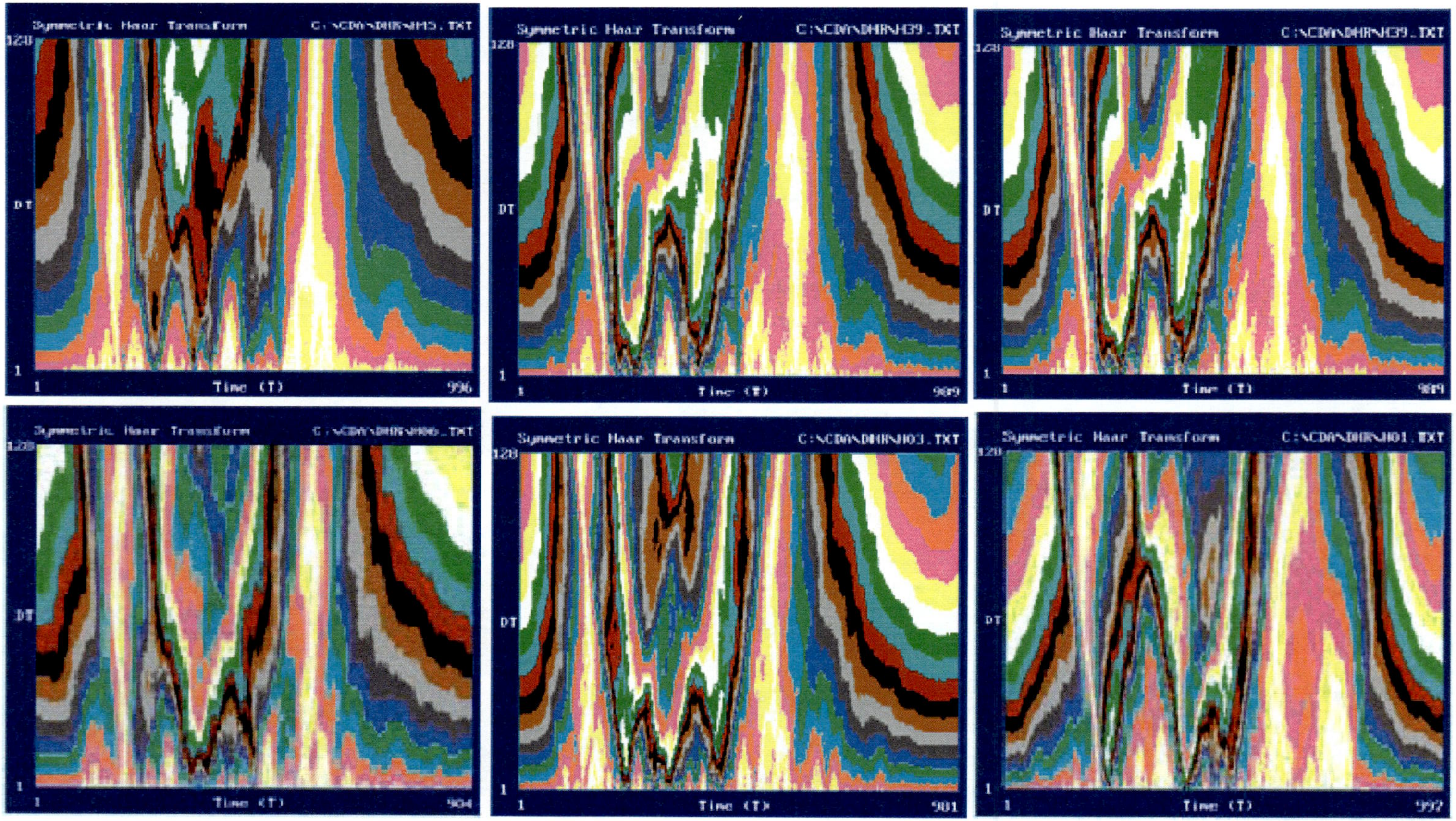

Figure 11. Wavelet plots of galactic distances in six different directions.

Recurrence plots of temporal series (Figure 12) show episodic patterns (complexes) as observed with other creative processes. Shuffling erases this organization in separate complexes. Similar results are obtained for continuous strips and for the non-continuous set of all galaxies in each survey. Shuffling also increases the number of isometric recurrences (Figures 13 to 17). This is an indication of novelty, a property of creative time series.

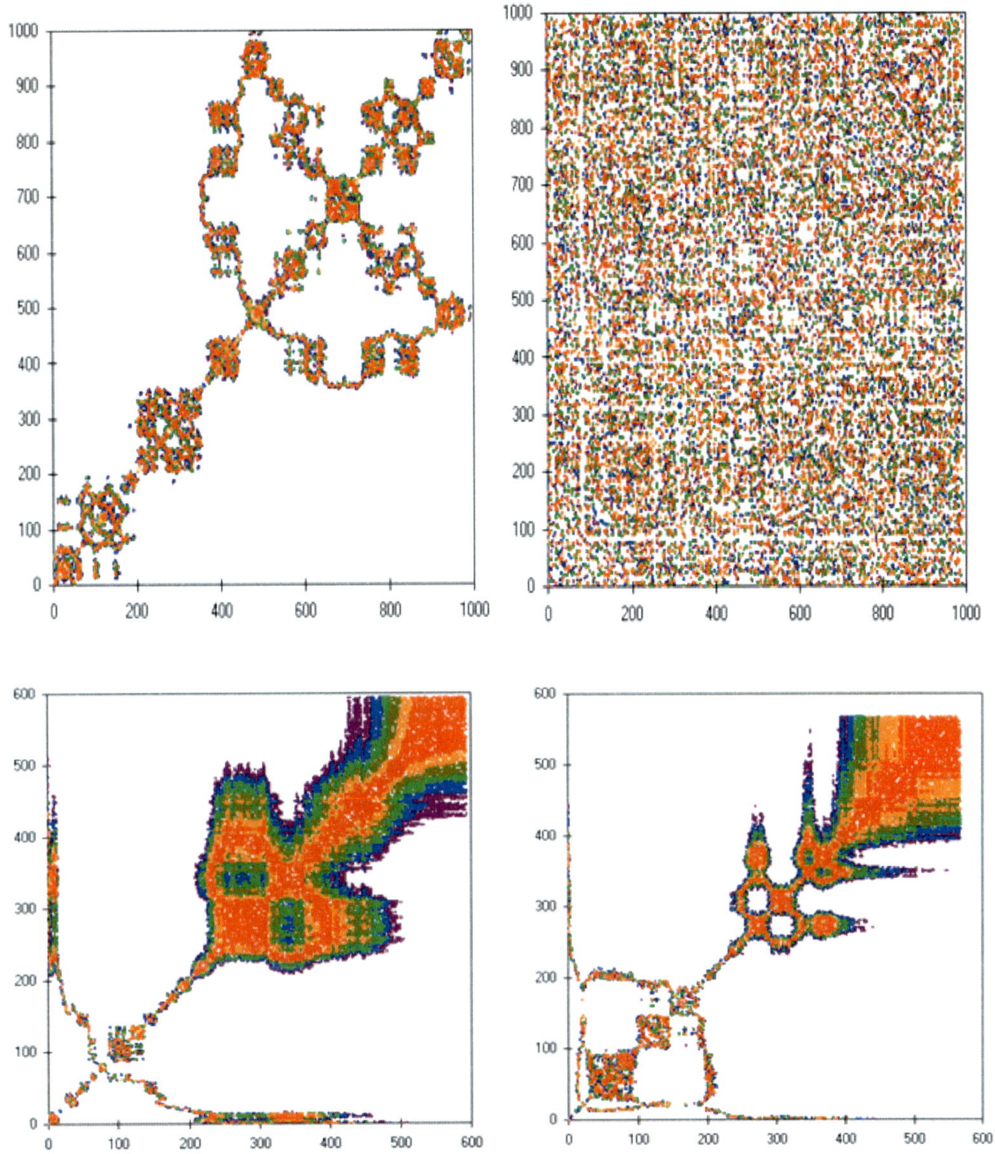

Figure 12. Recurrence plots at 5 embeddings. Top left: mathematical Bios. Top right: randomized (shuffled) copy of mathematical Bios. Shuffling eliminates complexes. Novelty is evident: the number of recurrences of isometric vectors is much increased by randomizing the data. Bottom left: Galaxies (Sloan Digital Sky Survey). Bottom right: Quasars. The pattern of these series is very complex and changes in time, as observed in Bios.

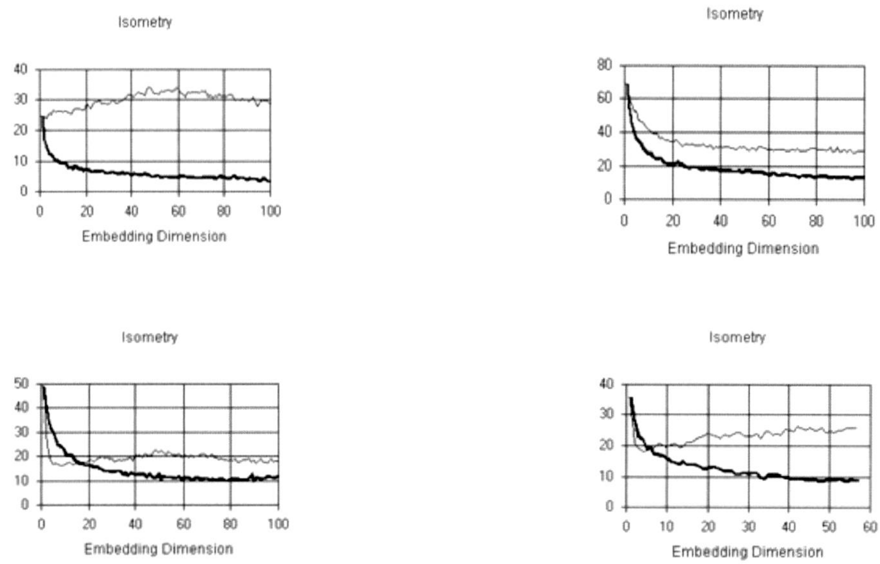

Figure 13. Embedding plots. Top left: mathematical Bios. Top right: chaos. Bottom left: Galaxies. Bottom right: Quasars.

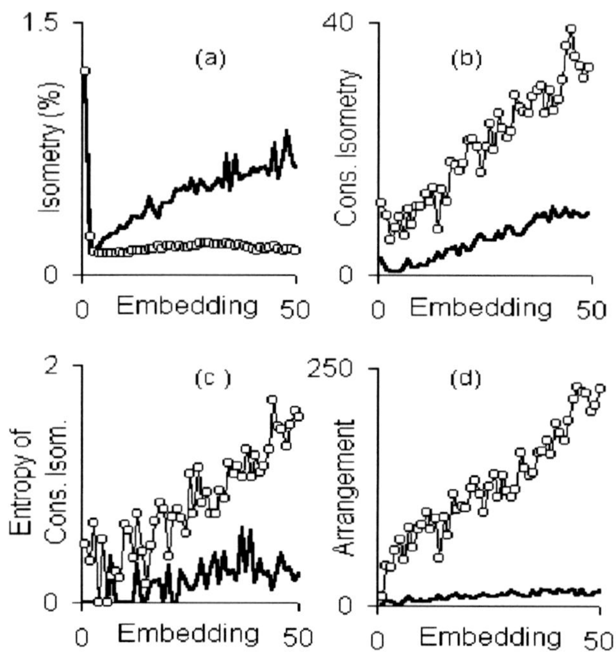

Figure 14. Embedding plots of temporal series of galaxies in Las Campanas.

Consecutive recurrence is an indication of determinism [Webber and Zbilut, 1994]. We observe that a high proportion of consecutive isometries at low embeddings indicates nonrandom causation. Galactic series show a higher proportion of consecutive isometries than their shuffled copies at low as well as at high embedding dimensions. This is also observed with mathematical Bios, but not with random series. A stochastic process such as Brownian noise may not show consecutive isometry at low embeddings because it is generated by the addition of random changes, but it shows consecutive isometry at high embeddings, revealing the deterministic accumulation of random changes. In Biotic series, one can demonstrate determinism in the series itself at low and high embeddings. The distinction between Bios and noise is most evident with a small radius of tolerance for the detection of isometries (Figure 18).

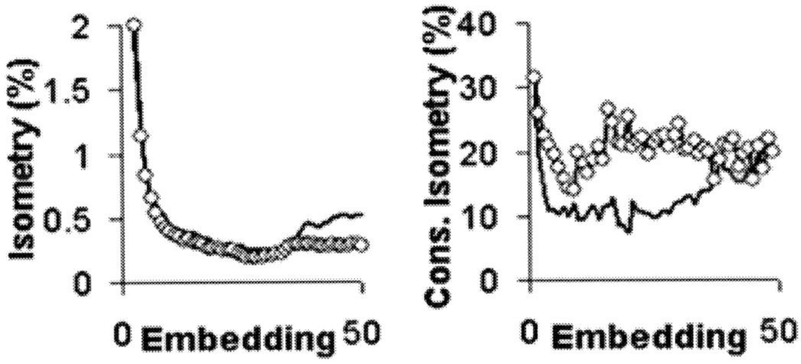

Figure 15. Embedding plots of temporal series of galaxies in 2dFGRS, Right Ascension, 6-7 radians. Number of galaxies per 0.5 Mpc. The thin line with circles shows the data and the heavier line is the shuffled copy. Cut-off radius 0.1.

The ratio of consecutive isometries to total isometry, which we call arrangement, is higher in galactic series than in their shuffled copies, indicating nonrandom complexity (Figure 16).

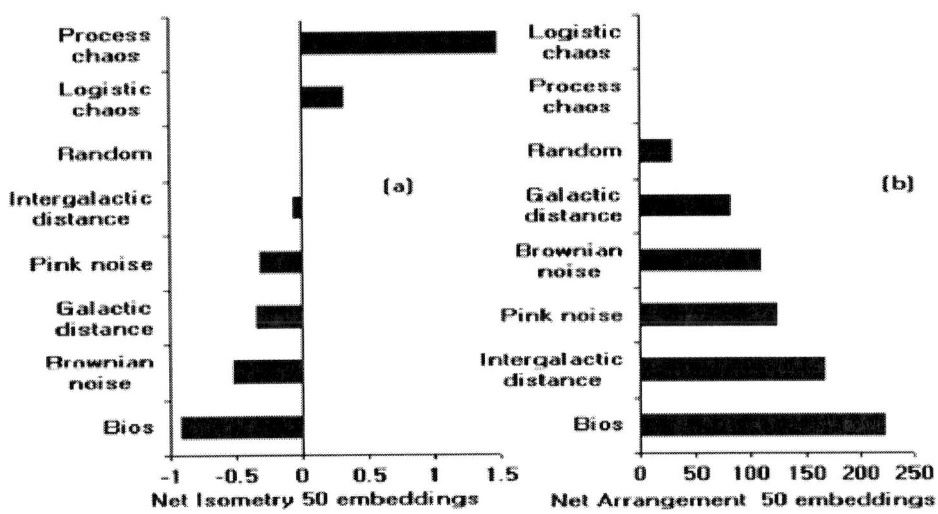

Figure 16. Isometry and arrangement in series of galactic distances and intergalactic distances compared to various prototypes of aperiodic patterns. 'Net' means values obtained in the original series minus shuffled copies. Net isometry is negative for galactic and intergalactic distances for Bios and for Pink and Brownian noise and positive for Chaos, while arrangement is higher for cosmological series and for Bios and Random Walk (Brownian), but not for Chaos.

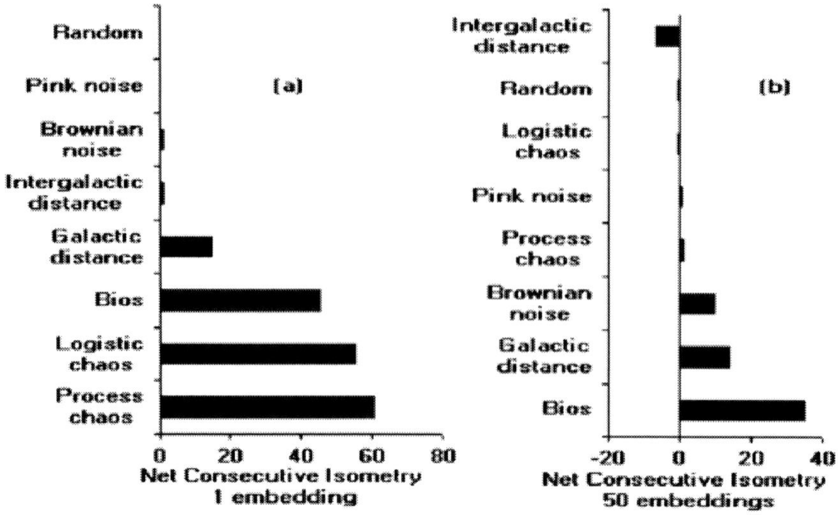

Figure 17. Consecutive isometry. At 1 embedding (Left) galactic distances show net consecutive isometry, as also observed with Bios and Chaos but not with Brownian and Pink noise. At 50 embeddings (right) galactic distance shows significance net consecutive isometry in Bios and Brownian Noise, but not in Chaos.

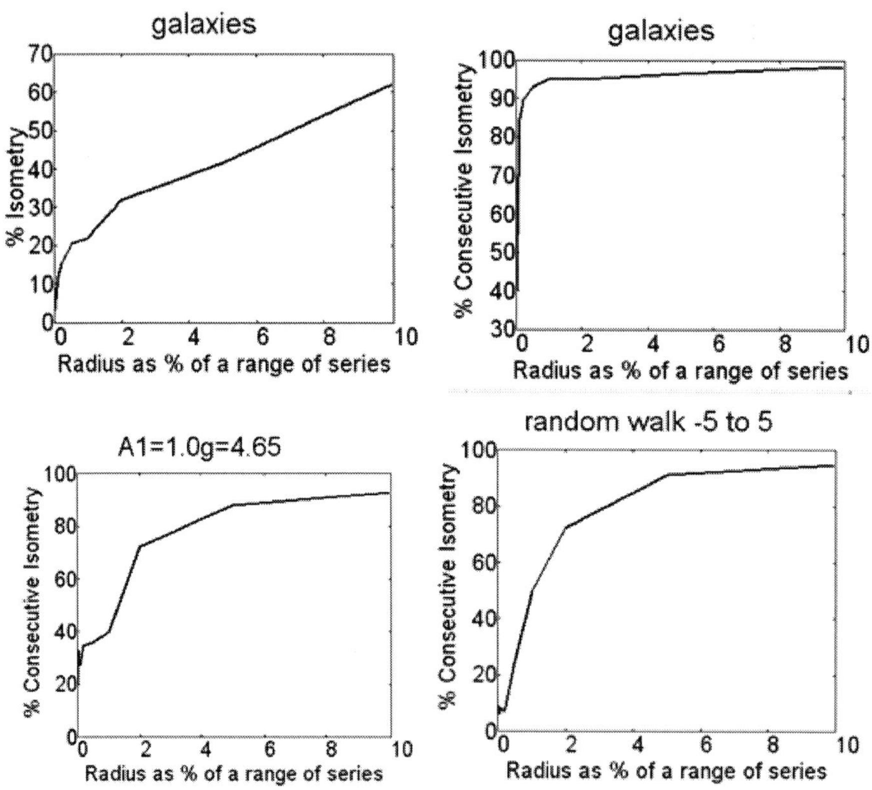

Figure 18. Embedding plots of consecutive isometry as a function of the radius of tolerance shows higher values for galaxy distribution series and for Bios than for Random Walks when the radius is small.

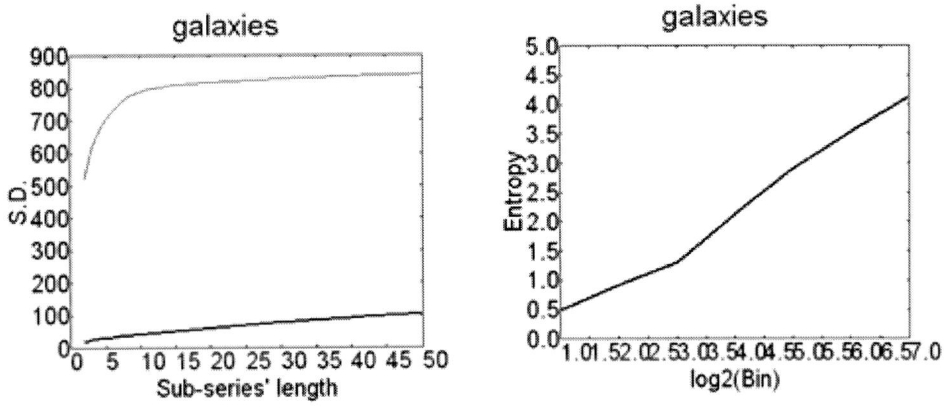

Figure 19. Diversification (left) and entropy (right) of galaxy distribution series.

Analyses of these series show diversification (Figures 19, 20, and 21) increase in S.D. and in Mean Squared Displacement from the point of origin as a function of duration (global diversification) and of embedding (local diversification) as observed with Bios. Chaos does not (Figures 20 and 21).

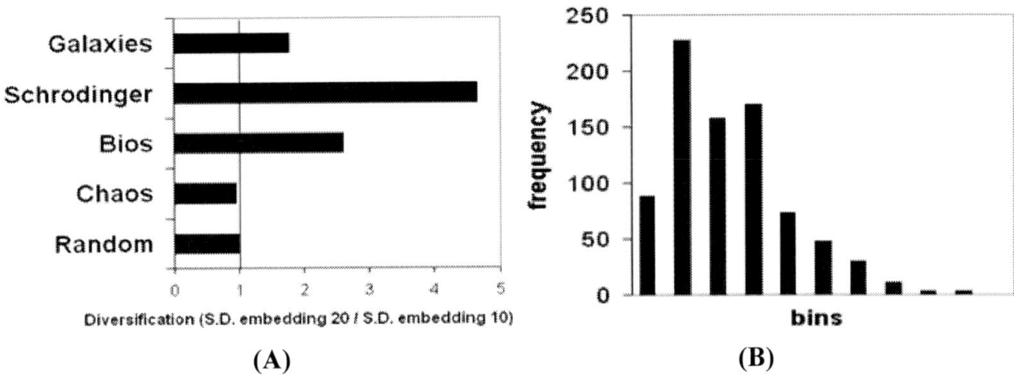

Figure 20. Diversification (A) and asymmetry (B). (A) Histogram of -12^0 declination strip of the Las Campanas survey. Distribution of galactic distances is asymmetrical in all samples and often multimodal. Asymmetry and multimodality fits Bios and contradicts randomness.

In summary, the time series of galactic distances thus shows a biotic pattern: (A) Features of creativity: diversification, novelty, and nonrandom complexity; and (B) Evidence of causality: consecutive recurrence, partial autocorrelation, patterned series of differences. These properties can be observed in biotic series generated deterministically or in stochastic processes. To distinguish between these two cases, we also studied the series of the differences between consecutive terms in these series. The difference between consecutive terms in a random walk series is of course random. In the simple biotic series generated by the process equation, the time series of the difference is chaotic, but in more complex computer models of Bios, the time series of the differences is biotic. For both galaxies and quasars, the series of differences between consecutive terms also showed diversification, novelty, complexes, and consecutive isometry. This shows that these series are generated deterministically, not stochastically. These results support the view that creative patterns spring from the interaction of preexisting simple processes, and not only as meaningful consequences of accidental events. This is also demonstrated by partial autocorrelation (Table 2, Figure 22).

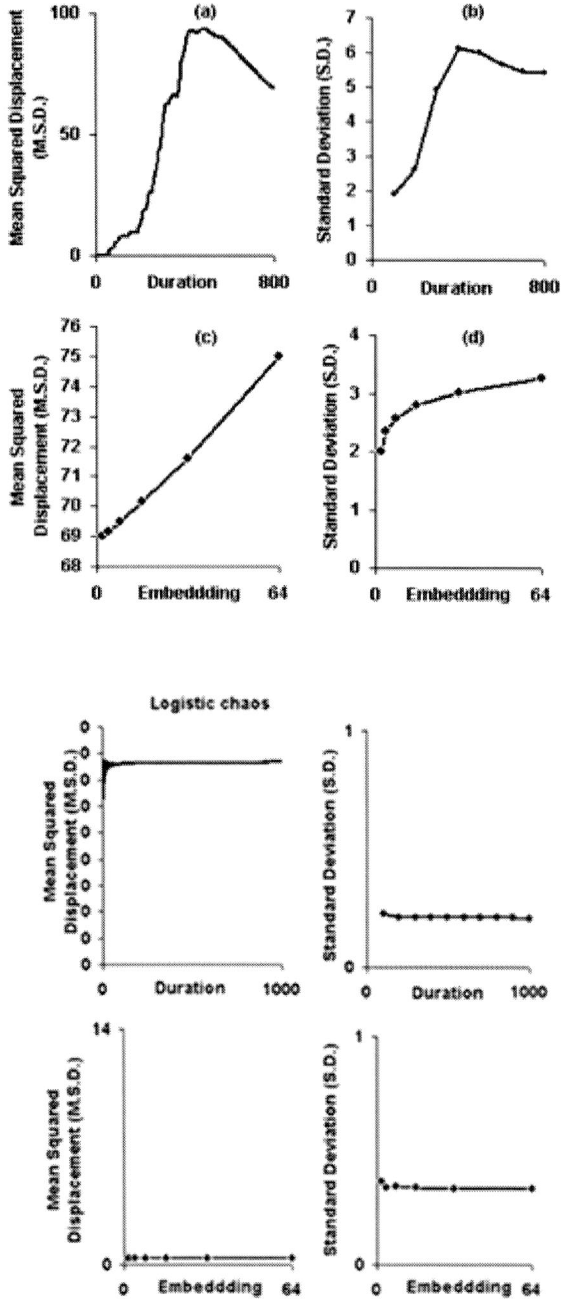

Figure 21. Diversification of galactic distribution that shows diversification (increase in S.D. and in Mean Squared Displacement from the point of origin) as a function of duration (global diversification) and of embedding (local diversification). Bios but not Chaos generates diversification.

Transitivity and Partial Autocorrelation as Evidence for Causality

Causality is transitive in the mathematical sense. Such transitivity can be detected by statistical correlation. Time series of galactic distances as well as mathematical Bios, and statistical noise (pink, brown) show high positive autocorrelation for at least 16 lags. In contrast, there is a negative autocorrelation between successive terms for periodic series and for chaotic series generated by the process equation; chaotic series generated by the logistic equation resemble random data in showing no autocorrelation. These results can be interpreted to mean that in periodicity and Chaos change embodies opposition, while in Bios and in cosmological processes, change connects similar entities.

The partial autocorrelation coefficient measures the association between A_t and A_{t+k} when the effects of other time lags (1, 2, 3..., up to k-1) are partialled out [Kendal, 1973]. Cosmological time series show extended partial autocorrelations, indicating deterministic rather than probabilistic origin. Positive and negative partial autocorrelations are also observed in biotic series generated by the process equation for several lags. In contrast 1/f noise and brown noise correlate for only one lag as change is random in these series.

Table 2. Partial Autocorrelation: The partial autocorrelation coefficient measures the association between A_t and A_{t+k} when the effects of other time lags (1,2,3..., up to k-1) are partialled out

			Lag 2	Lag 3	Lag 4	Lag 5
Mathematical prototypes		Random	0.0409 P=0.197	-0.0473 P=0.136	-0.0086 P=0.786	0.0108 P=0.735
		Expanding Random	.0312 P=0.325	-0.0561 P=0.077	0.0136 P=0.668	0.0496 P=0.119
		Brownian	-0.034 P=0.278	0.0104 P=0.743	-0.0072 P=0.820	-0.0219 P=0.419
		Bios	0.3940 P<0.001	0.2400 P<0.001	0.1338 P<0.001	0.0832 P=0.009
Galaxies by Mpc	LCRS Degree Declination	45 degrees	0.425 P<0.001	0.243 P<0.001	0.1470 P<0.001	0.1035 P=0.001
		42 degrees	0.4158 P<0.001	0.199 P<0.001	0.1981 P<0.001	0.1408 P<0.001
		39 degrees	0.3988 P<0.001	0.3176 P<0.001	0.1992 P<0.001	0.1659 P<0.001
		12 degrees	0.4293 P<0.001	0.2681 P<0.001	0.1268 P<0.001	0.1029 P=.001
		6 degrees	0.4197 P<0.001	0.2376 P<0.001	0.1642 P<0.001	0.1508 P<0.001
		3 degrees	.4787 P<0.001	0.2243 P<0.001	0.1541 P<0.001	0.1260 P<0.001
	2dFGRS Radian Ascension	0-1 radians	0.3581 P<0.001	0.1512 P<0.001	0.0964 P<0.001	0.1171 P<0.001
		5-6 radians	0.3377 P<0.001	0.1828 P<0.001	0.1171 P<0.001	0.0166 P=.602

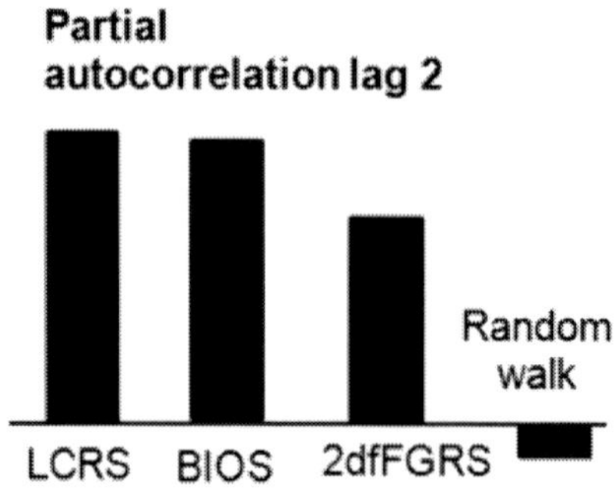

Figure 22. Partial autocorrelation for temporal series of galactic distribution in 2 samples, 2dFGRS (95000 galaxies) and LCRS (25000 galaxies).

Frequency Analysis and the Spatial Distribution of Galaxies as a Function of Time

Another way to examine the temporal evolution of galaxies is to examine their spatial distribution at different ages of the universe. There is a huge hierarchy of structure and motion in the observable universe. The galaxies –some 100 billion of them in the observable universe– journey through space in clusters bound by gravity, which in turn form superclusters of thousands of galaxies stretching many hundreds of millions of light years across space. These superclusters are arranged in filaments or sheet-like structures, between which there are gigantic voids of seemingly empty space. Stars, galaxies, clusters of galaxies, and superclusters, demonstrate an irregular hierarchical structure, which in a way continues the sequence of quarks, nuclei, atoms and molecules at the small end of the scale. The universe appears to be a hierarchical system, at variance with random stationary process models and with the notion of a uniform distribution of matter. These observations have led to a weaker reformulation of the hypothesis, namely that the universe is uniform, homogenous and isotropic only in a large scale and only in a statistical sense.

While at large distances (earlier in time), the distribution of galaxies appears random-like, nearer it shows fractality [Mandelbrot, 1975; Coleman and Pietronero, 1987, Pietronero, et al. 1997; Baryshev and Teerikorpi, 2002]. Also fractality has been ascribed to stationary randomness [Mandelbrot, 1975]; a fractal distribution is compatible with, but not predicted by, stationary randomness. Fractality is a characteristic of chaotic and biotic processes.

Recurrence plots show complexes separated by brief interruptions in the epochs from 100 to 600 Mpc, and become less evident at greater distance when compared with shuffled copies. Embedding plots of z in 100 Mpc epochs show two distinct patterns. In the range between 100 and 600 Mpc, the recurrence plot shows complexes and the quantification of isometries show high dimensional novelty, and nonrandom complexity (arrangement); the pattern is biotic. In contrast, up to 100 Mpc and beyond 600 Mpc, embedding and recurrence plots resemble 1/f noise; there is high dimensional novelty and nonrandom complexity.

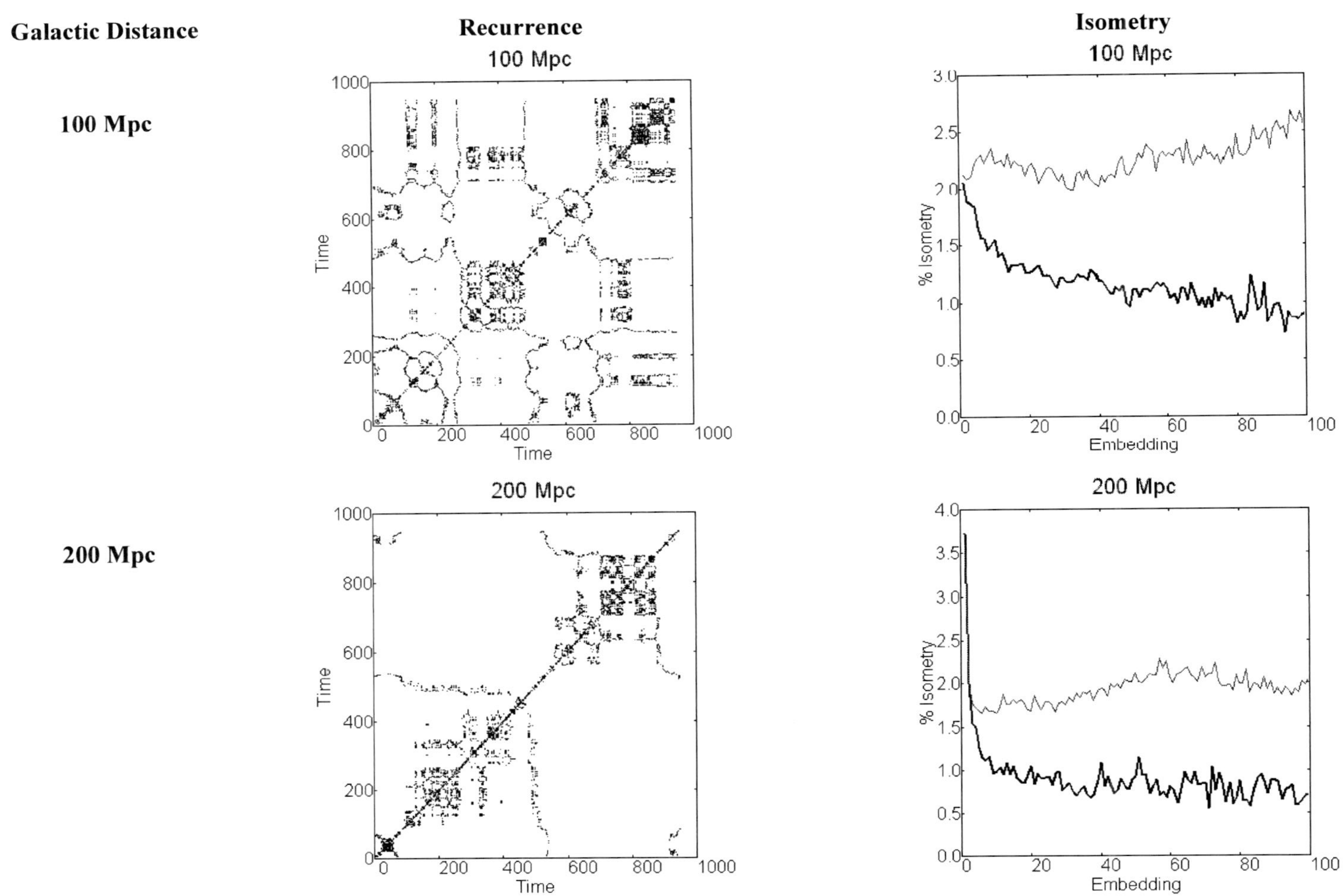

Figure 23. Continued on next page.

300 Mpc

400 Mpc

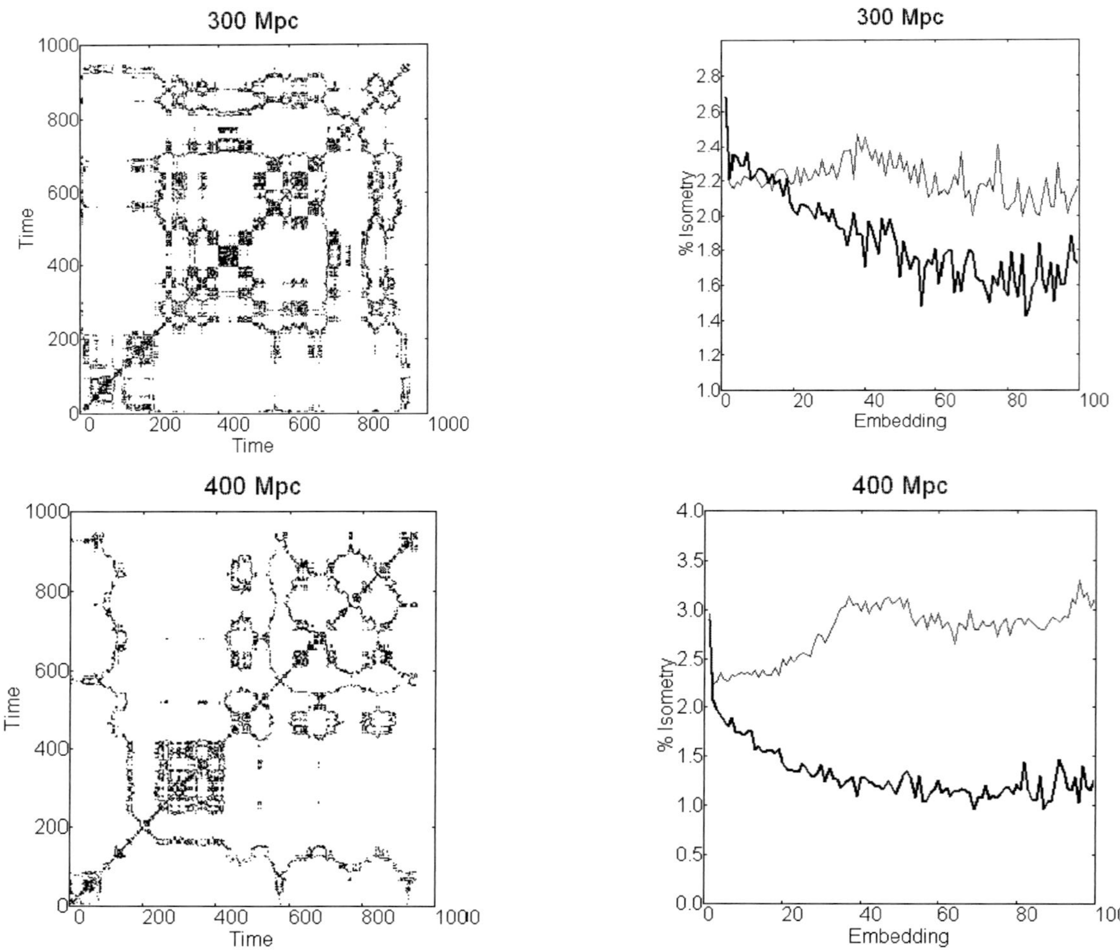

Figure 23. Continued on next page.

500 Mpc

600 Mpc

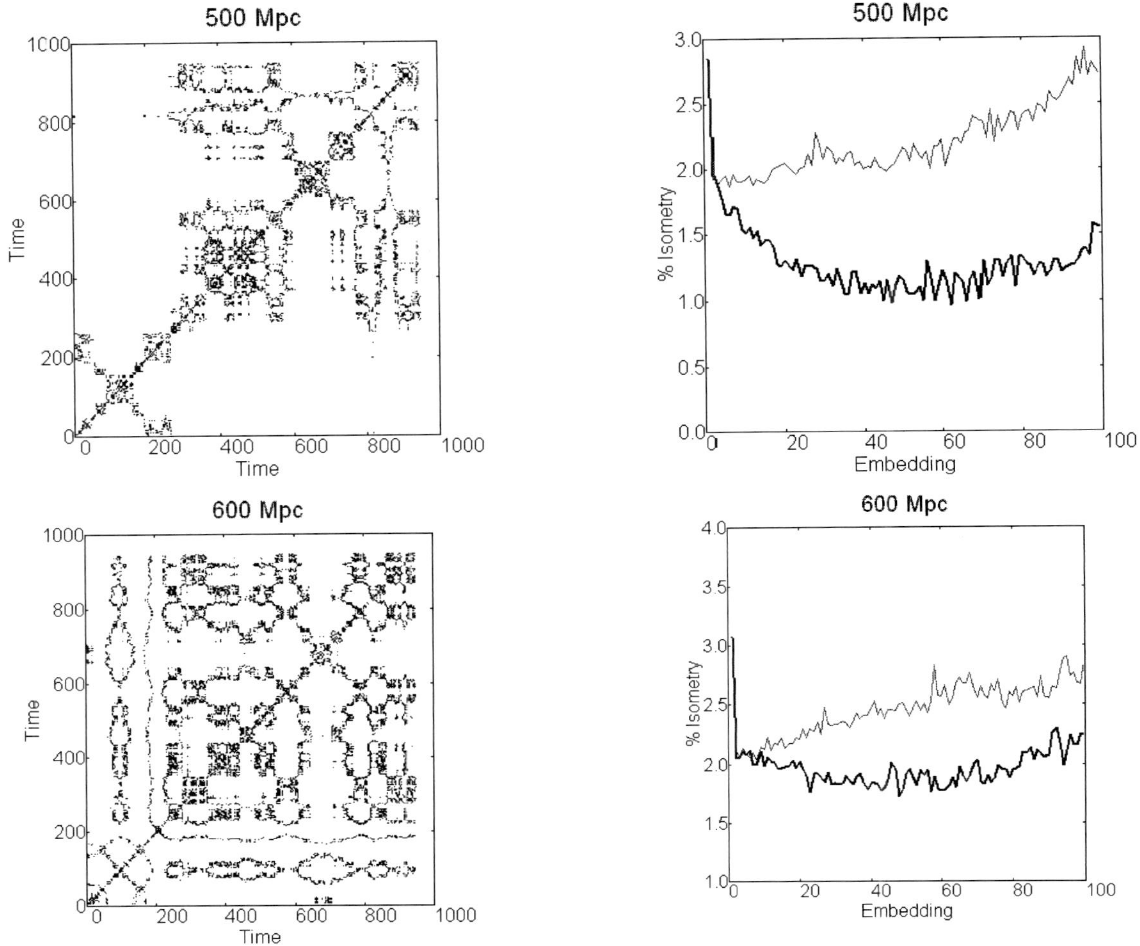

Figure 23. Continued on next page.

700 Mpc

800 Mpc

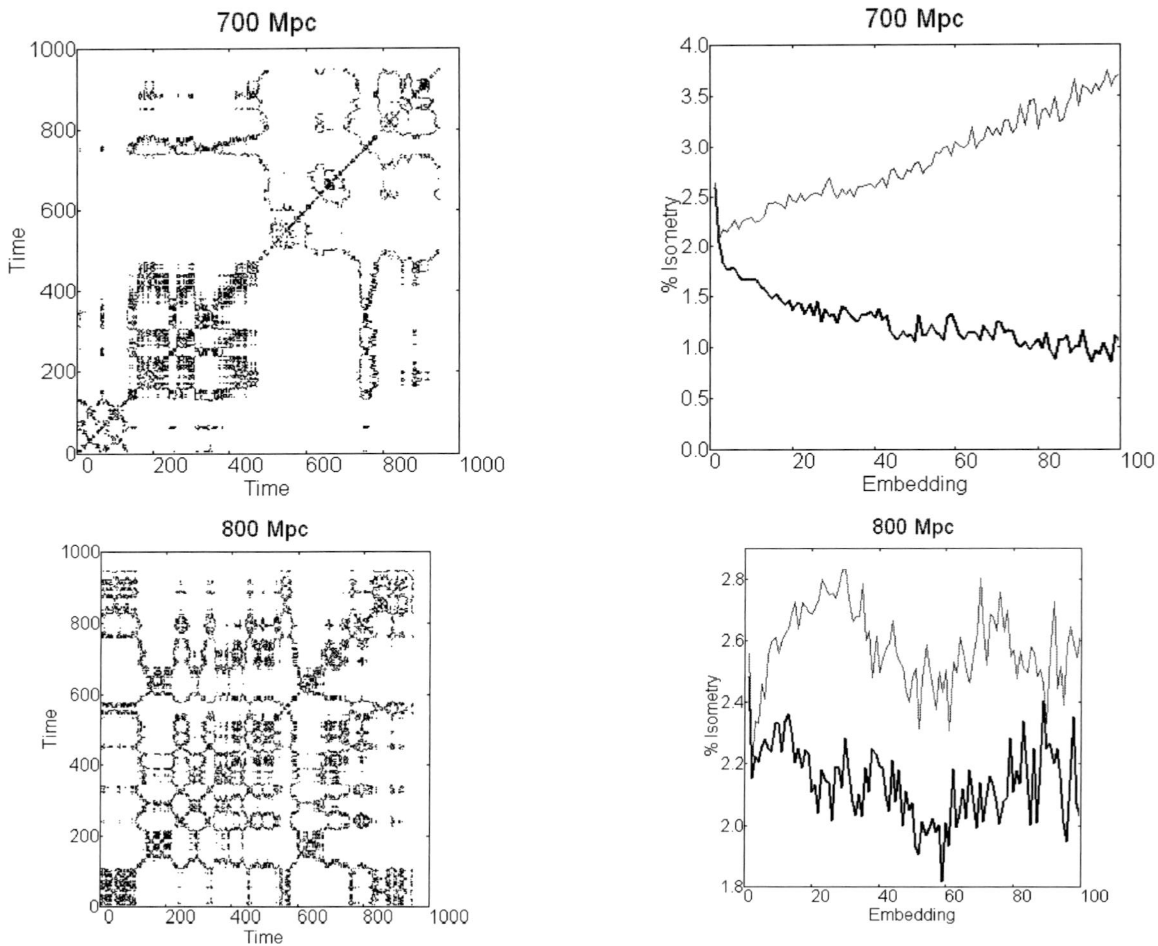

Figure 23. Continued on next page.

900 Mpc

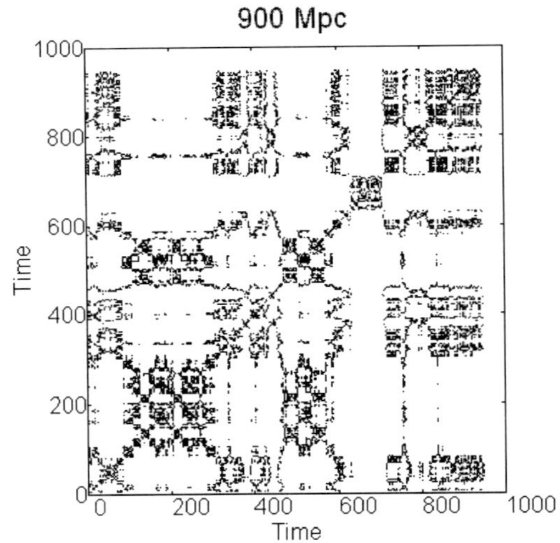

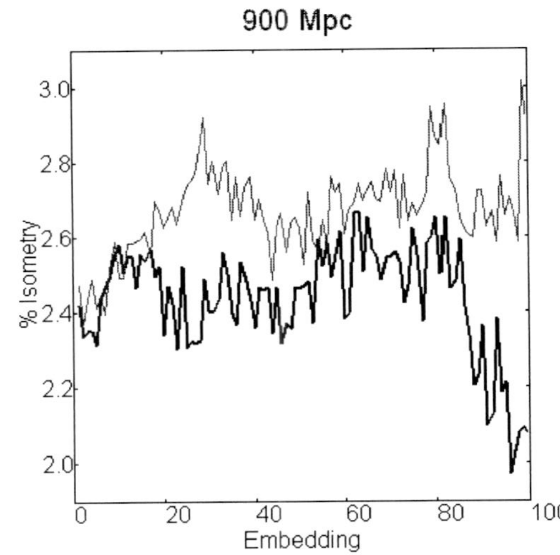

Figure 23. Power spectrum of the distribution of galaxies.

Wavelet plots (Figure 23, left) show structure at intermediate distances, as observed with different methods by previous investigators. Wavelet plots show nonrandom organization in the range between 100 and 700 Mpc (Figure 23). The phase identified as biotic by recurrence analysis is within these limits (100 to 600 Mpc).

The power spectrum of the distribution of galaxies shows wildly different results at different distances (Figure 23, right). A 1/f spectrum is often obtained, while in other cases there is a complex plot. The time series of differences has a positive value for β cosmological series as it is also observed in Bios, and Pink Noise. There also is significant partial autocorrelation for the distribution of galaxies in space from 100 to 800 Mpc.

Many investigators agree that galactic structures are fractal up to a distance scale [Baryshev and Teerikorpi, 2002], i.e. fractality may be limited to the local universe (less than 500 Mpc). Since Bios has fractal features [Sabelli, 1999], the demonstration of Bios is consistent with the observation of fractal features in the distribution of galaxies. Empirical data reveals heterogeneity and hierarchical clustering at all observable scales, and some models suggest that hierarchical clustering continues through to the very largest scales [Pietronero, 1987; Pietronero et al, 1997]. These Italian cosmologists maintain that the apparent homogeneity of the universe is an artifact of the way galaxy surveys are analyzed. A biotic pattern looks random when analyzed with many tests. It is the identification of complexes and the measurement of diversification and novelty that differentiates Bios from random data.

These results show that the spatial distribution of galaxies as a function of the age of the universe show diversification, separate complexes, novelty and causality, confirming the finding of biotic pattern in the two smaller galaxy surveys (LCRS and 2DFGRS) analyzed previously.

The significance of finding a biotic pattern in the distribution of galaxies lies in the expansive nature of Bios. Finding biotic features for at least 2.5 billion years indicates that biotic expansion must have been operative for a significantly large era of cosmic evolution. The process that generates Bios may cause or contribute to the expansion of the universe.

Cosmic Background Radiation

The universe is much older than 2.5 billion years. We can learn about the earlier stages in the evolution of the Universe by observing the radiation that originated in the early universe. The cosmic background radiation (CBR) originated in a universe that was much smaller than the one we inhabit. The standard model postulates the generation of an extremely hot radiation that cools off as the universe expands. This CBR currently is 2.726 ±0.010 °K above absolute zero. The CBR portrays photons bouncing off a hot gas cloud of free electrons and nuclei that filled the universe when it was about 400,000 years old, some 13.7 billion years ago. The CBR photons thus bear a unique imprint of the state of the Universe at 'recombination' time .

While the observable universe displays a rich hierarchical pattern of galaxy clusters and superclusters, the CBR is remarkably isotropic, indicating that the early Universe had a spatially homogeneous and isotropic geometry. This isotropy of the CBR indicates a fundamental oneness of the universe. The CBR also provides evidence for a fundamental twoness (Figure 6,16), a *dipole anisotropy*. This dipole results in part from a Doppler effect as the Earth moves with respect to the microwave background at about 600 km/s. Physicists assume that in addition to this dipole due to our motion, there is a true "cosmic dipole". Observations also demonstrate the existence of a quadrupole, and octapole, all the way to extremely small fluctuations; there is a sequence 1, 2, 4, 8... many as in bifurcation processes.

The small anisotropies in the CBR temperature provide us with an image of the early Universe. "Gravitational instability" models propose that a Gaussian random field of initial fluctuations seeded the formation of the large-scale structure we see today (galaxies, stars and planets). These slight 'ripples' were magnified by the effect of gravity as the small, initially overly dense, areas attracted additional mass.

The dodecahedral structure found by Luminet and coworkers (2001) in the CBR data indicates causal processes rather than stochastic ones. In addition, the disproportion between the observed heterogeneity in the distribution of galaxies and the remarkable isotropy of the CBR suggests that causal and creative processes, rather than a simple magnification of initial non-homogeneity, have operated throughout the history of the universe. Consider then the hypothesis that the early Universe embodies, since its origin, asymmetry and harmonic feedback that would create periodic, chaotic and biotic fluctuations. These initial fluctuations in the radiation could nucleate additional mass thereby generating large and heterogeneous structures.

Sabelli and Kovacevic (2006) set out to test this concept by sampling linear numerical series derived from the WMAP sky temperature maps that were made available for public use by NASA in 2003. The data provided in each frequency map includes the thermodynamic temperature (in mK) for each pixel. We selected a number of iso-latitude rings from the maps converted into the RING numbering scheme, and we constructed numerical series of temperatures.

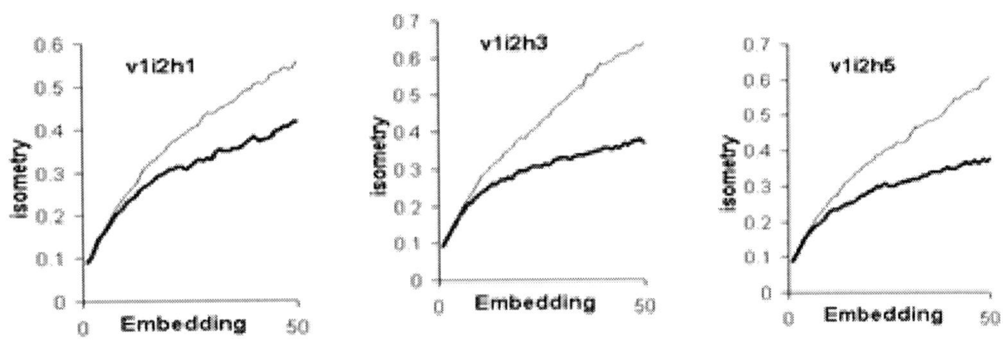

Figure 24. Embedding plots of cosmic background radiation, v1 band relatively near and far from the equator; in all cases we observed novelty.

Wavelet plots show significant structure, which is erased by shuffling. Recurrence plots of series of CBR show episodic patterns (complexes) that are erased by shuffling. Shuffling also increases the number of isometric recurrences, indicating novelty. Also the time series of the differences between consecutive terms shows novelty. The percentage of consecutive recurrence is the same in the original data and their shuffled copies, as observed with stochastic noise but also in some causally generated series such as DNA sequences [Sabelli,

2005]. The electromagnetic cosmic microwave background shows 1/f power spectrum, novelty and partial autocorrelation [Sabelli, 2005], as observed with Bios, and at variance with noise generated by random change. Numerical series are constructed using iso-latitude rings at several latitudes show small partial autocorrelation (except for those near the equator) yet larger than observed in shuffled copies.

In brief the CBR shows novelty but not all the features of Bios. Of course the way in which we have sampled the data as circles may not be useful to actually depict the actual pattern of the process. On the other hand, one may also attribute the occurrence of novelty without other characteristics of Bios to the fact that photons have no mass, as will be discussed later.

Gravitational Wave Background

Another relevant data source is the recordings of presumed gravitational waves by LIGO. Einstein predicted the existence of these gravitational waves in his general theory of relativity. Gravitational waves are waves in space-time that travel at the speed of light. Their emission by relativistic processes is a strong prediction of General Relativity. Since Einstein predicted the existence of gravitational waves in 1916, there has been a heated debate regarding the existence of such waves, and, more recently, numerous attempts to obtain empirical evidence supporting their existence. In the late 1950s, Bondi (1959) rigorously proved that gravitational radiation should in fact be a physically observable phenomenon; as gravitational waves carry energy, a system that emits them should lose energy. Joseph Taylor and Russell A. Hulse were awarded the 1993 Nobel Prize in Physics for their discovery of a binary pulsar (two neutron stars orbiting each other) in which the orbit period is declining by about 75 millionths of a second per year, rotating faster about each other in an increasingly tight orbit, presumably because the system is emitting energy in the form of gravitational waves. The theoretically calculated value from relativity theory agrees to within about one half of a percent with the observed value. As a result, there is increasing confidence that gravitational waves exist [Thorne and Kenan, 1997] Available technology allows one to detect the small changes in the energy of astronomical bodies predicted by the theory, and a number of stations have been set all over the world to detect gravitational waves. Such interferometers are plagued by seismic vibrations of the ground beneath the interferometer and thermally-induced vibrations of the test masses and of the wires that suspend them. Yet the empirical analysis of the data obtained in such observations may guide the design of methods for the study of gravitational waves.

With the construction of the LIGO facility at Hartford, Washington, recordings have been obtained that may include evidence for such waves originating from binary stars, colliding black holes, and other such massive events. Whether such data truly provides evidence for gravitational waves is still undetermined. Yet since extreme care has been made to eliminate known man-made effects, we believe that they provide enormously rich data reflecting non-linear natural phenomena.

The LIGO data includes four different categories of gravitational wave sources: millisecond bursts, inspiralling binary neutron stars, periodic waves from a known pulsar, and a background of gravitational waves analogous to the electromagnetic cosmic microwave background. This may appear to be noise in each detector, but the signal could be extracted through a correlation of the outputs of two detectors [Christensen, 1992; Allen and Romano, 1999; Pietronero et al, 1997]. Gravitation defines the spacetime in which actions take place. The force of gravity generates a curvature of spacetime. When massive objects move, the curvature changes, causing gravitational waves in spacetime that spread outward just like ripples on the surface of a pond. A gravitational wave is a fluctuation in the curvature of spacetime which carries energy (gravitational radiation). In a different conceptual framework, gravitational interactions are mediated by gravitons ((hypothetical particles having no mass and no charge that mediate the gravitational force) instead of being described in terms of curved spacetime) and gravitational waves are viewed as coherent states of many gravitons. A gravitational wave propagates outward from a moving object whose motion involves acceleration, provided that the motion is not perfectly spherically or cylindrically symmetric. A binary star system is expected to emit gravitational waves carrying off energy as it was observed by Hulse and Taylor. These waves are in the low-frequency end of the gravitational-wave spectrum (10-7 to 105 Hz); high-frequency waves (above 105 Hz and probably 1010 Hz) are theorized to be faint imprints of the Big Bang [Grishchuk, 1976] but may be artifacts.

Theoretically, it is expected that the present universe should be permeated by a diffuse background of gravitational waves. The gravitational wave background dates from the trillionth-of-a-second mark, while the CMB sets in around 380,000 years later when the first atoms formed. The GWB stems from three different production processes at work in the inflationary era: waves stemming from the inflationary expansion of space itself; waves from the collision of bubble-like clumps of new matter at reheating after inflation; and waves from the turbulent fluid mixing of the early pools of matter and radiation, before equilibrium among them ("thermalization") had been achieved. The gravity waves would never have been in equilibrium with the matter (since gravity is such a weak force there wouldn't be time to mingle adequately) [Schewe et al, 2007]. Garcia-Bellido and Figueroa (2008) predict that the GWB waves would have a broad spectrum (distinct from gravity waves from point sources such as the collision of two black holes, that release waves with a sharp spectral signal) and a larger red-shifting than in CBR. It is believed that these waves are a *stochastic* background produced by the same early-universe processes as might radiate at low and high frequencies: cosmic strings, phase transitions, and the Big Bang. The background gravitational waves have been labeled as noise prior to measurement.

Figure 25 (top, left) shows the pattern observed in all the time series examined. The data cluster at low values and demonstrate dispersion at higher values. There are no significant variations with time. The time series of the differences between consecutive terms in the original data show [Figure 25 (top, right)] a clustering at zero with scattered positive and negative data. Current analysis of the data has focused, and rightly so, on whether the data supports gravitational waves. As may often happen however, the data may provide other compelling information, especially if analyzed in a novel way. We have looked at such data with recurrence plots.

Figure 25 (second row) shows representative examples of the recurrence plots and suggests why this structure reflects an underlying process that is generating novelty and diversity. Recurrence plots of LIGO series display complex structure. The most outstanding feature of these plots is the clustering of recurrences (*complexes*), separated by recurrence-free intervals, with both geometric and organic features. Also the time series of differences shows complexes. These recurrence plots are similar to those obtained with mathematical Bios and with Random Walks.

In nearly all the series examined the number of isometries is very high at low embeddings and then rapidly decreases to a rate of two to five percent of isometries (Figure 25, third row). This rate was fairly constant in most observations and it is larger than the number of isometries observed after the data had been shuffled in 70 % of cases, indicating novelty. Novelty is greater and more consistent in the time series of differences. The fact that there is less recurrence in the original data than in the randomized copy suggests that there is a process that generates variation greater than expected by random change.

The percentage of isometries that are consecutive is high (Figure 25, fourth row); it is higher than the percentage observed in shuffled copies of the data only at low embeddings and only in 60% of the cases. Similar results are obtained in the time series of differences. Consecutive isometry at low embeddings indicates nonrandom causation ("determinism").

Both the original time series and the time series of the differences show higher arrangement than shuffled copies, local diversification and, most strikingly, asymmetry in the distribution as shown by an intercept below 1 in the entropy lot.

The standard deviation of vectors of consecutive points, starting at each point, increases with embedding to achieve a fixed value between ten and thirty embeddings in random and chaotic data. In contrast, as shown in Figure 25, LIGO series show an increase in the S.D. (vertical axis) with the duration of the sample (global diversification) or with duration of the vector (local diversification) used to calculate it. Global and local diversification are also observed in mathematical Bios and in Random Walks. The demonstration of diversification shows that the observed data does not represent the product of an ordered process that tends to an attractor –either linear, periodic, or chaotic. Diversification indicates the continual generation of variance.

Entropy of LIGO series is 0.15 $\pm$0.14 at 2 bins (asymmetry) and 3.48 $\pm$0. at 64 bins, indicating that the entropy increased 0.054 $\pm$.007 per bin (low variety). These values are widely different from those obtained with random data which, being symmetric, have an entropy of 1 as measured with two bins, and being maximally variable, have a slope of 45 degrees.

The demonstration of pattern in time series of differences between consecutive terms demonstrates causation. Differencing the LIGO data generated series that were bipolar but otherwise showed the same characteristics (novelty, diversification, low symmetry as measured by entropy) as observed in the original data; these features were observed even after differencing the data four times. Figure 26 shows analyses of the difference between consecutive terms in two recordings from the same source. Note the extreme complexity of the recurrence plots.

The interpretation of these data is difficult because it has not been established that the phenomena recorded are gravitational waves, and if they are, what is their origin? In fact much research is currently addressing these issues, and trying to identify patterns that will indicate particular events, such as rotating binary stars, that could originate gravitational waves. Exactly because such events have not been identified so far in the data, and instead the processes recorded have the same pattern at all times and in many directions of space, we are inclined to think that the patterns observed correspond to what cosmologists regard as the gravitational wave background. We propose, then, that the gravitational wave background is Biotic rather than random. This is particularly significant because it originates within one trillionth of a second after the Big Bang. We are, however, aware that there may be other sources for the data observed, which after all are obtained from the surface of our planet. So it may even include artifacts, in spite of the fact of the meticulous and sophisticated work of the LIGO scientists. Whether or not the LIGO recordings portray the initial GWB, as we perceive today the vibrations that surround our solar system are biotic and thereby indicate a causal, creative process at work.

Power spectrum analysis shows a low slope beta between white noise, such as observing random and chaotic series (beta = 0) and -1 as observed in 1/f processes such as pink noise. This is observed in some biotic series. The beta for the series of differences is usually around 1.5 and for the cumulative sum of the terms around -1.8, as observed with typical biotic series. The low but non zero beta of gravitational waves (Table 4) may indicate a mixture of white noise (Random or Chaos) with Bios, pink noise or Brownian noise. These studies were continued (Sabelli 2009, in press).

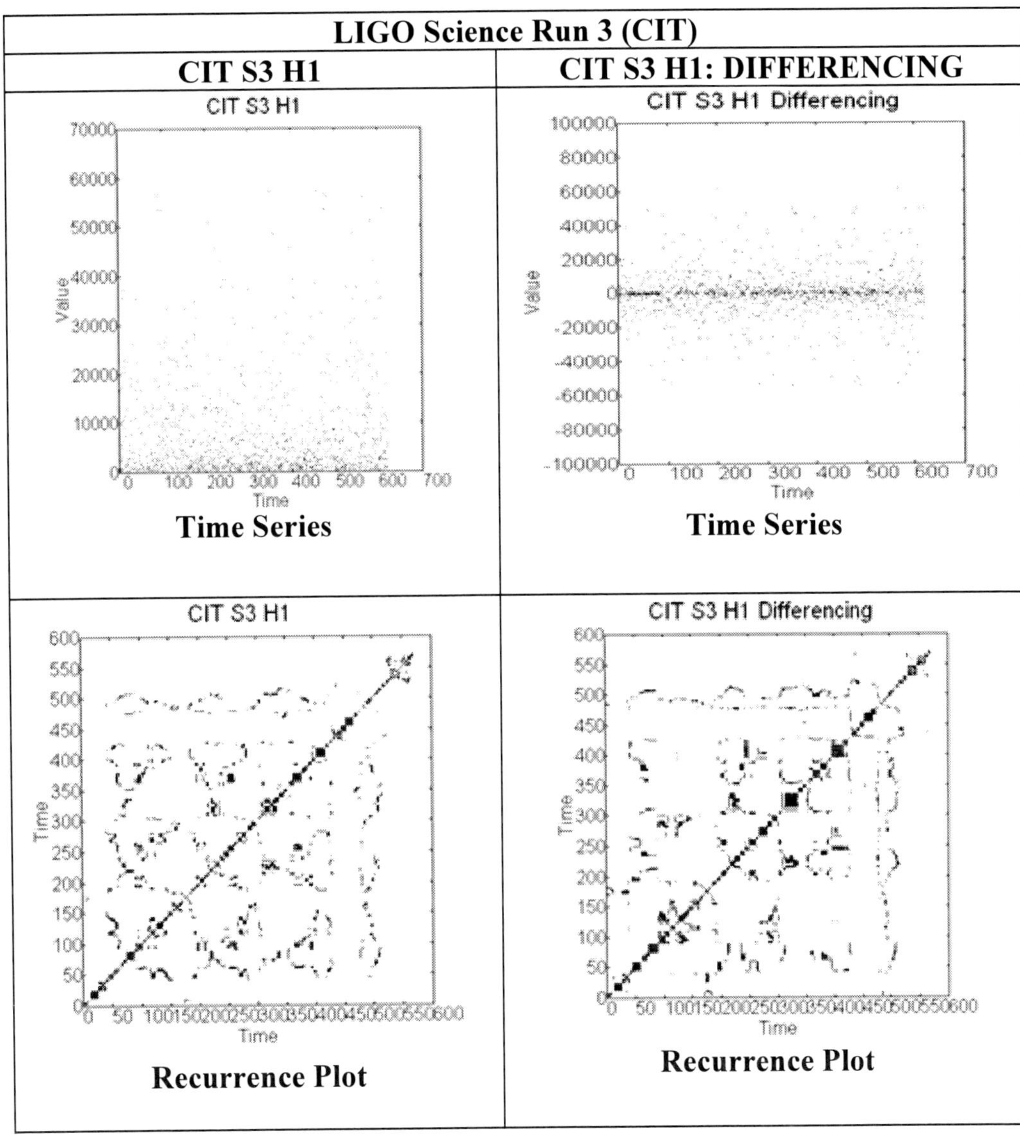

Figure 25. Continued on next page.

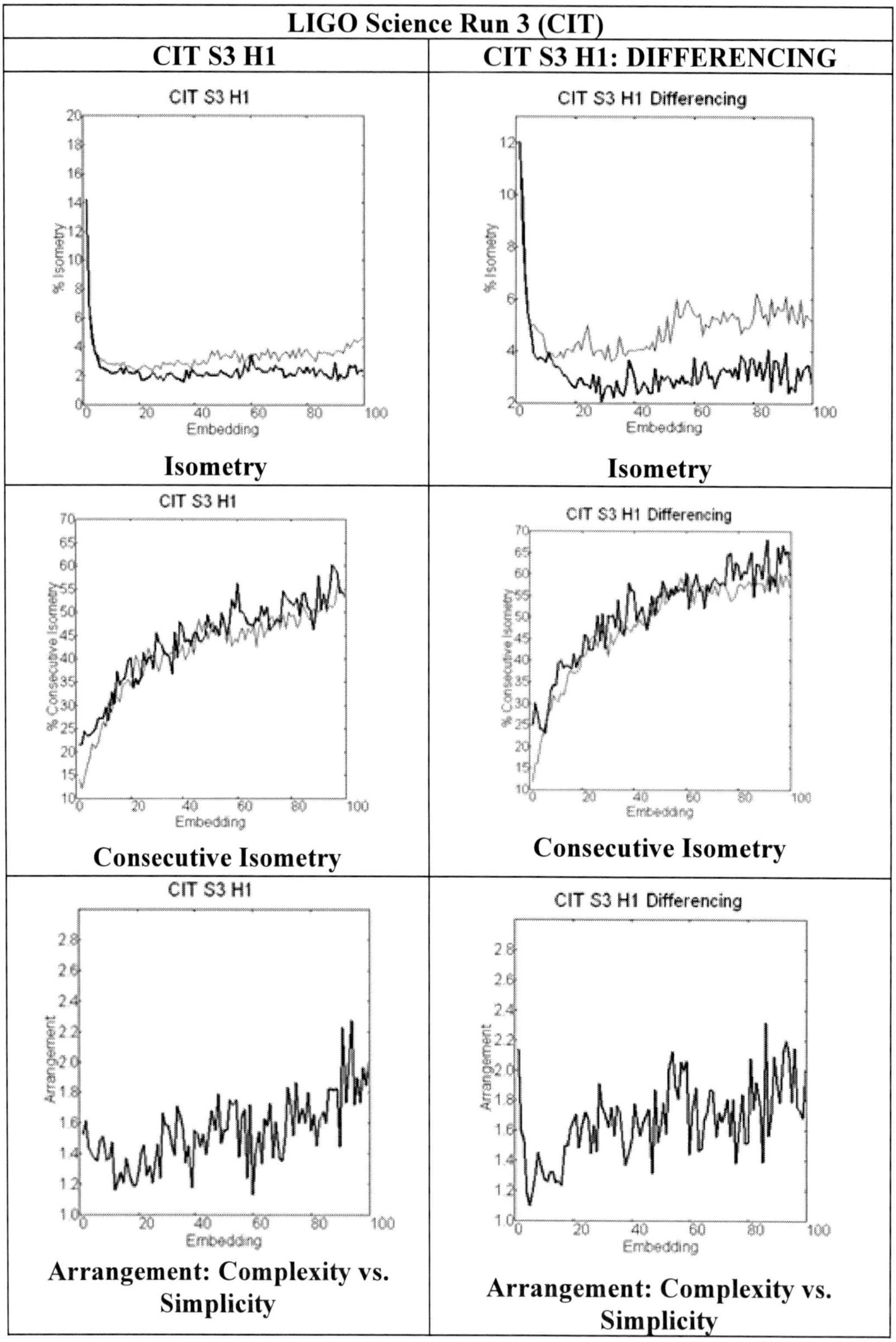

Figure 25. Continued on next page.

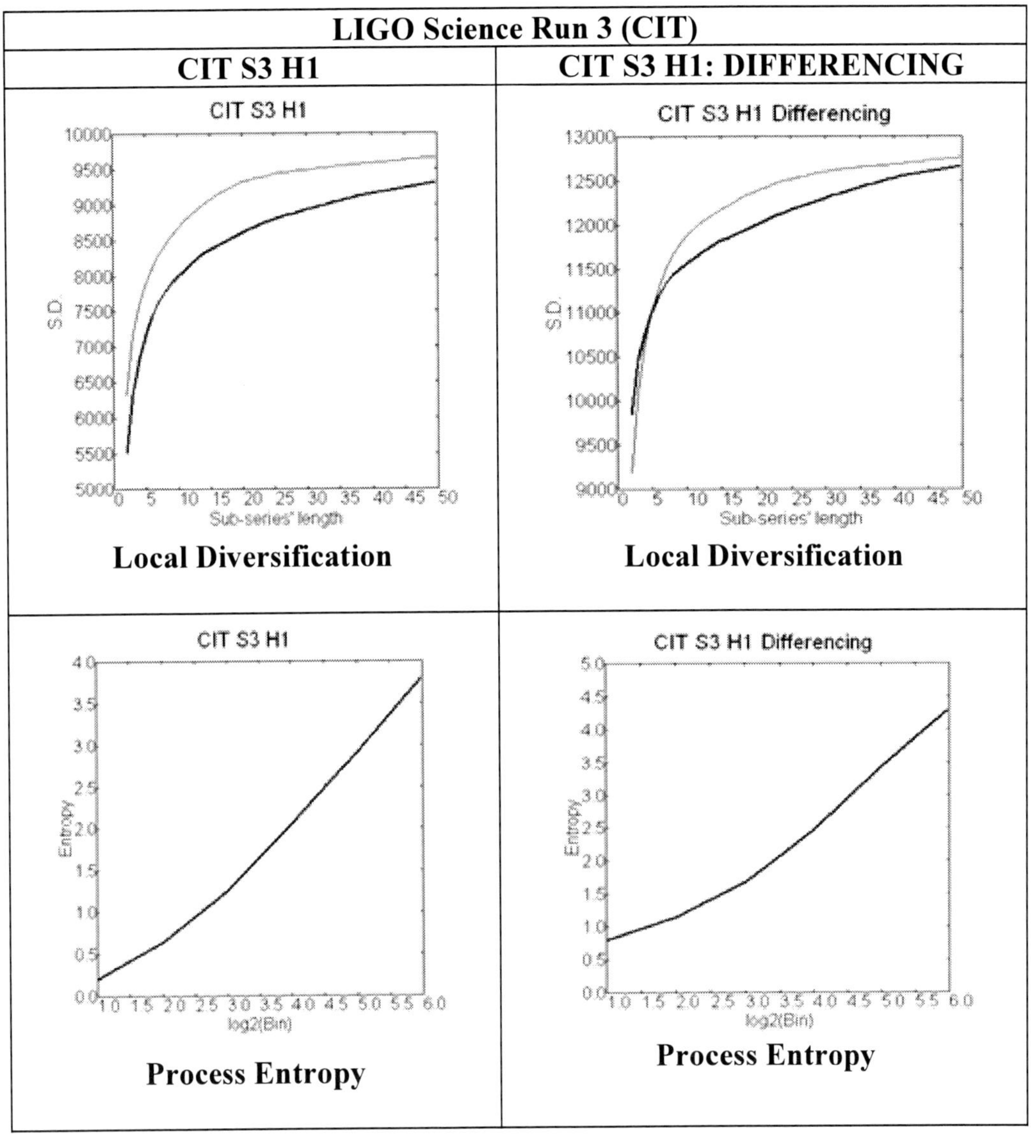

Figure 25. Time series recurrence plots and isometry quantification, local diversification, and process entropy of a recording from LIGO and the time series of differences between consecutive terms.

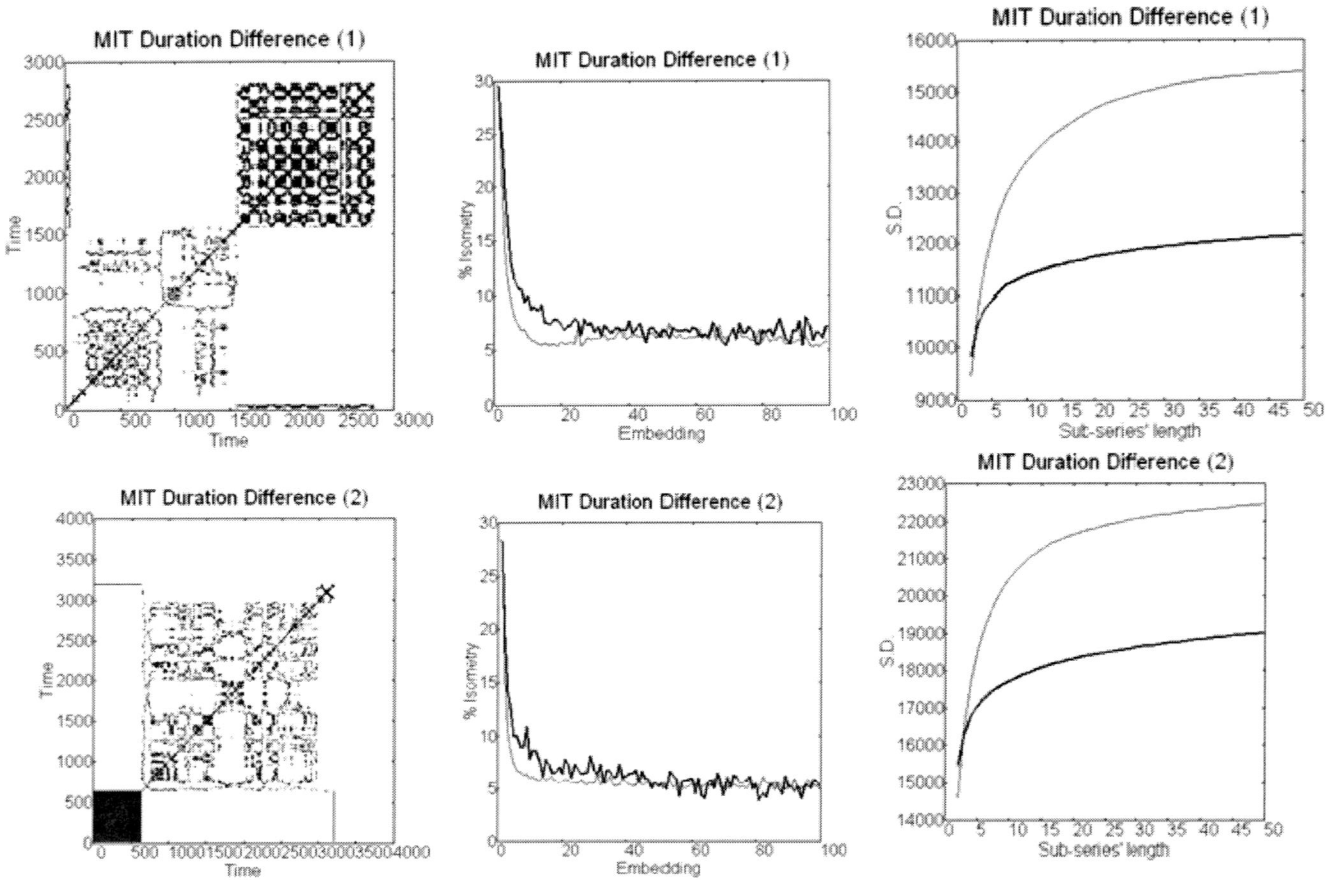

Figure 26. Data obtained from two different Science Runs from MIT. Pattern in the series of differences between consecutive terms of a series, indicates causal rather than random change.

Table 3. Slope β of the log-power/log-frequency plots

Sample	Beta +/- standard error		
	Series	Difference	Cumulative
CIT S5 H1RDSC03L2H	-0.1690 +/- 0.0294	1.6746 +/- 0.0266	-1.8082 +/- 0.0042
CIT S5 H2RDSC03L2H	-0.1694 +/- 0.0318	1.6823 +/- 0.0271	-1.8081 +/- 0.0044
CIT S5 L1RDSC03L1L	-0.2860 +/- 0.0317	1.4934 +/- 0.0293	-1.8076 +/- 0.0043
CIT S5 L1RDSC03L2L	-0.2776 +/- 0.0284	1.5541 +/- 0.0268	-1.8112 +/- 0.0039
CIT S5 H2RDSC03L1H	-0.1694 +/- 0.0318	1.6823 +/- 0.0271	-1.8081 +/- 0.0044
CIT S5 H1RDSC03L1H	-0.1989 +/- 0.0307	1.6695 +/- 0.0273	-1.8082 +/- 0.0042
CIT S4 L1RDSC04LX	-0.2395 +/- 0.1510	2.1098 +/- 0.0000	-1.7920 +/- 0.0246
CIT S3 H1RDSC01LX	-0.3224 +/- 0.0672	1.5623 +/- 0.0389	-1.7971 +/- 0.0103
CIT S3 L1RDSC01LX	-0.3538 +/- 0.0922	1.3723 +/- 0.0521	-1.7871 +/- .0123
CIT S3 H2RDSC01LX	-0.2038 +/- 0.0761	1.7077 +/- 0.0390	-1.7977 +/- 0.0114
LLO S5L1RDSC02LXl	-0.2199 +/- 0.0355	1.6221 +/- 0.0318	-1.8053 +/- 0.0049
UWM S5L1RDSC03L2L	-0.3303 +/- 0.0343	1.4780 +/- 0.0324	-1.8062 +/- 0.0047
UWM S5H2RDSC03L2H	-0.1694 +/- 0.0318	1.6823 +/- 0.0271	-1.8081 +/- 0.0044
UWM S5H1RDSC03L2H	-0.1952 +/- 0.0312	1.6878 +/- 0.0276	-1.8082 +/- 0.0042
MIT S5H2RDSRL3L	-.2309 +/- 0.0301	1.5876 +/ 0.0278	-1.8109 +/ 0.0041
MIT S5H2RDSC02 LXH	-0.2066 +/- 0.0328	1.6152 +/- 0.0299	-1.8103 +/- 0.0046
LHO S5H2RDSC03L2H	-0.1343 +/- 0.0322	1.6739 +/- 0.0297	-1.8075 +/- 0.0044
LHO S5H1RDSC03L2H	-0.1729 +/- 0.0343	1.6464 +/- 0.0309	-1.8071 +/- 0.0047

β = 0 randomness and chaotic series (white noise).
B = -1 (1/f or pink noise), β = -2 (Brownian noise), -2 < β < -4 (black noise).

<table>
<tr><th colspan="4">Birmingham, Alabama</th></tr>
<tr><th>Graph</th><th>1995</th><th>1996</th><th>1997</th></tr>
<tr><td>Time Series</td><td>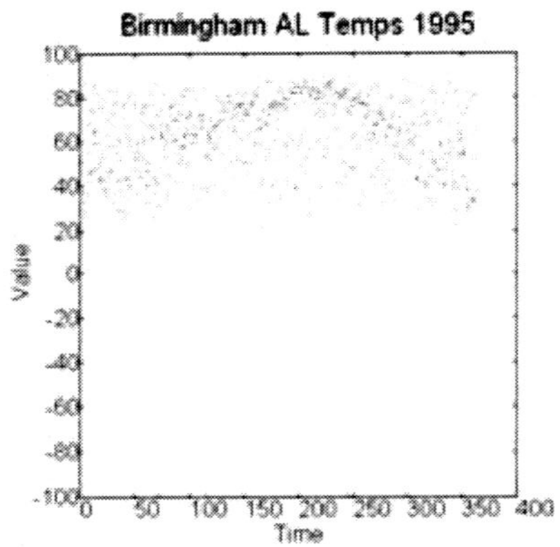</td><td>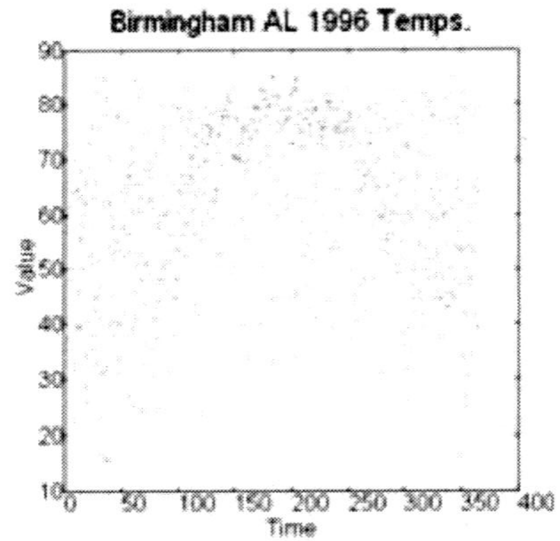</td><td>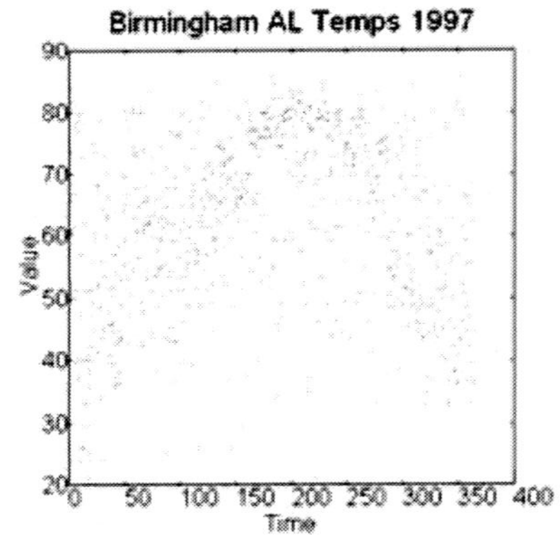</td></tr>
<tr><td>Recurrence Plot</td><td>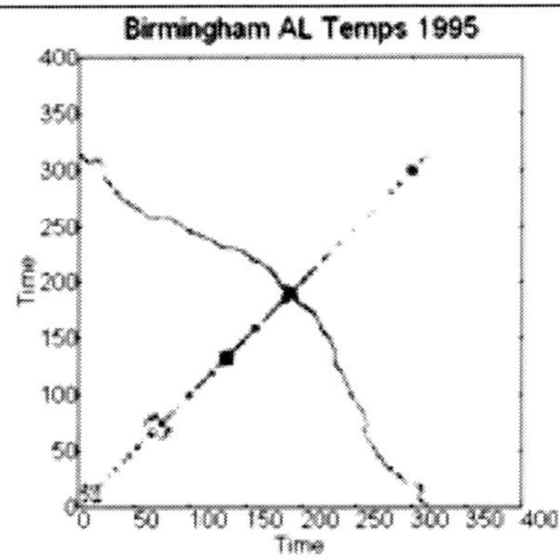</td><td>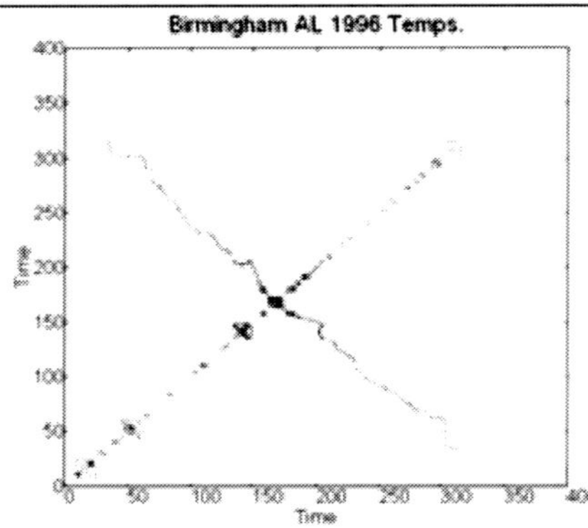</td><td>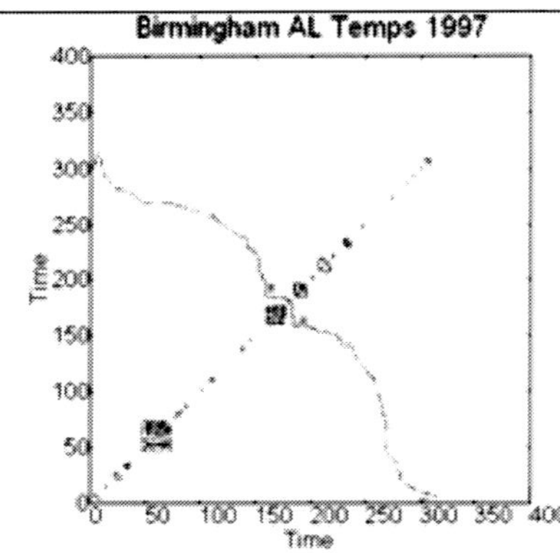</td></tr>
</table>

Figure 27. Continued on next page.

<table>
<tr><th colspan="4">Birmingham, Alabama</th></tr>
<tr><th>Graph</th><th>1995</th><th>1996</th><th>1997</th></tr>
<tr><td>Isometry</td><td>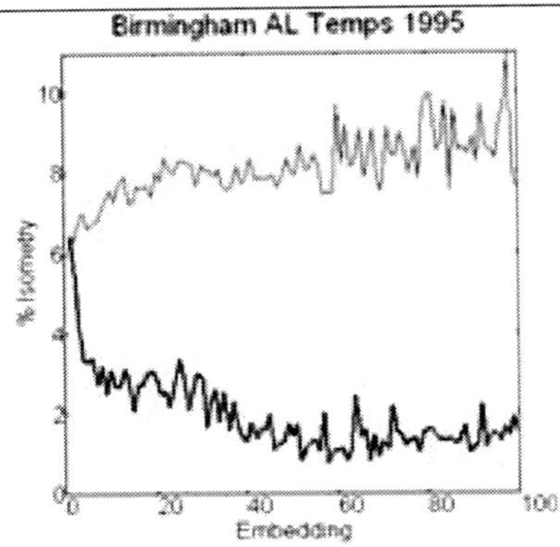
</td><td>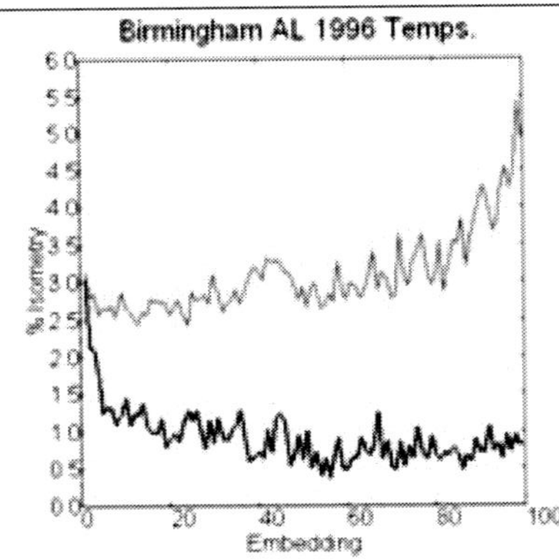
</td><td>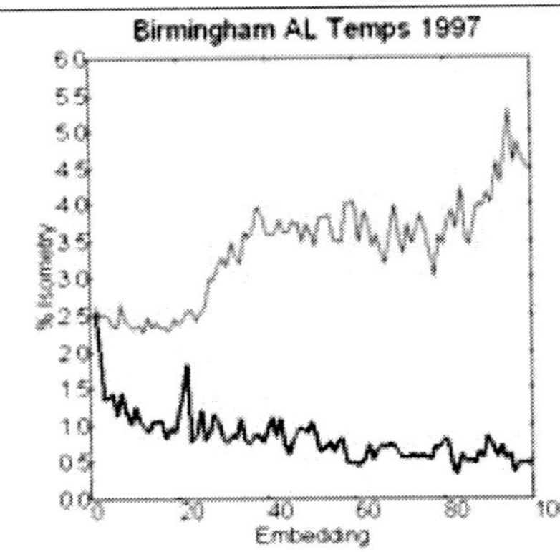
</td></tr>
<tr><th colspan="4">San Francisco, California</th></tr>
<tr><td>Time Series</td><td>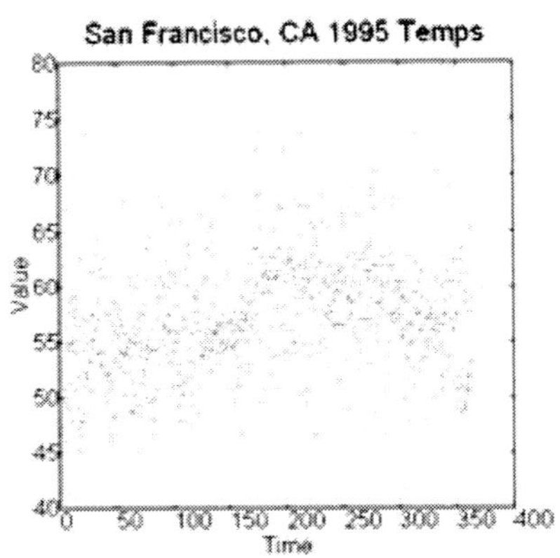
</td><td>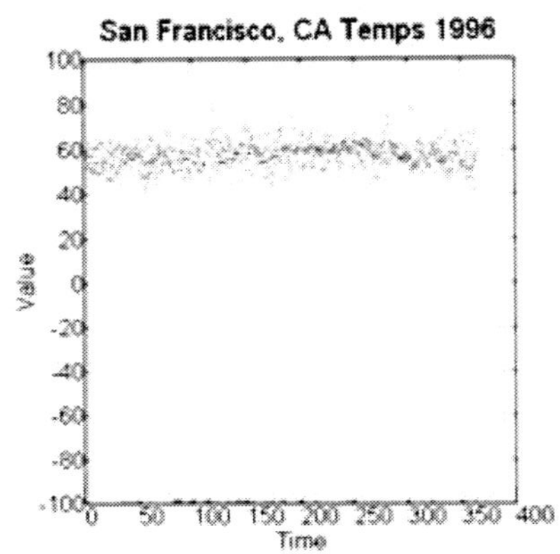
</td><td>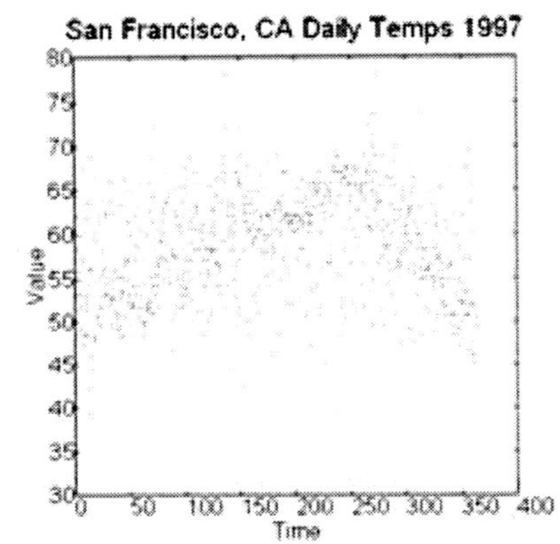
</td></tr>
</table>

Figure 27. Continued on next page.

San Francisco, California			
Graph	**1995**	**1996**	**1997**
Recurrence Plot	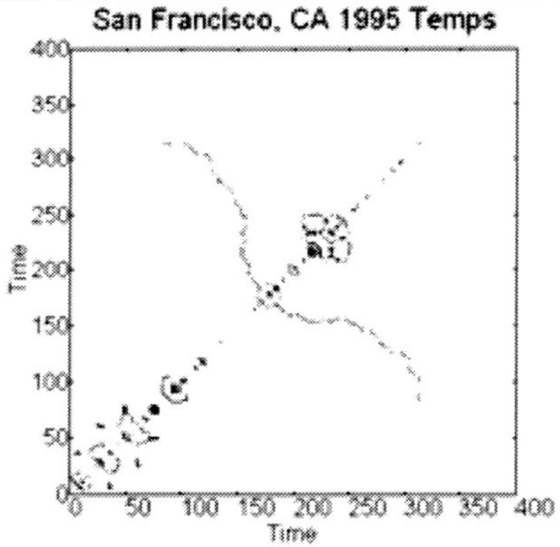	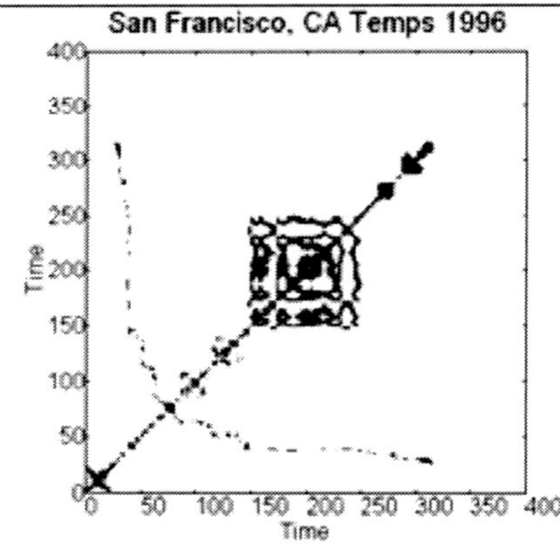	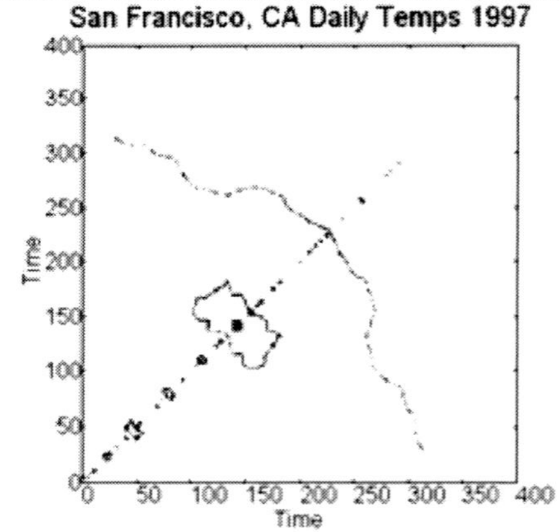
Isometry	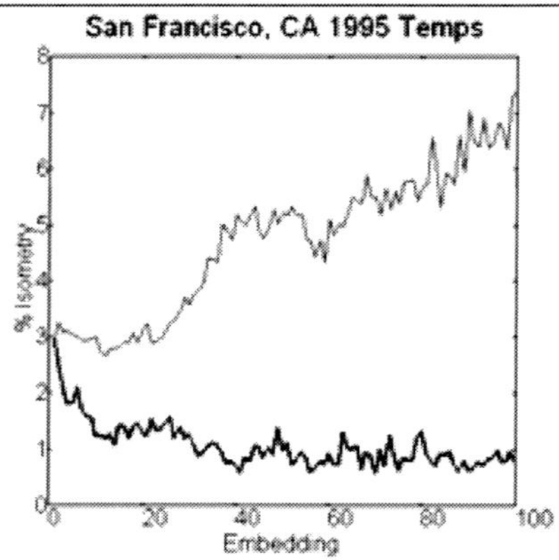	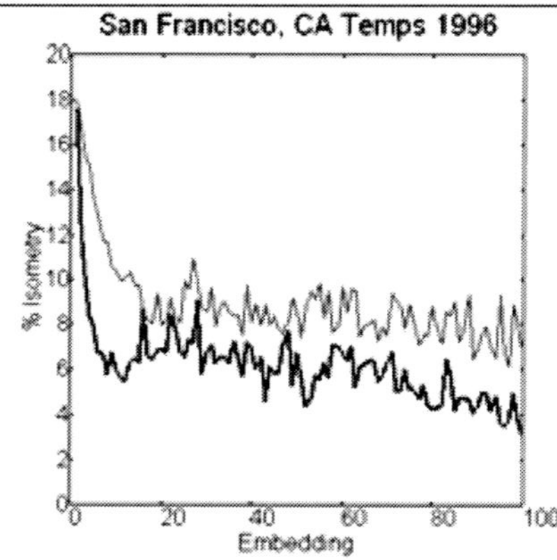	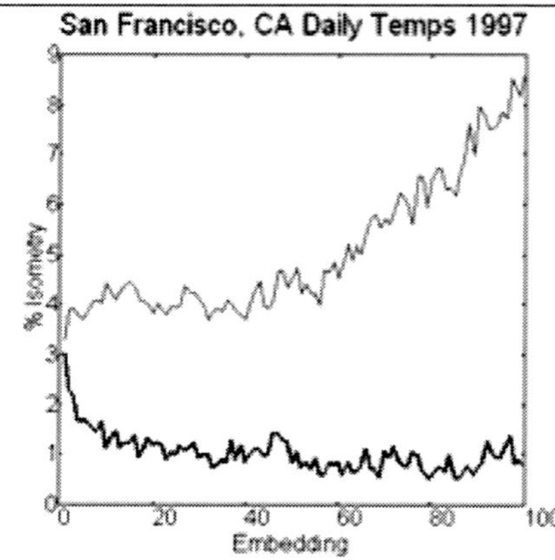

Figure 27. Continued on next page.

Graph	1995	1996	1997
	Chicago, Illinois		
Time Series	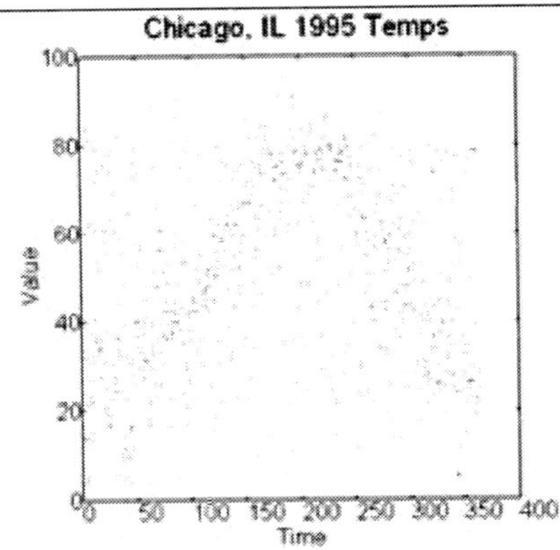	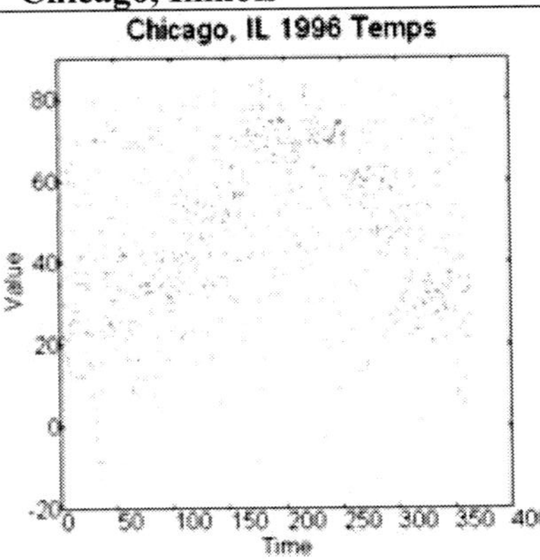	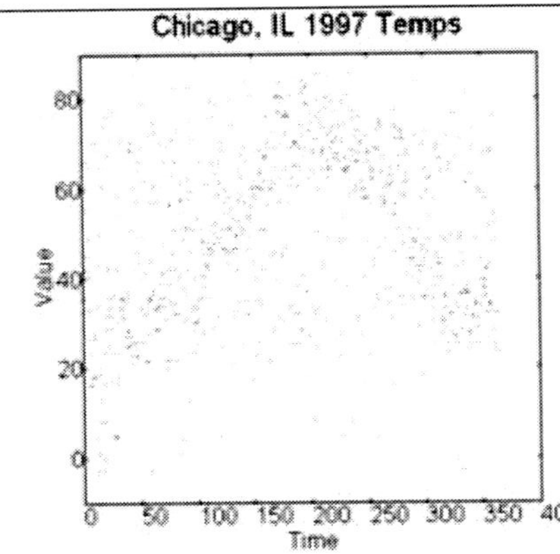
Recurrence Plot	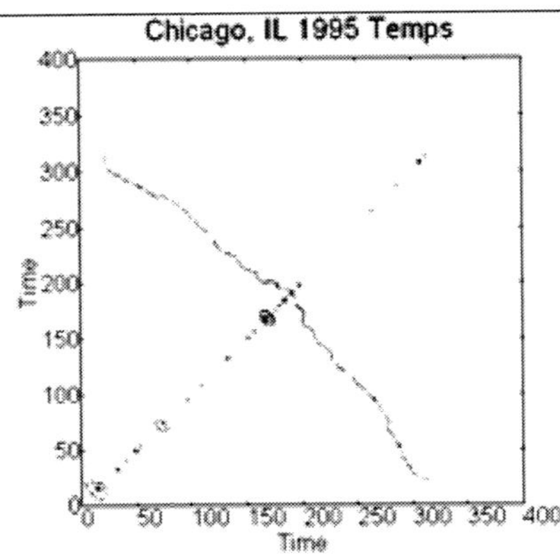	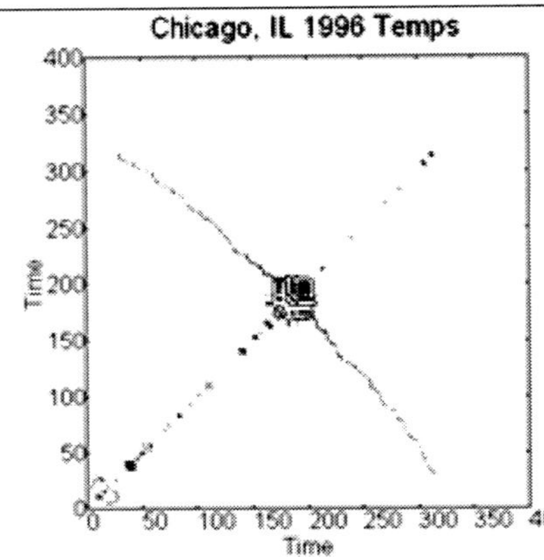	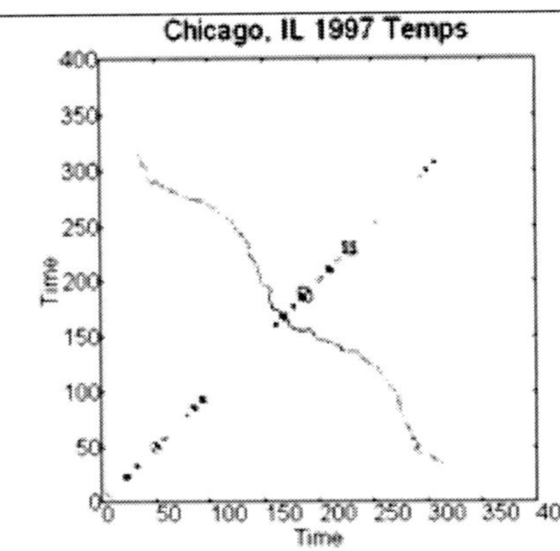

Figure 27. Continued on next page.

Graph	1995	1996	1997
	Chicago, Illinois		
Isometry	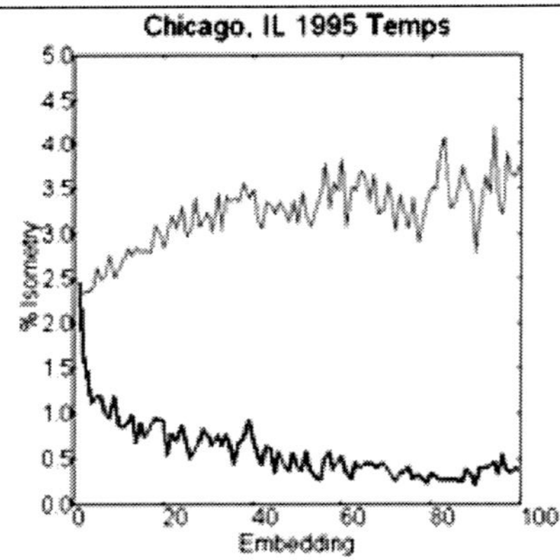	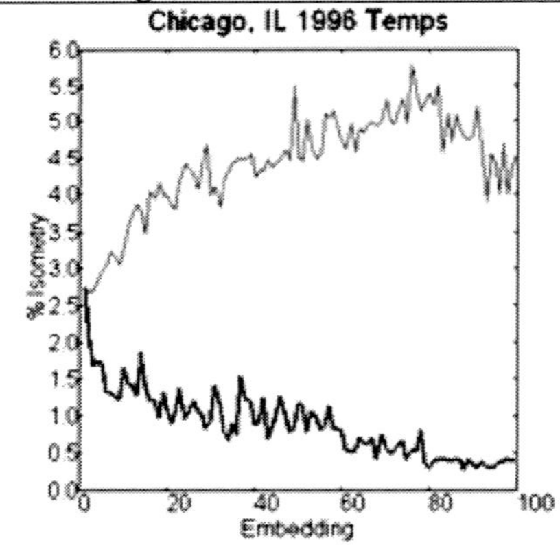	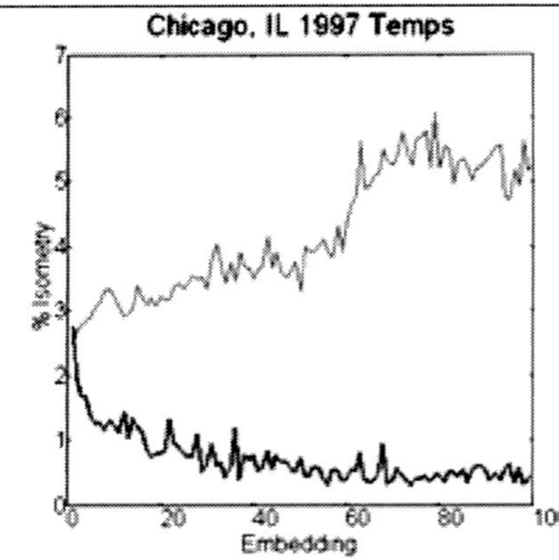

Figure 27. Annual temperature changes in three American cities and the corresponding recurrence and embedding plots of isometries.

Contemporary Planetary Processes: Temperature

Figure 27 shows the temperature recorded in a one year period in American cities. While the overall change in temperature across many years show global warming and yearly recording shows the expected cycle of seasons, recurrence plots show few recurrences clustered in highly organized complexes and isometry quantification shows novelty and other indications of Biotic pattern.

The Generation of Bios by Quantum Processes

The Schrödinger equation, and its relativistic generalizations, describes quantum processes whether they be atoms or molecules, nuclei or quarks. Presumably it also describes the quantum oscillations in the primordial atom. The standard interpretation appeals to concepts such as probability and uncertainty but the wave function as described by Schrödinger provides a complete and consistent description, a description that is deterministic to the extent that the wave function evolves in time in a unique way given appropriate boundary conditions at some initial instant of time.

In an earlier publication we reported that the Schrodinger equation for an electron restricted to a box generates a biotic pattern [Sabelli and Kovacevic, 2006]. In another set of experiments [Thomas et al, 2006] we computed the exact continuous Schrodinger equation, its relativistic equivalent the Klein-Gordon equation, and considered three different distance scales: atomic, nuclear and Planck (time and distance scale obtaining shortly after the Big Bang).

We start with the equations for a particularly simple one-dimensional Schrödinger equation for an electron trapped in a well by its electric charge. In three dimensions the Schrödinger equation has the following form for a general potential V :

$$i\hbar\frac{\partial\psi}{\partial t}=-\frac{\hbar^2}{2m}\nabla^2\psi+V\psi .$$

The equation is solved by considering a general superposition of energy *eigensolutions* to this equation where the time dependence is extracted: The general form isolates the spatial dependence in the factor ψ_n

$$\psi=e^{-iE_nt/\hbar}\psi_n .$$

For the special case of a one-dimensional problem where the *particle* is trapped inside a well, the equation for the spatial factor is an ordinary differential equation in the one-dimensional direction x and depends on the eigenenergy E_n:

$$E_n\psi_n=-\frac{\hbar^2}{2m}\frac{d^2\psi_n}{dx^2} .$$

At the boundary of the well, the wave function is required to be zero to reflect the concept that a well is by definition an area from which a particle cannot escape. The resultant equation is:

$$\left(\frac{2mE_n}{\hbar^2}+\frac{d^2}{dx^2}\right)\psi_n=0\,.$$

In a well of size a, as noted the wave function must vanish at both ends. The above equation is the differential equation for the *sine* and *cosine* functions.

Because of the boundary condition at $x=0$ and $x=a$, the energy eigensolution must be proportional to

$$\psi_n=\sin\left(\pi n\frac{x}{a}\right).$$

Furthermore, for this solution to satisfy the Schrödinger equation, the possible energy states are discrete and satisfy:

$$\frac{2mE_n}{\hbar^2}=\frac{\pi^2n^2}{a^2}$$

Thus there is no need for numerical approximation for such a solution. The general solution is a superposition of the above:

$$\psi=\sum_{n=1}^{\infty}a_n e^{-iE_nt/\hbar}\sin\left(\pi n\frac{x}{a}\right).$$

To explore the behavior of this solution, we express the distance and time in terms of dimensionless variables:

$$x=a\overline{x}$$

$$t=\frac{ma^2}{\hbar}\overline{t}\,.$$

The resultant solution is then a superposition of *sine* functions modulated by a complex time-dependent factor that changes the coherence over time:

$$\psi=\sum_{n=1}^{\infty}a_n e^{-i\frac{1}{2}\pi^2n^2\overline{t}}\sin\left(\pi n\overline{x}\right).$$

The time dependence is a function of the dimensionless energy $\overline{E}_n = \frac{1}{2}\pi^2 n^2$.

We analyze the possible behaviors of the system given an initial normalized packet for the wave function at time $\overline{t} = 0$:

$$packet = \sum_n a_n \sin\left(\pi n \overline{x}\right) = N e^{-\left(\overline{x}-\overline{x}_0\right)^2/\sigma^2} e^{i\overline{v}\overline{x}}$$

$$\int_0^1 d\overline{x} \left|packet\left(\overline{x}\right)\right|^2 = 1$$

We have chosen a Gaussian packet as exemplary, positioned inside the well. We have added a phase that is proportional to distance that does not change the normalization. The packet is normalized by requiring the area of the modulus squared to be unity. Note that the well is of unit length in terms of the reduced distance.

The coefficients a_n are the Fourier coefficients of the packet, defined as the complex numbers:

$$a_n = 2\int_0^1 d\overline{x} \sin\left(\pi n \overline{x}\right) packet\left(\overline{x}\right).$$

The coefficients depend on the initial packet. Given the exact form of the solution, and the above set of coefficients, we compute the time series for the Schrödinger equation by taking a fixed point in space and looking at the wave function for successive points in time:

$$\psi\left(\overline{x},\overline{t}\right) = \sum_{n=1}^{\infty} a_n e^{-i\frac{1}{2}\pi^2 n^2 \overline{t}} \sin\left(\pi n \overline{x}\right).$$

The classical limit corresponds to very short time intervals. The packet is supposed to be at rest, and its center does not move, though the wave form spreads out in time; since the area does not change, the height decreases slowly with time. The initial packet is at rest towards the initial end of thc well.

Well past the classical limit, the packet will spread out until it reaches the sides of the box. Now interesting things happen. To capture the complexity, we fix our position at a fixed point in space (near the initial maximum for example), and consider the time series that results. The recurrence plot provides a rather amazing pattern. The result (Figure 28) shows that the packet is at rest, and its center does not move, though the wave form spreads out in time; the area does not change and the height decreases with time. As the wave packet encounters the sides of the well, it is reflected back and like a wave, interferes with waves still coming toward the wall. The time series can be provided for the wave function's real part, imaginary part, and absolute value or phase. Here we present only the series of absolute value of the wave function as representative.

The series generated shows diversification (Figure 29), novelty and complexity (Figure 30). The complement plot (*sine* vs. *cosine* of the terms of the series) shows a Mandala pattern.

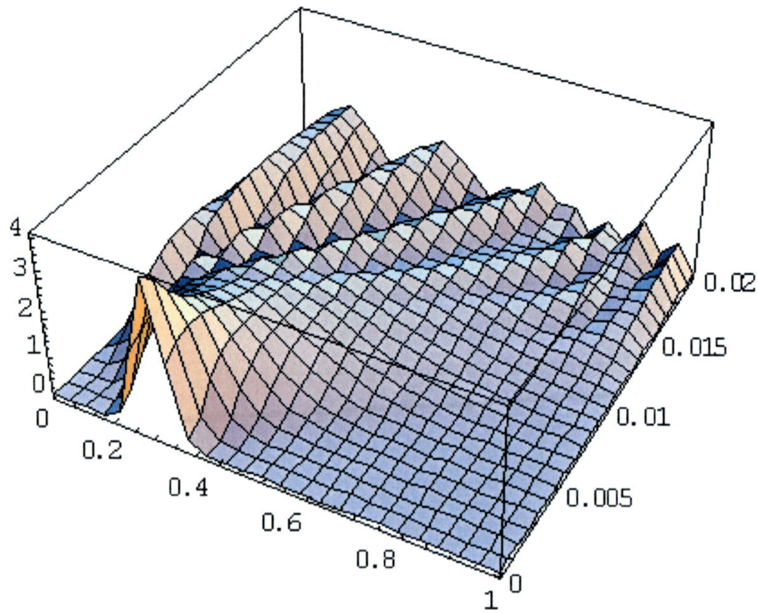

Figure 28. Evolution in time and space of wave packet generated with Schrödinger wave equation.

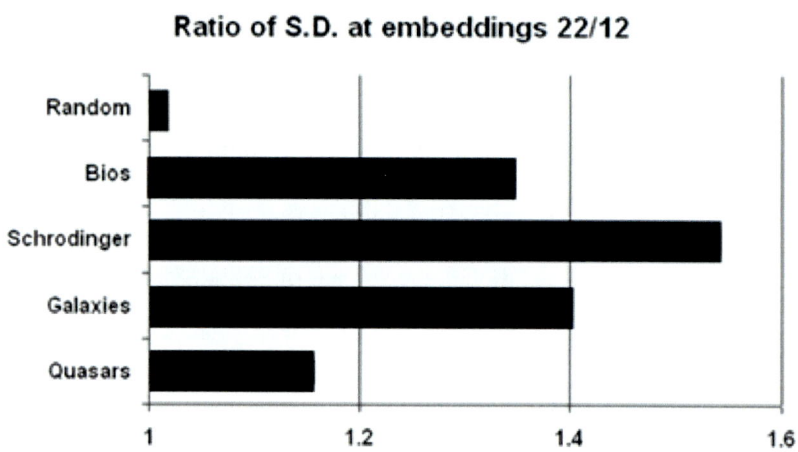

Figure 29. Diversification of Schrödinger series, galaxies and mathematically-generated Bios: Increase in Standard Deviation with increase in embedding from 12 to 22.

Biotic patterns are observed not only in the series generated by the Schrödinger equation but also in the series of differences between consecutive terms of the series. This is expected in deterministically generated Bios and incompatible with stochasticity.

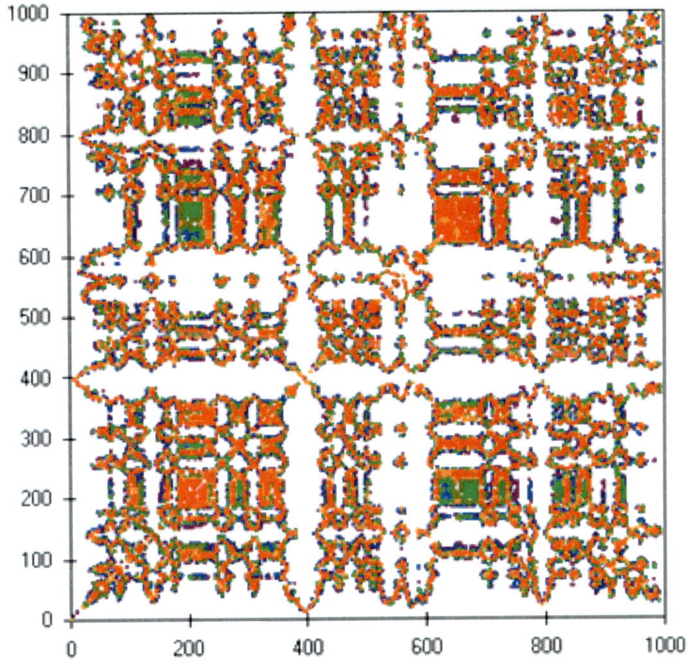

(A)

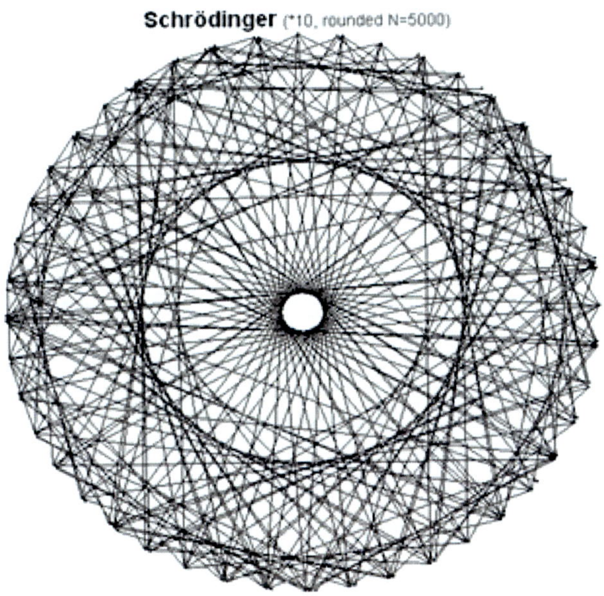

(B)

Figure 30. Schrodinger wave equation. (A) Recurrence plots at 5 embeddings. (B) complement plot (sine vs. cosine of the time series) shows a Mandala pattern as also observed with mathematically generated Bios and other biotic series.

Atomic Scales

The atom is of the order $10^{-8}cm$ and the typical time unit is $10^{-16}s$. Compare this with the time it takes the speed of light to traverse the atom $10^{-18}s$. We see that the atom is essentially non-relativistic: the speeds are much less than the speed of light. It is acceptable to use the Schrödinger equation as an approximation, as opposed to a relativistic form such as the Dirac equation.

Nuclear Scales

The nucleus is of the order $10^{-13}cm$ and the typical time unit is of order $10^{-24}s$. The time it takes light to traverse the nucleon is of order $10^{-23}s$.

Planck Scales

The size of the universe before inflation is $10^{-33}cm$ and so the typical time unit is $10^{-43}s$. It is assumed that matter is highly relativistic at this point. After inflation, which perhaps took 300,000 years ($10^{13}s$) the size of the universe might have expanded to $10^{-13}cm$, or the size of a nucleus. However the masses are now very large, so it is reasonable to assume that now the trapped particles are non-relativistic. Thus the structure we see today might well result from this early universe. Even if this universe were homogeneous, we see that the Schrodinger equation of a trapped particle in a well generates non-homogeneities simply because of the form of the evolution equation. It is interesting that we see no such effect when the particles are relativistic.

At relativistic energies, the Schrödinger equation is replaced with its relativistic counterpart, the Klein-Gordon equation. We illustrate this with the Klein-Gordon equation for spin 0 particles, which is representative of all relativistic particles with spin since we are concerned primarily with the effects due to the mass:

$$\frac{\partial^2\phi}{\partial t^2}-\frac{\partial^2\phi}{\partial x^2}+m^2\phi=0.$$

The form of the Schrödinger equation changes in such a way that the mass and momentum are now covariant with respect to reference frame:

$$\phi\left(\overline{x},\overline{t}\right)=\sum_{n=1}^{\infty}a_n e^{-i\overline{t}\sqrt{\pi^2n^2+m^2}}\sin\left(\pi n\overline{x}\right).$$

We make the important observation that in the relativistic regime where the mass contribution can be ignored, the solution can be expressed in terms of the initial packet $f\left(\overline{x}\right)$. The solution is proportional to $f\left(\overline{x}-\overline{t}\right)-f\left(\overline{x}+\overline{t}\right)$. At a fixed point in space, the time series is two packets offset from each other. The time series does not show biotic behavior unless there is a biotic behavior in the initial packet. The relativistic Klein-Gordon equation generates biotic patterns only when applied to entities with sufficient mass to slow down their velocity. Therefore the presence of mass is a necessary requirement for Bios in physical processes. This is consistent with the need for a conservation term in the process equation $A(t+1) = A(t) + g * \sin(A(t))$ [Sabelli, 2005]; recursions without the conservation term, such as $A(t+1) = g * \sin(A(t))$, generate only Chaos. For strictly massless particles such as photons, there is no Bios, though we explore below some interesting behaviors in microwaves. We note that biotic behavior has not been observed in the cosmic background radiation [Sabelli, 2005]. Thus both bipolar feedback and conserved mass appear necessary to generate causally biotic patterns.

In conclusion, both continuum and discrete models of the Schrödinger equation generate biotic patterns. Bios appears at three different distance scales: atomic, nuclear and Planck (time and distance scale obtaining shortly after the Big Bang). Quantum processes can generate biotic patterns with creative features. We also observed that biotic patterns are observed in the temporal distribution of galaxies. Finding Bios in both quantum equations and in the temporal distribution of galaxies suggests that the two biotic processes may be related. The Schrödinger equation generates Bios at atomic, nuclear and Planck distance scales [Thomas et al, 2006]. Since at its origin, the universe was of quantum size, quantum processes in the early universe may have been sufficient to initiate creative evolution.

Electromagnetic waves and gravity waves should be studied to make sure that structure effects seen in Klein-Gordon and Schrödinger equations are properly understood in context. Thomas has looked at electromagnetic waves in vacuum, and found that they show no change from when the wave form is created and at some distance after creation. They reflect a shape that moves through space without changing its form. This is similar to the Klein-Gordon equation solution when the momentum is significantly greater than the mass.

The situation changes when electromagnetic waves propagate in matter. Pure Silicon was considered, which has a conductivity $4.4\ 10^{-4}$ /meter-Ohm. The conductivity is small compared to metals. Metals respond to electric fields with currents that flow in the direction of the applied field. Conductance thus changes the propagation of waves. A wave (time series) was constructed, which has biotic-like properties—though this was not specifically intended. Waves of 10 meters were then considered at some distance from the source, moving along a single direction as in a wave guide. The second series is significantly more asymmetric and biotic-like. The attenuation of the electromagnetic wave due to the conductivity demonstrates effects for electromagnetic waves similar to the effects of matter in the Klein-Gordon equation. We thus have a physical and classical analog to the Schrödinger Equation (or Klein-Gordon equation with low momentum). In a similar way, one may speculate that gravity waves that travel in a vacuum might reflect the nature of their source, which could well be biotic. It is also possible that gravity waves traveling through space, which is not a perfect vacuum, will be affected in ways analogous to the above wave guide example. Over significant distances such waves might be distorted by Einstein's equations and create additional structure.

There are many points to consider having found such interesting patterns in physical equations. First, the number of terms in the Schrödinger series is quite large (we have kept 50 terms). The terms are interacting in a variety of ways, yet they combine to produce an interesting and simple pattern. The pattern reflects the type of complexity advocated by Bios, and we believe is outside of the classical limit. This we have already discussed at some length.

Another issue to contemplate is the common notion that quantum mechanics requires a probabilistic interpretation. We consider this question briefly along with the notion of classical action. How does classical action play a role in our understanding of biotic complexity? According to the view of modern physics the classical action determines the phase of the wave function, though the classical action does not satisfy the classical equations of motion. In the quantum formulation the phase of the wave function is constructed from Feynman's path integral as follows. For each possible path a particle could take, the classical action S is computed, and from this action the amplitude $e^{iS/\hbar}$ is computed. These amplitudes (which are complex numbers) are summed up. The resultant sum produces the wave function, which satisfies the Schrödinger equation and so this view is entirely equivalent to the previous discussion. This view also provides a clear explanation of what is meant by the classical limit. The classical limit is the special situation in which the action is large compared to the quantum of action, so that in the sum over paths, the only paths that count are those in which the action is least (strictly speaking, we mean that the action is maximal—either smallest or largest). This determines the classical equations of motion in exactly the same way as geometric optics is determined as an approximation of Maxwell's wave equations for light as the limiting case when the wave length of light is small compared to the size of the optical components.

The classical limit is more correctly a semi-classical limit, since one still has a wave function: $\psi \approx \sqrt{\rho} e^{iS/\hbar}$. The phase of the semi-classical wave function is given by the classical least action, and the modulus squared of the wave function is given by the density of the particle wave. The modulus squared has some properties associated with probability or frequency of occurrence for finding the particle, though at best this is a poor metaphor for the wave function, which is a complex number that reflects a more complicated behavior than classical probability. A better metaphor is matter density ρ for the square of the modulus and the corresponding current $\mathbf{j} = \rho \mathbf{v}$ determined by the density and the average velocity. Matter density as defined by the Schrödinger equation and its current obeys the conservation law $\frac{\partial \rho}{\partial t} + \nabla \bullet \mathbf{j} = 0$. We emphasize however that this is an interpretation of the quantum theory. This is not the quantum theory itself. Just as in optics the proper understanding of light moving through matter is based on Maxwell's Equations, so here the proper understanding of quantum mechanics is based on Schrödinger's Equation or Feynman's path integrals.

In particular, the matter density metaphor does not consider the phase of the wave function. We elucidate this issue by considering the semi-classical approximation in more detail. We start by separating the wave function into two factors, the modulus (which will be related to the classical matter density) and phase (which will be related to the classical action):

$$\psi = f \exp\left(\frac{iS}{\hbar}\right).$$

In this form, the phase S is defined in terms of the wave function and the Schrödinger equation, which as an equation for complex numbers, consists of two separate equations. Without going through the derivation, one equation that results relates the phase (action) to the matter density:

$$\frac{\partial S}{\partial t} + \frac{1}{2m}\nabla S \bullet \nabla S + V = \frac{\hbar^2}{2m} f^{-1}\nabla^2 f$$

In the limit that the Planck constant $\hbar$ goes to zero, this reduces to the Hamilton-Jacobi equation for action S, that follows from Newtonian mechanics. Since Planck's constant is very small, it demonstrates that the semi-classical corrections to classical mechanics are "doubly" small. The second equation takes the form of matter conservation with $\rho = f^2$ and $\mathbf{v} = \nabla \frac{S}{m}$:

$$\nabla \bullet (\rho \mathbf{v}) + \frac{\partial \rho}{\partial t} = 0.$$

The Hamilton-Jacobi equation not only follows from Newtonian mechanics, but starting with this equation it reproduces the classical equation of motion. The full derivative of the phase demonstrates that it is equal to the difference between the kinetic energy and potential energy of the system (which is called the "Lagrangian"):

$$\frac{dS}{dt} = \frac{\partial S}{\partial t} + \frac{\partial S}{\partial \mathbf{v}_k}\mathbf{v}_k = \frac{\partial S}{\partial t} + m\mathbf{v} \bullet \mathbf{v}$$

$$\frac{\partial S}{\partial t} + \tfrac{1}{2} m\mathbf{v} \bullet \mathbf{v} + V = 0 \Rightarrow \frac{dS}{dt} = \tfrac{1}{2} m\mathbf{v} \bullet \mathbf{v} - V = L$$

The Lagrangian is equal to the integral over time of the action, so the phase is the classical action. Moreover, the laws of classical mechanics follow from the principle of least action, which is equivalent to the least phase variation that followed above from the path integral formulation, and this leads to the standard equations for the Lagrangian, which are a standard formulation of classical mechanics.

$$\frac{\partial L}{\partial \mathbf{x}} - \frac{d}{dt}\left(\frac{\partial L}{\partial \mathbf{v}}\right) = 0 \Rightarrow m\frac{d\mathbf{v}}{dt} = -\frac{\partial V}{\partial \mathbf{x}}$$

The second equation is Newton's equation stating that the force is equal to the gradient of the potential.

This detail helps identify the phase with the classical action, and makes the point that we need not identify quantum mechanics as a probabilistic theory, but rather as a wave theory in which the classical limit is a geometric optics approximation of a wave. According to Goldstein (1959) it was noted as far back as 1834 by Hamilton that surfaces of constant action propagate as if they were a wave. Hamilton noted that it was possible to view classical mechanics as the geometric optics approximation to a wave equation. Of course it took experimental evidence to make this leap acceptable, and such evidence did not occur until 1926 with Schrödinger. This puts in context the particle path approach by Feynman who at the outset takes the view that the wave is constructed from the action in exactly the same way as a wave in electromagnetism can be constructed by summing up the least time paths for photons, associating each path with an appropriate complex number that has a magnitude and phase: the phase is proportional to the time. Our view is that the wave function represents the reality, not the classical or semi-classical interpretations of that wave , including the probabilistic interpretations.

The consequence of taking the wave function description seriously is that the behavior at any point in time and position may depend on the values of the wave function at all past times and positions. In a relativistic theory, there is a limitation however that any dependence on past positions must "propagate" at speeds equal to or less than the speed of light. This behavior does not square with classical mechanics, though it is exactly what is required in the theory of Maxwell for electromagnetic radiation.

Discussion

All the series studied demonstrate plots with diversification, novelty, arrangement (nonrandom complexity) and temporal complexity (*complexes* of recurrences), indicating creativity. The complexes display equally complex forms at various scales, as observed with fractal series. Shuffling largely obliterates such organization. Very different forms are observed at different points in these series, and also the internal shape of these complexes continuously changes. The demonstration of these multifarious patterns points to creative processes; complexes are observed in biotic series and in brown noise, while random and chaotic series show more uniform recurrence plots, similar to those of their shuffled copies.

These observations show that the evolution of the universe represents a creative process, Bios, in physical processes ranging in age from the early universe to the present, and in size from the Schrodinger equation at Planck length 10^{-35} m to cosmological dimensions. Finding Bios in cosmological series is at variance with random stationary process models that are the current view in cosmology as well as with the hypothetical decay towards entropic disorder. Finding Bios in the Schrödinger and the Klein-Gordon-Fock equations is likewise at variance with the probabilistic interpretation of quantum mechanics, but supported by more modern views of quantum field theory such as Feynman. Our studies support causality in all the processes studied. This conclusion fits well with recent studies that showed that the formulation of quantum gravity requires causality (Ambjørn and Loll, 2004).

Finding biotic features for billions of years, possibly since the first second after its Big Bang origin, indicates that biotic expansion must have been operative for a significantly large era of cosmic evolution. The process that generates Bios must thus contribute to, or cause, the expansion of the universe. In turn the process of expansion may play a fundamental role in the evolution of the universe, and its complexification.

Spontaneous, apparently random, erratic fluctuations are observed at all levels of organization. Empirically, the distribution of galaxies is highly heterogeneous and dynamic for at least 1000 Mpc. Stars, galaxies, clusters of galaxies, and superclusters demonstrate an irregular hierarchical structure, which in a way continues the sequence of quarks, nuclei, atoms and molecules at the small end of the scale. The universe appears to be a hierarchical system rather than a random distribution.

Quoting Einstein, most models of the Universe assume that the present distribution of matter is uniform on the very large scale. Actually Einstein understood that the universe is highly heterogeneous. He likened it to the surface of the earth. The standard cosmological model [Peebles, 1993] interprets the cosmological principle to mean that the universe is homogeneous and isotropic. A random stationary origin is assumed and support for uniform distribution is provided by the correlation function method [Peebles, 1993] although this method gives correlation lengths of 5-10 Mpc, in contrast to the observation in galaxy maps of structures with over 50 Mpc. These and other observations indicating a far-from-uniform distribution of galaxies have led to a formulation of a weaker hypothesis, namely that the universe is uniform, homogeneous and isotropic only on a large scale and only in a statistical sense. The observable hierarchical structure in the distribution of matter is explained as local fluctuations, which are assumed to be confined to scales less than a few hundred million light years.

Mandelbrot (1975) proposed an alternative model: a self-similar fractal geometry generated by stationary stochastic processes. A fractal model of the universe has been developed by Pietronero and coworkers (1997).

Big Bang models suggest that the initial distribution of matter was uniform. How does gravity turn an initially smooth distribution of matter to hierarchical clustering? Within the "Hot Big Bang " model, the formation of the large-scale structure we see today could be explained as the result of the growth of gravitational instabilities during an early period of rapidly accelerating expansion (inflation). "Gravitational instability" models propose that a Gaussian random field of initial fluctuations seeded the structure-formation process. These slight "ripples" were then magnified by the effect of gravity as the small, initially overly dense, areas attracted additional mass. It is further speculated that these ripples would evolve independently during the early stages of this expansion, but that as the structures grow in mass, they would interact with each other in nonlinear ways by phase correlations [Coles and Chiang, 2000]. In the beginning, the phases are random. This corresponds to a state of minimal information that is regarded as maximum entropy. As information flows into the phases, the information content increases and entropy decreases. (Models that assume random origin require evolution against the law of maximization of entropy.) Some physicists point out that this process may not be sufficient to account for the formation of the observed hierarchical structure. Using distance measurements other than redshift, such as the one Tully and Fischer (1977) devised, it has been calculated that galaxies as a rule do not move much faster than a thousand kilometers per second (about one-three-hundredth as fast as the speed of light), indicating that if the universe started out with matter spread smoothly through space, galaxies could have moved only sixty-five million light-years, too short a time to make up Tully's complexes. The alternative is the formation of large inhomogeneities very early in the evolution of the universe as result of quantum processes in a Planck size universe obtaining shortly after the Big Bang .

Both the standard and the fractal models regard the galaxy space distribution as the result of stationary stochastic processes. The pattern of the process does not evolve with time. In this sense, these are static concepts. Yet, intergalactic distance is convoluted with time. The study of the macroscopic distribution of matter is not geometry but history. The expansion of space, the synthesis of elements in the core of stars, and the generation of complex aggregates of matter ranging from atoms to galaxy superclusters indicate that the universe is a process of creative evolution, that is to say, non-stationary and largely causal. Evolution speaks against stationarity, and the hierarchical organization of the groupings of matter from atoms to superclusters of galaxies speaks against random distribution. Other evidence for nonrandom organization is the form and mass of galaxies. Galaxies have many features in common, such as a dense nucleus, a faint outer region or envelope, and a dark matter halo. Many galaxies show spiral-shaped arms, which emanate from or near the nucleus and gradually wind outward to the edge. Many galaxies are of the same order of magnitude as the Milky Way; seldom, if ever, are they four times bigger. Also, the dodecahedral structure found by Luminet [2001] and coworkers in the CBR data indicates causal processes rather than stochastic ones.

Since mechanism cannot account for the generation of novelty, it is often assumed that innovations are due to chance. As random models dominate scientific discourse, the heterogeneous distribution of galaxies and quasars is explained as the result of random distribution of energy in the early universe. The gravitational waves generated in the early universe have been labeled as stochastic before being observed. However, the distribution of galaxies and quasars as well as the LIGO data demonstrate a causal origin: consecutive isometry, partial autocorrelation, and most important, pattern in the series of differences between consecutive terms.

What generates novelty and complexity in cosmological processes? Is it random flux or directed action? Planck discovered that radiation can be emitted only in multiples of a quantum h that has the dimensions of action = energy x time. Within its limits, change is continual and unpredictable; the uncertainty principle excludes absolute rest. Above this limit, determined, directed action (not energy alone) appears to be the sole constituent of the universe. Energy and time are inseparable; they are quantum conjugates. Physical action is physical causation.

Notwithstanding, in formulating cosmological theories, many mathematical physicists postulate random change first. How action and causation emerge from randomness is never made clear. How randomness emerges from causal action is likewise not explained. Randomness is assumed rather than demonstrated. It is however possible to distinguish causal from random change. The characteristic features of deterministic action are: (1) temporal asymmetry; (2) transitivity (continuity, causation); and (3) a direct relation between energy E and frequency f (Planck's law E =h f) where h is the Plank constant. We propose that the widespread 1/f power spectrum found in natural processes represents the integration of the fundamental relation between frequency and energy stated in Planck's law [Sabelli, 2005]. In contrast, random flux is: (1) symmetric; (2) acausal, composed of independent events; and (3) variable along the entire frequency spectrum with no prominent peaks, and with no relation between frequency and energy (white noise). These differences can be detected with recurrence analysis and statistical methods. In this manner we have analyzed empirical and mathematical time series to investigate the role of deterministic action and probabilistic flux in creative processes.

The time series of cosmological processes, as well as mathematical Bios, show asymmetry, positive autocorrelation, and extended partial autocorrelation. Random, chaotic and stochastic models show symmetric statistical distributions, and no partial autocorrelation. The percentage of continuous proportions is high in biotic series, and scarce in pink noise and Chaos. These findings differentiate creative biotic processes from chaotic and stochastic series. Natural creativity must then emerge from determined interactions rather than from the accumulation of accidental random changes.

The significance of finding a biotic pattern in the distribution of galaxies lies in the expansive nature of Bios. Biotic patterns model deterministic diffusion [Chirikov et al., 1985; Geisel and Nierwetberg, 1982] and creative processes [Kauffman and Sabelli, 1998]. As biotic processes have been operative for a significantly large era of cosmic evolution, the process that generates Bios must thus contribute to, or cause, the expansion of the universe.

Bios Theory of Cosmological Evolution

Mathematical Model: Biotic Equations

How does causal action create diversity and complexity? Mathematical models for Bios help us to search in mathematical data for causal processes that can generate patterns with creative features. Biotic series can be generated deterministically by recursions of sinusoidal functions (such as those embodied by electromagnetic energy) but not by random processes.

Bios is generated by recursions (action) in which change results from positive and negative feedback [$g * \sin(A_t)$] (bipolar opposition) and that preserve the previous state (A_t) (conservation). Temporal change, bipolar opposition, and conservation of matter and energy are universal features of physical processes. As they are sufficient to generate Bios in mathematical recursions, they may be expected to generate Bios in the physical universe. Recursions in which the feedback is bipolar and the successive terms accumulate, such as the *process equation* [Kauffman and Sabelli, 1998]

$$A(t+1) = A(t) + g * t * \sin(A(t))$$

and the *diversifying equation* (Sabelli, 2005)

$$A(t+1) = A(t) + \sin(g * t * A(t))$$

show a sequence of phases: steady state, periodicities, Chaos and Bios, which alternates with leaps in quantity (infinitations) and fractal repetition of the fundamental period 2 (Figure 37).

We coined the term 'process equation' to point to what we consider a fundamental process in nature, not only because of the empirical demonstration of Bios in many processes, but also because of a theoretical interpretation of the terms of the equation as abstractions for (1) action (recursion) ; (2) bipolar opposition, as in the sinusoidal pattern of electromagnetic energy and other waves, and in harmonic feedback, which we interpret as a dialectic; and (3) accumulation of change by a conserved term A(t).

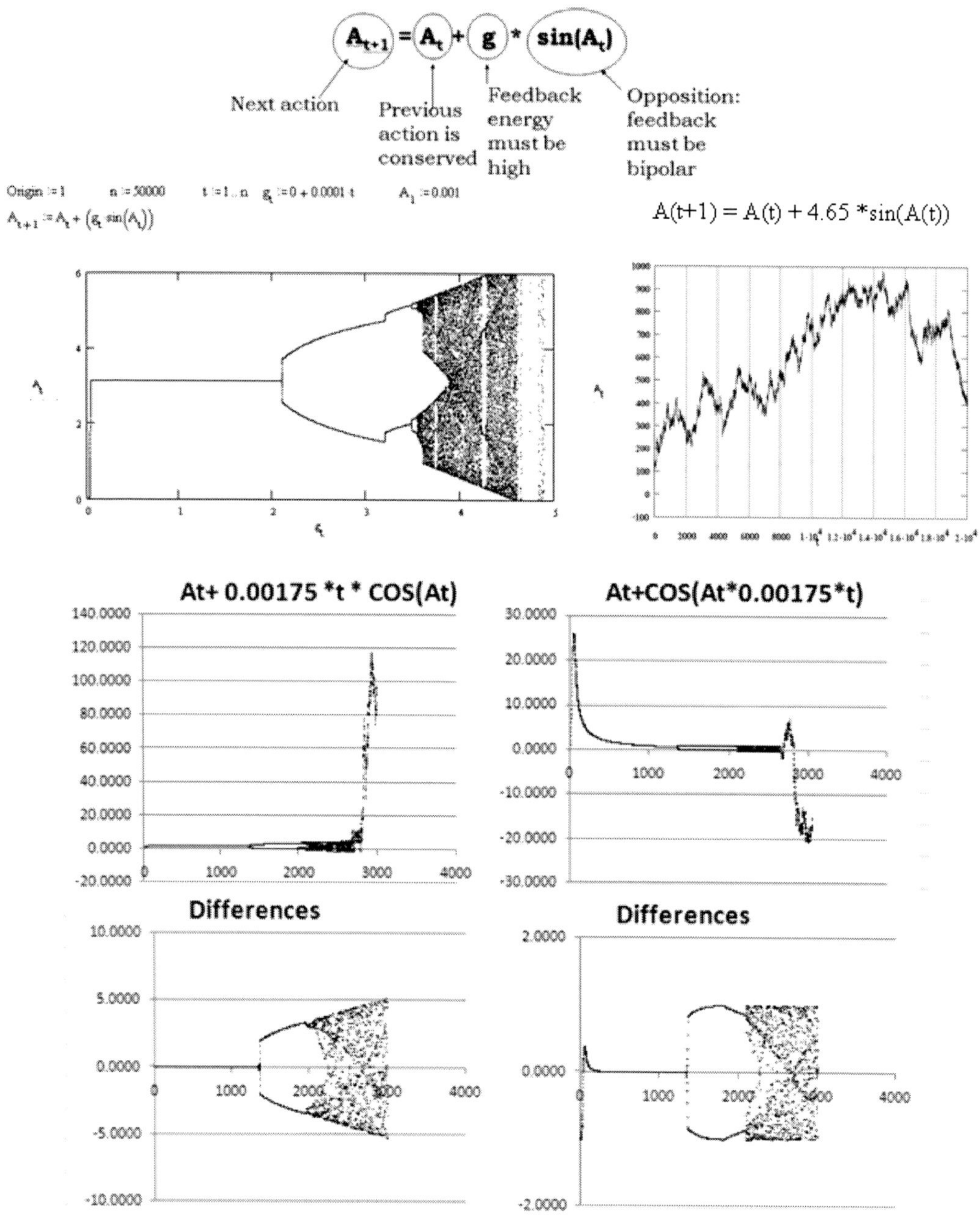

Figure 31. Biotic equations. Top: The process equation $A_{t+1} = A_t + g \sin(A_t)$ illustrates the requirements to generate Bios. The equation embodies the basic postulates of process theory: (1) iteration models the linear flow of action (i.e. time); (2) the sine function models the cycling of complementary opposites generating positive and negative feedback; (3) the conserved term models the conservation of matter. Middle: Its iteration generates biotic patterns with dynamic features (right) as observed in empirical recordings of cosmological and physiological data. Before Bios (left), the process equation generates steady state, bifurcations, and Chaos. Bottom: Process At+ 0.00175*t*COS(At) and diversifying At+COS(At*0.00175* t) equations. Time series and time series of differences.

Just as biotic patterns are generated by the relativistic Klein-Gordon equation only for entities with sufficient mass to slow down their velocity, mathematical experiments demonstrate that bipolar harmonic motion without the conservation term, such as A(t+1) = g * sin (A(t)), generates only Chaos [Sabelli, 2005]. In contrast, unipolar feedback, as in the logistic recursion, which portrays scarcity, and the progression of patterns in its outcome is accelerated but not qualitatively changed by adding a term that allows for accumulation.

The generation of Bios requires a rough symmetry of opposites (Figure 38), although a small degree of asymmetry is favorable.

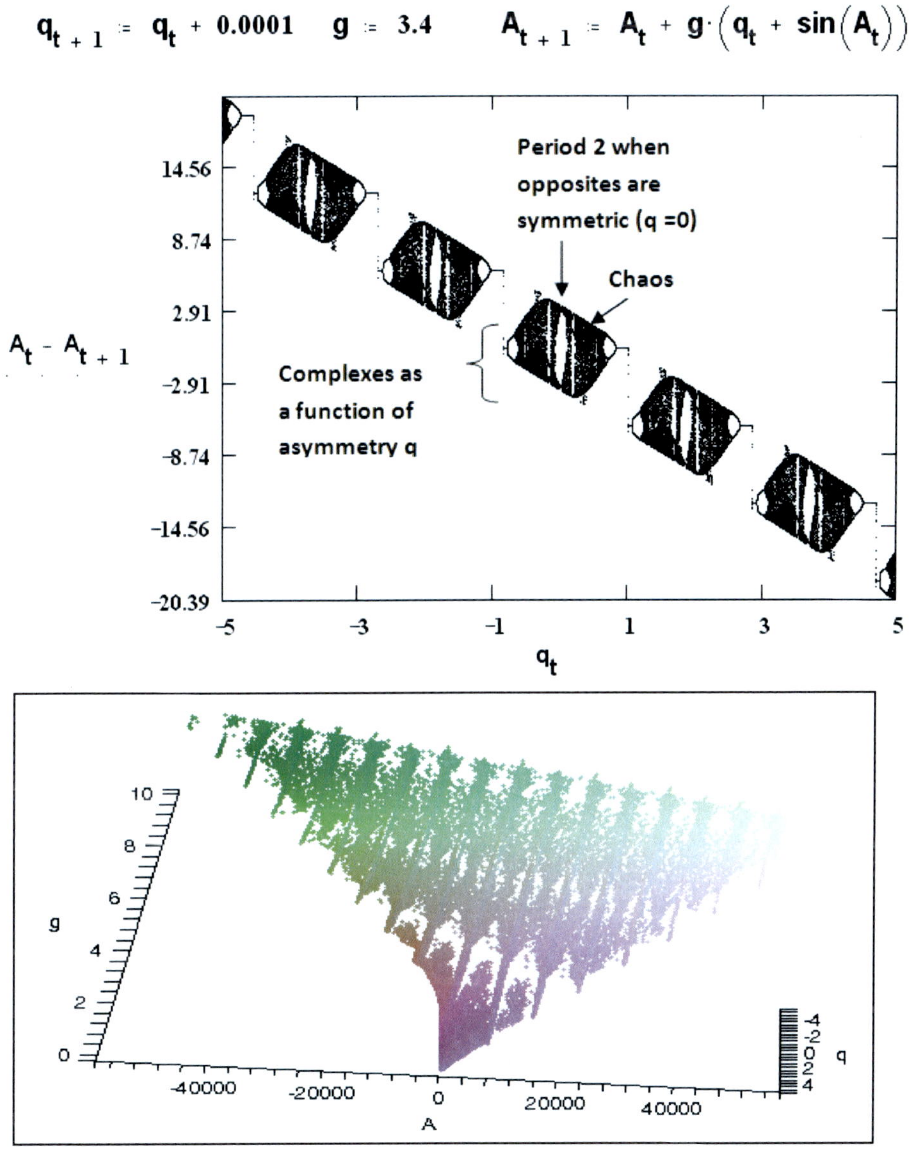

Figure 32. Continued on next page.

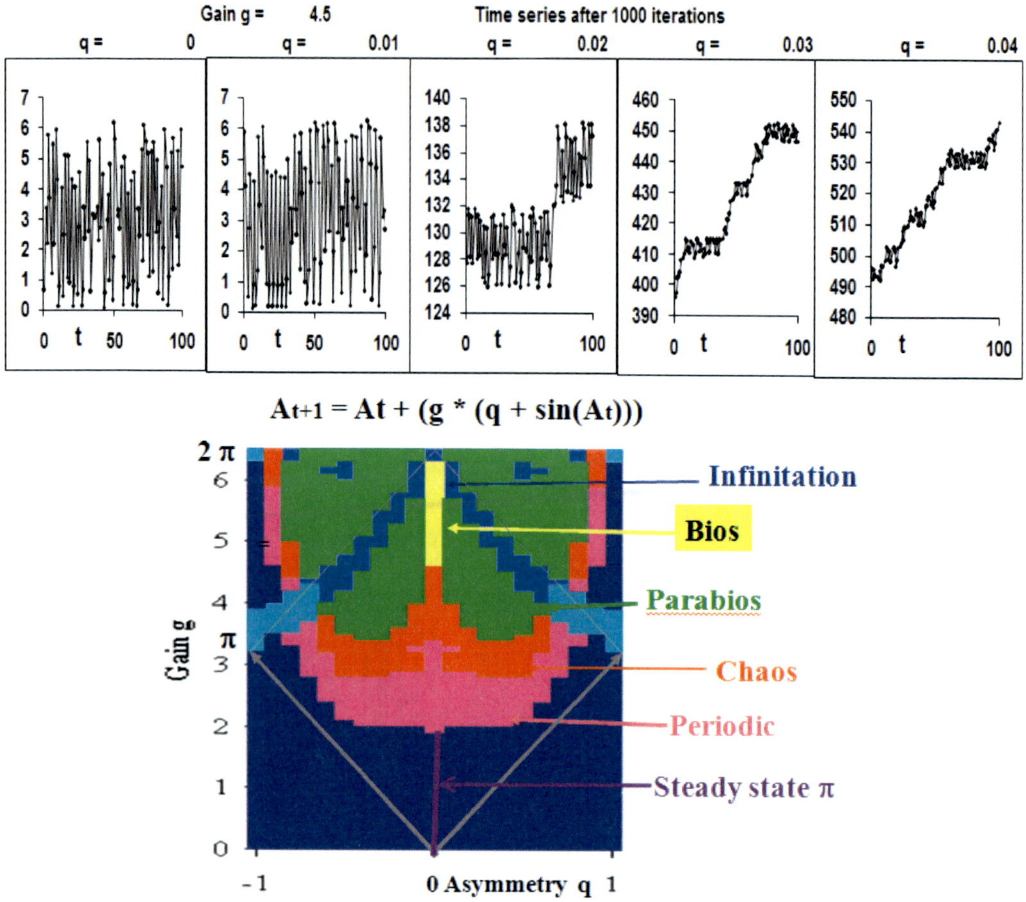

Figure 32. Time series generated by $A_{t+1} = A_t + g*\sin(A_t) + q$, where the gain g represents energy and q represents asymmetry. Bios is generated by intense and quasi- symmetric processes. A small asymmetry favors the development of Bios at lower gain.

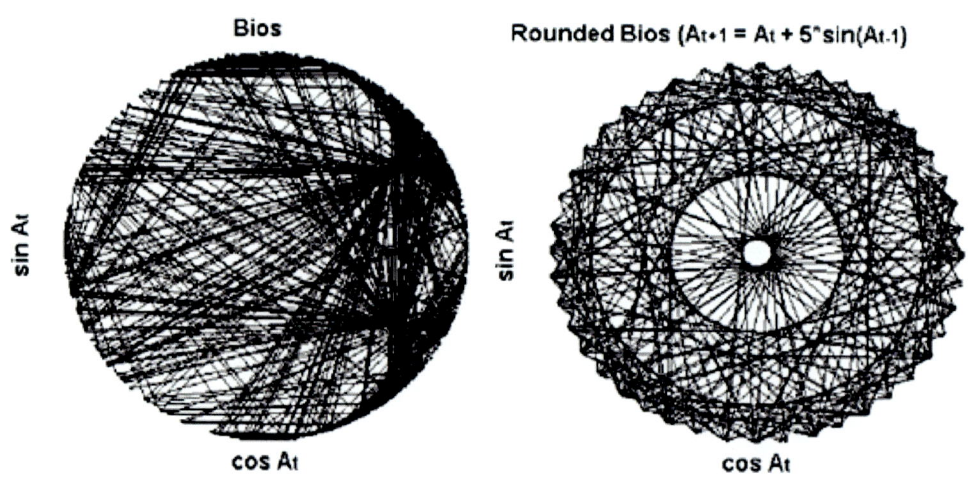

Figure 32. Continued on next page.

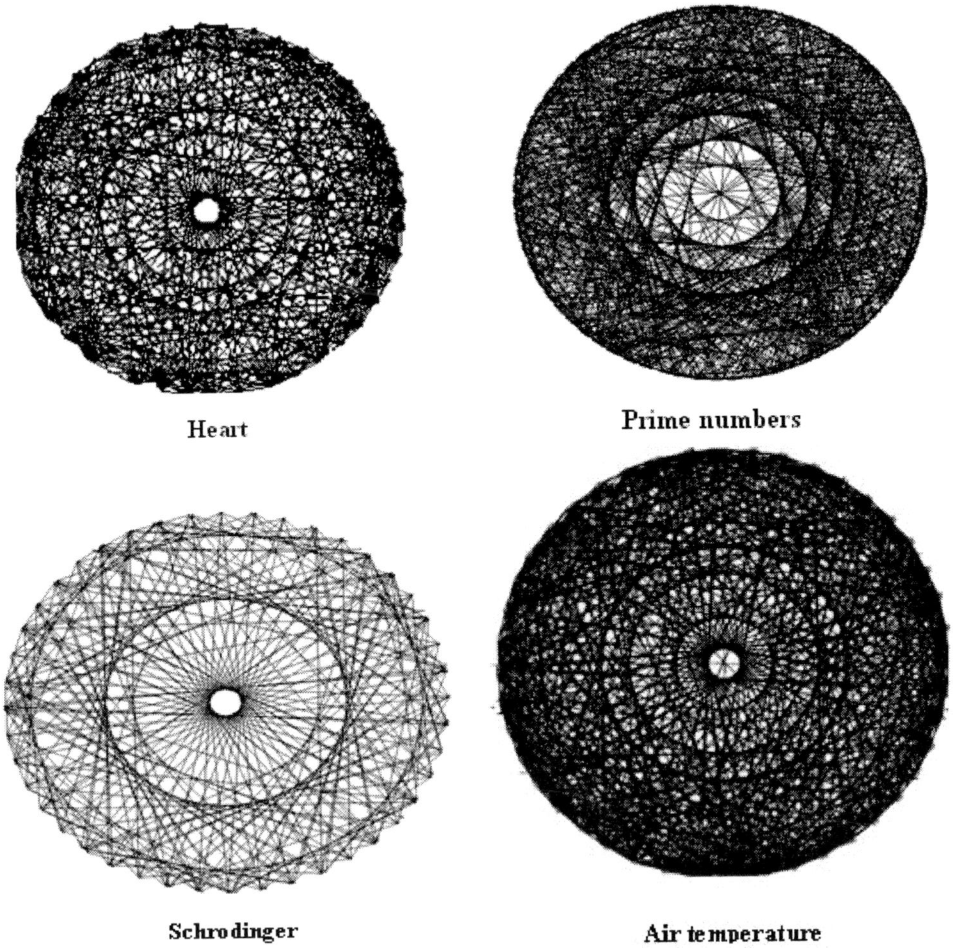

Figure 33. Complement plots (sine versus the cosine of a time series) reveal Mandala pattern in some biotic series.

The process equation generates Bios with the recursion of a sine wave. Conversely, trigonometric transformations of biotic empirical data often display a striking Mandala pattern (Figure 39). The deterministic generation of biotic patterns by simple equations gives credence to the hypothesis that the generation of diverse, novel, and complex processes can be the necessary consequence of interactions among simple processes.

Bios Theory: Lattice, Group and Topology as Primary Processes

These studies support the view of the universe as a process that is both creative and causal, as contrasted to deterministic stability and to entropic decay (Figure 40). Finding non-random pattern and causation in cosmological data supports the notion of natural creation as opposed to stationary random models of galaxy distribution, supernatural intervention, non-creative mechanism, thermodynamic decay towards entropy, or the neo-Darwinian portrait of meaningless evolution driven by chance.

Spontaneous fluctuations are present in the void and at all levels of integration. Tryon (1973) suggested that the universe arose from a random quantum fluctuation. Slutzsky proposed that order in nature could be produced by lawful operations on totally random fluctuations. Sabelli (1989, 2003) proposed that unidirectional action, bipolar opposition and conservative transformation mold the ever-present flux into organized processes and structures (Figures 34- 35).

The mathematical experiments described above led us back to Heraclitus' Process Theory and its development in Hegel's dialectics and in evolutionary theories in which the interaction of opposites is regarded as primary in the generation of complexity. We developed a theory of creative processes founded on the discovery of Bios in empirical data and the demonstration that Bios was produced by the recursion of bipolar opposites, such as in a sinusoidal wave [Sabelli, 2005]. This is a 'trigonometric' formulation of the notion of development by the interaction of opposites. These ideas are abstracted and generalized by fundamental mathematical structures.

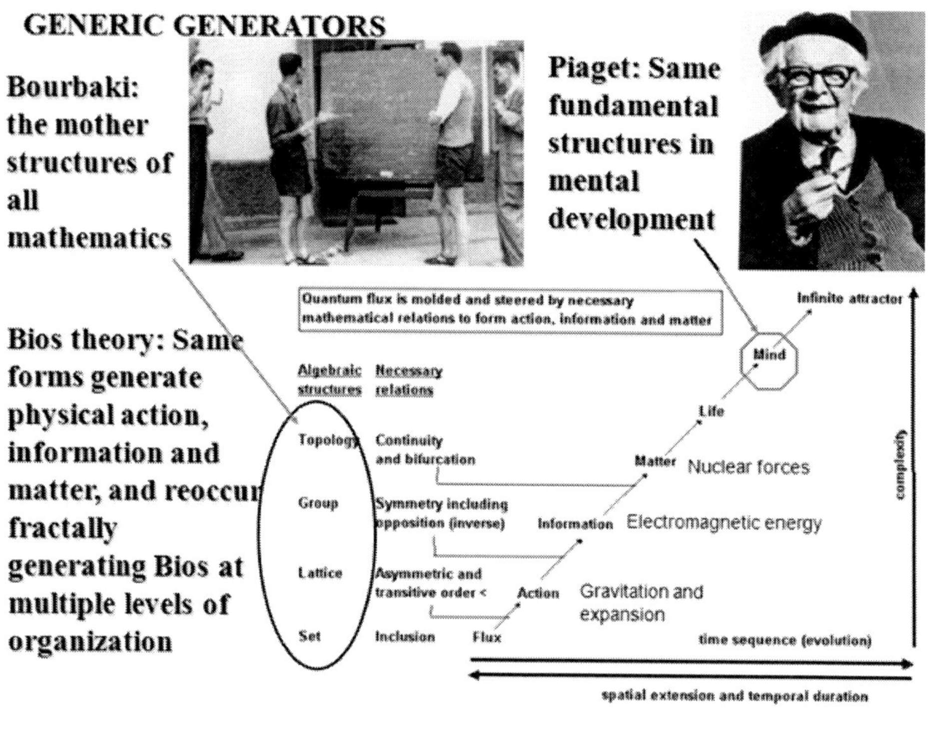

Figure 34. The three fundamental mathematical structures identified by Bourbaki, lattice, group, and topology, which Piaget encountered them in the cognitive processes, and we regard as primary processes.

Bourbaki (1948) demonstrated that mathematics can be generated by three "mother structures", namely lattice order (asymmetry and transitivity), group symmetry, and topological transformation with continuity. These generic forms are homologous (in the biological sense) to the number forms 1, 2 and 3: order is uni-directional, group symmetry hinges on the existence of opposites, and physical topology involves tri-dimensional structure. From a static perspective, lattices portray hierarchical order, groups describe symmetry, and topology deals with continuity of form. From a process perspective, lattices describe temporal sequence, groups focus on opposition, and topology deals with transformation. Piaget (1968) discovered these same generators in the psychological development of children; thus these forms are embodied in the human brain. Sabelli (1989) proposed that these forms also are embodied in primordial processes at all levels of organization. The fact that the same fundamental structures generate all of mathematics and guide mental development suggests that **lattice, group, and topology also generate physical and biological evolution**. "Mother structure" would thus be an appropriate name. Being logically necessary, mathematical relations operate at all levels of organization. The universe itself may be the necessary consequence of the embodiment of these forms by the ever-present flux in the void [Sabelli, 2003].

Lattice, group and topological properties are evident at multiple levels of organization. The formation of multiple structures from a single origin likewise represents lattice organization. The physical universe displays group properties at multiple levels, beginning with the well-known symmetries exhibited by elementary particles and including the rotational movements of astronomical bodies. The conservation and continual transformation of energy represents topological transformation with continuity. Specifically, mathematical order (lattice theory) is materialized by the temporal flow of energy (action) and causation in unidirectional time; charge embodies opposition and more generally group symmetry is embodied in groups of physical entities (such as elementary particles), energy waves, and pairs of informational opposites; and topological formation (transformation, continuity and conservation) is materialized by tridimensional matter and energy fields. These three forms, together, may constitute a "cosmic genome" that organizes processes at all levels of integration [Sabelli and Carlson-Sabelli, 1996]. At the quantum level, action, charge, and 'color' are exemplary of the embodiment of these mathematical forms [Sabelli, 2005]. Physical processes embodying Bourbaki's fundamental forms generate Bios and thereby initiate creative evolution.

	Mother Structures of Mathematics	Aspects or "components" of natural objects	Physical forces	Quantum mechanical phenomena
1. Uni-directionality	Lattice order: asymmetry and transitivity	Energy flow in time (action)	Unipolar gravitational attraction and space expansion	Action quanta
2. Twoness: opposition	Group: inverse and other symmetries	Physical symmetries; waves; two-valued information	Double bipolar electromagnetic force	Symmetric (electron/positron) and asymmetric (electron/proton) pairs of particles
3. Threeness: tridimensional space	Topology: continuity, transformation	Spatial structures (matter)	Tripolar strong forces nucleate matter	Three color charges, and three flavors. Three quarks constitute a proton and a neutron. Tridimensional matter.

Figure 35. The three fundamental mathematical structures identified by Bourbaki, lattice, group, and topology, which Piaget encountered them in the cognitive processes, and we regard as primary processes.

A mathematical view of creation is consistent with a long scientific tradition that regards mathematics as the account of nature. It may seem unlikely that the universe emerges simply from the molding of flux by three mathematical forms as here described, but who dares to assert that it evolves independently from, or at variance with, mathematical form? Form and transformation determine existence, as energy is contained in both void and matter. Form underlies existence. Absolute uniformity is the "void." The "vacuum state" needs not be emptier than matter.

What Are the Forms of Action that We See in the Universe?

First, the uni-verse embodies the unidirectional arrow of time and causation that makes all actions part of a single process. An asymmetric order (≤) defines *lattices* (one of the three fundamental components of mathematics), and it is the core of logic (if then) and of arithmetic.

Second, waves oscillate and the universe expands in (roughly) symmetric manner. Just as we know directly the order of time, we also see the circular form of the universe in the appearance of the sky and the shape of the sun and the moon. Symmetry is described by *group theory*, which is the second of the three fundamental components of mathematics. The sinusoidal form of electromagnetic waves embody the arrow and the circular symmetry.

Third, energy and matter are always conserved. The transformation of entities without loss of substance is portrayed by *topology*, the third fundamental mathematical structure. All real structures are at least tri-dimensional.

1. Asymmetric Action

The first principle of Bios Theory is that **asymmetric action is the one and sole constituent of the universe**. The physical dimension of action is energy multiplied by time. Energy and time do not vary independently from each other. Time and energy are inseparable; they are quantum conjugates just as position and momentum.

Action embodies one as oneness of substance and as the unidirectionality of time and of causation. Both action and causation embody lattice asymmetry and transitivity. Action involves temporal unidirectionality, as contrasted to the time reversibility of classic, relativistic and quantum mechanics. Matter displays asymmetry. Although often described as neutral, matter actually is bipolar, and each atom embodies an asymmetry between the positive nucleus and the peripheral negative electrons. Biological molecules are asymmetric, deviating light in one direction. Based on his discovery of biomolecular asymmetry, Pasteur postulated the existence of a fundamental physical asymmetry; indeed asymmetry has been found in physical, biological and psychological processes [Anderson and Stein, 1987; Corballis and Beale, 1976; Haldane, 1960; Sabelli, 1989]. One can interpret the unidirectionality of time and causality as Pasteur's cosmic asymmetry [Sabelli, 1989; 2005].

The universe does not show symmetry under time reversal, as evident in the uncertainty principle (at quantum scales) and thermodynamic entropy (at larger scales). Maxwell's equations with material absorption or Newtonian mechanics with friction are not time-reversal invariant at the macroscopic level. Time reversal is not a symmetry.

Electromagnetism, gravity and the strong interaction all display charge-symmetry, but weak interactions does not [Gerber, 2004]. Parity is violated in weak interactions. Weakly-interacting particles have a definite "handedness." If nuclear beta decay is reflected in a mirror (such a mirror reflection is called a parity operation, P), a right-handed particle becomes left-handed. The weak force is the only one that connects to asymmetry regarding opposition, time, and charge, but it embodies a triadic symmetry. It is transformative of matter. It transforms three dimensional protons into three dimensional neutrons.

Matter displays asymmetry first in the distribution of charges in the atom, positive in the massive nucleus, and negative in the electronic shells, and second in the greater abundance of matter than of anti-matter.

The degree of asymmetry of time series is portrayed by its statistical distribution. The distribution of galactic distances, heartbeat intervals, economic series, and meteorological data all are asymmetric. Asymmetry is also present in biotic series generated by the process equation. In contrast, symmetry is the rule for the distribution of random, periodic, and chaotic series.

Some physicists attribute the asymmetry of the universe to its initial conditions rather than to the laws of physics. In their view, the asymmetry of time results from the onset of the Universe was in a smooth and organized state, and it became more irregular and disorganized only as it expanded. The description of the Big Bang as a smooth and organized state is a counterintuitive hypothesis. While some physicists regard thermal equilibrium as a more "elegant" portrait for the initial universe, it seems difficult to imagine how equilibrium could become evolution. Even in symmetric physical systems there is temporal asymmetry.

The classical laws of physics discovered by Galileo, Newton and Einstein are time-symmetric, but the CPLEAR experiments at CERN collided antiprotons and hydrogen atoms to make kaons and their antimatter counterparts, antikaons. As they travel, antikaons can transform into kaons and vice versa. The rate for antikaons transforming into kaons is higher than that for kaons becoming antikaons. This experiment has observed a microscopic arrow of time [Gerber, 2004].

The evolution of complexity requires and generates asymmetry. In contrast, many physical theories posit a primordial symmetry, and Thermodynamics postulates a continual increase in symmetry. Thermodynamics has led to Clausius' notion of decay towards disorder (as entropy is interpreted); thus thermodynamics not only fails to account for evolution but in fact predicts its opposite. The de Moivre equation employed to quantify entropy in statistical mechanics and information theory actually measures diversity and symmetry [Sabelli, 2005], not disorder.

Taking asymmetric action rather than symmetric equilibrium as the simplest condition differentiates Bios theory from other conceptions of physics and of dynamics. In formulating physical and cosmological theories, a majority of scientists place symmetry first, building and attempting to account for the evolution of the universe by postulating that symmetry somehow is broken, without explaining how this happens. In contrast, symmetry can readily emerge from asymmetry as a case of well-known processes of equilibration.

2. Opposition, a Universal Symmetry

Actions are asymmetric (unidirectional time), but the symmetry of opposite actions and structures is a leading characteristic of fundamental physical processes. Symmetries are described by group structure, in which every element has an opposite. Opposites are orthogonal, synergic and additive in one dimension and antagonistic and subtractive in another. This requires representing opposites in at least two separate dimensions (i.e. a plane). **Opposites are orthogonal, not polar.** Opposition is bidimensional, not linear. The quaternity of orthogonal pairs of bipolar opposites is contained in a circle, and in the sinusoidal pattern of electromagnetic waves. Opposition is a cyclic engine of change. Opposition is the physical substrate of information.

As predicted by Dirac's relativistic quantum mechanics, every particle has a corresponding antiparticle that has exactly the same mass and lifetime, and exactly opposite charge. Yet there are no significant concentrations of antimatter in the observable universe. Symmetry coexists with asymmetry. The coexistence of asymmetry and symmetry is a fundamental union of opposites [Sabelli 1989, 1997, 2005], a concept recently repeated, without quoting these previous publications, as "synsymmetry", which is defined as "the paradoxical synthesis of symmetry and asymmetry" (Rosen, 2008). In our view there is nothing paradoxical in the coexistence of these and other opposites; their coexistence is the rule, not the exception. Change and evolution arise from the interplay of symmetry and asymmetry (Sabelli, 1989, 2005), a concept recently repeated by Rosen.

Opposition is the simplest symmetry, but **fundamental oppositions are asymmetric** – e.g. positive massive nucleons and negative electrons, horizontal sympathetic and vertical parasympathetic innervations, simple and complex.

Noether's Theorem states that in a multi-component field (such as the electromagnetic field of light, or the metric field of spacetime), symmetry is associated with a conservation law, and conversely conservation laws are associated with symmetries. Stress on symmetry over asymmetry thus corresponds to a preference for static states over evolutionary processes.

Field theories such as general relativity and quantum field theory stress local and quite general symmetries that are expected to be obeyed for all dynamics. Thus even Schrödinger's equation would satisfy that, yet it produces Bios. Local symmetries provide only a general guideline like conservation of energy and momentum, charge, baryon number. They leave relatively unconstrained the dynamic behavior. Only the ground state does not decay, but dynamics keeps systems constantly excited.

3. Tridimensional Materialization

Processes create structures, starting with tridimensional matter, the third fundamental component of nature. Matter is the minimal of complexity; it involves unipolar gravitational attraction, electromagnetic energy (equilibrium of opposites) and three-dimensional structure which forms around protons and neutrons constituted by the tri-polar nuclear force. As discovered by Gell-Mann (1964), all known baryons (protons and neutrons that make the atoms' nuclei, as well as particles created artificially in the laboratory) are nucleations of three quarks. They are held together by the strong nuclear force, which is the strongest. While gravitational and electromagnetic attraction decrease with distance, the interaction between quarks becomes arbitrarily weaker as distance converges to zero (asymptotic freedom) [Wilczek and Gross, 1973]. The energy of a quark in isolation is infinite; thus we never observe individual quarks. The color charge can be canceled if a quark and an antiquark coexist at exactly the same location, but this is energetically costly, as quarks tend to spread out. The color charge is more readily controlled by the nucleation of three quarks. According to Wilczek (2003) , as a result of this nucleation, mass is generated, in fact most of mass, with the exception of the small mass of quarks and electrons.

Charge, Parity and Time (CPT) symmetry could be another example of ternary stability. The quantum field theory has CPT symmetry, i.e., the laws are invariant under simultaneous transformations that involve time reversal and the inversions of charge and parity simultaneously. A mirror-image of our universe — with all objects having momenta and positions reflected by an imaginary plane (corresponding to a parity inversion), with all matter replaced by antimatter (corresponding to a charge inversion)— would evolve like the real universe. CPT symmetry is postulated to be a property of physical processes. In order to preserve this symmetry, every violation of the combined symmetry of two of its components (such as charge and parity) must have a corresponding violation in the third component, time. However time is unidirectional in reality, if not in some physical models, so the asymmetry of charge or parity is real.

Einstein established the interconversion of energy (E) and matter (M). Energy creates matter and matter transforms into energy, but, in the evolution of the universe, energy has formed matter to a larger extent than matter becomes energy. As these processes also increase information (I), and hence complexity, we may write the inequality

$$E \overset{\longrightarrow}{\leftarrow} I * M$$

The process equation shows that conservation is necessary to generate Bios; in the physical universe, energy-matter is conserved. The Klein-Gordon-Fock equation shows that Bios is generated only when velocities are slowed down below the speed of light; this is the result of mass. Conservation and mass are fundamental features of matter. Matter appears to be necessary to generate Bios. The CBR is the only one of the cosmological series examined that does not have a full biotic pattern, perhaps because photons have no mass. The concordance of these three disparate lines of evidence – an abstract equation, the quantum equations, and empirical data--, support the hypothesis that matter is necessary to generate Bios.

4. Creative Processes

The coexistence of asymmetric action, symmetric opposition, and conservation appears necessary and sufficient to generate biotic complexity. Simple and complex processes originate from simple deterministic mathematical structures (cause), not from maximally complex random change (chance).

Simple forms are included in the complex ones that they generate; thus the overall structure is fractal, and complex processes contain crucial simple components. The mathematical generators of Bios involve time asymmetry (as lattice asymmetry), trigonometric symmetric opposition (as in groups), and conservation with changes of form (as in topology). We thus propose that logically necessary mathematical relations shape cosmological evolution; that they represent the original heterogeneity that caused the heterogeneous distribution of energy and matter in the universe; and that they continue to generate structure –e.g. the expansion of the universe is accelerating in our times, and galaxy formation is a process extending from early times to the present.

The embodiment of lattice order, group opposition, and topological connectivity by quantum processes may generate biotic processes that play a significant role in cosmic evolution. Biotic diversification, novelty and expansion may explain how the universe evolved from an initial asymmetric, non-equilibrium state, and continues becoming more asymmetric, diverse, and complex. Cosmologists speculate that certain regions of the early universe expanded, became huge, and account for most of the volume of the universe. One may further speculate that the regions that expanded were those in which the pattern was biotic, while those regions that remained in steady, periodic or chaotic states did not inflate, and thus became insignificant.

We thus propose that the universe evolves **from order to Bios** (complex, biological-like organization) rather than from random or Chaos to order. In other words, cosmological processes are, as biological development, causal (deterministic) in their origin, but creative in their evolution. In any case, static, crystal-like order is not a convincing model for the current physical organization of the universe, which, as physics is fundamental, must also account for the generation of life, as Dirac remarked.

Mathematical order is arguably more "natural" than randomness. In particular, unidirectional action is far simpler and more evident, as time and as causation, than randomness. Occam's razor guides us to assume the simplest possible pattern for this primordial process. Occam's razor has been interpreted to imply that we must start by assuming stationarity and randomness. However, randomness has maximal algorithmic complexity [Chaitin, 1987], stationarity does not produce change, and explosions like the Big Bang do not generate a smooth and organized state; it seems paradoxical to regard stationary randomness as the simplest process. Moreover, randomness has not been observed and the universe expands and displays hierarchical organization. The observed data are accounted for by Bios, and Bios does not emerge from randomness. The cosmological principle asserts that laws of nature are the same everywhere. Parsimony suggests considering biotic feedback processes as responsible for the hierarchical distribution of matter in the universe. The term "order" is used to describe simple, static order (e.g., crystals), and complex, evolving organization (e.g., organisms). This has promoted the acceptance of order and symmetry as "natural." From a different perspective, symmetric randomness has been regarded as natural.

Leading physicists such as Carroll [2006] and others have pointed out that our notions of what is natural are historically determined. That in recent times physicists have come to regard equilibrium, time reversibility, and high entropy states as "natural" do not make them so. In fact this view forces us to regard the early universe as being in an extremely delicately specified state, and to consider the evident directionality of time as a problem. Why should we regard randomness as natural and organization as requiring explanation, when everything we see is organized and we cannot even produce a truly random series in a computer? It seems to us that, on the contrary, mathematical order is logically necessary, and therefore anything that exists physically by necessity must be organized mathematically.

The Physical Generation of Biotic Patterns: The Four Fundamental Forces

These three primary forms are embodied in the process equation, in the quantum equations, and in many others. They are also embodied in the physical forces –unipolar gravitation that generates unidirectional action, bipolar electromagnetic waves that embody circular symmetry in its sinusoidal shape, and tripolar nuclear forces that create atomic nuclei, the core of all structures.

We thus explore the hypothesis that the differentiation, coexistence and interaction of three forms of energy -- unidirectional inertial action (including gravitation), bipolar waves (including electromagnetic radiation), and triadic nucleation-- constitute an engine for the causal creation –i.e. the generation of diversity, novelty, and temporal complexity, the features that define the class of patterns that we call biotic.

This hypothesis is testable. It is falsifiable [Popper, 1965] and confirmable --that is to say, it makes new predictions that can be verified [Smolin, 2006]--, as required for a hypothesis to be scientific. In contrast, attributing evolution to random events makes no predictions, implies stationarity, claims to have explained the observations discouraging further search for pattern [Berry, 1987], and, regarding human processes, it provides no tools for action and demands no responsibility. The hypothesis predicted that the temporal distribution of galaxies and of the gravitational waves would display biotic patterns, as observed.

Bios Theory May Be Summarized in Physical Terms as the Priority and Universality of the Forms of Gravitational, Electromagnetic and Nuclear Forces Which Together Generate Periodicities, Chaos and Bios

Energy is the common "substance" of action, opposition and matter, but the concept of "energy" is an abstraction. Leaving aside the mysterious "dark energy", there are four types of energy: gravitational, electromagnetic, the weak nuclear force, and the strong nuclear force. Together they embody the factors required to generate Bios. Gravity embodies linear action and acceleration. Electromagnetic energy embodies coexisting opposites. The nuclear forces form the atomic nuclei of matter which is conserved. The neutron excess of the stable nuclei reflects the interaction between the strong and electromagnetic forces via the weak force.

It is assumed by most physicists that all four forces were unified in the earlier hotter universe, and gradually differentiated. Gravitational, electromagnetic and nuclear forces are ordered according to a number of properties:

Strength: gravity (1) < weak force (10^{25}) < electromagnetic force (10^{36}) < strong color force (10^{38}).

Domain: gravity (infinite in range and universal in quality, acting on radiation, baryonic matter, and dark matter) > electromagnetic energy (it also has an infinite range but acts only on charged matter, and it does not affect dark matter or neutrinos) > weak force (left-handed leptons and quarks, range 10^{-18}) > strong nuclear forces (act on protons and neutrons, and does not affect leptons, such as electrons; it has a very small range 10^{-15}).

Spin: 1 for gluons, 1 for w, z mediating weak force, 1 for photons, 2 for gravitons.

Polarity: gravity is unipolar, the electromagnetic force is bipolar (charge), and the weak and the strong nuclear forces are tripolar ("color" for the strong force). The strong nuclear force is more complex not only by being tripolar but also having an opposite to each of the three colors. Gravity is also more complex as it reflects possible accelerations in space which can occur in any one of three directions, giving rise to stresses that are represented by a tensor of values. For this reason physicists think of gravitons as spin 2.

The properties of the fundamental physical forces suggests an evolution of energy from gravitational to electromagnetic to nuclear forces (Figure 40). The simplest physical actions should operate on undifferentiated entities and should be universal, of lowest possible strength, unidirectional in time (unipolar), and inseparable from the entities it affects. Inertial movement is intrinsic to the object. Being the weakest, simpler (unipolar), and most universal form of interaction, gravity, which is inertial, has priority. The strong nuclear force, being stronger, restricted, and tripolar, has local supremacy regarding both complexity and strength. The existence of dark matter --an entity of unknown composition whose presence can be inferred from its gravitational effects on visible matter, it does not emit or reflect electromagnetic radiation-- indicates that unipolar gravitation is older than electromagnetic bipolarity. The greater domain of gravitation in space and in the types of entities it affects, and the greater strength and complexity of the strong force, suggests an evolution from gravity to the strong force.

A current understanding of the evolution of physical forces assumes that at very high energies, the four forces are unified. At the highest energies achievable with present-day particles accelerators (which correspond to temperatures circa 10^{15} degrees Kelvin), the weak and electromagnetic forces merge into a single electro-weak force. According to the Grand Unified Theories, the strong and electro-weak forces will behave as a single unified force at particle energies and temperatures that are about a trillion times higher still. Many physicists believe that at even higher energies, gravitation will unite with the others. Thus, during the first instants of the Big Bang, the Universe was dominated by the unified super-force. Thereafter, as the Universe expanded and cooled, the forces separated out and acquired their individual identities, gravitation at around 10^{-43} seconds after the beginning of time, the strong force at around 10^{-35} seconds, and the weak and electromagnetic forces at around 10^{-11} seconds.

We conjecture that the **polarity of a process may be a basic aspect of its complexity**; correspondingly, evolution may involve the differentiation from universal unipolarity (as in unidirectional time, inertial movement, gravitational attraction, entropy increase, and evolution from simple to complex) to bipolarity (as the electromagnetic force, mathematical groups, nature's cycles and logical oppositions), to tripolar polarities (as the strong and weak nuclear forces and conceptual triads) and the corresponding hexadic structures (e.g. snowflakes and the phenyl rings of organic molecules).

Mathematical recursions generate cascades of bifurcations. We thus conjecture that the *differentiation* of the four forces may have been generated by a similar process of repeated *bifurcations* from expansive inertial energy to bipolar electromagnetic radiation and tripolar nuclear forces, just as mathematical recursions generate periods 2^N, Chaos (which often includes period 3), and Bios. The cascade of bifurcations of physical forces proceeds may be regarded as a creative development, 'development' because it is determined by initial and simpler causes, and 'creative' because it generates diversity, novelty and complexity.

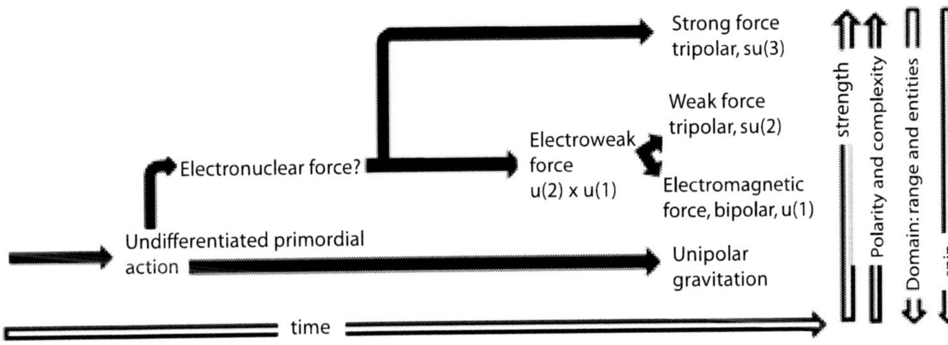

Figure 36. Hypothetical cascade of bifurcation of the physical forces. The electronuclear force has never been detected.

The concept of a hierarchy of forces and their sequential development during evolution is here connected with the idea of numbers as measures of complexity. But this is not the only way in which we can look at polarity and complexity. Gravity is 10-polar as it couples to the 10 components of the energy momentum tensor. Mercury orbit demonstrates the tensor character of the force and that the force does not, as Newton suggested, arise from a scalar potential. The conservation law associated with the 10 energy momentum charge is the energy momentum conservation. Weak couples to SU(2) making it bipolar (or even tripolar, given the group rotations of SU(2) are in the complex space of two dimensions in which one length is held fixed). Electromagnetic energy is unipolar being the symmetry U(1). Actually, both descriptions of polarity capture some aspect of reality.

A Gravitational Dialectics?

In his gravitational dialectics, Gal-Or (1987) argues that gravitational processes, as described by general relativity, provides a framework for understanding evolution. Gravity drives galaxy and star formation, and thereby nucleosynthesis, which is the first step in the evolution of molecules, life, and intelligence [Gal-Or, 1987]. As part of the physical process of global expansion and gravitational attraction, the universe increases in complexity.

Current models of the universe describe it as having no center, although the origin of the universe in a primeval "Big Bang " represents a universal origin to space-time. The expansion of the universe represents a diversification of time and of space, a bifurcation tree, from one universal origin to multiple local relativistic times as separate frames of reference. As result of expansion, processes become separated also in time, so their interactions, when possible, become delayed and acquire the properties of feedback processes. At even greater distances, there are no interactions. Relativity further asserts that simultaneity cannot be defined for separate spatial regions, and further asserts that there is no preferential frame of reference. The expansion of the universe thus produces an evolution from a compact universe to multiple local separate frames of reference –from one to many. Diversification is a major feature of creative processes.

There is a diversification of time and space. Time becomes fractal and space becomes multicentered. Space is locally curved by gravitation at each center. Mass determines its local geometry. Cosmic space is thus shaped by the coexistence and overlap of these spheroid regions that overlap with each other generating complex geometry.

Any entity (including observers, but also purely physical entities) "stands" at the surface of a spheroid (planet, star, galaxy), not at a "point" in space. Each frame of reference thus includes three different local asymmetries in the three macroscopic spatial axes: (z) away from the entity, an axis in which time and space are conflated; (x) in the direction of rotation of the spheroid; (y) orthogonal to it. There is a significant asymmetry between the north and south galactic caps in galaxy density and in the cosmic microwave background radiation. Each asymmetry is local but the existence of asymmetry is universal: matter nucleates throughout the universe. Everywhere in the universe we see local centers: the atomic nucleus, the sun surrounded by planets, galaxies spiraling around an axis, black holes at their center. A finite universe by necessity has a geometrical center. (Copernicus placed the sun at the center of his system, so the notion that there are no special places in the universe cannot be properly called the "Copernican principle".)

It is expected that the integration of the relativity theory of gravitation with quantum mechanics will provide a theory of universal evolution. Quantum mechanics per se does not provide a matrix for unification [Gal-Or, 1987]. It does not predict the expansion of the universe, and it cannot account for evolution because it portrays time as reversible. Although Schrödinger equation (and its equivalent, Heisenberg's matrices) are deterministic, quantum mechanics is often given a probabilistic interpretation. And the uncertainty introduced by interactions ("the observer") is brought as a deus ex machina to account for physical events. In fact the Schrödinger equation is both causal and creative – it generates Bios, not stationary states or stochastic noise.

The integration of quantum mechanics with gravity ("quantum gravity") requires causality to make meaningful models of the data [Ambjørn et al., 2004, 2008]. Causal, realistic physics offers predictions. The Soviet physicist Friedman (1999) demonstrated that general relativity predicts the global expansion of the universe. Einstein highlighted that physical processes are real and causal, rather than undetermined or dependent on observation. Relativity theory describes gravitational processes as actions in time, not only space, transmitted at the speed of light. In general relativity, gravitational interactions are described as the curving of space-time by gravitational fields, and objects moving freely simply undergo inertial motion in curved space-time. A gravitational field is an abstraction of a real global process of interactions ("forces") composed of inertial movement (including universal expansion and gravitational attraction), radiation and matter. Sometimes gravitational changes are described as oscillations of the space-time coordinates used to label a point, but this description has no physical consequence. Time and space are abstractions of fundamental features of these processes and structures, not a stage where they occur. Space-time does not exist independent from energy and matter. Only energetic and material changes are physical.

Gravitational dialectics focuses on mechanical movement and on opposition. The expansion of the universe and gravitational attraction are universal and unidirectional, they represent the simplest level. Quantum Electrodynamics and Chromodynamics address bipolar and tripolar processes, the next two more complex levels. Correspondingly, a theory of gravitation such as Relativity theory should be the foundation for the unification of physics and of science. The study of unidirectional inertial movement must come first. Unidirectional movement in time and space has logical, and possibly temporal, priority, so it must be incorporated into quantum and macroscopic mechanics. Unidirectional movement in time and space is a fundamental case of the universal asymmetry postulated by Pasteur.

Gravitation is inertial. While static medieval thinking adhered to Aristotle's notion that equilibrium was natural and change required an external propelling force, Galileo inaugurated modern science by demonstrating empirically that movement is inertial. Movement is the simplest action, and just as movement is self-sustained, all forms of energy are conserved. Action, not equilibrium, is spontaneous. At the simplest level of mechanical movement, inertia is spontaneous. As energy is conserved, action continues, its flow generating sequential temporal order and accumulating as action, the integral of changes of energy in time.

Electromagnetic Dialectics: From Helices To Bios

Gal-Or speaks of dialectics, a term that refers to change fueled and guided by the interaction of united opposites. Electromagnetic energy is the physical prototype of the interaction of united opposites. Action contains opposites: time and energy, the two inseparable conjugates that are the dimensions of physical action, are opposites insofar as energy is embodied in frequency, which is the inverse 1/t of time. Action is thus a union of opposites [Sabelli, 1989]. Group inverses are mathematical portraits of an important form of opposition. But opposition takes multiple forms. Each physical system (atom, star, galaxy) is held together by attractive forces, and prevented from collapse by repulsive actions, such as rotation for planetary systems --actions oppose forces. In a similar manner, the expansion of the universe is opposed by gravitational attraction (Figure 41, left). Expansion predominates globally while gravitation predominates locally, generating multiple black holes. Expansion increases with time and distance in the three directions of space x, y and z, while gravitation decreases with the square of distance.

Opposition, and in fact two dimensional opposition involving at least four "poles", is thus observed in all the physical forces. Gravitation, which is only attractive, is connected to the expansion of the universe; in this sense action and force are opposites. Regarding the basic forces, the opposition is two dimensional as represented by the sinusoidal wave form of the electromagnetic force, and the existence of colors and anti-colors regarding the nuclear force.

Electromagnetic radiation involves more complex opposition, the opposition between positive and negative polarities (Figure 41, right), the dual opposition of electrical and magnetic axes, and the 2^{∞} bipolar opposites of the circular rotation embodied in its helical form. The coexistence of linear change and circular symmetric feedback with conservation generates Bios as illustrated by the process equation. Thus Bios is the necessary, unavoidable product of simple, fundamental and widespread features of the physical universe.

Harmonic models have inspired the natural and human sciences from antiquity to Fourier and our times. Sinusoidal functions describe electromagnetic energy and other wave phenomena. At the abstract level, group inverses represent opposition. Our model for simple but fundamental forms of opposition is trigonometry, and this trigonometric representation changes our concept of dialectics and rendering it mathematical, physical (electromagnetic energy), and practical. The equations that generate Bios involve trigonometric functions. Trigonometric functions offer noteworthy portraits of complex time series (harmonic decomposition [power spectrum analysis], trigonometric walks [Dekking and Mendès-France,

1981], complement plots [Sabelli, 2005]. A trigonometric dialectics represents a significance transformation of philosophical dialectics. Trigonometric feedback generates periodicity, Chaos, Bios and infinitation.

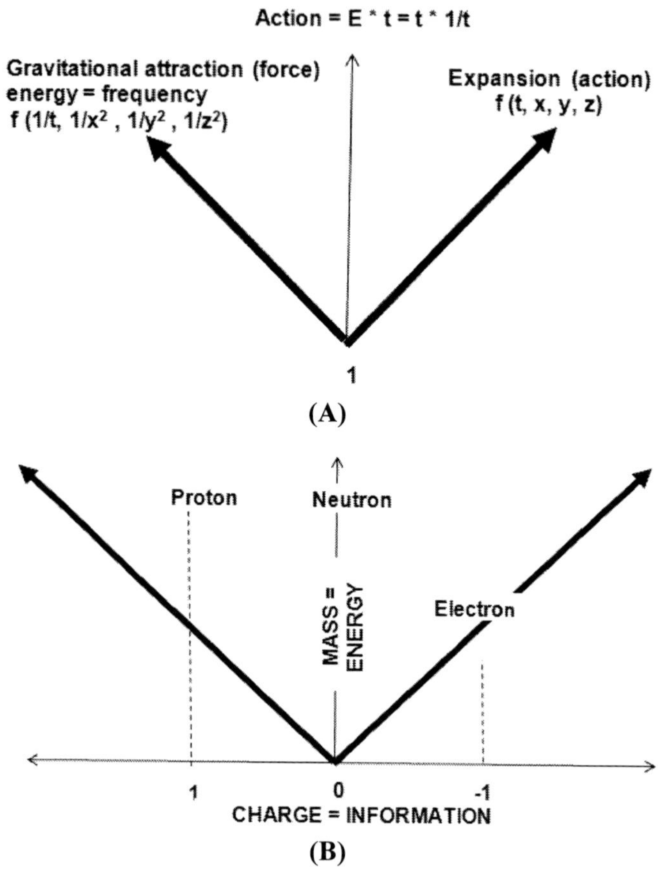

Figure 37. Diamond of opposites portrays fundamental opposites at macroscopic (A) and quantum (B) dimensions.

Tripolar Dynamics

Triadicity is fundamental. Beyond dyadic dialectics, macroscopic space is tridimensional and nuclear forces are tripolar (e.g. chromatic). Tripolar local interactions may generate a global tridimensional space.

Every interaction between two entities involves a third, such as the system that contains them. Space-time coordinates are quaternions with time in the real axis and three "imaginary' spatial axes (Einstein). Poincare's three attractors generate Chaos. Sarkovskii's period three at the end of the sequence of periodicity is generated by recursions. "Period three implies Chaos" [Li and Yorke, 1975] is a hypothesis about the universe not explicit in Sarkovskii's theorem. Unipolar feedback generates period 3 Chaos.

Starting with the formation of protons and neutrons, nucleation generates mass and mass curves space-time, establishing the paths for least action. Thus the generation of mass by complex nuclear forces feedbacks into simpler inertial movement, which thereby becomes complex gravitation. These feedback actions of complex processes transforming the simpler ones that generated them and then coexist with them may be one of the engines of evolution.

The curvature of space-time around each center of nucleation results in a complex global structure, which is metric (it has numerical properties) and fluid (topological). Every action occurs in such a tridimensional space structured by multiple masses. Gravitational attraction does not require mass, as it affects also massless photons, but mass affects inertial movement by creating a global field.

Regardless of what process underlies the differentiation of the primordial energy, the unipolar, bipolar and tripolar forces coexist. Indeed cosmological processes involve not only linear causality and dialectic oppositions but also tripolarity. In short the physical forces embody action, opposition, and conservation; thus they may be expected to generate Bios as indeed it is observed.

While some still speculate on the improbability of life as a random molecular accident, there is an increasing awareness that biomolecules may be cosmologically abundant products of the clouds forming young stars leading to cosmologically derived RNA molecules performing both catalysis and replication.

Bios Theory of Cosmological Evolution

In summary, biotic processes are physical, and contribute to cosmological evolution. The evolution of the universe is a creative development. The development of a biological organism creates a unique individual starting from a set of structures (the genes) that determine its overall features largely independently from external inputs. In a similar manner, the evolution of the universe would be a creative development from a "cosmic gene" (lattice order, group symmetry, and topological form) rather than as the product of chance and selection. Lattice, group and topological organization are embodied in the four basic physical forces and are evident at multiple levels of organization. Concretely, cosmological structure originated via a cascade of bifurcations of the original undifferentiated action through unipolar, bipolar, and tripolar forces to Chaos and Bios.

More generally, the Bios hypothesis predicts that biotic complexity will emerge whenever complementary opposites interact, such as in the Schrödinger equation when a moving quantum entity is confined to a box, in economic processes because there is both abundance and scarcity, and in animal populations where there is reproduction and predation. In fact biotic patterns are widespread in natural and human processes, but they have been described as noise or Chaos. Biotic patterns have also been found at many levels of organization:

(1) Mathematics: prime number series [Sabelli, 2007] and the Riemann equation [Kauffman and Sabelli, 2007].
(2) Cosmology: temporal distribution of galaxies and quasars; LIGO recordings.
(3) Planetary: air and ocean temperature, river levels, river and shore fractal forms [Sabelli, 2005].
(4) Molecular biology: sequences of bases in DNA [Sabelli, 2005].

(5) Physiology: heartbeat series and respiration [Carlson-Sabelli et al, 1994; Sabelli and Carlson-Sabelli, 2005].
(6) Biology: population size of several animal species [Sabelli and Kovacevic, 2007].
(7) Economic series [Patel and Sabelli, 2003; Sabelli, 2005].

Like Chaos, Bios has a simple origin – autodynamic, self-generating, "deterministic" interactions rather than random, independent interactions. Like statistical noise, Bios generates diversity, novelty and complexity. It is this combination of determined causation and generativity that makes a process creative.

We interpret Bios as a prototype for natural creation, and we identify the bipolar feedback that generates it as a dialectic interaction of opposites. Bios thus provides an anchor for developing a theory of creative processes. Indeed Bios is demonstrable in the prime number series (Sabelli, 2007; Kauffman and Sabelli, 2007), and the prime numbers are paradigmatic of novelty, as they are not multiples of any other, and they are generated mathematically, not by random chance. Several studies have demonstrated the relevance of periodic functions in the generation of prime numbers.

What are the opposites that generate biotic patterns in cosmological processes? There certainly is evidence for multiple interactions of opposites such as the coexistence of intergalactic expansion and gravitational attraction, electric and magnetic charge, and the formation of pairs of virtual particles in "empty space". We thus conclude that the universe is a creative process driven by the interaction of complementary opposites and triads.

The empirical analyses of a number of different cosmological series reveals the existence of common features of pattern characteristic of Bios, namely diversification, novelty, temporal complexity, and evidence for causal generation. This commonality between disparate series points to the existence of profound homologies in the universe. Bios is present in physical processes at multiple levels, from quantum to cosmological dimensions. If the LIGO recordings portray the GWB that stems from the waves emerging from the inflationary expansion of space itself, waves from the collision of bubble-like clumps of new matter at reheating after inflation, and waves from the turbulent fluid mixing of the early pools of matter and radiation, before equilibrium among them, then Bios is already present in the universe one trillionth-of-a-second after the Big Bang .

There is Bios everywhere. Where does it come from? How is it generated? Bios is not generated by random processes. Quantum processes generate an expansive pattern characterized by novelty, diversification, non-random complexity, asymmetric statistical distribution, non-uniform recurrence plots and evidence for causal generation. This characterizes Bios. Bios is generated mathematically by the recursion of three simple and universal processes, temporal (asymmetric) action, symmetric opposition, and conservation, as demonstrated by the process equation. Notably these are primary forms, as made evident by mathematics, physics and even simple inspections of the world we inhabit. Thus cosmological Bios may have originated with quantum processes. Indeed, at its origin, the universe was a 'primeval atom', as Lemaître described what became known as the Big Bang theory of the origin of the Universe.

In summary, the demonstration of Bios indicates an ongoing increase in diversity, the appearance of novel forms, and the existence of time limited patterns that follow each other. Bios is the signature of creative processes that may generate the evolution of matter, and potentially, of life and mind.

Methodological Considerations

While the data is too limited to validate these hypotheses, one must consider them as worthy alternatives to the widespread assumption of random changes which is repeated incessantly by many scientists, without the least statistical evidence, and with the same eagerness as others proclaim divine creation. Assuming that changes are random discourages the search for causes, does not provide tools to act, has no more explanatory power than attributing processes to an act of God, removes responsibility, and is hardly demonstrable.

Creative natural processes are known to exist; causation is demonstrable with analytic methods such as those illustrated here. Causal novelty is greater than random change; hence assuming random change is not necessary to account for biological or physical variation.

Isometry quantification demonstrates novelty while the quantification of standard recurrences do not; isometry measures the magnitude of the vector while recurrence takes into consideration also its direction, which is irrelevant because the changes that generate Bios are bipolar and diverse.

Behind the concrete results, our research represents a concerted effort to understand the creation of complexity, including life and mind, in a natural and causal way. All natural processes obey principles that can be codified by mathematics exemplified by physics, chemistry and biology. There is no need to invoke supernatural forces to explain what we all see. Because this is so ingrained in physical disciplines, we sometimes lose sight of the fact that creative processes are often viewed as rare. Creative processes are regarded as exemplified by living organisms that did not make their appearance until many billions of years after the Big Bang. Even biological evolution is explained by a process of natural selection.

Equally ingrained in the physical disciplines is the notion that natural processes are deterministic. This causes vast confusion however because from a philosophical point of view, determinism means that nothing new is generated. Determinism also seems to be the opposite to a strongly held view that life forms such as humans, as well as at least some animals, have freedom of choice. We each can make decisions that are not predetermined by a set of conditions. The Newtonian view was that the universe was like a clock, so that it would continue to exist in a regular and repetitive way. Once you start with a specific form, that form never changes. For centuries this view has persisted in one form or another. It is however untenable. Old simpler forms are not predictive, but new forms are generated. It is clear that such physical models can't conform to the real world unless they allow for creative processes, and that such creative process production can't be rare. Models that enshrine the form-deterministic models of past centuries (including many standard economic models, though Thomas, 2006 provides the first economic model, physical decision theory, which produces Bios) cannot describe reality.

Given this state of affairs, it becomes paramount to understand whether creative processes are indeed rare in physical theories, or rather simply forms that have not been studied. If creative processes are in fact not rare, they should occur in cosmological events as well as in such fundamental equations as the Schrödinger equation.

The distribution of matter in galaxies, the CBR, the gravitational background and even the very existence of galaxies and stars are examples of creative processes long before life appeared on our small segment of the universe. Biotic processes are also generated by quantum equations. In this chapter we verify that the pattern is not an artifact of the discrete approximations used in earlier publications but follows whenever a quantum particle is contained in a finite area. We show this by looking at a simple model in which the solutions are exact, sine and cosine. In fact the pattern follows from the fact that the spatial behavior and the time behavior don't evolve in the same way so that an initial simple structure is not stable but evolves into new and interesting complexes. The behavior is present for both the non-relativistic case and the relativistic case as long as the speeds are not too high. It is interesting that once speeds are significantly relativistic, the behavior is that of pure radiation showing no creative processes at all.

These studies support causation over random chance. Laplace championed a strict determinism. The demonstration of biotic processes in physical processes enhances causation beyond the limits of determinism to evolution. As biotic processes occur in these physical processes, and if we accept that such processes characterize what we can agree are creative processes, then we must conclude that creative processes are not in fact rare.

We found Bios because we developed methods to detect diversification, novelty, and non-random complexity in what appears to be random or simply chaotic. We focused on actual time series rather than on averages that eliminate variation. We also developed ways to distinguish causal Bios from Random Walks. The assumption of causality, and the corresponding rejection of random change as an explanation, does not stem solely from a philosophical stance but it is also based on physical facts: **Every action is transmitted, and energy is conserved, so every action has a consequence and a cause**. And every system interacts with others. "Isolated random events" are simply not possible.

The statistical assumption of randomness unless causality is demonstrated is unwarranted. There are many cases believed to be random and later on shown to be Chaos or Bios. **Seeking causality <u>unless</u> randomness can be <u>demonstrated</u>** promotes the discovery of causality and of creative processes. A causal hypothesis is better than none. The notion of Bios led to new methods for time series analysis and to the demonstration of non-random pattern and causation in empirical data (cosmological, biological populations, physiological functions, economic data).

While reductionism still supports the description of a very general physical theory as a "theory of everything", actual science often requires a multidisciplinary approach rather than reductionism. The analytic methods that lead to the identification of the Bios model were originally developed in the course of a study of variations in heart rate [Sabelli et al, 1997]. As this chapter illustrates, our research strategy also involves going from methods and ideas developed in the context of studying complex processes (e.g. heart rate variation) to their application to studying and understanding simple ones (e.g. physical processes). This "complexity inference" [Sabelli, 2005] complements the usual strategy of going from the simple to the complex, which we regard as fundamental, as simple processes create complex ones, and in turn complex processes always contain the simple processes that generated them. Process Theory based on philosophy (Sabelli, 1989) and psychology (Sabelli and Carlson-Sabelli, 1989) became Bios Theory of cosmological evolution.

Physics often appears dominated by the notion of invariant laws, symmetry, and stationarity. Biological (Lamarck, Darwin) and social (Comte, Marx, Morgan) evolution were established in the nineteenth century, while thermodynamics asserted inexorable decay and physicists up to Einstein postulated a non-evolving universe. Only in the twentieth century did Hubble establish the evolution of the universe.

The study of cosmological, quantum, and meteorological processes extends the notion of causal and creative Bios to the physical realm. Bios research includes: the development of methods to demonstrate creativity and causality [Sabelli et al., 2005]; the demonstration of Bios at all levels of organization, suggesting a mechanism that may contribute to biological [Sabelli and Kovacevic, 2008] and physical [this chapter], and human [Sabelli, 2009, in press] evolution; the formulation of equations that produce Bios; the interpretation of the component of these equations as embodied in basic physical processes (unipolar gravitation, electromagnetic oppositions, and tri-polar nuclear forces); the connections of these basic forces to the fundamental mathematical forms described by Bourbaki and found in cognitive processes by Piaget; their links to the small number forms (1,2,3) already highlighted by Pythagoras and in our times by Gamow (1967); and the application of these ideas to develop logic (including the logic of quantum computers [Sabelli and Thomas, 2008]), psychotherapeutic methods [Sabelli, 2008], innovation methods [Sabelli, 2008], and, in this chapter, to the formulation of physical theory.

Acknowledgments

This research was supported by the Society for the Advancement of Clinical Philosophy and by generous gifts from Ms. Maria McCormick.

References

Allen, B., Romano, J. 1999. *Phys. Rev.* 59[10]: 44. Detecting a stochastic background of gravitational radiation: Signal processing strategies and sensitivities.

Ambjørn, J., Jurkiewicz, J., Loll, R. 2008. *Scientific American*. 42-49. The Self-Organizing Quantum.

Anderson, P.W., Stein, D.L. In *Self-Organizing Systems*; Yates, F.E.; Ed.; Broken symmetry, emergent properties, dissipative structures, life: Are they related? Plenum: New York, NY, 1987; pp 459-474.

Aref'eva, I. Y., Volovich, I. V. (2007). Quantization of the Riemann Zeta-Function and cosmology [arXiv:hep-th/0701284v2].

Baryshev, Y.; Teerikorpi, (2002)P. *Discovery of Cosmic Fractals*. World Scientific: Hackensack, NJ.

Bennett, C. L. *Nature*. 2006, 440, 1126-1131. Cosmology from start to finish.

Berry, W. *Home Economics*. North Point Press, San Francisco, CA, 1987.

Beth, E. W., Piaget, J., *Relaciones entre la lógica formal y el pensamiento real*; Editorial Ciencia Nueva, S. L. Madrid, Spain, 1968.

Bondi, H., Pirani, F.A.E., Robinson, I. *Proc. Roy. Soc. London Set.* 1959, 251, 519-533. Gravitation waves in general relativity. III. Exact Plane Waves.

Bourbaki, N. In *Les grands courants de la pens´ee math´ematique*; Le Lionnais, F.; Ed.; L'architecture desMath´ematiques; Cahiers du Sud: Paris, France, 1948. Translated by Dresden, A. *Amer Math Month*. 1950, 57, 221–232.

Carroll, S. M. *Nature*. 2006, 440, 1132-1136. Is our Universe natural?

Chaitin, G.J. *Information, Randomness, and Incompleteness*. World Scientific: Hackensack, NJ, 1987.

Chirikov, B.V., Lieberman, M.A., Shepelyansky, D.L., Vivaldi, F. *Physica*. 1985, 14, 289-304. A Theory of Modulational Diffusion.

Christensen, N. *Phys. Rev.* 1992, 46, 5250-5266. Measuring the stochastic gravitational-radiation background with laser-interferometric antennas.

Coleman, P.H.; Pietronero, L. *Phys. Rep.* 1987, 213 (6), 311-389. The Fractal Structure of the Universe.

Coles, P. and Chiang, L-Y. (2000). Characterizing the nonlinear growth of large-scale structure in the Universe. *Nature* 406: 376 - 378.

Colless, M.M. (1999). First Results from the 2dF Galaxy Redshift Survey. [http://arxiv.org/abs/astro-ph/9804079]

Colley, W.N. (1996). Two dimensional topology of large scale structure in the Las Campanas Redshift Survey. Astrophysics Journal [http://arxiv.org/abs/astro-ph/9612106], 19:34:42.

Corballis, M. C.; Beale, I. *The psychology of left and right.* Erlbaum: Mahwah, NJ, 1976.

Dekking, M., and Mendès-France, M. *Journal Fur die Reine und Angerwandte Mathematik.* 1981, 239: 149-153. Uniform Distribution Modulo One.

Frey, R. E. (2007) LIGO: Status and Recent Results. http://www.ligo.caltech.edu/docs/P/P070079-01.pdf

Friedman, A. *Gen Rel Grav*. 1999, 31, 1991-2000. On the Curvature of Space.

Gal-Or, B. *Cosmology: Physics and Philosophy*. Springer Verlag: New York, NY, 1987, pp 41-46, 154.

Gamow, G. *Proc Nat Acad Sc*. 1967, 57, 192. Does gravity change with time?

Garcıa-Bellido, J., Figueroa, D.G. (2008). A new gravitational wave background from the Big Bang [arXiv:0801.410]. http://arxiv.org/PS_cache/arxiv/pdf/0801/ 0801.4109v1.pdf

Geisel, T., Nierwetberg, J. *Physic. Rev. Letters* 1982, 48, 7-10. Onset of Diffusion and Universal Scaling in Chaotic Systems.

Gell-Mann, M. *Phys Lett*. 1964, 8, 214-215. A Schematic Model of Baryons and Mesons.

Gerber H. *The European Physical Journal C - Particles and Fields*. 2004, 35, 195-196. Evidence for time-reversal violation?

Gilden, D. L. *Psych. Rev*. 2001, 108, 33- 56. Cognitive Emissions of 1/f Noise.

Grishchuk, L. P. *JETP Lett.* 1976, 23, 293. Primordial Gravitons and the Possibility of Their Observation.

Guth, A.H. *Phys Rev*. 1981, 23, 347. The Inflationary Universe: A Possible Solution to the Horizon and Flatness Problems.

Haldane, J.B. *J Genet*. 1960, 57, 351-360. More precise expressions for the cost of natural selection.

Handel, P. H., Chung, A. L. *Noise in physical systems and 1/f fluctuations*; American Institute of Physics: New York, NY, 1993.

Hubble, E.; *The Observational Approach to Cosmology*; Clarendon Press: Oxford, England, 1937.

Kauffman, L., Sabelli, H. *Cybernetics and Systems*. 1998, *29* (4), 345-362. The Process equation.

Kendal, M. *Time Series*. Charles Griffin: London, England, 1973.

Li, T. Y., Yorke, J. *Am Math Month*. 1975, 82, 985-992. Period Three Implies Chaos.

Loll, R. (2008). The Emergence of Spacetime. arXiv [gr-qc]

Luminet, J.P. *L'Univers Chiffonné*. Fayard: Paris, France, 2001.

Mandelbrot, B. B. *The Fractal Geometry of Nature*; W.H. Freeman and Company: New York, NY, 1975

Patel, M., Sabelli, H. *Kybernetes. 2003,* 32: 692-702. Autocorrelation And Frequency Analysis Differentiate Cardiac And Economic Bios From 1/F Noise.

Nicolis, G., Prigogine, I. *Self-Organization in Non-Equilibrium Systems*; Wiley: New York, NY, 1977.

Peebles, P. J. E. *Principles of physical cosmology*; Princeton University Press: Princeton, NJ, 1993, pp 209–210.

Pietronero, L. *Physica A.* 1987, 144: 257-284. The fractal structure of the universe: correlations of galaxies and clusters and the average mass density.

Pietronero, L., Montuori, M., Sylos, F. L. In *Proceedings of the Conference Critical Dialogues in Cosmology*; Turok, N.; Ed.; World Scientific: Hackensack, NJ, 1997.

Popper, K. *Conjectures and Refutations: the growth of scientific knowledge.* Routledge and Kegan Paul, New York, NY, 1965.

Press, W.H. *Comments on Modern Physics, Part C- Comments on Astrophysics*; 1978, 7, 103-109. Flicker noises in astronomy and elsewhere.

Prigogine, I. *From Being to Becoming: Time and Complexity in the Physical Sciences*; Freeman: New York, NY, 1979.

Rosen, R. *The Self-Evolving Cosmos*; World Scientific: Hackensack, NJ, 2008.

Sabelli, H. *Union of Opposites: A Comprehensive Theory of Natural and Human Processes*; Brunswick Publishing: Lawrenceville, VA, 1989.

Sabelli, H., Carlson-Sabelli, L. In *Sustainable Peace in the World System, and the Next Evolution of Human Consciusness*; Hall, M.L.W. Ed.; Proc Inter Systems Soc: Louisville, KY; 1996. A cosmic gene? A biological model of complex systems.

Sabelli, H. In *Systems thinking, globalization of knowledge, and communitarian ethics.* Rhee, P.Y. Ed.; The union of opposites: North and South, East and West, Korea and America. *Bailey Proc. International Systems Society*: Seoul, Korea; 1997; 434-445.

Sabelli, H. In *Toward a New Paradigm of System Science.* Rhee, P.Y. Ed.; Process theory: mathematical formulation, experimental method, and clinical and social application. Seoul National University Press: Seoul, Korea, 1999, 159-201.

Sabelli, H. In *Understanding Complexity*. Ragsdell, G., J. Wilby, J. Eds.; The Co-Creation Hypothesis. Kluwer Academics/Plenum Publishing: London, England, 2000.

Sabelli, H. *Nonlinear Dyn Psych and Life Sci.* 2001, 5, 89-113. Novelty, a Measure of Creative Organization in Natural and Mathematical Time Series.

Sabelli, H. *Kybernetes*. 2003, 32: 752-766. Mathematical development: A theory of natural creation.

Sabelli, H., Kovacevic, L. *VIP Forum of the IPSI-2003*, Montenegro. Biotic Expansion of the Universe. Sabelli, H., Kovacevic, L. *Complexity*, 2006, 11, 14-25. Quantum Bios and Biotic Complexity in the Distribution of Galaxies.

Sabelli, H. and Kovacevic, L. "Biotic complexity of population dynamics." *Complexity*, 2008, 13 (4), 47-55.

Sabelli, H., Carlson-Sabelli, L., Patel, M., Sugerman, A. *J. Mind and Behavior*. 1997, 18, 305-334. Dynamics and psychodynamics: Process Foundations of Psychology.

Sabelli, H., Sugerman, A., Kovacevic, L., Kauffman, L., Carlson-Sabelli, L., Patel, M., Konecki, J. *Nonlinear Dyn Psyc and the Life Sci.* 2005, 9, 505-538. Bios Data Analyzer.

Sabelli, H. Biotic patterns in LIGO recordings point to the creativity of gravitational interactions. *Complexity*, 2009 (in press).

Sabelli,H. Social theory beyond the collapse of communism and capitalism. The International Journal of Interdisciplinary Social Sciences, 2009 (in press).

Sabelli, H. *Bios. A Study of Creation*; World Scientific: Hackensack, NJ, 2005.

Sabelli, H., Thomas, G. The Future Quantum Computer: Biotic Complexity. In (Ed.) F. Orsucci, N. Sala, *Reflecting Interfaces: The Complex Coevolution of Information Technology Ecosystems*. Idea Group, Anaheim, CA, 2008.

Sabelli, H. *Bios Theory of Innovation*. (Forthcoming).

Shectman, S.A., Landy, S.D., Oemler, A., Tucker, D.L., Lin, H., Kirshner, R., & Schechter, P. 1996. The Las Campanas Redshift Survey. *Astrophysics Journal* [http://arxiv.org/abs/astro-ph/9604167], 470 (172).

Schewe, P. F., Stein, B., Turner B., Castelvecchi, D. (2007). Gravitational wave background and cosmic microwave background [www.aip.org/pnu].

Schröeder, M. *Fractals, Chaos, Power Laws*; Freeman: New York, NY, 1991.

Smolin, L. *The Trouble With Physics*; Houghton Mifflin Company: Boston, MA, 2006.

Springel, V., Frenk, C. S., White, S. D. M. *Nature*. 2006, 440: 1137-1144. The large-scale structure of the Universe.

Sprott, J., Rowlands, G. *Nonlinear Dyn. Psych. and Life Sci.* 2000, 4, 195-199. Software review Chaos Data Analyzer, Professional Version.

Starobinsky, A.A. *Phys Lett.* 1980, 91, 99-102. A new type of isotropic cosmological models without singularity.

Sugerman, A. In *Bios. A Study of Creation*; Sabelli, H. Bios Data Analyzer. World Scientific: Hackensack, NJ; 2005; 114-252.

Thomas, G. H., Sabelli, H., Kauffman, L. H., Kovacevic, L. (2006). Biotic patterns in Schrödinger's equation and the evolution of the universe. Inter Journal Complex Systems, 1787. [http://www.interjournal.org/manuscript_abstract.php?1161328888]

Thorne, K. S., Kenan, W.R. (1997). Gravitational Radiation: A New Window onto the Universe. http://arxiv.org/abs/gr-qc/9704042v1

Tomozawa, Y. (2007). The CMB Dipole and Circular Galaxy Distribution [arXiv:astro-ph/0701151v5].

Tryon, E. P. *Nature*. 1973, 246: 396-397. Is the Universe a Vacuum Fluctuation.

Tully, R. B.; Fisher, J.R. *Astron and Astrophys.* 1977, 54, 661-673. A new method of determining distances to galaxies.

Webber, C.L.; Zbilut, J.P. *J Appl Phsiol.* 1994, 76, 965-973. Dynamical assessment of physiological systems and states using recurrence plot strategies.

Wilczek, F. and Gross, D.J. *Phys Rev D.* 1973, 8: 3633. Asymptotically Free Gauge Theories.

Wilczek, F. (2003). The Origin of Mass, Annual Physics at MIT, 24-35. [MIT-CTP-3410].

In: Black Holes and Galaxy Formation
Editors: A.D. Wachter and R.J. Propst, pp. 87-100
ISBN: 978-1-60741-703-3

Chapter 2

TURBULENT FORMATION OF PROTOGALAXIES AT THE END OF THE PLASMA EPOCH: THEORY AND OBSERVATIONS

Rudolph E. Schild[1*] and Carl H. Gibson[2**]

[1] Center for Astrophysics, 60 Garden Street, Cambridge, MA 02138, USA

[2] University of California San Diego, La Jolla, CA 92093-0411, USA

Abstract

The standard model of gravitational structure formation is based on the Jeans 1902 acoustic theory, neglecting nonlinear instabilities controlled by viscosity, turbulence and diffusion. A linear instability length scale equal to the sound speed times the gravitational freefall time emerges. Because the Jeans scale L_J for the hot primordial plasma is always slightly larger than the scale of causal connection *ct*, where *c* is the speed of light and *t* is the time after the big bang, it has been assumed that no plasma structures could form without guidance from a cold (so $L_{J\ CDM}$ is small) collisionless cold-dark-matter CDM fluid to give condensations that gravitationally collect the plasma. Galaxies by this CDM model are therefore produced by gradual hierarchical-clustering of CDM halos to galaxy mass over billions of years, contrary to observations showing that well formed galaxies existed much earlier. No observations exist of CDM halos. However, Gravitational instability is non-linear and absolute, controlled by viscous and turbulent forces or by diffusivity at Schwarz length scales smaller than *ct*. Because the universe during the plasma epoch is rapidly expanding, the first structures formed were at density minima by fragmentation when the viscous-gravitional scale L_{SV} first matched *ct* at 30,000 years to produce protosupercluster voids and protosuperclusters. Weak turbulence produced at expanding void boundaries guide the morphology of smaller fragments down to protogalaxy size just before transition to gas at 300,000 years. The size of the protogalaxies reflect the plasma Kolmogorov scale with a linear and spiral morphology predicted by the Nomura direct numerical simulations and confirmed by deep field space telescope observations. On transition to gas the kinematic viscosity decreases by ten trillion so the protogalaxies fragment at Jeans scale clouds, each with a trillion earth-mass planets, as predicted by Gibson 1996 and observed by Schild 1996. The planets promptly form stars near the cores of the protogalaxies, but as the universe cools the

* E-mail address: rschild@cfa.harvard.edu

** E-mail address: cgibson@ucsd.edu, http://sdcc3.ucsd.edu/~ir118

planets gradually freeze to form the galaxy dark matter. Space telescope observations of the most distant galaxies confirm the Kolmogorov scale size and Nomura geometry of the protogalaxies. High resolution images of planetary nebula and supernova remnants reveal thousands of frozen hydrogen-helium dark matter planets with large atmospheres evaporated by such events. Galaxy mergers show frictional trails of young globular clusters formed in place, proving that dark matter halos of galaxies consist of dark matter planets in metastable clumps. Galaxy evolution is guided by friction from this galaxy dark matter.

Introduction

The standard model of gravitational structure formation and the formation of galaxies is based on a linear acoustic instability theory proposed by James Hopwood Jeans in 1902 in a lengthy monograph titled "The stability of a spherical nebula" transmitted to the Philosophical Transactions of the Royal Society of London Series A (Vol. 1999, pp. 1-53) through G. H. Darwin, son of Charles Darwin. The purpose of the work was to understand how the solar system might have formed from a gas cloud. Galaxies and the expansion of the universe were not known at that time. The alternative theory due to G. H. Darwin in his 1889 paper (Vol. 180, pp. 1-69) was titled "On the mechanical conditions of a swarm of meteorites, and on theories of cosmogony". Neither of the theories are correct as a description of how either the solar system or galaxies were formed. Darwin's theory focusing on meteorites [1] is a better guide than Jeans' in both cases since accretion within clumps of primordial planets and effective viscosity from planet collisions are crucial and the collapse of gas clouds following Jeans 1902 [2] is irrelevant. The Jeans acoustic scale has only one known application, which is to set the mass of globular star clusters at the plasma to gas transition (decoupling), but it is for different reasons than described by Jeans. Modern fluid mechanical concepts [3-19] applied to cosmology are termed hydro-gravitational-dynamics HGD. Relevant length scales are summarized in Table 1.

The fluid mechanics of Jeans 1902 is that of the 19th century. Jeans started with the Euler equations that neglect viscous forces, used linear perturbations stability analysis that neglects turbulence forces, and took no account of diffusivity effects arising from weakly collisional non-baryonic dark matter that probably constitutes about 97% of the rest mass of the universe where the universe is flat with density approximately equal to the present critical density of $\rho_{Crit} = 10^{-26}$ kg m^{-3} with no dark energy or cosmological constant Λ. Jeans' assumptions lead to linear acoustic momentum conservation equations for the gas, so the Jeans 1902 acoustic instability scale is $L_J = V_S / (\rho G)^{1/2}$, where V_S is the speed of sound, $\tau_g = (\rho G)^{-1/2}$ is the gravitational free fall time, ρ is the density and G is Newton's constant of gravitation.

The Jeans criterion for gravitational structure formation is that a large cloud of gas with density ρ is unstable to gravitational structure formation only on scales larger than J_H. By this criterion it is impossible to make any structures at all in the plasma epoch of the universe because the sound speed $V_{S-plasma} = c / 3^{1/2}$ is so large that $L_J > L_H$, where $L_H = ct$ is the horizon scale or scale of causal connection at time t after the big bang and c is the speed of light. Even after the plasma turns to gas at $t \approx 10^{13}$ seconds (300,000 years) the Jeans mass is an enormous million solar masses so the first stars cannot form until much later and planets can never form from gas at average temperatures because the density of the expanding universe decreases more rapidly than the sound speed with time. The problem with the Jeans criterion is that it is simply wrong [10]. Gravitational instability is absolute not linear [11], meaning that in a fluid with density fluctuations, structure will immediately begin to form at all scales of available density fluctuations unless prevented by the molecular diffusivity of the nearly collisionless non-baryonic dark matter or by viscous or turbulence forces. Mass moves toward density maxima and away from density minima at all scales unless prevented by forces or diffusivity.

When the plasma turns to gas it becomes a fog of planets [10] that become galaxy dark matter, as observed [20]. The only known form of non-baryonic dark matter NBDM is neutrinos, but there may be other massive neutrino-like particles formed in the early universe at temperatures unavailable in laboratories or there may be more ordinary neutrinos than known from present estimates. The evidence indicates the existence of something non-baryonic and quite massive that prevents the disintegration of galaxy clusters by gravitational forces, just as baryonic dark matter is needed to inhibit the disintegration of individual galaxies by centrifugal forces, but at this time no one knows what it is.

Density perturbations exist in the plasma because the big bang was triggered by a form of turbulent combustion at Planck temperature 10^{35} K termed the Planck-Kerr instability [12,13]. Planck particles and anti-particles appear by quantum tunneling and form the Planck scale equivalent of positronium. Positronium is produced by supernova temperatures of 10^{10} K sufficient to cause electron positron pair production, where the antiparticle pairs form a relatively stable combinations in orbit. Prograde captures give 42% release of the Planck particle rest mass energy which can only go into producing more Planck particles to fuel similar interactions. This gives a turbulent big bang with Reynolds number of about a million, terminated when cooling permits formation of quarks and gluons and a strong increase of viscosity and viscous stress that can cause an exponential increase in space to approximately a meter size in 10^{-33} seconds from 10^{-27} meters (10^{8} Planck lengths). The speed is about 10^{25} c. This inflation epoch produces the first fossil turbulence because all the turbulent temperature fluctuations of big bang turbulence are stretched beyond the scale of causal connection $L_H = ct$.

These fossils of big bang turbulence then trigger the formation of density fluctuations in the hydrogen and helium formed in the first three minutes by nucleosynthesis. The density fluctuations then seed the first gravitational structure formation [14]. The mechanism is not by the Jeans criterion but under the control of viscosity, turbulence or particle diffusivity, depending on which of three Schwarz scales is largest for the fluid in question. Because the collision length for non-baryonic dark matter is larger than $L_H = ct$ for the plasma epoch, only viscous forces or turbulence forces can prevent structure formation. Fragmentation of the non-baryonic dark matter occurs after the end of the plasma epoch, so this material is nearly irrelevant to the first structure formation, which begins when the horizon scale matches the Schwarz viscous scale at about 10^{12} seconds after the big bang (30,000 years) and the mass scale is that of superclusters. The plasma epoch is dominated by viscous forces with only weak turbulence because the photon viscosity is very large. Photons scatter from the free electrons of the plasma and transmit momentum because the electrons strongly couple to the ions to maintain electrical neutrality. Reynolds numbers are close to critical values so the Schwarz viscous and turbulent scales are nearly identical. The first structures to form are voids because the universe is expanding. Condensations occur only after decoupling. The voids expand at the sonic speed of the plasma epoch which is $c / 3^{1/2}$, so we have significantly larger supervoids than superclusters as discussed in the following sections [15].

Because observations showed early structure must have occurred, cosmologists resorted to a *deus ex machina* solution (magic): they invented cold dark matter in order to create the present standard cosmological model. Following Jeans, it was assumed that a cold non-baryonic material must exist with sound speed sufficiently small that gravitational condensations in the plasma epoch could occur; that is, with $L_{J_{CDM}} < L_H$. Seeds of CDM were assumed to condense to provide gravitational potential wells into which the baryonic plasma would fall. The seeds would magically merge and stick over time, collecting more and more of the baryonic material until stars could form, and the clumps could cluster to larger and larger scales to produce galaxies and galaxy clusters and last superclusters of galaxies. This is hierarchical clustering, so the standard model is often called CDMHC. No CDM halo has ever been observed. In the following we suggest they do not exist and that CDM does not exist. Neither does the latest addition to the standard cosmological model; that is, dark energy and the cosmological constant Λ [16–19].

Instead, the plasma fragmentation beginning at 10^{12} s continues to smaller and smaller scales. Galaxies represent the smallest mass gravitational structures formed in the plasma before decoupling at 10^{13} s (300,000 years). Galaxy morphology reflects the turbulence existing in the plasma when they were formed. Protogalaxies fragmented along turbulent vortex lines at the Kolmogorov scale of the plasma and with a chain and spiral morphology demonstrated by the direct numerical simulations of Nomura. This morphology and a consistent protogalaxy length scale termed the Nomura scale is observed in a variety of Hubble Space Telescope observations discussed in the rest of this chapter [15].

At decoupling the kinematic viscosity ν decreased from photon viscosity values of $\sim 10^{26}$ m^2 s^{-1} to hot gas values of 10^{13} m^2 s^{-1}, a factor of ten trillion. This decreased the mass scale of fragmentation from that of protogalaxies to that of planets. Simultaneously the gas protogalaxies fragmented at the Jeans mass of globular star clusters, giving hot gas clouds of planet mass in million solar mass clumps. As the universe expands it cools. The gas clouds of hydrogen and helium eventually begin to freeze to form the galaxy dark matter, which is proto-globular-star-cluster PGC clumps of primordial-fog-particle PFP planets [10] as observed [20]. All stars form in PGCs from the trillion planets contained by a binary accretion process that gives larger and larger Jovian planets. The first stars to form were therefore small and long-lived, as observed in ancient globular star clusters. Structure formation from fluid mechanics naturally explains globular star clusters, which have always been a mystery. In standard ΛCDMHC cosmology, the first stars to form are superstars that re-ionize the universe to explain why so much hydrogen is missing in high red-shift quasar spectra. In our hydro-gravitational-dynamics HGD cosmology the missing hydrogen is tied up in frozen planets, the superstars (Population III) never happened, and Reionization of the Universe never happened [15].

Table 1. Length scales of gravitational instability

Length Scale Name	Definition	Physical Significance
Jeans Acoustic	$L_J = V_S/(\rho G)^{1/2}$	Acoustic time matches free fall time
Schwarz Viscous	$L_{SV} = (\gamma\nu/\rho G)^{1/2}$	Viscous forces match gravitational forces
Schwarz Turbulent	$L_{ST} = (\varepsilon/[\rho G]^{3/2})^{1/2}$	Turbulent forces match gravitational forces
Schwarz Diffusive	$L_{SD} = (D^2/\rho G)^{1/4}$	Diffusive speed matches free fall speed
Horizon, causal connection	$L_H = ct$	Range of possible gravitational interaction
Nomura, smallest plasma piece	$L_N = 10^{20}$ m	Proto-galaxy-fragmentation on vortex lines

V_S is sound speed, ρ is density, G is Newton's constant, γ is the rate of strain, ν is the kinematic viscosity, ε is the viscous dissipation rate, D is the diffusivity, c is light speed, t is time.

Theory

Figure 1 shows the sequence of gravitational structure formation events leading to the present time according to hydro-gravitational-dynamics HGD cosmology.

The big bang mechanism is one of turbulent combustion limited by gluon viscosity, where the maximum Taylor microscale Reynolds number $Re_\lambda = 1000$ occurred at $t = 10^{-33}$ seconds at 10^8 L_P, where $L_P = 10^{-35}$ is the Planck scale of quantum-gravitational instability [12,13]. The only particles possible during this epoch of high temperatures were Planck particles and Planck anti-particles that interact in the manner of electron-positron pair production to efficiently produce more Planck particles until the event cools enough for quarks and gluons to appear. Large negative Reynolds stresses expand space according to Einstein's theory of general relativity. During the big bang turbulence epoch, indicated in Figure 1a by the red star, the kinematic viscosity ν is the mean free path for particle collisions L_P times the speed of the particle. Even though the particle speeds were light speeds c at $t = 0$, the collision distances were small, giving a small initial kinematic viscosity $\nu_0 = 3 \times 10^{-27}$ m^2 s^{-1}. The Planck scale Reynolds number is near critical so L_P matches the Kolmogorov and Batchelor scales [3] of big bang turbulence and turbulent mixing.

The end of the big bang turbulence epoch occurred when the temperatures and particle speeds decreased as the Reynolds numbers increased till quarks and gluons could appear. This phase change is termed the strong force freeze out, with length scale $L_{SF} = 10^8$ L_P, or 10^{-27} m. Gluons are particles that carry momentum for the quarks, similar to photons carrying forces from momentum differences between the electrons and protons of the plasma epoch. It is assumed that the gluon viscosity is dramatically larger than the Planck viscosity, just as the photon viscosity is dramatically larger than the proton-proton collision viscosity of the plasma epoch, so that large negative gluon viscous stresses will accelerate the expansion of space exponentially during the inflation epoch indicated by the blue triangle. Other mechanisms such as the Guth false vacuum may be important for driving inflation. Whichever mechanism is dominant, many observations support an inflationary event.

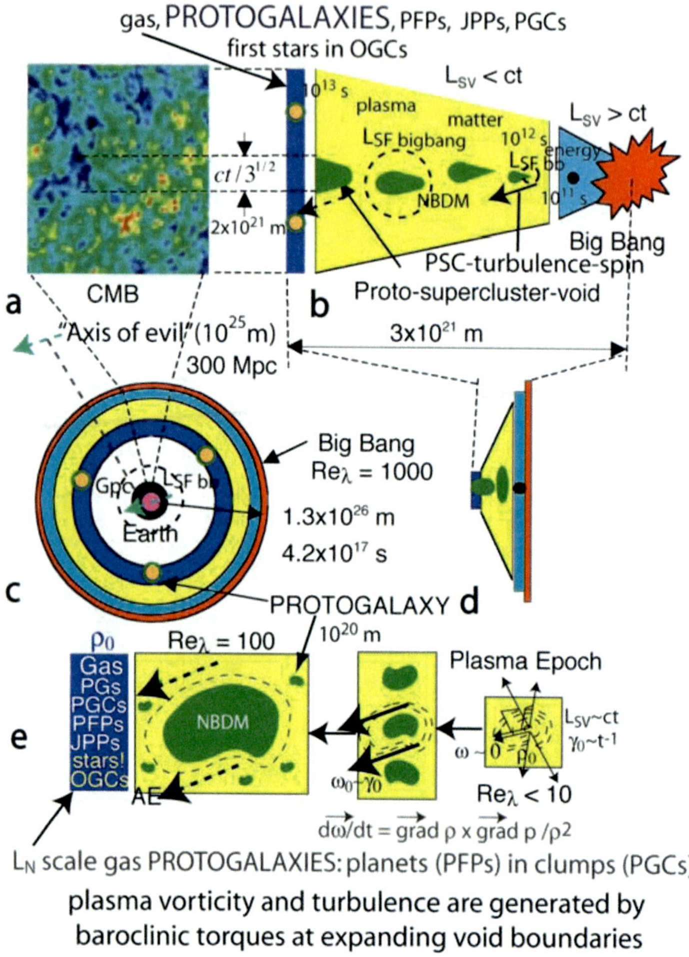

Figure 1. Protogalaxy formation at the end of the plasma epoch by hydro-gravitational-dynamics HGD theory. a. Cosmic Microwave Background temperature anisotropies reflect structures formed in the plasma epoch. b. From HGD the photon viscosity of the plasma epoch prevents turbulence until the viscous Schwarz scale L_{SV} becomes less than the Hubble scale (horizon scale, scale of causal connection) $L_H = ct$, where c is the speed of light and t is the time. The first plasma structures were proto-super-cluster voids and proto-super-clusters at 10^{12} seconds (30,000 years). c. Looking back in space is looking back in time. Protogalaxies were the last fragmentations of the plasma (orange circles with green halos) at 10^{13} seconds. d. The scale of the gravitational structure epoch is only $3x10^{21}$ m compared to present supercluster sizes of 10^{24} m and the largest observed supervoid scales of 10^{25} m. e. Turbulence in the plasma epoch is generated by baroclinic torques on the boundaries of the expanding super-voids.

During inflation, turbulent temperature fluctuations are fossilized by stretching beyond the scale of causal connection L_H. The mass-energy and entropy vastly increase, presumably by the rapid stretching of superstrings with Planck tension $c^4/G = 1.2x10^{44}$ kg m s^{-2} [21]. The black dot in the inflation epoch symbolizes fossilized turbulent temperature fluctuations that can guide nucleosynthesis and preserve big bang turbulence information for testing using CMB temperature anisotropy patterns. In a period of 10^{-27} seconds the size of the big bang universe increases from 10^{-27} m to a few meters with density exceeding 10^{80} kg m^{-3}. The mass-energy of the universe within our present horizon scale $L_H = 10^{26}$ m is about 10^{52} kg, about 10^{-40} of the mass-energy created by the big bang. The mass-energy of the universe before inflation was less than 10^{16} kg so the increase was by a factor of 10^{74}. Guth describes this as the ultimate free lunch [21].

The transition between energy domination and mass domination occurred at about 10^{11} s, the beginning of the plasma epoch shown in Figure 1b. The plasma is absolutely unstable and will fragment once the Schwarz viscous scale L_{SV} matches the horizon scale L_H. This occurs at time $t = 10^{12}$ s when the density of a flat universe is 10^{-15} kg m^{-3}. Most of this density, about 97%, is non-baryonic dark matter, NBDM, presumably some combination of neutrinos. The rest is plasma. None of it is dark energy. This gives a primordial density ρ_0 of 4×10^{-17} kg m^{-3}, which matches the density of old globular star clusters OGC. Because the universe continues to expand, fragmentation of the plasma at density minima is favored over condensation at density maxima. Voids appear in the plasma at 10^{12} seconds and proceed to expand as rarefaction waves at speeds limited by the speed of sound $c/3^{1/2}$ for 10^{13} seconds until decoupling. The material between the protosupercluster voids is protosuperclusters with mass 10^{46} kg, that of a thousand galaxies. The proto-super-clusters never collapse gravitationally but continue to expand with the universe. This is a key point; instead of a messy chaotic formation scenario of mergers the expansion was tranquil and preserved primordial alignments and densities. The present scale of superclusters is observed to be 10^{24} m.

Smaller scale fragmentations occur in the plasma epoch at density minima up to the time of plasma to gas transition (decoupling). The last fragmentation is to produce protogalaxies at 10^{13} seconds, as shown in Figure 1b. The morphology of the protogalaxies is determined by weak turbulence generated by baroclinic torques on the surfaces of the supervoids as they expand. The direction of the spins is determined by the fossil strong force freeze-out density gradient at length scale L_{SF}. This spin direction determines the small wavenumber spherical harmonic directions for the CMB, the spins of galaxies in the local group, and even the spin of the Milky Way. All are directed toward the "axis of evil" [17], which has length scales in quasar polarizations to a Gpc, or 3×10^{25} m, as shown in Figure 1c.

Figure 2 illustrates erroneous concepts of cold dark matter hierarchical clustering (CDMHC) theory, which is at present the standard cosmological model along with the equally erroneous concept of dark energy and a cosmological constant Λ (ΛCDMHC). The standard model has no turbulence in the big bang but predicts Gaussian scale independent fluctuations and white noise spectrum of CMB fluctuations from the big bang and inflation epochs. All structure formation in the plasma epoch is attributed to the non-baryonic dark matter NBDM.

According to CDMHC seeds or halos of CDM form because their Jeans scale L_J is less than L_H. This is physically impossible because the diffusivity of any form of NBDM will cause the diffusive Schwarz scale L_{SD} of the material to exceed L_H. Even if a seed could form it could not stick to another seed because sticking requires collisions of the CDM particles, no matter how cold they might be. In Figure 2 each erroneous CDM concept is indicated by a red X.

It is claimed by CDMHC that the baryonic matter plasma is inviscid and falls into the CDM halos where it oscillates without viscous damping to give the acoustic peaks observed in the CMB temperature anisotropy spectrum. However, the plasma is far from inviscid, with a photon viscosity of $\nu = 4\times10^{26}\ m^2\ s^{-1}$ [11]. Even if CDM halos existed, the plasma is far too viscous to enter the halo and oscillate acoustically.

Other erroneous concepts of CDMHC are listed in Figure 2. The CDM halos gradually cluster and collect gas for 300 My of dark ages (10^{16} s) in mini-galaxies which form the first superstars, one per mini-galaxy, in an enormous set of supernovae that re-ionizes all the gas formed at decoupling. The re-ionization concept is unnecessary to explain the lack of UV-absorbing gas as observed in quasar spectra because the gas is sequestered as dark matter planets in PGC clumps.

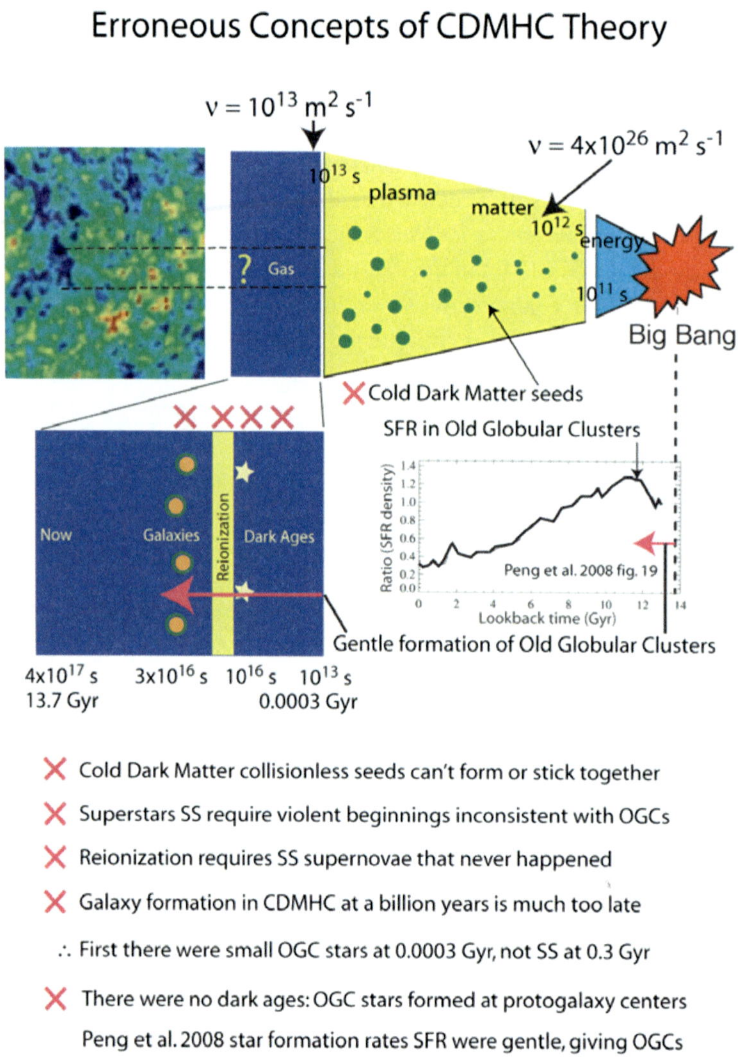

Figure 2. Galaxy formation by the standard ΛCDM model is impossible to reconcile with observations and fluid mechanical theory (see text). Failed aspects of the model are indicated by red Xs.

The first CDMHC galaxies to form occur much too late to be consistent with observations, at about a billion years after the big bang (3×10^{16} s). The star formation rate SFR in Peng et al. 2008 Figure 19 [23] cannot be superstars because superstars can only occur supported by violent turbulence stresses that would prevent formation of the very small stars observed to exist in old globular star clusters OGC.

From HGD and the mass of small stars it is easy to show from the Schwarz length scale criteria of Table 1 that the maximum viscous dissipation rate ε to permit the formation of a small OGC star is 10^{-12} m^2 s^{-3}. However, strong turbulence is required with ε values of $5x10^{-4}$ m^2 s^{-3} to permit formation of superstars and the re-ionization of the universe. If such large ε values existed before the first star formed, no old-globular-clusters OGCs could have formed because their small stars would be inhibited by turbulence produced as NBDM CDM halos filled with baryonic gas. Since OGCs are observed in all galaxies, the superstars and re-ionization concepts of CDMHC must be ruled out.

Observations

The fifth year WMAP observations of the CMB have been released, as shown in Figure 3. Figure 3a shows the Dunkley et al. 2008 5th year WMAP spherical harmonic temperature anisotropy data, with un-binned samples as light dots in the background [24].

The range of the un-binned samples in Figure 3a is a measure of the intermittency (non-Gaussianity) of the turbulence associated with the samples. The range indicated by circles and arrows is smaller at the sonic peak than at higher wavenumbers. This is attributed to the weak turbulence produced in the plasma at the boundaries of the expanding supervoids. The wider scatter at high wavenumbers k is attributed to the greater intermittency of big bang turbulence, which has a higher Reynolds number. An Obukhov-Corrsin turbulent mixing spectrum $\beta\chi\varepsilon^{-1/3}k^{-5/3}$ is fitted to the lowest wavenumber CMB spectrum, where β is a universal constant, χ is the dissipation rate of temperature variance and ε is the dissipation rate of velocity variance. It is assumed that the measured spherical harmonic CMB spectrum multiplied by $\ell(1+\ell)$ is not much different than a turbulent dissipation spectrum $k^2\phi$.

The red solid curve labeled CDM+ν=0 in Figure 3a is a CDM model without viscosity forced by ad hoc (GIGO) numerical assumptions to fit the data. Photon viscosity will not permit the baryonic plasma to enter the CDM potential wells even if they were physically possible, and will not permit any sonic oscillations because of viscous damping. Therefore the green dashed curve from big bang turbulence theory labeled CDM+ν should apply.

Evidence of turbulence from the expansion of super-voids producing baroclinic torques and vorticity at void boundaries is shown by a series of turbulence statistical parameters computed by Bershadskii and Sreenivasan from the low wavenumber CMB spectrum in Figure 3b, 3c, and 3d and compared to the same parameters computed for atmospheric, laboratory and numerical simulations of turbulence over a wide range of Reynolds numbers.

Figure 3b compares Weiner-filtered CMB data with extended self similarity turbulence data for various order structure functions. The agreement clearly demonstrates the CMB temperature anisotropies in the indicated angular size range 0.5 to 4 degrees studied were produced by turbulence. From HGD the turbulence reflects baroclinic torques at gravitationally driven expanding void boundaries.

Figure 3c compares the same CMB and turbulence data using a statistical parameter to test for Gaussianity. Again, the agreement with turbulence is remarkable. Both sets of CMB and turbulence data deviate from Gaussianity in the same way in the angular size range studied.

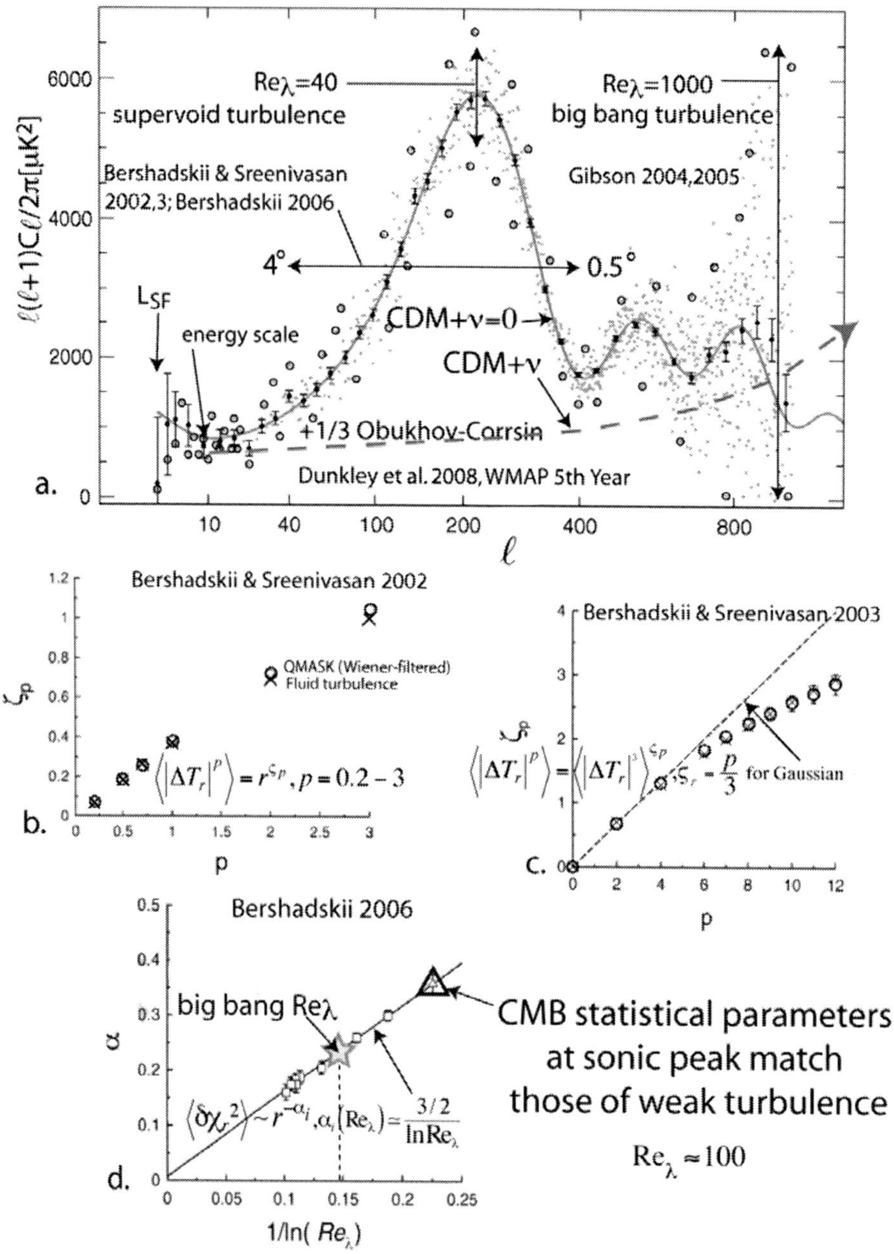

Figure 3. a. Spherical harmonic spectrum of CMB temperature anisotropies from the fifth year WMAP observations, Dunkley et al. 2008 [24]. The large sonic peak at wavenumbers near 200 reflects weak turbulence produced by gravitationally driven protosupervoid expansion according to Figure 1, not inviscid CDM (solid red). The dashed green line represents a viscous CDM model; that is, with no sonic signal and should match the +1/3 Obukhov-Corrsin turbulent mixing dissipation spectrum of big bang turbulence. Turbulence parameters are compared to CMB statistics in b. [25], c. [26] and d. [27] as described in the text.

Figure 3d shows an estimate of the sonic peak Reynolds number estimated by Bershadskii 2006 to be about $Re_\lambda \sim 100$. A more precise estimate is $Re_\lambda = 40$ as shown near the peak, using the expression derived for the temperature anisotropy variance as a function of separation distance r between sampling points. The power law exponent is a function of Re_λ. The value of Re_λ for the CMB is extracted using a series expansion of $\alpha(Re_\lambda)$ for high Reynolds number as a function of $1/\ln Re_\lambda$. It turns out for turbulence that only the first term is important, as shown, so that the CMB data near the sonic peak gives $Re_\lambda = 40$ as indicated by the bold triangle. For comparison, the big bang turbulence Re_λ value of 1000 is shown by a bold star. There can be little doubt from the combination of evidence shown in Figure 3 abcd that the plasma epoch was weakly turbulent in the range of scales including the sonic peak. As noted by Alexander Bershadskii (2001 personal communication to CHG) "the fingerprints of Kolmogorov are all over the sky".

As shown in Figure 1e, the last stage of plasma fragmentations guided by weak turbulence is the formation of proto-galaxy-mass objects with a linear morphology reflecting vortex tubes of the turbulence where the rates-of-strain that enhance fragmentation of the plasma to form voids is maximum.

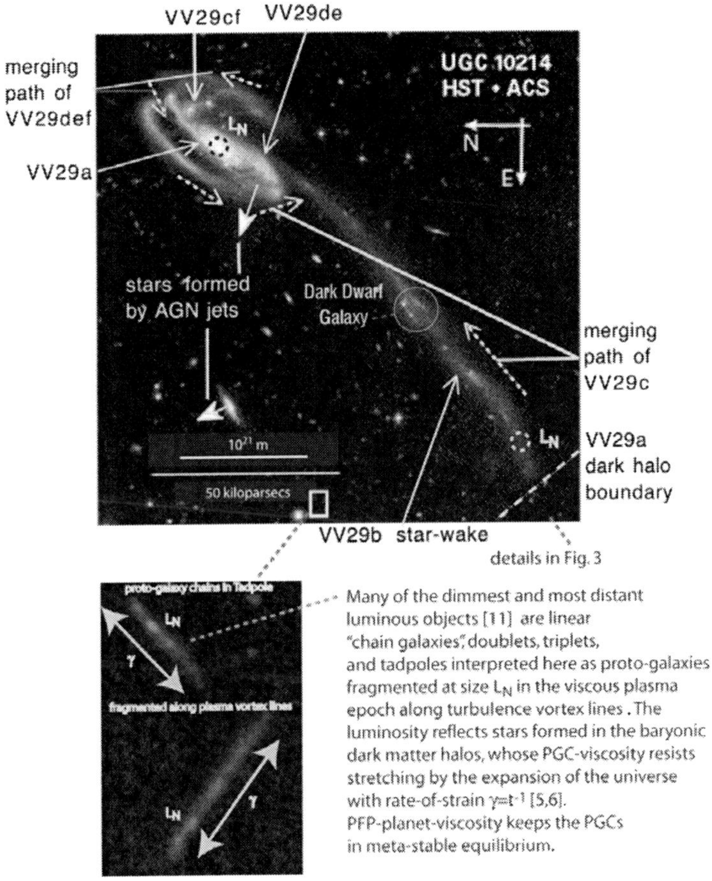

Figure 4. Tadpole galaxy merger (top), with background chains of protogalaxies shown in enlargement of small area at bottom center (bottom).

Figure 4 (bottom) is an image from the Hubble Space Telescope Advanced Camera for Surveys (HST-ACS) showing two proto-galaxy-chain clusters in the dim background of the Tadpole galaxy merger. Virtually all (94%) very dim galaxies from HST-ACS; that is, with magnitudes greater than 24, have such a linear morphology as expected from HGD [15]. The objects consist of bright nuclei separated by a dimmer glow that we interpret as proto-galaxies with a concentration of first-stars near the cores, with star-forming galaxy dark matter (planets in clumps) in between.

The star-clumps have a uniform size set by the turbulence and fluid mechanics to be the Nomura scale 10^{20} m (Table 1) with the Nomura morphology of weak turbulence [15].

Figure 4 (top) shows the formation of star trails and dust trails in the baryonic dark matter halo of the Tadpole central galaxy VV29a as galactic objects VV29cdef merge in frictional spirals on the VV29a disk plane [19]. A sharp VV29a dark halo boundary is clear in the high-resolution HST-ACS images. The eponymous VV29b spiral filament shows numerous young globular star clusters [28] aligned in a direction pointing precisely at the point of frictional merger as the objects move through the dark matter halo triggering star formation by tidal forces in a L_N scale diameter star wake. The size of VV29a dark matter halo diffused from the L_N scale central core of old small stars is 8×10^{21} m, or 0.9 Mpc. From this evidence, the concept of frictionless tidal tails in galaxy mergers appears to be obsolete [29].

Conclusion

A fluid mechanical analysis of gravitational structure formation termed hydro-gravitational-dynamics HGD shows that the CDMHC model is inconsistent with theory and observations and must be abandoned. From HGD, gravitational structure formation is driven entirely by the baryonic matter of the plasma epoch. The first structures are proto-super-cluster-voids and proto-super-clusters, followed by proto-galaxies at decoupling. The proto-galaxies are formed in very weakly turbulent plasma, but possess the morphology of weak turbulence in that the protogalaxies are stretched along the vortex lines of the plasma turbulence with a diameter reflecting their viscous-inertial-vortex-gravitational origin at the Nomura scale 10^{20} m. Although our description of HGD seems to emphasize turbulence processes, the universe described is in fact more tranquil than previously envisioned, and can preserve primordial alignments seen in many cosmological structures, in particular proto-galaxies.

References

[1] Darwin, G. H. 1889. On the mechanical conditions of a swarm of meteorites, and on theories of cosmogony, *Phil. Trans.*, **180**, 1-69.

[2] Jeans, J. H. 1902. The stability of spherical nebula, *Phil. Trans.*, **199A**, 0-49.

[3] Gibson, C.H. (1991). Kolmogorov similarity hypotheses for scalar fields: sampling intermittent turbulent mixing in the ocean and galaxy, *Proc. Roy. Soc. Lond. A*, **434**, 149-164.

[4] Gibson, C. H. (2006). *Turbulence, update of article in Encyclopedia of Physics*, R. G. Lerner and G. L. Trigg, Eds., Addison-Wesley Publishing Co., Inc., pp.1310-1314.

[5] Gibson, C. H. (1981). Buoyancy effects in turbulent mixing: Sampling turbulence in the stratified ocean, *AIAA J.*, **19**, 1394.

[6] Gibson, C. H. (1968a). Fine structure of scalar fields mixed by turbulence: I. Zero-gradient points and minimal gradient surfaces, *Phys. Fluids*, **11**: 11, 2305-2315.

[7] Gibson, C. H. (1968b). Fine structure of scalar fields mixed by turbulence: II. Spectral theory, *Phys. Fluids*, **11**: 11, 2316-2327.

[8] Gibson, C. H. (1986). Internal waves, fossil turbulence, and composite ocean microstructure spectra," *J. Fluid Mech.* **168**, 89-117.

[9] Gibson, C. H. (1999). Fossil turbulence revisited, *J. of Mar. Syst.*, **21**(1-4), 147-167, astro-ph/9904237.

[10] Gibson, C.H. (1996). Turbulence in the ocean, atmosphere, galaxy and universe, *Appl. Mech. Rev.*, **49**, no. 5, 299-315.

[11] Gibson, C.H. (2000). Turbulent mixing, diffusion and gravity in the formation of cosmological structures: The fluid mechanics of dark matter, *J. Fluids Eng.*, **122**, 830–835.

[12] Gibson, C.H. (2004). The first turbulence and the first fossil turbulence, *Flow, Turbulence and Combustion*, **72**, 161–179.

[13] Gibson, C.H. (2005). The first turbulent combustion, *Combust. Sci. and Tech.*, **177**: 1049–1071, arXiv:astro-ph/0501416.

[14] Gibson, C.H. (2006). *The fluid mechanics of gravitational structure formation*, astro-ph/0610628.

[15] Gibson, C.H. (2008). Cold dark matter cosmology conflicts with fluid mechanics and observations, *J. Applied Fluid Mech.*, Vol. 1, No. 2, pp 1-8, 2008, arXiv:astro-ph/0606073.

[16] Gibson, C.H. & Schild, R.E. (2007). *Interpretation of the Helix Planetary Nebula using Hydro-Gravitational-Dynamics: Planets and Dark Energy*, arXiv:astro-ph/0701474.

[17] Schild, R.E & Gibson, C.H. (2008). Lessons from the Axis of Evil, axXiv[astro-ph]:0802.3229v2.

[18] Gibson, C.H. & Schild, R.E. (2007). *Interpretation of the Stephan Quintet Galaxy Cluster using Hydro-Gravitational-Dynamics: Viscosity and Fragmentation*, arXiv[astro-ph]:0710.5449.

[19] Gibson, C.H. & Schild, R.E. (2002). *Interpretation of the Tadpole VV29 Merging Galaxy System using Hydro-Gravitational Theory*, arXiv:astro-ph/0210583.

[20] Schild, R. 1996. Microlensing variability of the gravitationally lensed quasar Q0957+561 A,B, *Ap J*, **464**, 125.

[21] Greene, B. 1999. *The Elegant Universe, Superstrings*, Hidden Dimensions, and the Quest for the Ultimate Theory, W. W. Norton & Compant, New York.

[22] Guth, A. H. 1997. The Inflationary Universe, *The Quest for a New Theory of Cosmic Origins,* Helix Books, Addison-Wesley Pub. Co., Inc., New York.

[23] Peng, E. W. et al. 2008, The ACS Virgo Cluster Survey. XV. The Formation Efficiencies of Globular Clusters in Early-Type Galaxies: The Effects of Mass and Environment, *Ap J*, **681**, 197-224.

[24] Dunkley, J. et al. (2008). Five-year Wilkinson Microwave Anisotropy Probe (WMAP) Observations: Likelihoods and Parameters from the WMAP data, subm. *Ap JS*; arXiv:0803.0586v1.

[25] Bershadskii, A., and Sreenivasan, K.R. 2002. Multiscaling of cosmic microwave background radiation, *Phys. Lett. A*, **299**, 149-152.

[26] Bershadskii, A., and Sreenivasan, K.R. 2003. Extended self-similarity of the small-scale cosmic microwave background anisotropy. *Phys. Lett. A*, **319**, 21-23.

[27] Bershadskii, A. 2006. Isotherms clustering in cosmic microwave background, *Physics Letters A*, **360**, 210-216.

[28] Tran, H. D., Sirianni, M., & 32 others 2003. Advanced Camera for Surveys Observations of Young Star Clusters in the Interacting Galaxy UGC 10214, *Ap J*, **585**, 750.

[29] Toomre, A., & Toomre, J. 1972. Galactic Bridges and Tails, *Ap J*, **178**, 623.

In: Black Holes and Galaxy Formation
Editors: A.D. Wachter and R.J. Propst, pp. 101-120
ISBN: 978-1-60741-703-3

Chapter 3

TRACKING ORDER AND CHAOS IN A BINARY QUASAR DYNAMICAL MODEL

N.D. Caranicolas
Department of Physics, Section of Astrophysics,
Astronomy and Mechanics, University of Thessaloniki
54124 Thessaloniki Greece

Abstract

A dynamical model is presented for the study of the behavior of orbits in a binary quasar. The two quasars are hosted in a pair of interacting disk galaxies. Two cases are investigated: the time independent model and the evolving model. Our numerical experiments show that a large fraction of orbits are chaotic in the time independent case. A careful analysis suggests that several Lindblad resonances are responsible for the chaotic motion of stars in both host galaxies. In the time dependent case, we follow the evolution of the model as one of the two interacting galaxies develops a quasar, by mass transport from the disk to the nucleus. In this interesting case there are orbits that change character from regular to chaotic. This result, strongly indicates, that the regular or chaotic nature of orbits depends, not only in the galactic interaction but also in the presence of the quasar. The outcomes derived from our dynamical model are compared with data available from observations.

1. Introduction

Quasars were first discovered by Maarten Schmidt in 1963 as distant, energetic objects with a star like appearance. Since the discovery of the first quasar, astronomers have been trying to understand the conditions of birth and existence of these immensely powerful objects. Today most astronomers believe that quasars are the luminous supermassive cores at the center of distant galaxies. Recent observations indicate that a large number of bright quasars are hosted in massive spiral galaxies with really prominent disks(Letawe et al. 2006).

In an earlier paper (Papadopoulos and Caranicolas 2005) we have presented a simple dynamical model with a rotating disk and a massive nucleus in order to study the properties of motion in a quasar. On the other hand, there is evidence (Letawe et al. 2007) that interacting galaxies host more powerful quasars with strong emission lines. It was also found that about 50% of the observed host galaxies display signs of interaction, which are compatible with several previous studies(see Hutching & Neff 1992, Bahcall et al. 1997, Boyce et al. 1999, Kaspi et al. 2000, Floyd et al. 2004, Jahnke et al. 2004).Therefore, it seems challenging to built a model for a pair of interacting quasars and study its dynamical properties. The article is organized as follows: In Section 2 we present the dynamical model. In Section 3 we study the nature of orbits in the time independent potential, as well as in the evolving model. In the case of the evolving model we follow the evolution of orbits as the host galaxy develops a quasar in its centre. Some interesting semi-numerical results are also presented in the same Section. Finally, in Section 4, a discussion and the conclusions of this research are given and a comparison of the results with observational data is made.

2. The Dynamical Model

Our model consists of a pair of disk galaxies, each having a dense massive nucleus. The potential describing the motion in each galaxy is given by the equations

$$V_1(r) = -\frac{M_{1d}}{\left(r^2+a^2\right)^{1/2}} - \frac{M_{1n}}{\left(r^2+c_1^2\right)^{1/2}},$$

$$V_2(r) = -\frac{M_{2d}}{\left(r^2+b^2\right)^{1/2}} - \frac{M_{2n}}{\left(r^2+c_2^2\right)^{1/2}}, \qquad (1)$$

where $r^2 = x^2 + y^2$, while M_{1d}, M_{1n}, a, c_1 and M_{2d}, M_{2n}, b, c_2 is the mass of disk, the mass of nucleus, the scale length of disk and the scale length of nucleus for the galaxies 1 and 2 respectively. In our study we shall use the theory of the circular restricted three body problem (see Caranicolas and Innanen 2009, Caranicolas and Papadopoulos 2009). We study the motion in a clockwise rotating frame with the centre at the centre of mass of the system and constant angular velocity $\Omega_p > 0$ given by Kepler's third law

$$\Omega_p = \left[\frac{GM_t}{R^3}\right]^{1/2} \qquad (2)$$

where $M_t = M_{1d} + M_{1n} + M_{2d} + M_{2n}$ is the total mass of the system, while R is the distance between the two galaxies.

The total potential describing the motion of a star of a unit mass is

$$U(x,y) = U_A(x,y) + U_B(x,y) \qquad (3)$$

where

$$U_A(x,y) = V_1(r_1) + V_2(r_2) \tag{4}$$

$$U_B(x,y) = -\frac{\Omega_p^2}{2}\left[\frac{M_2}{M_t}r_2^2 + \left(1-\frac{M_2}{M_t}\right)r_1^2 - R^2\frac{M_2}{M_t}\left(1-\frac{M_2}{M_t}\right)\right] \tag{5}$$

and

$$r_1^2 = (x-x_1)^2 + y^2, \quad r_2^2 = (x-x_2)^2 + y^2 \tag{6}$$

with

$$M_2 = M_{2d} + M_{2n}, x_1 = -\frac{M_2}{M_t}R, \quad x_2 = R - \frac{M_2}{M_t}R \tag{7}$$

where $x_1, y_1 = 0$ and $x_2, y_2 = 0$ are the fixed positions of the centres of the two galaxies.

In this article we shall use a system of galactic units, where the unit of length is 20 kpc, the unit of mass is $1.8\times10^{11} M_{\odot}$ and the unit of time is 0.99×10^{8} yr. The velocity unit is equal to 197km/s, while G is equal to unity. Figure 1 shows the contours of constant potential $U(x,y) = const$. The values of the parameters are M_{1d}=2.0, M_{2d}=0.6, M_{1n}=0.08, M_{2n}=0.04, α=0.3,b=0.2, c_1=0.0125, c_2=0.01,R=2.5. L_1, L_2, L_3, L_4, L_5 are the five Lagrange equilibrium points, where we have

$$\frac{\partial U}{\partial x} = \frac{\partial U}{\partial y} = 0. \tag{8}$$

L_1, L_2, L_3 are saddle points while L_4, L_5 are maxima of $U(x,y)$.

The outcomes of this research are mainly based on the numerical integration of the equations of motion

$$\begin{aligned} \ddot{x} &= -2\Omega_p\dot{y} - \frac{\partial U}{\partial x}, \\ \ddot{y} &= 2\Omega_p\dot{x} - \frac{\partial U}{\partial y}, \end{aligned} \tag{9}$$

where the dot indicates derivative with respect to the time. The system of differential equations (9) admits the well known Jacobi integral

$$J = \frac{1}{2}\left(p_x^2 + p_y^2\right) + U(x,y) = E_J \tag{10}$$

where p_x, p_y are the momenta per unit mass conjugate to *x,y*, while E_J is the numerical value of the Jacobi integral.

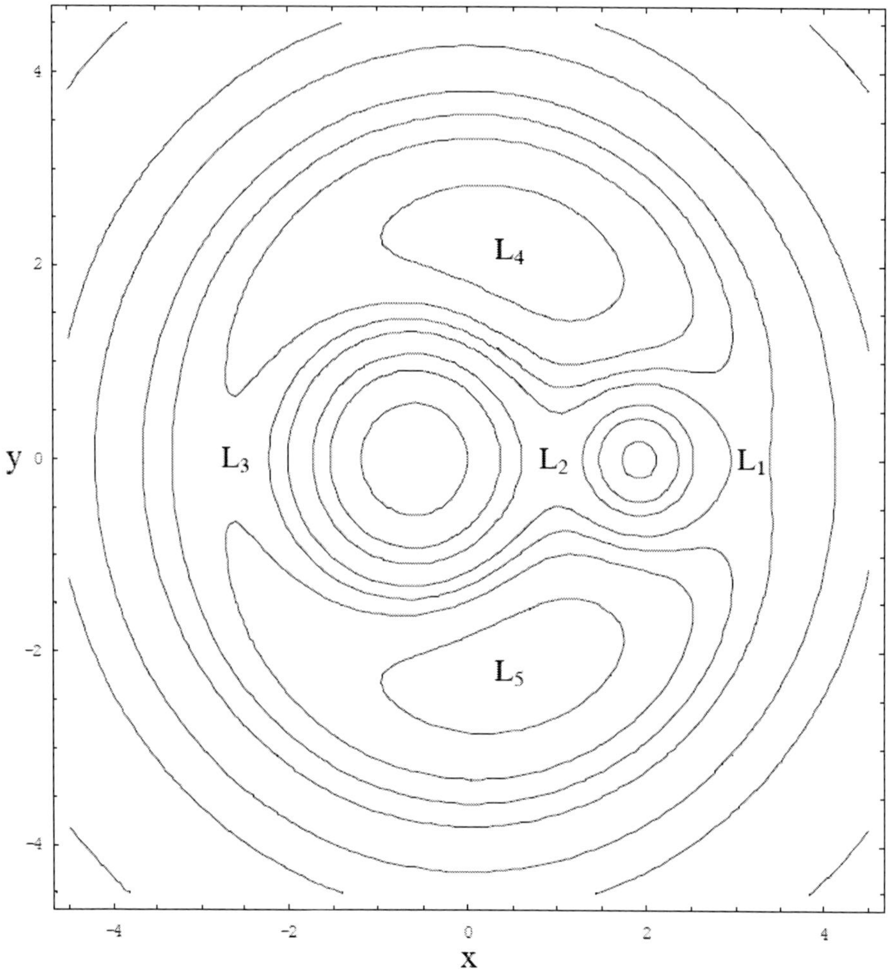

Figure 1. Contours of constant potential $U(x,y) = const.$ The values of the parameters are M_{1d}=2.0, M_{2d}=0.6, M_{1n}=0.08, M_{2n}=0.04, α=0.3, b=0.2, c_1=0.0125, c_2=0.01, R=2.5. The five Lagrange equilibrium points are indicated as L_1, L_2, L_3, L_4, L_5.

3. The Library of Orbits

A. Time Independent Model

In the first part of this Section we shall deal with the time independent model. We shall use the x-p_x y=0, p_y>0 Poincare phase plane in order to find the main families of orbits and their regular or chaotic character. Figure 2a shows x-p_x phase plane, when R=2.5, E_J=-2.0.The values of the other parameters are M_{1d}=2.0, M_{2d}=0.6, M_{1n}=0.08, M_{2n}=0.04, α=0.3,b=0.2, c_1=0.0125,

c_2=0.01, Ω_p=0.417229. We see that the regular motion is confined around the retrograde periodic point, in the larger galaxy, while in the smaller galaxy there are regular regions around both, the direct and the retrograde periodic points. The rest of the phase plane is covered by chaotic orbits. Some small islands produce by secondary resonances are also present.

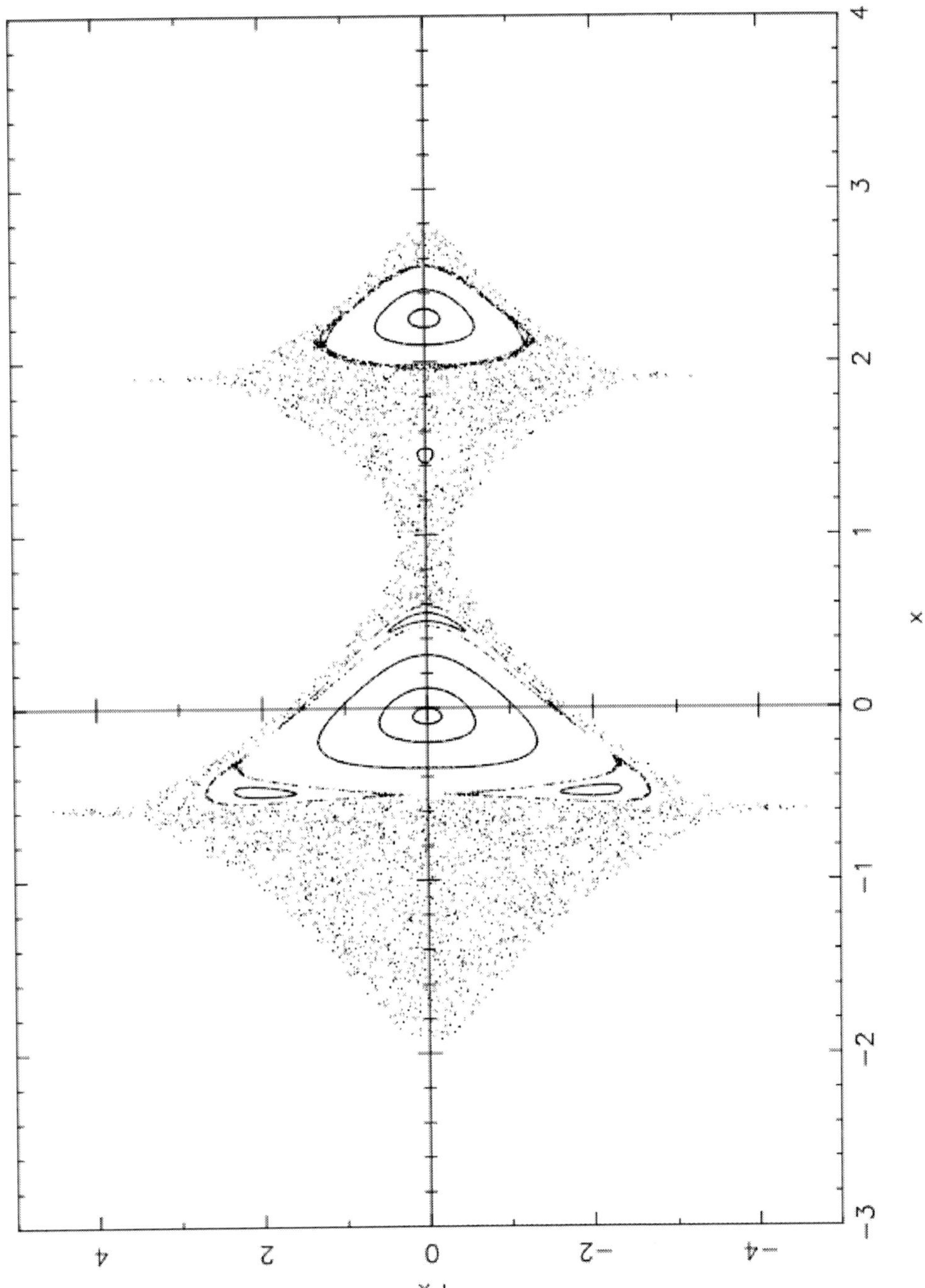

(A)

Figure 2. Continued on next page.

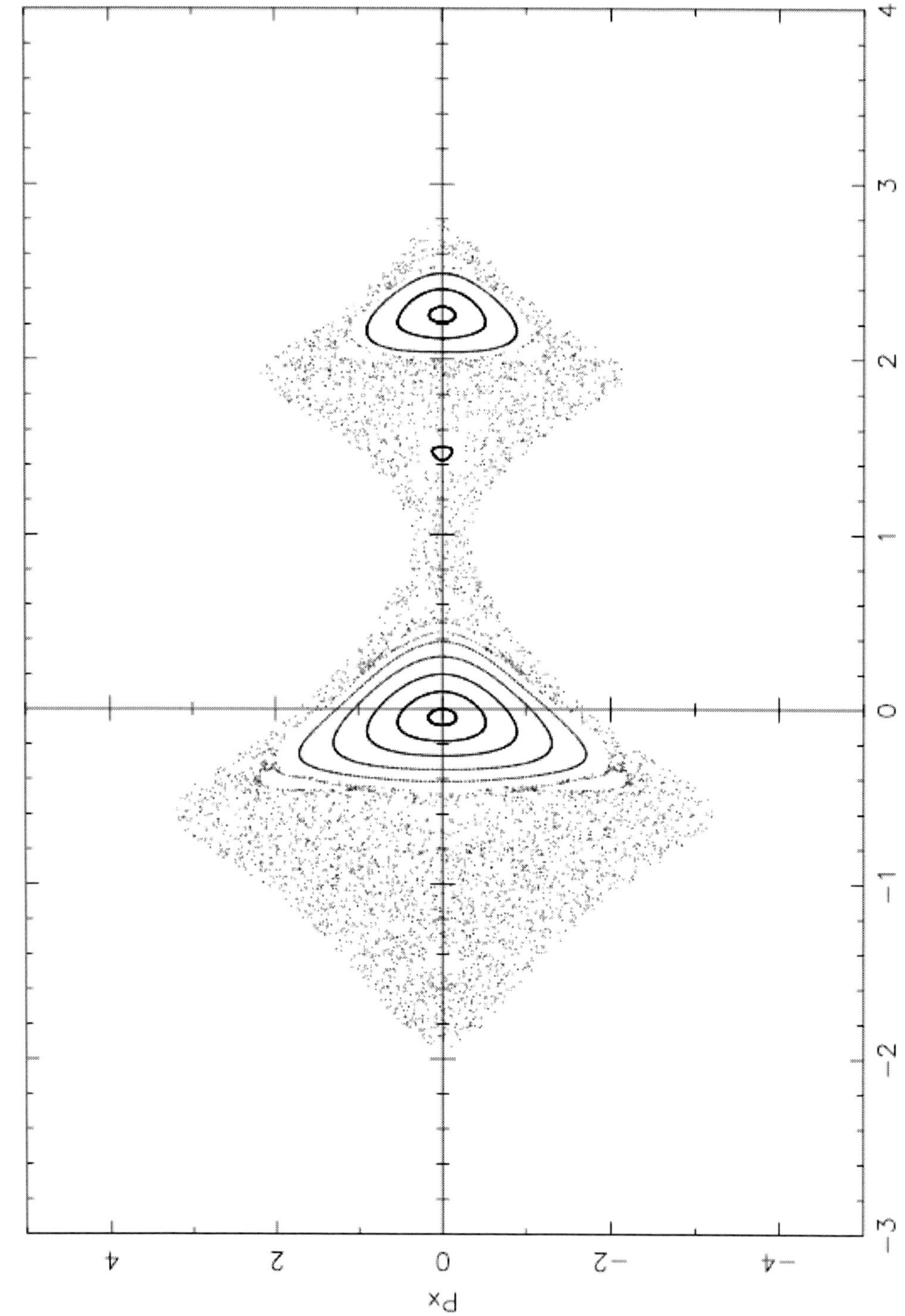

(B)

Figure 2a-b. The x-p_x Poincare phase plane corresponding to the model (3) when R=2.5, E_J= -2.0. The values of the other parameters are (a) M_{1d}=2.0, M_{2d}=0.6, M_{1n}=0.08, M_{2n}=0.04, α=0.3, b=0.2, c_1=0.0125, c_2=0.01, Ω_p=0.417229. (b) M_{1d}=2.08, M_{2d}=0.64, M_{1n}= M_{2n}=0.

Another interesting point in Figure 2a, is the high velocities, observed near the centres of the two galaxies. Figure 2b is similar to Figure 2a but when M_{1d}=2.08, M_{2d}=0.64.This means that all the mass is on the disks of the two galaxies. The pattern has one main difference from the pattern of Fig 2a. Here the chaotic motion near the centre of both galaxies is made at much lower velocities. Therefore, one can say, that the structure of chaos in galaxies is not only a result of galactic interaction but it is also a result of galactic activity. In other words the presence of the quasars affects drastically the character of motion in the host galaxies. Moreover the absence of the quasars reduces the velocities near the centres of the two galaxies.

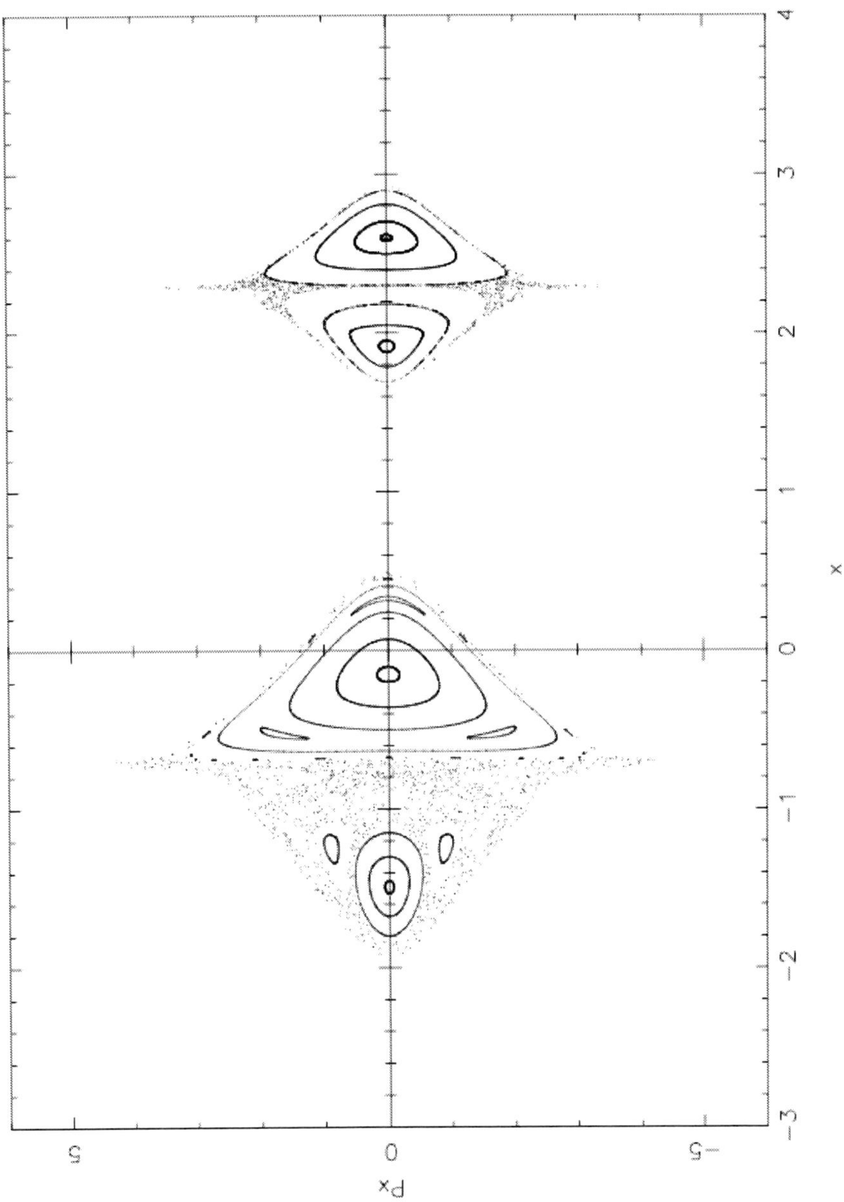

(A)

Figure 3. Continued on next page.

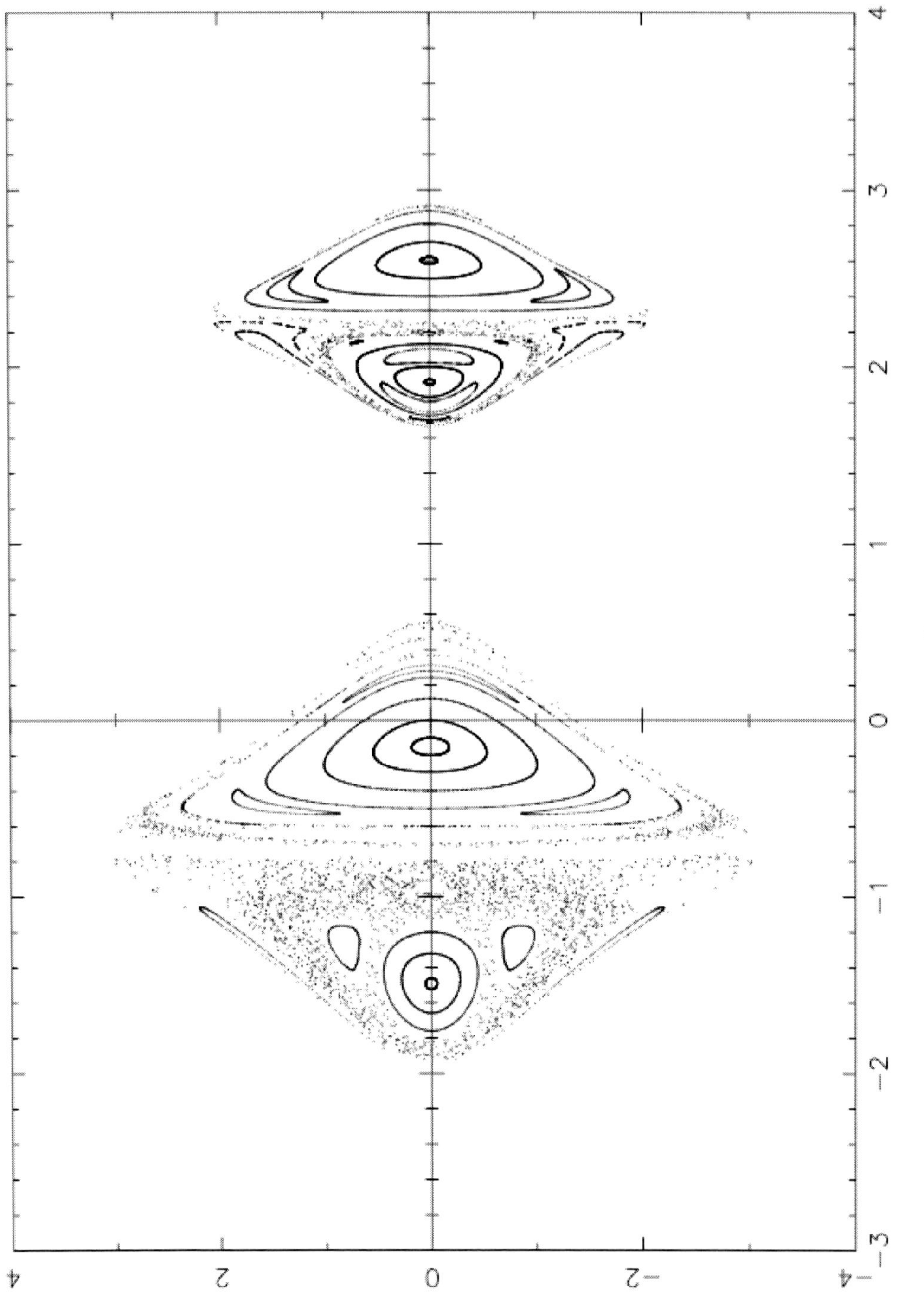

(B)

Figure 3a-b.Similar as Figure 2a-b when R=3.0, Ω_p=0.317397.

Figure 3a is similar to Figure 2a but when R=3, Ω_p=0.317397. In this case we can see two separate regions of motion around each galaxy while the majority of orbits are regular.

There is a considerable chaotic region, mainly near the central region of each host galaxy. High velocities are also observed in the central regions of the two host galaxies. Figure 3b is similar to Figure 2b but when R=3, Ω_p=0.317397. Here the picture is the same as that shown in Figure 2a. As the quasars are absent the velocities near the centre of the two galaxies are smaller than those observed in Figure 3a.

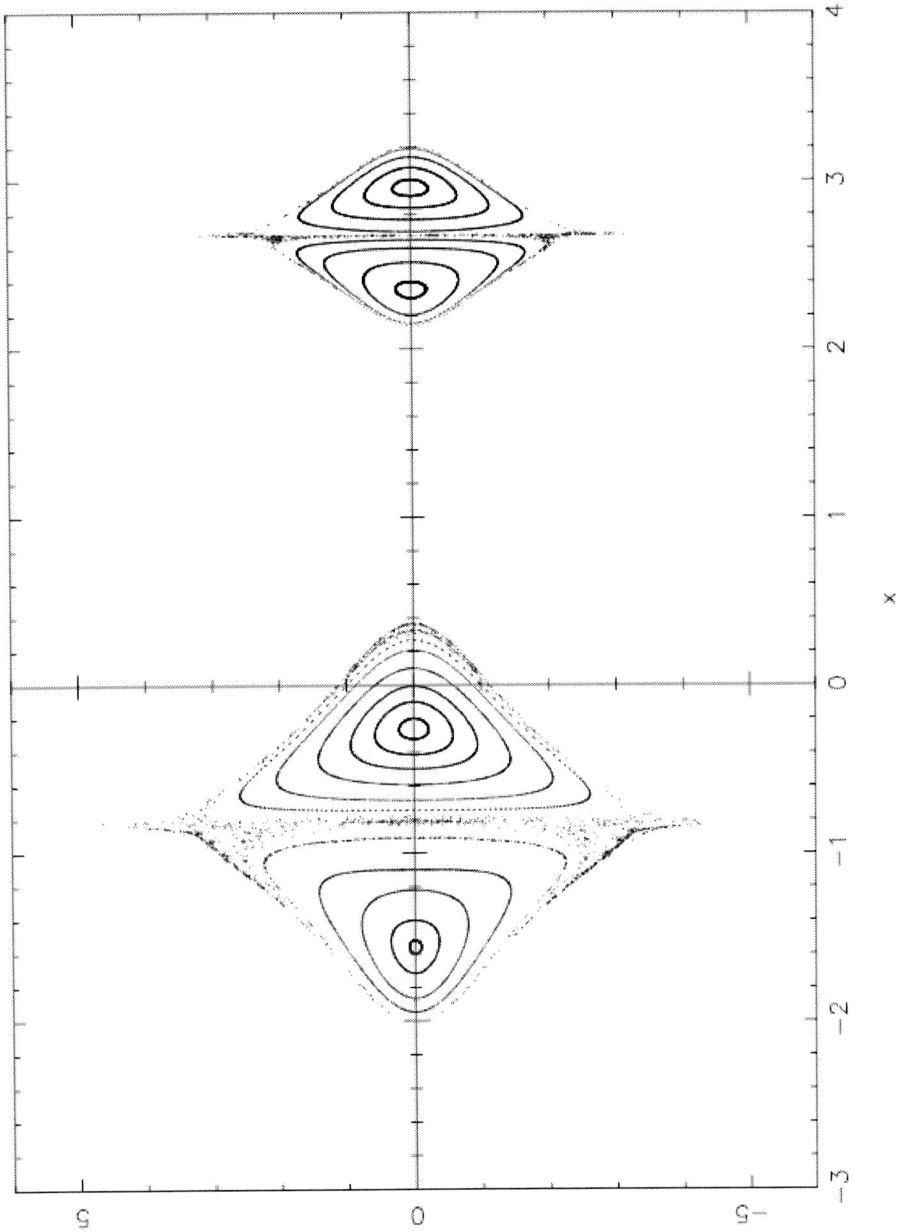

(A)

Figure 4. Continued on next page.

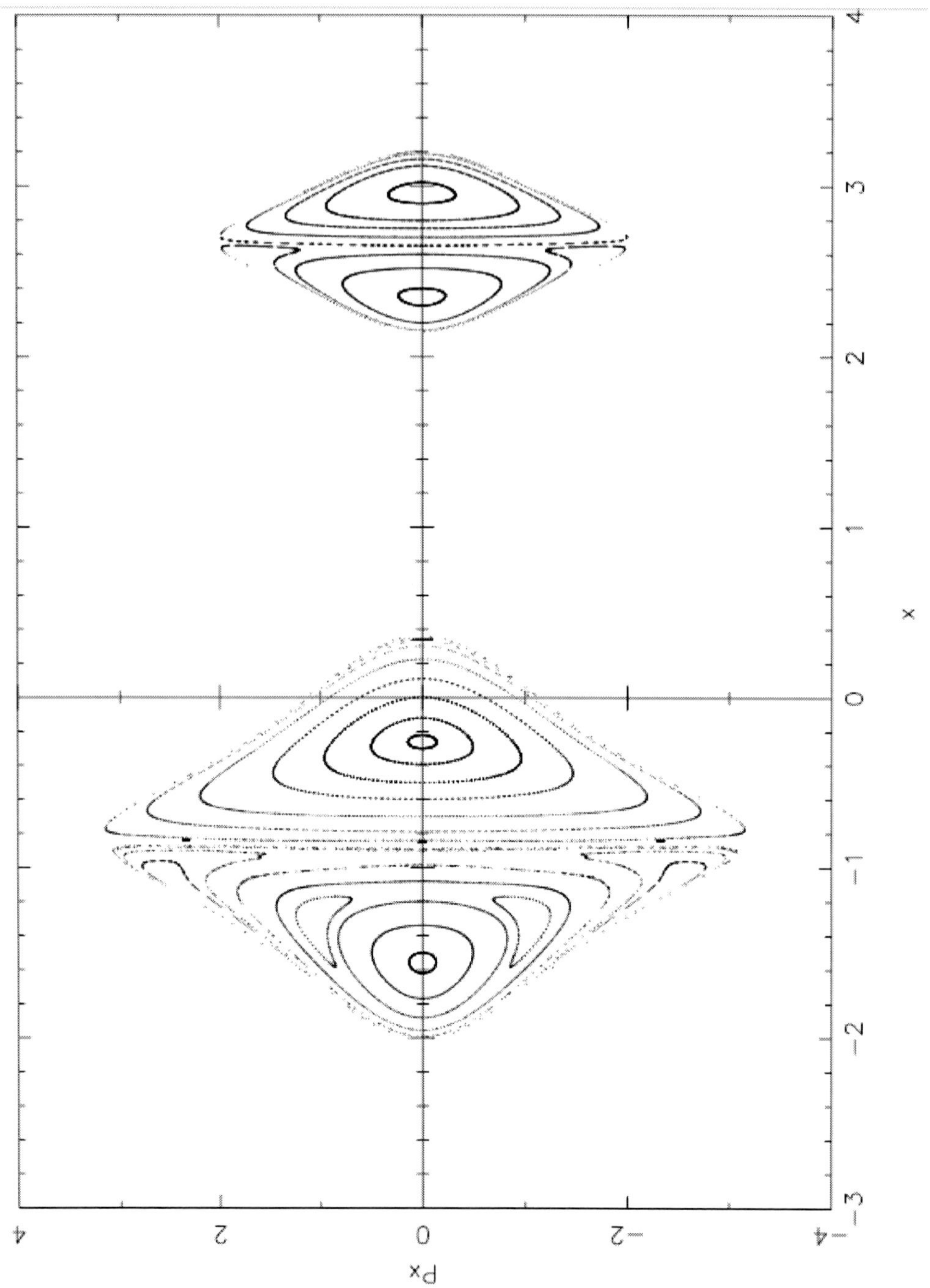

(B)

Figure 4a-b.Similar as Figure 2a-b when R=3.5, Ω_p=0.251830.

Figure 4a is similar to Figure 2a but when R=3.5, Ω_p=0.251830.Here, again, there are two separate regions of motion around each galaxy.There is a relatively small chaotic layer, mainly near the centre of each galaxy, while the rest of the phase plane is covered by regular orbits. Some small sticky regions are also present in both galaxies.Figure 4b is similar to Figure 2b but when R=3.5, Ω_p=0.251830. All orbits in both galaxies seem to be regular.

Chaotic motion was not observed and, if present, it is negligible. Therefore, we can say that, the two interacting galaxies, for large separations, do not show chaotic motion when the quasars are not present.

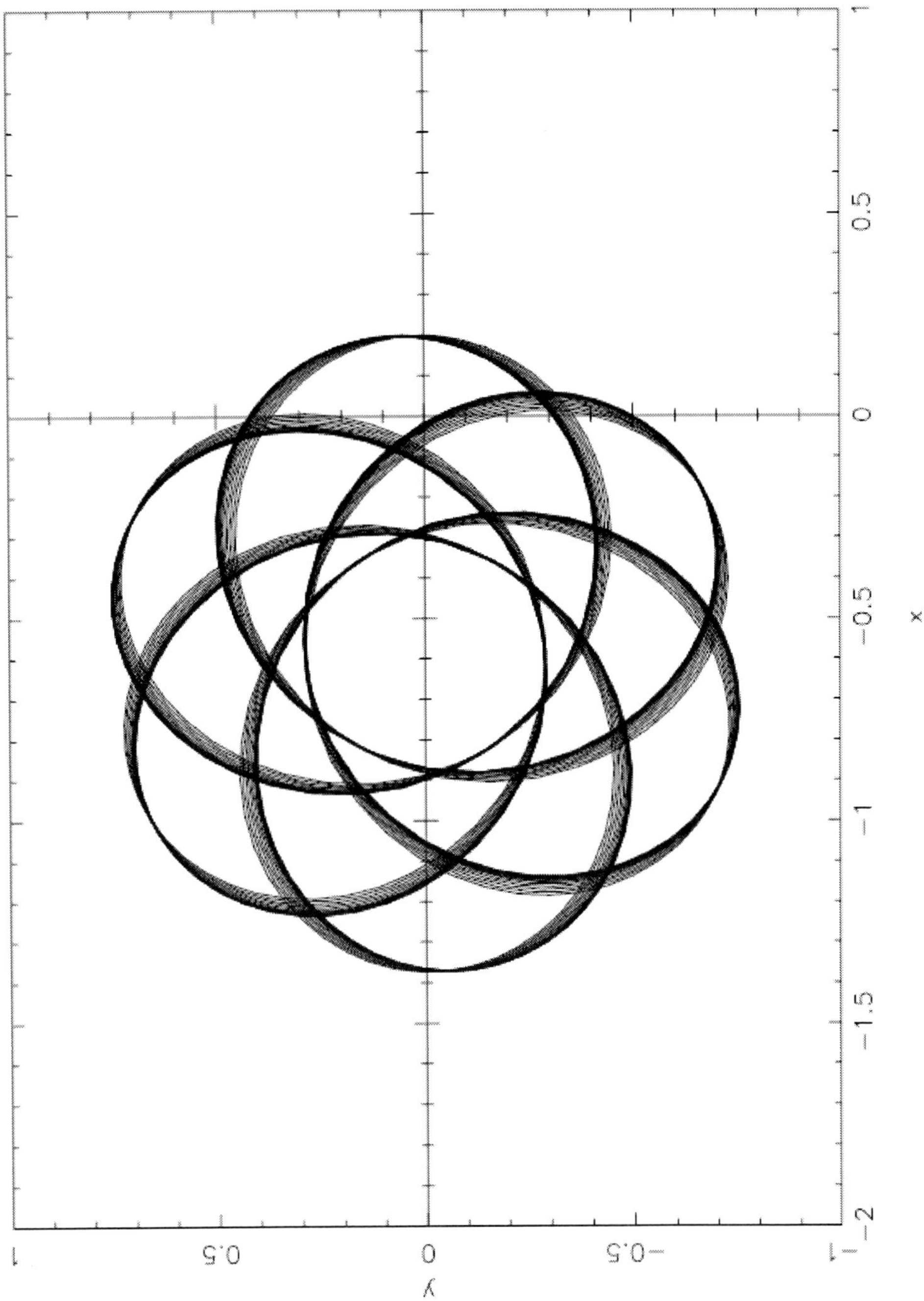

(A)

Figure 5. Continued on next page.

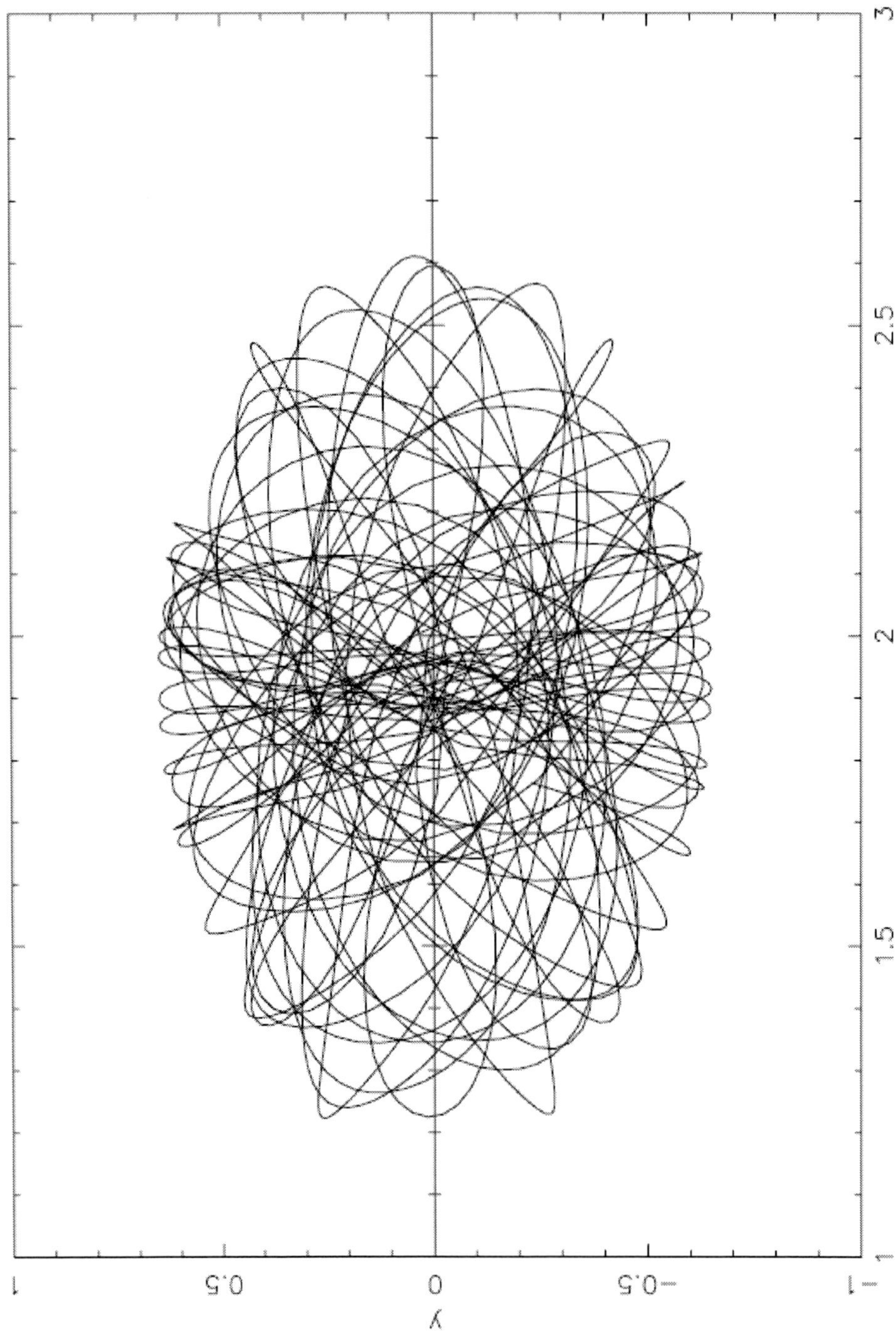

(B)

Figure 5a-b (a) A quasi periodic orbit in the galaxy 1. Initial conditions x=0.2,y= p_x =0.The value of p_y is found from the Jacobi integral.(b) A chaotic orbit in the galaxy 2. Initial conditions x=1.75,y= p_x =0. The value of the Jacobi integral is E_J=-2.0.

Figure 5a shows a quasi periodic orbit in the galaxy 1. Initial conditions x=0.2,y= p_x =0.The value of p_y is found from the Jacobi integral in all cases. A chaotic orbit in the galaxy

2 is shown in Figure 5b. Initial conditions x=1.75,y= p_x =0.The value of the Jacobi integral is E_J=-2.0 for both orbits, while the values of all other parameters are as in Figure 2a.In Figure 6a-b we can see two regular orbits in the galaxies 1 and 2 respectively. The initial conditions are x = -0.8235, y= p_x =0 for Figure 6a and x=3.0,y= p_x =0 for Figure 6b. The value of the Jacobi integral is E_J=-2.0 for both orbits, while the values of all other parameters are as in Figure 4b. All orbits were calculated for a time period of 150 time units.

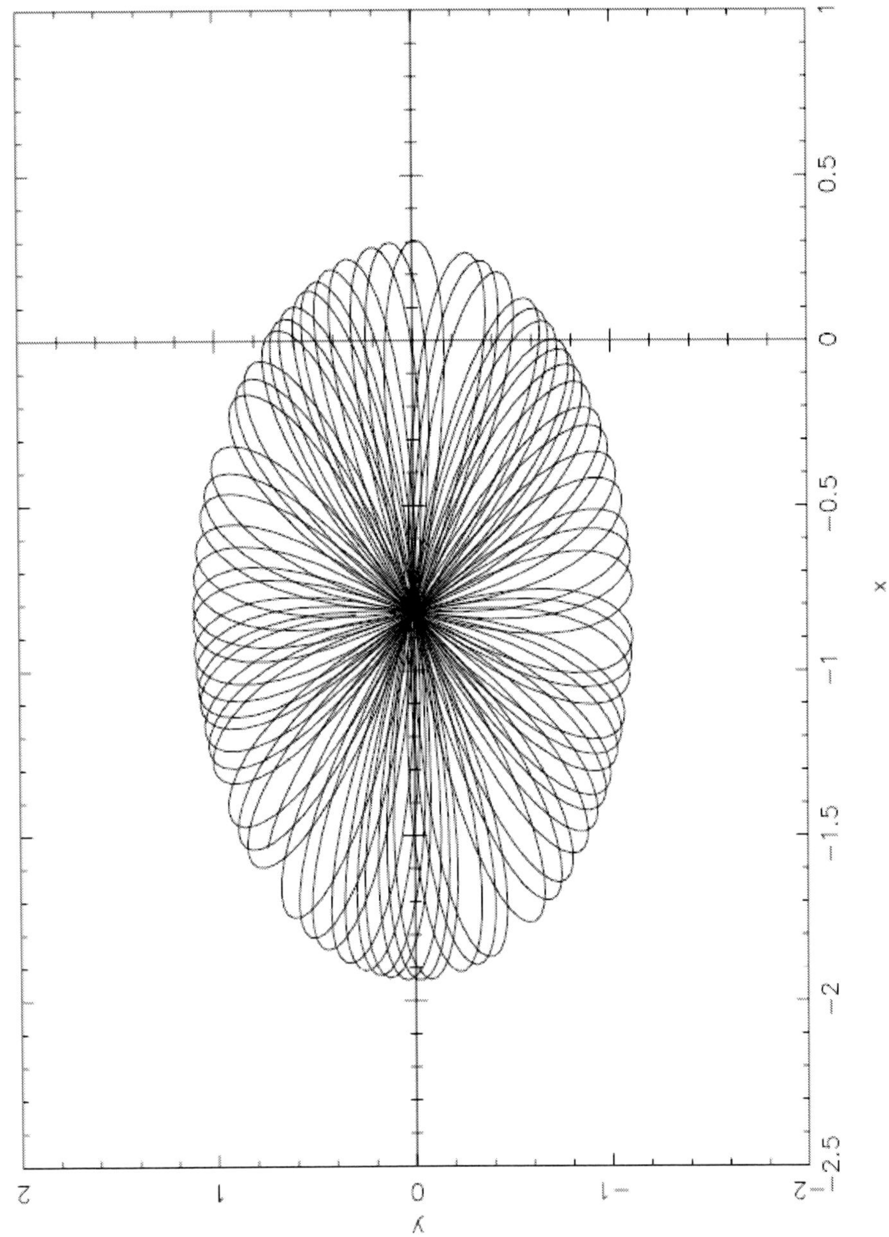

(A)

Figure 6. Continued on next page.

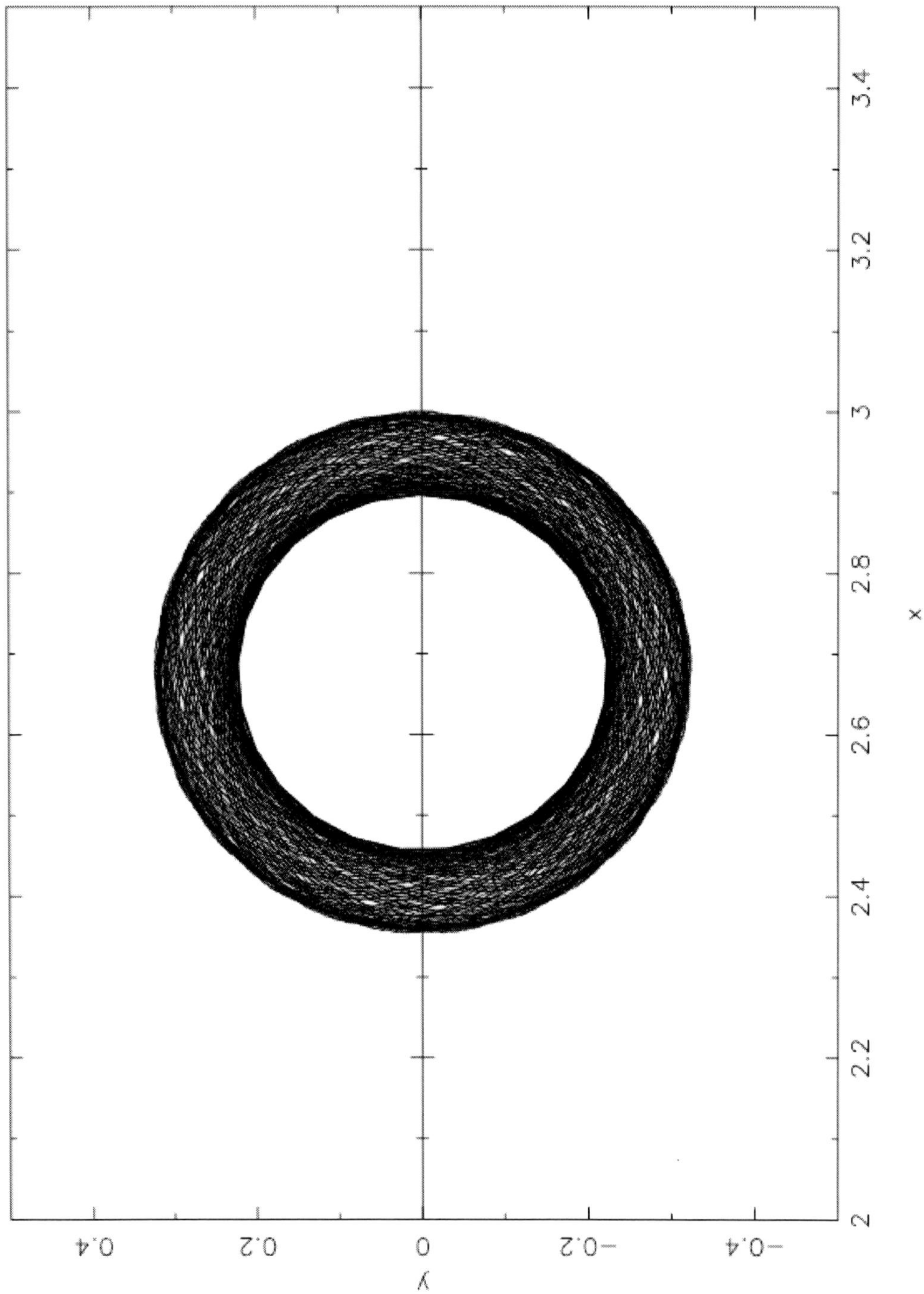

(B)

Figure 6a-b. Two regular orbits in the galaxies 1 and 2 respectively. The initial conditions are (a) x = -0.8235, y= p_x =0 and (b) x=3.0,y= p_x =0. The value of the Jacobi integral is E_J=-2.0.

In Figure 7 we can see the curves (1) $\Omega - 2\kappa/3$, (2) $\Omega - 5\kappa/8$, (3) $\Omega - 3\kappa/4$, (4) $\Omega - 7\kappa/9$, vs r for the large galaxy. Here κ is the epicyclic frequency and the two straight lines are the curves Ω_p=±0.417229. All the parameters are as in Figure 1. Figure 8 is similar to Figure 7 but for the small galaxy. As we can see, there is a considerable number of resonance radii for both the direct and the retrograde orbits in both galaxies. In other words, the above four Lindblad resonances are responsible for the chaotic motion in the two galaxies. It is interesting to note that the above resonances produce large chaotic regions for small values of R, while for larger R, as in Figure 4a, the chaotic regions are small although the resonance radii are still present.

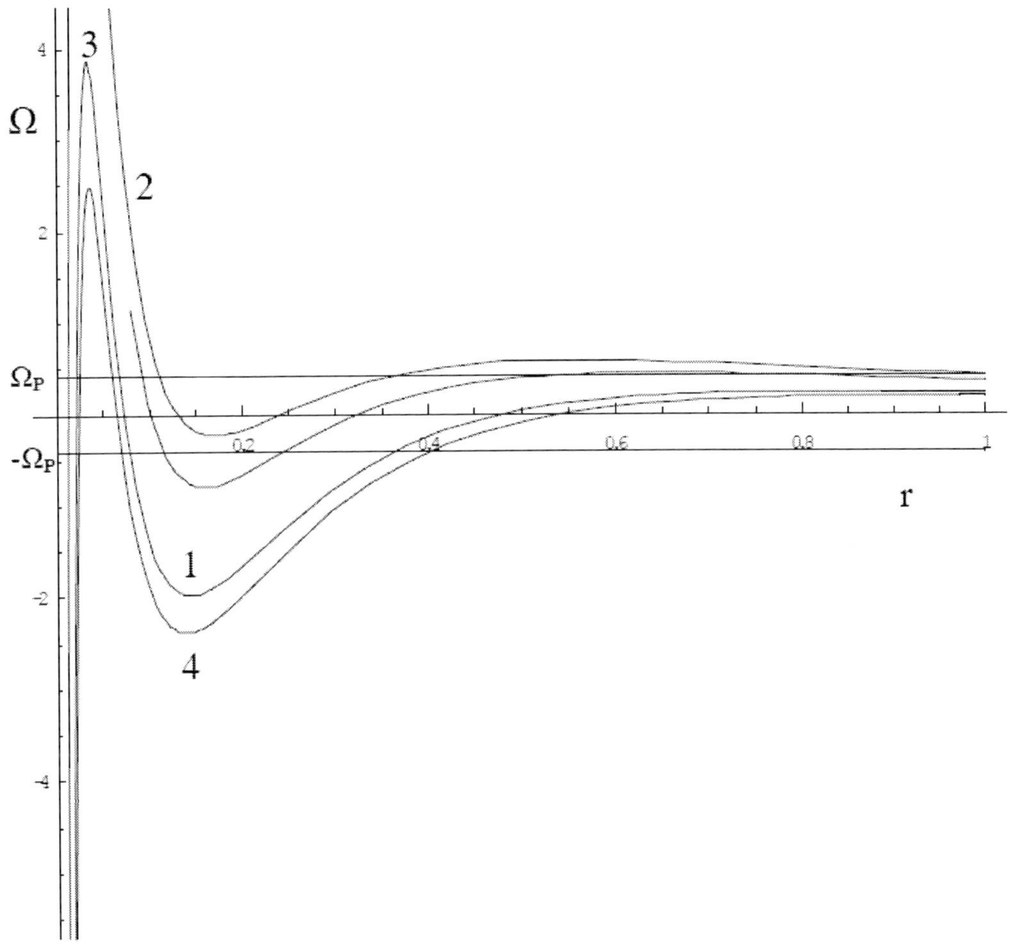

Figure 7. The curves $\Omega - n\kappa/m$ vs r for the host galaxy 1. The values of the corresponding parameters are as in Figure 1.

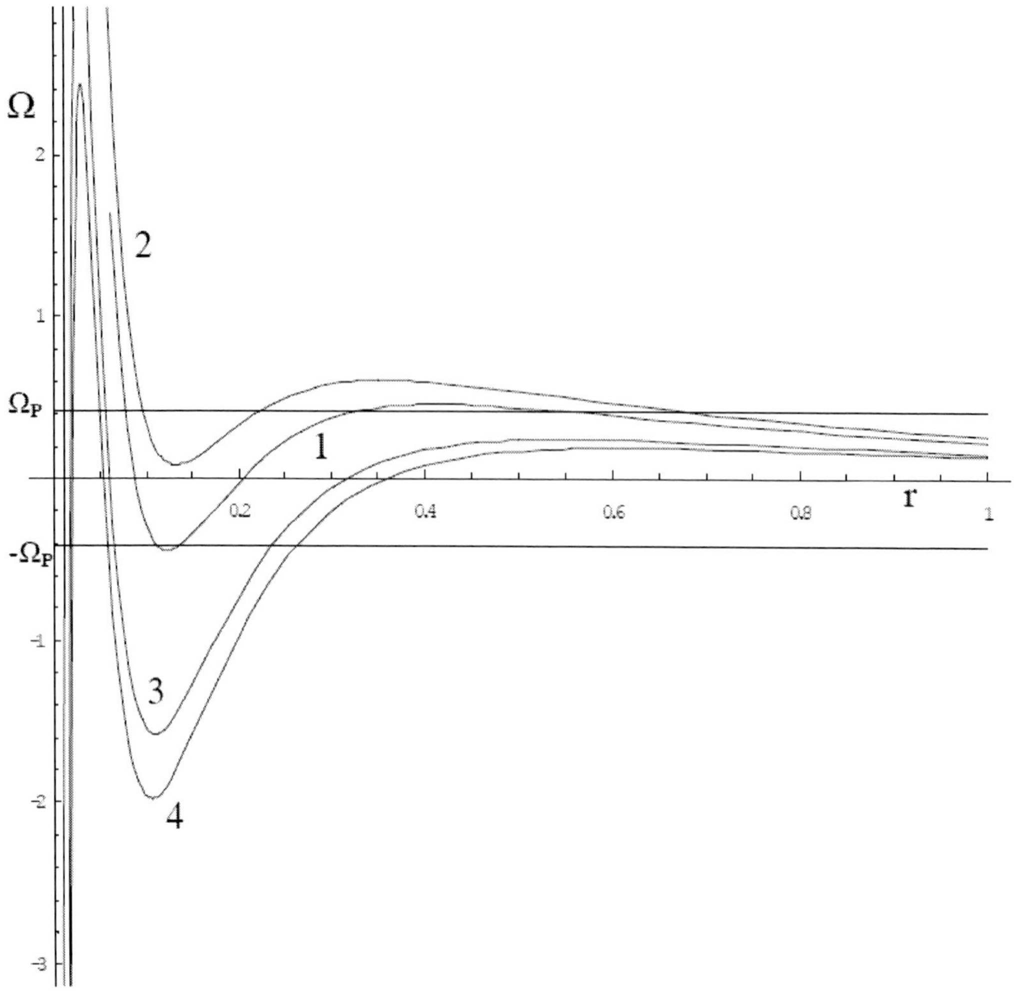

Figure 8. Similar as Figure 7 for the host galaxy 2.

B. Time Dependent Model

Let us now come to follow the evolution of orbits as mass is transported from the disk of the galaxy 1 to its centre. By this procedure a quasar is developed in the central regions of galaxy 1. The mass transport is linear following the equations

$$M_n(t) = M_{ni} + kt, \quad M_d(t) = M_{di} - kt, \tag{11}$$

where $M_{ni} = 0$, and $M_{di} = 2.08$, are the initial values of the mass of the nucleus and disk respectively, while k is a positive parameter. As in Caranicolas and Innanen (2008) we assume that the linear rate (11) is slow compared to the orbital period therefore it is adiabatic.

It is also assumed that the transportation stops as the mass of the quasar takes the value $M_{nf} = 0.08$.

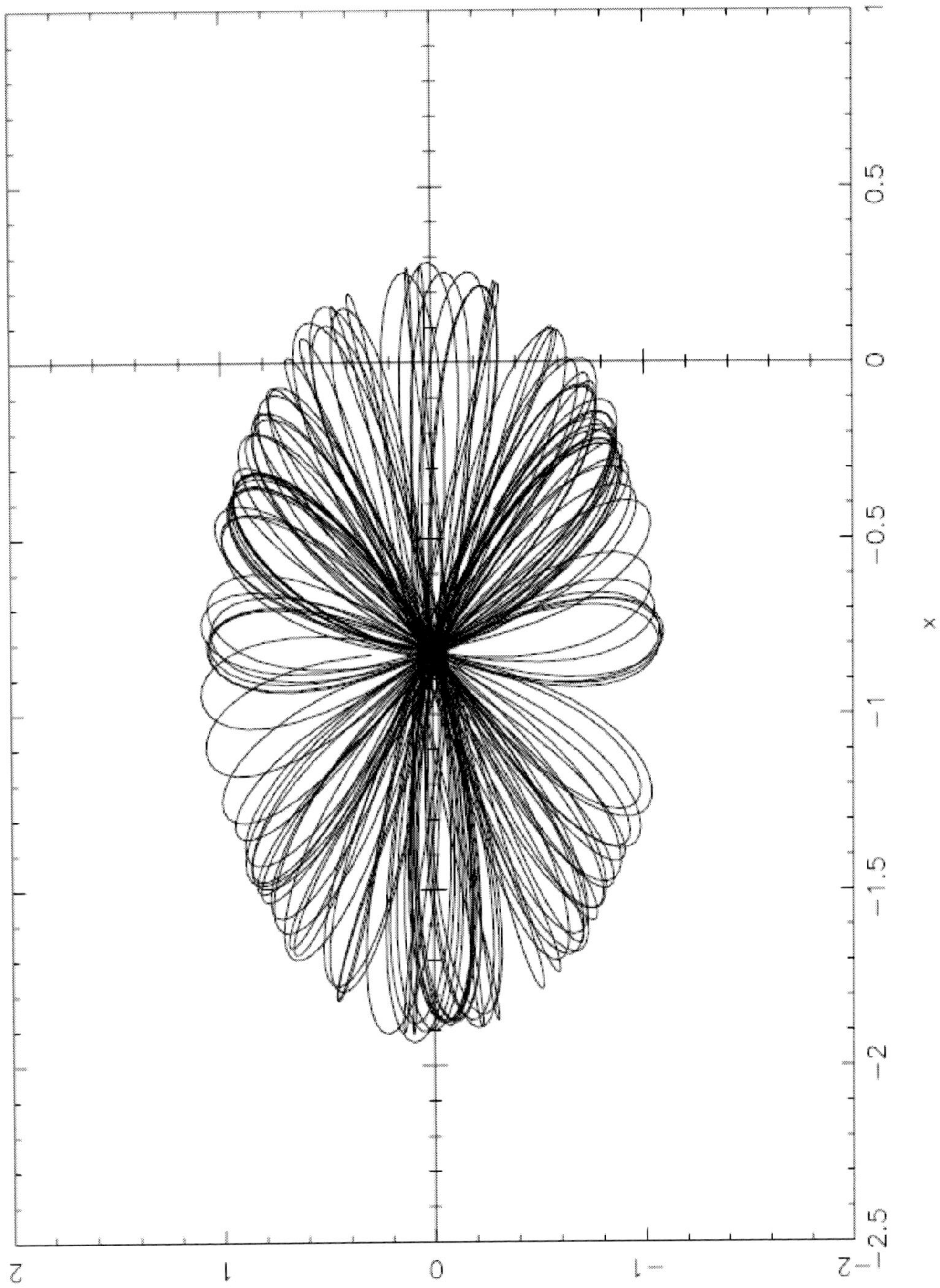

(A)

Figure 9. Continued on next page.

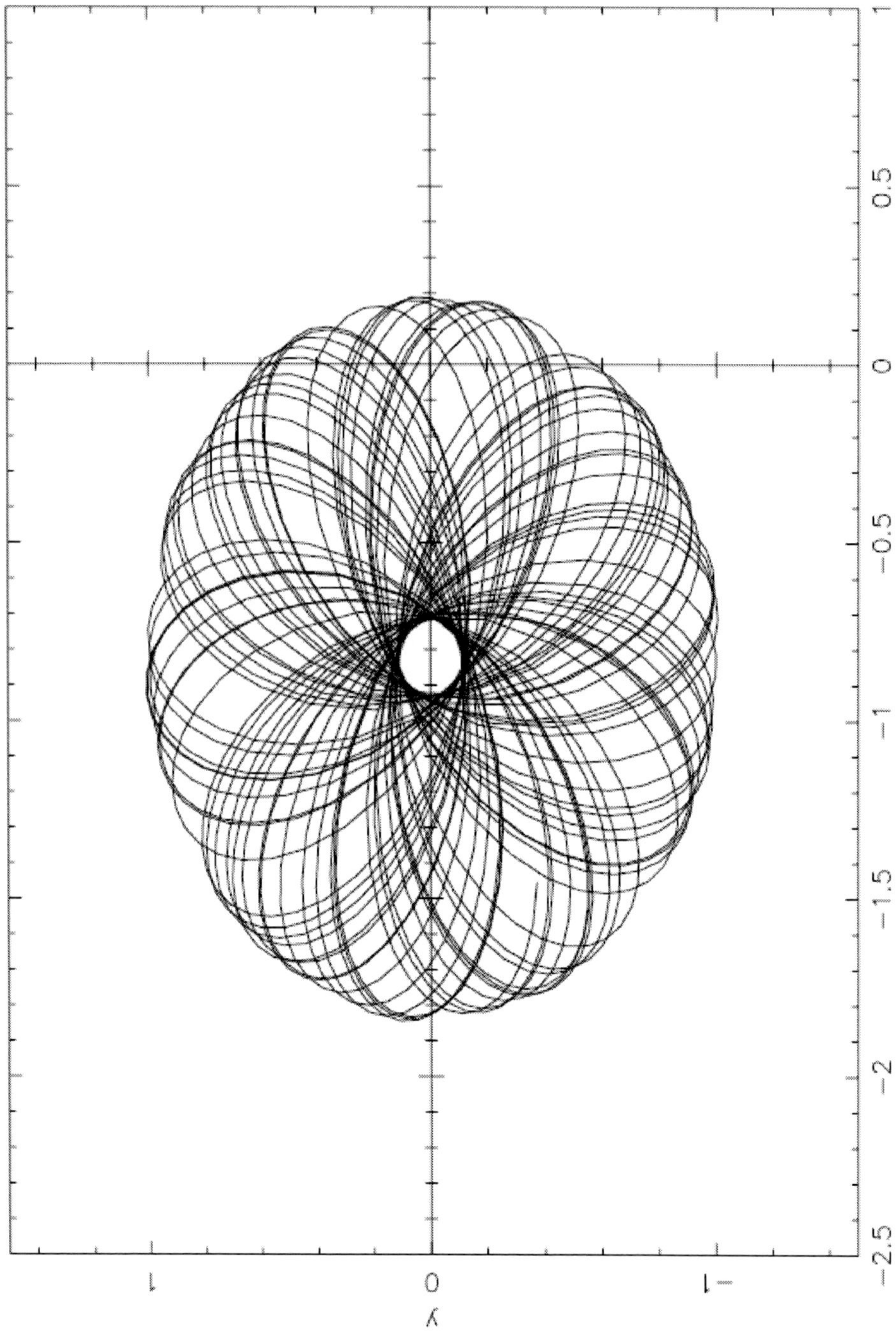

(B)

Figure 9a-b. (a) An orbit in the evolving model. Initial conditions are x=-0.8235, y= p_x =0, while k is equal to 0.0008. The initial value of the Jacobi integral is E_J=-2, while the values of all other parameters are as in Figure 4b. The presence of the quasar changes the character of the orbit, from regular to chaotic.(b) An orbit with the same values of parameters as in Fig 9a but with different initial conditions x=-0.2, y= p_x =0.The orbit starts as regular and remains regular during the galactic evolution.

Figure 9a shows an orbit as the galaxy evolves following equations (11). The initial conditions are x=-0.8235, y= p_x =0,while k is equal to 0.0008. The initial value of the Jacobi integral is E_J=-2, and the values of all other parameters are as in Figure 4b. When t=100 time units, the mass of the quasar is $M_{nf} = 0.08$ and the evolution stops. The value of the Jacobi integral is now E_J = -2.092 and we continue the calculations for 150 more time units. It is evident that, if mass transport was not present, we would observed the regular orbit shown in Figure 6a.The presence of the quasar has change the character of the orbit, from regular to chaotic. Moreover, as the change of the value of the Jacobi integral is negligible we can say that in the phase plane of the galaxy 1 shown in Fig 4b chaotic regions appear and it is transformed in that shown in Figure 4a, as the quasar is formed in the centre of the galaxy 1. Figure 9b shows an orbit with the same values of parameters as Figure 9a but for different initial conditions. The initial conditions are x=-0.2, y= p_x =0.The orbit starts as regular and remains regular during the galactic evolution. The orbit was calculated for 150 time units and the change in the value of the Jacobi integral was from E_J = -2 to E_J = -2.033.

Numerical experiments, not given here, suggest that the character of orbits can change from regular to chaotic or not change at all, as mass is transported and the quasar is developed in the centre of the galaxy 1.The final nature of the orbit depends on the initial conditions. In all cases we use the parameters given in Figure 4b and relations (11).

4. Conclusion

In this article we have built a simple dynamical model in order to study the nature of motion in a binary quasar dynamical model. We have chosen the host galaxies, in our model, to be disk galaxies because the assessment that bright quasars reside in massive ellipticals (Dunlop et al. 2003) must be revised as recent observations show that a large number of bright quasars are hosted in disk galaxies (Letawe et al 2006). Observations of radial velocities have shown that the masses of the large spirals are about $2x10^{11}$ solar masses. This mass is compatible with the mass of our disk galaxy 1.As observation data show that about 50% of the host galaxies show signs of interactions, we have decided to present a quasar model based on a pair of two interacting host galaxies.

Two cases were investigated: the time independent model and the evolving model, that is the case when a quasar is formed in the centre of galaxy 1.Our numerical calculations show that two factors produce the observed chaotic motion in the time independent model: (i) the galactic interaction and (ii) the galactic activity, that is the presence of the quasars. Furthermore, the presence of the quasars increases the velocities near the centres of the host galaxies. On the other hand, it was found that the two interacting galaxies, for large values of R, do not show chaotic motion, when the quasars are absent.

It is of interest to follow the evolution of orbits as the quasar is formed in the centre of the host galaxy 1. In order to keep things simple we have started from the phase plane of galaxy 1 shown in Figure 4b, where all orbits are regular and we went to the phase plane of galaxy 1 shown in Figure 4a, following equations (11). We see that the final character of orbits depends on the initial conditions. Therefore regular orbits can turn to chaotic or remain regular during the quasar formation and since the quasar has been formed.

An interesting question is whether interaction is essential for triggering the mass transportation and, therefore, the galactic activity. This question remains unsolved as observational astronomers have found disk galaxies (see Letawe et al. 2006) showing no signs of interaction at all but harbouring active nuclei.

References

[1] Bahcall, J.N.,Kirhakos,S., Saxe,D.H., 1997,*ApJ* **479**, 642.
[2] Boyce,P.J.,Disney,M.J.,Bleaken,D.G., 1999, *MNRAS*, **302**, L39.
[3] Caranicolas,N.D., Papadopoulos, N., 2005, *Astron. Nach.* **326** (2), 96.
[4] Caranicolas,N.D., Innanen K. A., N., 2009, *Astron. Nach.* **330** (1), 20.
[5] Caranicolas,N.D., Papadopoulos, N., 2009, *New Astron.,* **14**,207.
[6] Dunlop,J.S.,Mclure,R.J.,Kukula,M.J., et al. 2003, *MNRAS*, **340**, 1095.
[7] Floyd,D.J.E., Kukula,M.J., Dunlop,J.S., et al. . 2004, *MNRAS*, **355**, 196.
[8] Hutching,J.B., Neff,S.J.,1992, *AJ* **104**, 1.
[9] Jahnke,K.,Kuhlbrodt,B.,Wisotzki,L., 2004, *MNRAS*, **352**, 399.
[10] Kaspi,S., Smith,P.S., Netzer,H., et al. 2000, *ApJ* **533**, 631.
[11] Letawe, G., Magain,P., Courbin, F., et al. *New Astron. Rev.* 2006,50, 779.
[12] Letawe, G., Magain,P., Courbin, F., et al.. 2007, *MNRAS*, **378**, 83.

In: Black Holes and Galaxy Formation
Editors: A.D. Wachter and R.J. Propst, pp. 121-147
ISBN: 978-1-60741-703-3

Chapter 4

BLACK HOLES AND OTHER EXOTICA AT THE LARGE HADRON COLLIDER

Arunava Roy [*] ***and Marco Cavaglià*** [≠]
Department of Physics and Astronomy,
University of Mississippi, University, MS, USA

Abstract

Black holes and other exotica might be produced at the forthcoming Large Hadron Collider if the fundamental scale of gravity is near the TeV scale. Simulations of black holes and other proposed physics beyond the Standard Model, supersymmetry and string theory are presented here. Black hole events are simulated using the CATFISH Monte Carlo generator, simulations of string resonances use PYTHIA and supersymmetric simulations use a combination of ISAJET and PYTHIA. Our analysis shows that black holes can be discriminated from supersymmetry and string resonances can be distinguished from Standard Model processes and black hole events. In particular, isolated high transverse momentum leptons allow discrimination between black holes and supersymmetry. Black holes and string resonances can be distinguished using Z^0 bosons and photons with high transverse momentum. Additional discrimination is provided by the analysis of energy, momenta and event-shape variables.

PACS Numbers 04.50.-h, 04.60.Bc, 04.70.Dy, 04.65.+e, 11.30.Pb, 12.60.Jv, 11.25.Wx

1. Introduction

At CERN's Large Hadron Collider (LHC) [1], protons will soon collide to provide experimental evidence for new physics at the TeV scale. Supersymmetry (SUSY) [2, 3] is widely considered to be one of the best candidates for physics beyond the Standard Model (SM). It provides an explanation for the Higgs mass problem [4, 5], a candidate for cold dark matter, and unification of low energy gauge couplings by introducing superpartners to SM

* E-mail address: arunav@olemiss.edu
≠ E-mail address: cavaglia@olemiss.edu

fields (see Ref. [6] and references therein). Extra dimensional models such as Large Extra Dimensions (LEDs) [7], warped braneworlds [8] or universal extra dimensions [9] have also been proposed. In these models, gravity becomes strong at the TeV scale where radiative stability is achieved. The fundamental scale of gravity M_* is lowered to ~ 1 TeV by the presence of the extra n-dimensional space or by a warping of the metric. One of the most astounding consequences of the existence of extra dimensions would be non-perturbative production of subatomic Black Holes (BHs) [10, 11, 12] and real/virtual gravitons [13] in particle colliders and cosmic ray showers. (For reviews, see Refs. [14].)

Both SUSY and extra dimensions are essential ingredients of string theory [15, 16]. The string scale is defined as [16]

$$l_s = \sqrt{\alpha'}, \tag{1}$$

where α' is the slope parameter with units of inverse energy squared. The strength of string interactions is controlled by the string coupling g_s. The Planck scale M_{PL} is related to the string scale M_s by

$$M_s = g_s M_{PL}. \tag{2}$$

The string coupling is generally assumed to be of order one since string effects are expected to appear just before quantum gravity effects set in. A small coupling also justifies the use of perturbative analysis [17]. In this scenario, the string scale is close to the Planck scale. However, in presence of LEDs gravity becomes strong at the TeV scale; the Planck scale and the string scale would both be ~ 1 TeV i.e. $M_s \approx M_*$. If this is the case, string resonances [18] would be observed at the LHC before the onset of non-perturbative quantum gravity effects such as BHs. Detection of string events through corrections to SM amplitudes [19, 20, 21] at the LHC would be the most direct evidence of string theory and LED.

The ATLAS [22] and CMS [23] detectors at the LHC are entrusted with the task of studying events with large transverse momentum (P_T), a signature common to SUSY, extra dimensions and string-mediated interactions. Comparisons of SUSY and universal extra dimensions/little Higgs models in colliders have been investigated by various authors [24]. Discrimination of SUSY and BH events by means of dilepton events was recently discussed by the authors in Ref. [25, 26]. Potential for discovery of string resonances at the LHC was discussed in Ref. [27, 28]. In this paper, we review these analyses and show that SUSY, BH and string resonances can be clearly distinguished at the LHC. BH events tend to be more spherical than SUSY events because of the isotropic nature of BH decay. Thus event shape variables, such as sphericity, provide good discriminators of SUSY and BHs. The dilepton invariant mass is also an excellent discriminator; the SUSY invariant mass shows a sharp cutoff at ~100 GeV, which is absent in the BH model because most of the dileptons originate from uncorrelated events. String events in the $pp \rightarrow \gamma + jet$ channel [19] can be distinguished from the SM background using event shape variables, visible energy, momentum of photons and dijet invariant mass. Production of Z^0 bosons from BH decay allow the discrimination of string and BH events.

The remainder of this paper is organized as follows. In Sect. 2, 3 and 4 we briefly review the fundamentals of SUSY, TeV BHs and string amplitude calculations which are needed for our analyses, respectively. Event simulations are described in Sect. 5. The methods for discriminating SUSY and BHs are presented in Sect. 6. Discrimination of string events from the SM background is discussed in Sect. 7. Comparison of string resonances and BH events is presented in Sect. 8. Conclusions are presented in Sect. 9.

2. Supersymmetry at the LHC

According to the Minimal Supersymmetric extension of the Standard Model (MSSM) [2], all SM bosons (fermions) must have a fermionic (bosonic) partner. These superpartners have identical properties as their SM counterparts (masses, charges and quantum numbers), differing only in their spin. The MSSM allows for the unification of electromagnetic, weak and strong forces at $M_{GUT} \approx 10^{16}$ GeV. Since we do not observe superpartners of SM particles at low energies, SUSY must be a broken symmetry. The SUSY breaking scale is generally assumed to be around 1 TeV. A method of SUSY breaking which is mediated by gravitational interactions is supergravity (SUGRA). In its minimal version, the physics of mSUGRA is determined by a point in the five-dimensional moduli space with parameters: m_0, the common scalar mass at M_{GUT}, $m_{1/2}$, the common gaugino mass at M_{GUT}, A_0, the common trilinear coupling at M_{GUT}, $\tan\beta$, the ratio of the vacuum expectation values of the two Higgs fields and μ, the sign of the Higgsino mass parameter.

mSUGRA parameters for typical LHC points are given in Table 1 [29]. SUSY particle (or sparticle) masses, for example neutralino ($\widetilde{\chi}_i^0$), gluino ($\widetilde{g}$) and squark ($\widetilde{q}$), are determined by m_0 and $m_{1/2}$: $\widetilde{\chi}_1^0 \approx m_{1/2}/2$, $\widetilde{\chi}_2^0 \approx \widetilde{\chi}_1^{\pm} \approx m_{1/2}$, $\widetilde{g} \approx 3m_{1/2}$ and $m(\widetilde{q}) \approx (m_0^2 + 6m_{1/2}^2)^{1/2}$ [30]. The gluino and the first two generations of squark masses do not depend on A_0. Boson masses are dependent on m_0 and fermion masses on $m_{\frac{1}{2}}$ and A_0 [31]. Therefore, sparticle masses are generally dependent on the values of m_0 and $m_{\frac{1}{2}}$.

SUSY interactions conserve R-parity [2]:

$$P_R = (-1)^{3B+L+2s}, \tag{3}$$

where L (B) is the lepton (baryon) number and s is the particle spin. Superpartners (SM particles) have P_R=-1(1). As a consequence, R-parity implies that SUSY particles are always pair produced from SM particles. R-parity conservation ensures that a SUSY process ends in a state with SM particles and two lightest stable SUSY particles (LSPs), which are generally neutralinos. The LSPs are colorless and chargeless and escape the detector. They are the source of missing transverse momentum, an important signature of SUSY events [1].

In the following we will show that the invariant mass of dilepton events provides an effective discrimination of SUSY and BH events. A list of dominant SUSY interactions leading to dileptons at the LHC is described below for the LHC points of Table 1. The invariant mass of a dilepton event is

$$M_{12} = \sqrt{(E_1 + E_2)^2 - (p_1 + p_2)^2} = \sqrt{2 p_1 p_2 (1 - \cos\theta)}, \tag{4}$$

where θ is the angle between the two leptons.

Table 1. Parameters for typical LHC points. The first five points refer to ATLAS and LM1 refers to CMS. The scalar mass and the gaugino mass are given in GeV.

LHC point	m_0	$m_{1/2}$	A_0	**tan** β	μ
A	100	300	300	2.1	+
B	400	400	0	2	+
C	400	400	0	10	+
D	200	100	0	2	-
E	800	200	0	10	+
LM1	60	250	0	10	+

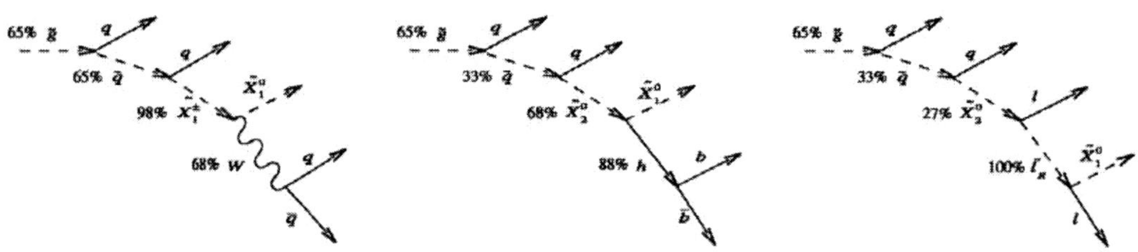

Figure 1. The top three SUSY decay chains at point A and their branching ratios.

The relevant decay chains for ATLAS point A are shown in Fig. 1. The third decay chain in Fig. 1 allows the separation of isolated leptons from the hadronic background [32]. Requiring that the leptons be isolated and have high P_T (in excess of 15 GeV) ensures that they were indeed produced from the above decay chain and not from the 2×2 hard interaction. In the rest frame of the second lightest neutralino, the dilepton invariant mass is

$$M_{ll} = \left[M^2_{\tilde{\chi}^0_2} + M^2_{\tilde{\chi}^0_1} - 2 M_{\tilde{\chi}^0_2} M_{\tilde{\chi}^0_1} \left(1 + \frac{P^2_{\tilde{\chi}^0_1}}{M^2_{\tilde{\chi}^0_1}} \right) \right]^{1/2} . \tag{5}$$

[1] If R-parity is not conserved, the missing transverse energy is reduced by the LSP decay.

Since the momentum of the LSP is not constrained, the invariant mass distribution shows an edge at ~ 100 GeV. M_{ll} attains its maximum value when the LSP is produced at rest: in that case the dilepton invariant mass is the difference between the masses of the first and second lightest neutralinos which is ~ 100 GeV for point A.

Heavy sparticles are produced at points B and C. $\widetilde{X}_2^0$ decay into sleptons is forbidden due to high mass of the slepton [5]. Therefore, the dominant decay mode at these points is $\widetilde{X}_2^0 \to \widetilde{X}_1^0 h$. Dileptons are produced from the decay of heavy charginos: the dilepton invariant mass plot does not have an edge as observed at point A. Due to the large value of $\tan\beta$ at point C there is increased prodction of Z^0 bosons and hence the number of dileptons.

At Point D there is production of light SUSY particles, thus sparticle production rates are large [33]. The dominant SUSY processes (branching ratios, BRs) are $\widetilde{g}g$ with $\widetilde{g} \to \widetilde{b}\bar{b}$ (89%)$\to \widetilde{X}_2^0 b\bar{b}$ (86%)$\to \widetilde{X}_1^0 b\bar{b}l^+l^-$ (33%) [5]. The above interactions imply that sparticle decays at this point are characterized by b jets and dileptons.

$M_{\widetilde{g}} < M_{\widetilde{q}}$ at point E, thus gluino production is predominant. Neutralino decay into sleptons is suppressed because of the high mass of the slepton. There is Z^0 production a feature common to models with large $\tan\beta$. Dominant decays (BRs) at this point are $\widetilde{X}_{3,4}^0 \to Z^0 \widetilde{X}_{1,2}^0$ (39%) and $\widetilde{X}_2 \to Z^0 \widetilde{X}_1$ (32%) [5]. Dileptons are produced from $\widetilde{X}_2^0 \to \widetilde{X}_1^0 l^+l^-$ (6% [5]) and from the leptonic decay of Z^0 bosons.

We end this section after studying the lepton production processes at point LM1 [34, 35]. The large value of $\tan\beta$ leads to a lower mass of $\widetilde{\tau}$ and $\widetilde{b}$ which results in the decay of the second lightest neutralino into $\widetilde{\tau}$. Dileptons may be produced via the process $\widetilde{X}_2^0 \to \widetilde{\tau}_1^{\pm}\tau^{\mp} \to \tau^+\tau^-\widetilde{X}_1^0$ and $\widetilde{X}_2^0 \to \tau^+\tau^-\widetilde{X}_1^0$, where the τ decays in the lepton channel [36, 37, 38]. Moreover, Z^0 production is enhanced due to higher sbottom production: $\widetilde{b}_1 \to \widetilde{X}_{3,4}^0 b, \widetilde{X}_{3,4}^0 \to Z_0 \widetilde{X}_{1,2}^0$ [36, 37]. $\widetilde{g} \to \widetilde{q}q$ is the dominant process as at point A since the mass of the gluino is larger than the mass of the squark. The BRs of the dominant decays are $\widetilde{X}_2^0 \to \widetilde{l}_R l$ (11.6%), $\widetilde{X}_2^0 \to \widetilde{\tau}_1^{\pm}\tau^{\mp}$ (46%) and $\widetilde{X}_1^{\pm} \to \widetilde{\nu}_l l$ (36%) [35]. Therefore, the invariant mass distribution of dileptons is expected to show an edge.

3. Black Holes at the LHC

In scenarios with LEDs, pp collisions at the LHC could produce TeV-mass BHs with expected characteristic lifetimes of 10^{-25} seconds [14]. BH production at the LHC occurs as follows. According to Thorne's hoop conjecture [39], a BH of mass M is formed when an object is compressed in all directions such that

$$C < 2\pi R_s(M), \tag{6}$$

where C is the circumference of the region where the object is compacted into and R_s is the Schwarzschild radius for a BH of mass M. Assuming no gravitational energy loss at formation provides the upper limit on BH mass, corresponding to the black disk (BD) cross-section $\sigma_{BD} = \pi R_s^2$. A more realistic model assumes some CM energy being lost as gravitational radiation (for a more detailed discussion, see Ref. [40]). To estimate the energy loss, the colliding particles are treated as two Aichelburg-Sexl shock waves [41]; the overlap of the shock waves forms a trapped-surface (TS) which sets a lower limit to the mass of the BH [42]. (For an alternative estimate of the collisional gravitational loss, see Ref.[43].) The cross section at the LHC involves summing up the contributions from all the initial partons. The cross section in the TS scenario is

$$\sigma_{pp\to BH}(s,n) = \sum_{ij} \int_0^1 2z dz \int_{x_m}^1 dx \int_x^1 \frac{dx'}{x'} f_i(x',Q) f_j(x/x',Q) F \sigma_{BD}(xs,n), \tag{7}$$

where Q is four-momentum transfer squared, $f_i(x,Q)$ are the parton distribution functions, z is the normalized impact parameter, and F is a form factor. The cutoff in x is related to the minimum allowed mass of the object, M_{min}, and to the fraction of center-of-mass energy trapped in the BH, $y(z)$, by $x_m = M_{min}^2/[sy^2(z)]$. TeV BHs may carry color or electric charge and angular momentum. They decay immediately after formation through loss of excess multipole moments (balding phase), gravitational + Hawking radiation [44] (evaporation phase) and final n-body decay or remnant production (Planck phase). SM particles are emitted on the brane and can be detected [45]. Simulations usually neglect the energy loss in balding phase since it is poorly understood. Since emissivities of rotating BHs are not known for all fields, the description of the evaporation phase is also approximated [47, 46]. Using averaged greybody factors, the total decay multiplicity is [48]

$$N = \frac{(n+1)S}{4\pi} \frac{\sum_i c_i R_i \Gamma_{R_i}}{\sum_j c_j P_j \Gamma_{P_j}}, \tag{8}$$

where c_i are the degrees of freedom of species i, Γ_{P_i} and Γ_{R_i} are the relative emissivities of Ref. [47], S is the initial entropy of the BH, and P_s and R_s are spin-dependent normalization factors. For a more detailed discussion of the evaporation and Planck phases of TeV BHs see Refs. [14].

Dilepton production in BH events differ greatly from dilepton production in SUSY. There is no single process of dilepton production for BHs. Dileptons are either produced by the BH directly or by the decay of heavier particles such as the Z^0 boson, $t\bar{t}$ pairs or a combination of the two. Therefore, the BH dilepton invariant mass does not show a sharp cut-off at high energy.

4. String Resonances at the LHC

The relevant process for string resonances is $pp \to \gamma + jet$, where the event is dominated by gluon-gluon scattering: $gg \to g\gamma$. The string amplitude for this process is [19]

$$|M(gg \to g\gamma)|^2 = g_s^4 Q^2 C(N) \left\{ \left[\frac{s\mu(s,t,u)}{u} + \frac{s\mu(s,u,t)}{t} \right]^2 + (s \leftrightarrow t) + (s \leftrightarrow u) \right\}, \quad (9)$$

where s, t and u are the Mandelstam variables and

$$\mu(s,t,u) = \Gamma(1-u) \left(\frac{\Gamma(1-s)}{\Gamma(1+t)} - \frac{\Gamma(1-t)}{\Gamma(1+s)} \right). \quad (10)$$

Here $Q^2 = \frac{1}{6}\kappa^2 \cos^2\theta_W \approx 2.55 \times 10^{-3}$, where κ=0.02 is a mixing parameter and θ_W is the Weinberg angle, N=3 is the number of D branes needed to generate the eight gluons of the SM and $C(N) = \frac{2(N^2-4)}{N(N^2-1)}$ is a constant parameter. The values of the parameters are chosen as in Ref. [19, 28].

In the limit $s \to nM_s^2$ the string amplitude possesses poles at $n = s/M_s^2$, where n is an integer. The amplitude for odd values of n is

$$|M(gg \to g\gamma)|^2 = g_s^4 Q^2 C(N) \frac{4}{(n!)^2} \frac{s^4 + u^4 + t^4}{M_s^4 [s - nM_s^2]} \left\{ \frac{\Gamma(t/M_s^2 + n)}{\Gamma(t/M_s^2 + 1)} \right\}^2. \quad (11)$$

The behavior for even values of n is obtained from Eq. (11) with the substitution $s \to t$ and $n \to m = t/M_s^2$ in the square bracket term. The singularities of the amplitude are smeared with a fixed width $\Gamma = 0.1$ for all $n > 1$ and as

$$|M(gg \to g\gamma)|^2 \approx \frac{g^4 Q^2 C(N)}{M_s^4} \left\{ \frac{M_s^8}{(s - M_s^2)^2 + (\Gamma^{J=0} M_s^2)^2} + \frac{t^4 + u^4}{(s - M_s^2)^2 + (\Gamma^{J=1} M_s^2)^2} \right\}, \quad (12)$$

for $n=1$ [19]. Equation (12) includes a correction for spin dependent widths: $\Gamma^{J=0}=0.75\alpha_s M_s$ and $\Gamma^{J=1}=0.45\alpha_s M_s$ where $\alpha_s=g_s^2/4\pi$ is the strong coupling constant. The presence of the poles indicates the formation of string resonances. The total cross section for the $pp\to\gamma+jet$ event is obtained by integrating the parton cross section over the parton distribution functions of the protons $f(x,Q)$ [49]

$$\sigma_{pp\to string\to\gamma+jet}=\int dx_1\int dx_2\int dt\; f_1(x_1,Q)f_2(x_2,Q)\frac{d\sigma}{dt}, \tag{13}$$

where x_i is the fraction of the total momentum carried by parton i, Q is the four-momentum transfer squared and

$$\frac{d\sigma}{dt}=\frac{|M(gg\to g\gamma)|^2}{16\pi s^2}. \tag{14}$$

The limits on x_1 and x_2 are $(0,1)$. The limits on t are fixed by the conditions $s+u+t=0$ and $|t||u|/s\geq P_{Tmin}^2$. The choice of CTEQ6D parton distribution functions allows direct comparison of our results to those of Ref. [19].

The left panel of Fig. 2 shows the differential cross section of the $pp\to\gamma+jet$ process with total center-of-mass energy E_{CM}

$$\frac{d\sigma}{ds}=\int dx_2\int dt\,\frac{2\sqrt{s}}{x_2E_{CM}^2}\,f_1(x_1,Q)f_2(x_2,Q)\frac{d\sigma}{dt}. \tag{15}$$

Resonances are evident for $s=nM_s^2$. The right panel of Fig. 2 shows the total cross section as a function of the minimum transverse momentum of the two outgoing particles of the 2×2 scattering, P_{Tmin}. The string cross section for M_s=1 TeV (solid red line) and the cross section for M_s=2 TeV (dashed green line) are $\sim10^3$ and 5×10^4 times less than the cross section for the SM processes respectively. Our sample run of 10^7 events produced ~ 10000 (200) string events for M_s=1 (2) TeV, with an integrated LHC luminosity of 1 fb^{-1}.

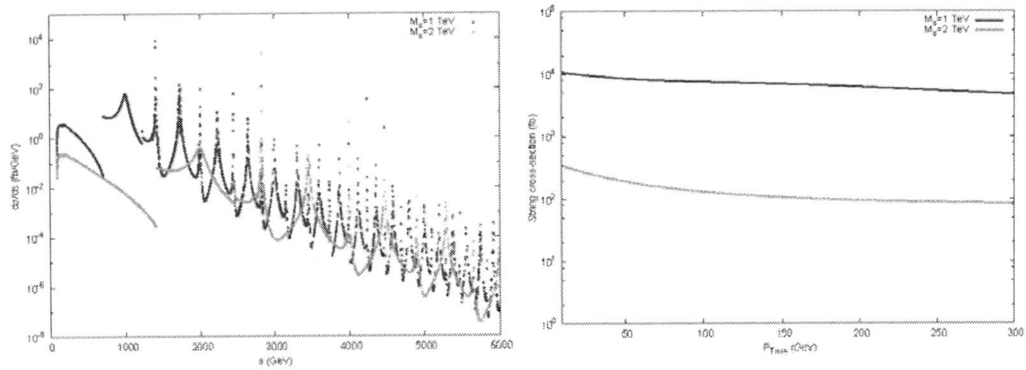

Figure 2. Left Panel: Differential cross section of string resonances for M_s = 1 TeV (red dots) and M_s = 2 TeV (green crosses) with P_{Tmin}=50 GeV. String resonances are clearly seen when $s = nM_s^2$. Right Panel: String resonance cross section for M_s = 1 TeV (red dots) and M_s = 2 TeV (green crosses) with P_{Tmin}=50 GeV. The cross section for M_s = 1 TeV is ~ 44 times larger than the cross section for M_s = 2 TeV.

We fix P_{Tmin}=50 GeV for both string and SM events. This choice leads to a signal-to-background ratio of ~73. For lower values of P_{Tmin} the string cross section is highly suppressed *w.r.t.* the SM cross section, for example $\frac{\sigma_{string}}{\sigma_{SM}} \approx 10^{-5}$ for P_{Tmin} = 10 GeV. Therefore, the discrimination of string events from the SM background is difficult for events with low P_{Tmin}. For higher values of P_{Tmin}, both the SM background and the signal are substantially reduced. For example, at 300 GeV they are reduced by a factor of ~98% and ~42% *w.r.t.* values at P_{Tmin} = 50 GeV, respectively. The optimal signal-to-background ratio is obtained for $P_{T\min} \leq 100$ GeV.

5. Event Simulations

SUSY simulations are carried out using a combination of ISAJET (ver. 7.75) [50] and PYTHIA (ver. 6.406) [51], with the former generating the mass spectrum. BH simulations are carried out using the CATFISH [48] Monte Carlo generator. String and SM simulations use PYTHIA (ver. 6.406). The setup for each simulation is summarized below:

- SUSY:
  - The MSSM mass spectrum is generated with ISAJET;
  - The mass spectrum in SLHA format is fed into PYTHIA;
  - SUSY processes except SM Higgs production are simulated;
  - Unstable SM particles and sparticles are hadronized or decayed with PYTHIA.

- BHs:
  - The cross section of the BH event is calculated in the center-of-mass frame;
  - The initial BH mass is sampled from the differential cross section;
  - The BH is decayed through Hawking mechanism and final n-body event (or remnant);
  - Unstable quanta are hadronized with PYTHIA.

- Strings:
  - The differential cross section for the $pp \rightarrow \gamma + jet$ process [19] is calculated;
  - The total cross section is obtained from Eq. (13);
  - The resonance decay is simulated with PYTHIA.

- SM background:
  - - $t\bar{t}, Z_0, W, WZ_0, WW$ *and* Z_0Z_0 SM processes are simulated with PYTHIA.

The parameters for the BH benchmark model are fundamental Planck scale $M_* = 1$ TeV, minimum BH mass $M_{min} = 2$ TeV, classical-to-quantum threshold $Q_{min} = 1$ TeV, six extra-dimensions ($n = 6$) and two-body final decay ($n_p = 2$). The benchmark for string simulations is defined by: M_s=1 TeV and P_{Tmin}=50 GeV. Particles produced in the initial-radiation phase are removed by imposing P_T cuts of 5 GeV and 15 GeV for leptons and photons+hadrons, respectively [48].

6. SUSY and BH Event Analysis

Our analysis is first based on visible energy, missing momentum of leptons and hadrons and event shape variables. This is followed by an analysis of high-P_T leptons. Figure 3 shows visible energy, missing transverse momentum and sphericity distribution for the five ATLAS points of Table 1, where all SUSY processes (except SM Higgs production) have been implemented. The difference between the five points is not significant for the purposes of the first part of our analysis. Therefore, we will consider point A as benchmark. Point A allows for Higgs production [33] in the supersymmetric decay chains. Since BHs may evaporate into Higgs (see Sect. 3), the presence of the Higgs channel in SUSY processes leads to a more effective comparison of SUSY and BH events. BH events are also characterized by up to several TeV of missing transverse momentum like SUSY point A.

BH events tend to be more spherical than SUSY events due to the spherical nature of the Hawking radiation. This is specially evident for high-mass BHs. The formation of a stable BH remnant at the end of the evaporation phase also helps to discriminate SUSY and BH events because of the large amount of energy which is carried away by the remnant. Analysis of isolated dilepton events is another powerful method to distinguish the two models. This is due to the fact that SUSY dileptons typically originate from a single decay chain whereas leptons are emitted less frequently by BHs (the hadron-to-lepton ratio is approximately 5:1) and are uncorrelated; they can be emitted at any angle *w.r.t.* beam axis.

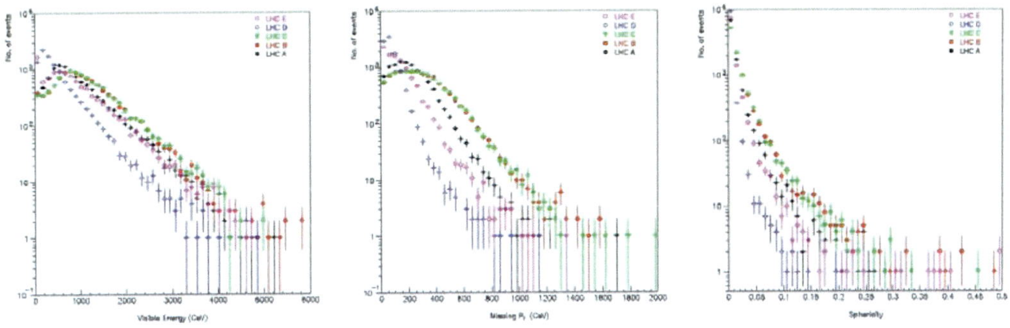

Figure 3. Comparison of visible energy (left), missing transverse momentum P_T (middle) and sphericity (right) for 10000 events for the five LHC points of Table 1 (A: filled black circles, B: filled red squares, C: filled green triangles, D: open blue circles and E: open pink squares).

Figure 4 shows visible energy, missing transverse momentum, and transverse momentum of leptons and hadrons+photons for 10,000 SUSY and BH benchmark events. The amount of visible energy and P_T is comparable for the two scenarios, even in the absence of a BH remnant. This is due to the presence of invisible channels in both models: neutrinos+gravitons for the BH and LSP for SUSY. Leptons (e and μ) provide better discrimination than hadrons and photons.

SUSY interactions do not produce leptons with energy above the TeV since isolated leptons are produced by the decay of sparticles with typical energy of less than a few hundred GeV. On the contrary, quanta produced in the BH decay are characterized by an average energy $E \approx M/N$, where the multiplicity N is less than 10 for typical BHs at the LHC. Since Hawking evaporation does not distinguish leptons from hadrons, hard leptons with energy up to several TeV are likely to be produced during the BH decay. This suggests that isolated leptons may provide a powerful means to discriminate the two models. This is indeed the case, as we shall see below.

The plots in the left panel of Fig. 5 show how variations in the BH Planck phase affect the observables of Fig. 4. The plots compare BH remnant, two-body and four-body decay. The BH is expected to have shed its electric and color charges by the time the remnant is formed[2]. The BH remnant is undetectable and a source of missing transverse momentum in addition to neutrinos and gravitons which are emitted during the Hawking evaporation phase. This leads to a larger difference in P_T between SUSY and BH models. The visible transverse momentum in hadrons+photons is sensibly reduced in the presence of a BH remnant. The BH remnant carries away energy which otherwise would have been emitted in visible channels (mostly hadrons) during the BH decay phase. Since leptons are rarer than hadrons in the BH decay phase, the leptonic channel is essentially unaffected by the presence of a BH remnant.

2 For an alternative scenario see, Ref.[52].

Variations in the energy distribution of the leptonic channel are thus suppressed compared to the hadronic channel. Changes in the number of final Planckian hard quanta do not produce significant differences in the distributions; more quanta of lower energy behave statistically like less quanta with higher energy. Provided that the BH decays at the end of the Hawking phase, it is thus safe to set the number of Planckian quanta to $n_p = 2$ (or $n_p = 4$), although BHs may decay in different numbers of particles on a event-to-event basis. Variations in the classical-to-quantum threshold Q_{min} are also not expected to cause significant differences in the energy/momentum distributions. A higher threshold increases the emission in the Planck phase while decreasing Hawking radiation. Since these phases differs only in relative greybody factors, the effect is too small to be detected.

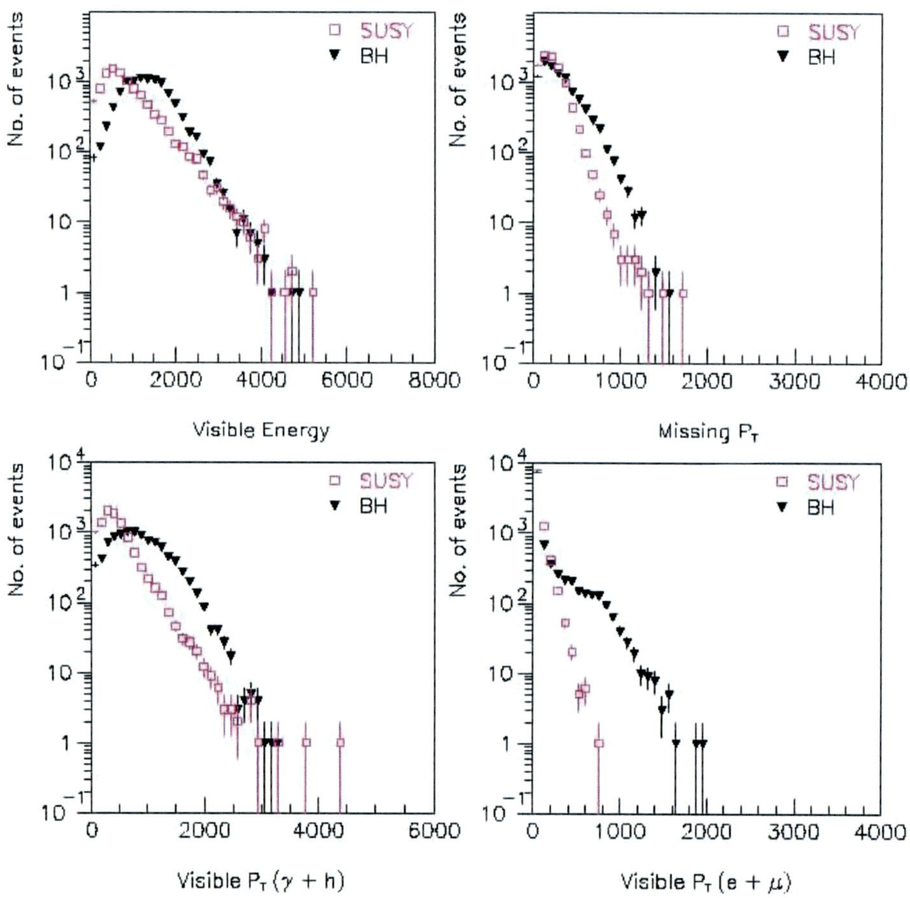

Figure 4. Comparison of 10,000 SUSY and BH benchmark events at the LHC. Visible energy and missing transverse momentum P_T (top panels) are comparable due to the presence of invisible channels in both models. Leptons with large transverse momentum provide instead an effective discriminator (bottom right panel).

The plots in the right panel of Fig. 5 show visible energy, P_T , and visible transverse momenta of leptons and hadrons+photons for different values of the fundamental Planck scale. Higher values of M_* lead to more massive BHs; i.e. higher multiplicity and more energetic quanta. This causes a significant increase in missing and visible momenta. If the value of the fundamental Planck scale is large, BHs are likely to be found and easily distinguished from SUSY through detection of highly energetic isolated leptons and hadronic jets. Missing transverse momentum of several TeV would also be observed [3].

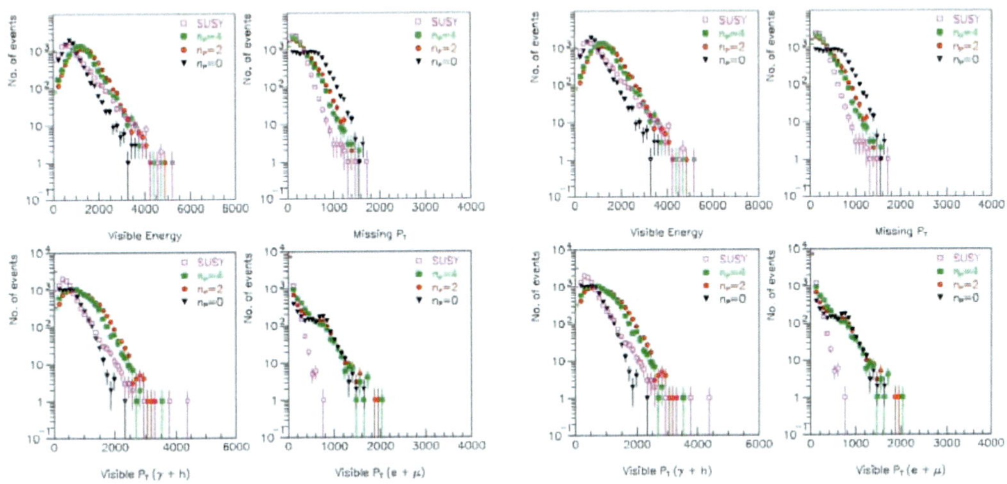

Figure 5. Distribution of visible energy, P_T and transverse momenta of leptons and hadrons+photons. SUSY plots are shown as pink open squares. The four plots in the left panel show the effect of different decay modes in the Planck phase of ten-dimensional BHs: remnant formation ($n_p = 0$, black filled triangles), two-body decay ($n_p = 2$, red filled circles) and four-body decay ($n_p = 4$, green filled squares). The fundamental Planck scale is $M_\star = 1$ TeV. The four plots in the right panel show the effect of varying the fundamental Planck scale: $M_\star = 1$ TeV (black filled triangles), $M_\star = 2$ TeV (red filled circles) and $M_\star = 3$ TeV (green filled squares). The ten-dimensional BHs decay in two hard quanta at the end of the evaporation phase.

Event shape variables such as sphericity and 2nd Fox-Wolfram moment can be used to complement the above analysis. BH events are more spherical because of the nature of Hawking radiation and the production of jets from SUSY decays. Formation of a BH remnant and high values of the fundamental scale lead to significant higher sphericity than SUSY (top panels of Fig. 6). The 2nd Fox-Wolfram moment (bottom panels of Fig. 6) is stable versus changes in the BH Planck phase and provides a good SUSY/BH discriminator. BH models with higher M_* can be differentiated more easily from SUSY events. Similar conclusions can be reached by looking at jet masses and number of jets. SUSY events generate more and

[3] Since the LHC is expected to produce very light BHs, this analysis may not provide the most effective discriminators [48, 53]. The study of leptonic final states alleviates this problem.

lighter jets than the BH model due to copious production of quarks (Fig. 7). The difference is again specially significant for high values of $M_\star$ and in the presence of BH remnants. Absence of sub-Q_{min} hard jets could provide strong evidence for BH remnant production. (See the suppression of heavy jets below $Q_{min} = 2$ TeV in the top leftmost panel of Fig. 7.)

We now focus our attention on the analysis of high-P_T leptons. The SM dilepton background contribution can be removed by applying suitable cuts on transverse momentum of the leptons [29]:

$$P_{Tl} \geq 15 \text{ GeV}, \; |\eta_l| < 2.5,$$

$$\text{Isolation cut, } \sum\nolimits_l P_{Tl} < 7 \text{ GeV in a cone of } R = 0.2,$$

$$\not{P}_T \geq 200 \text{ GeV},$$

where P_{Tl} is the transverse momentum of the leptons, $R = \sqrt{\Delta\eta^2 + \Delta\phi^2}$, $\eta = -\ln[\tan(\theta/2)]$ is the pseudorapidity, ϕ and θ are the azimuthal and polar angles of the lepton *w.r.t.* beam axis, and $\not{P}_T$ is the missing transverse momenta of a particular event.

Isolated dileptons as a SUSY signature have been discussed in the literature [33, 54]. Although the production of leptons is not as high as colored particles, high energy isolated leptons provide a cleaner environment by removing the SM background. The dominant SUSY interaction for opposite-sign, same-flavor (OSSF) dileptons at ATLAS point A is $\tilde{\chi}_2^0 \to l^\pm \tilde{l} \to l^\pm l^\mp \tilde{\chi}_1^0$ with a 27% BR (See Fig. 1). The maximum dilepton invariant mass for this interaction is

$$M_{ll}^{max} = m_{\tilde{\chi}_2^0} \left[\left(1 - \frac{m_{\tilde{l}}^2}{m_{\tilde{\chi}_2^0}^2} \right) \left(1 - \frac{m_{\tilde{\chi}_1^0}^2}{m_{\tilde{l}}^2} \right) \right]^{1/2} \approx 100\,GeV. \tag{16}$$

Isolated leptons in BH decays may come directly from the BH itself, from the decay of Z^0 bosons and top quarks, or from a combination of the two. Since top quark decay into leptons is rare [55] and the BR of Z^0 into leptons is small, $\Gamma(l^+l^-)/\Gamma_{tot} \approx 0.034$ [56], production of OSSF dileptons is less frequent in the BH model than in SUSY events. Our analysis shows that an OSSF dilepton event occurs approximately every 100 BH and 20 SUSY events, with a ~1:5 ratio of BH-to-SUSY dilepton events at fixed luminosity.

The left panel if Fig. 8 shows the invariant mass distribution for 2000 SUSY OSSF dilepton events and 2000 BH OSSF dilepton events with $M_* = 1$ TeV and two final hard quanta.

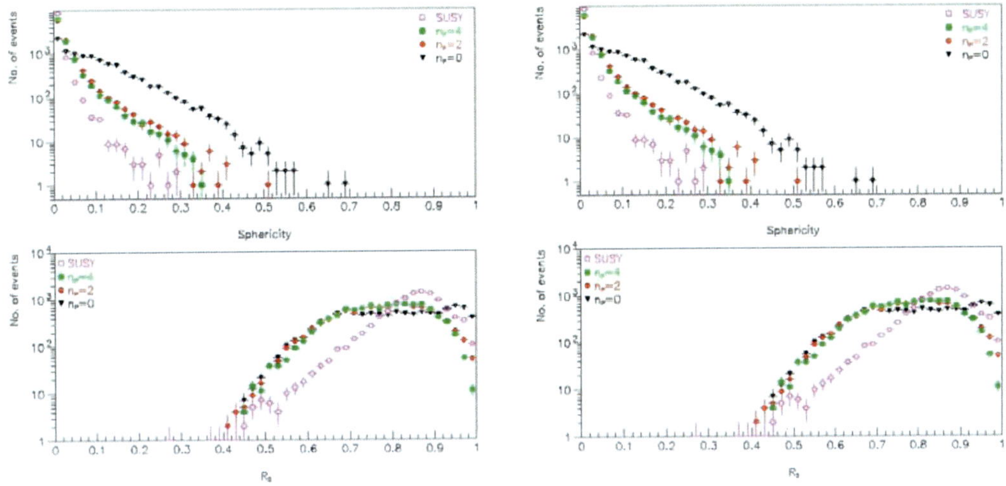

Figure 6. Sphericity (top panels) and 2nd Fox-Wolfram moment (bottom panels) for 10,000 BH and MSSM events (pink open squares). The left panels show the effect of different Planckian decay modes: BH remnant (black filled triangles), two-body decay (red filled circles) and four-body decay (green filled squares). The fundamental scale is $M_* = 1$ TeV and the number of extra dimensions is six. The right panels show the effect of different fundamental scales: $M_* = 1$ TeV (black filled triangles), 2 TeV (red filled circles) and 3 TeV (green filled squares). The ten-dimensional BHs decay in two quanta at the end of the Hawking phase.

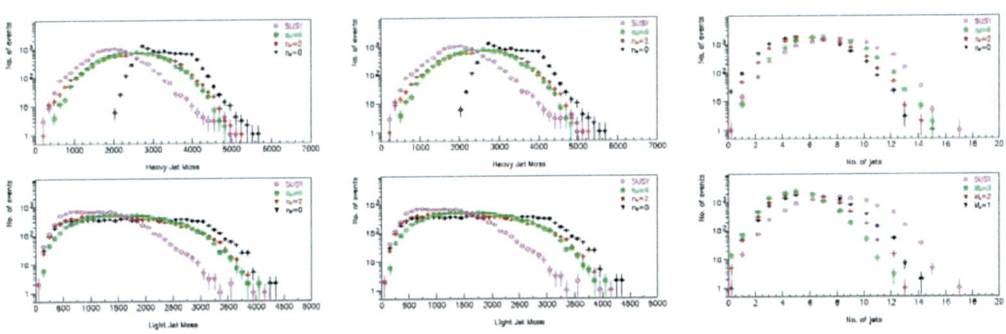

Figure 7. Light and heavy jet masses (four leftmost panels) and number of jets (right panels) for 10,000 BH and MSSM events. Symbols are like in previous figures.

As is expected, the SUSY invariant mass distribution shows a sharp edge at ~100 GeV. The BH invariant mass distribution is characterized by two peaks at ~90 GeV and 1 TeV, and a tail at high P_T. The first peak is due to dilepton events from single Z^0 bosons that are directly emitted by the BH. This is the dominant channel of OSSF dilepton production in BH events. The smaller peak at 1 TeV is due to dileptons emitted at the end of the Hawking phase [48]. Since the BH at the end of the Hawking phase is expected to be electrically neutral, isolated OSSF lepton events can occur: when the two final quanta are opposite-sign leptons or a $t\bar{t}$ pair. This peak is expected to be smeared out in a more realistic description of the final

BH phase [26]. The high-M_{ll} tail of the distribution is originated by pairs of uncorrelated leptons from the BH.

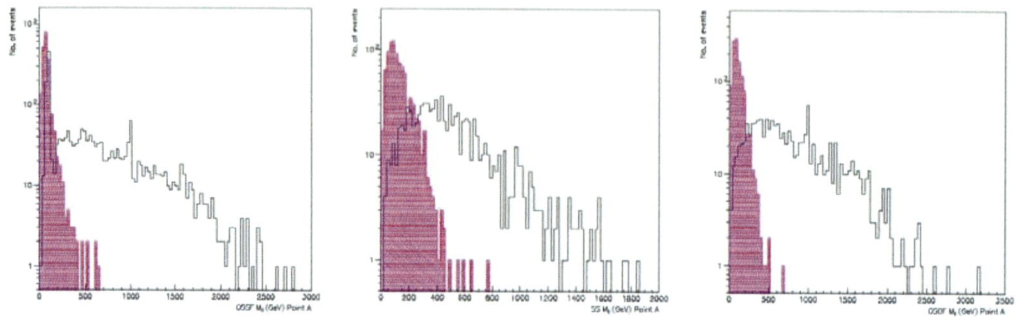

Figure 8. Left Panel: Invariant mass distribution (in GeV) for 2000 SUSY and BH OSSF dilepton events for Point A. The SUSY distribution (shaded pink histogram) shows the typical endpoint due to the presence of the LSP. The high-P_T tail of the BH distribution is originated by uncorrelated lepton pairs emitted during the Hawking evaporation phase. The final BH decay is in two-quanta. The SM background is negligible. Invariant mass distribution for 900 same-sign (1100 OSOF) dilepton events is depicted in the middle (right) panel.

The middle panel of Fig. 8 shows the invariant mass distribution for 900 same-sign (SS) SUSY and BH events. SUSY SS dileptons are produced by $\widetilde{\chi}_2^{\pm} \to W^{\pm}\widetilde{l}\,l^{\pm}, \widetilde{\chi}_1^{\pm} \to W^{\pm}\widetilde{\chi}_1^{0}$, and from top quark decay into leptons. SS dileptons from BHs are originated either from the BH itself or from the decay of heavier quarks and bosons ($t, \bar{t}, W, Z_0$). In both cases SS dilepton events are rare because of the low BR of W into leptons (~11% [56]). SS dileptons are selected using the same cuts of the OSSF channel. The high-M_{ll} tail of the BH invariant mass distribution can be clearly distinguished from the SS case. Isolation requirements reduce the SM background to virtually zero [57, 58].

Dilepton events with same sign and/or opposite flavor leptons can also be used as discriminators (right panel of Fig. 8). The "democratic" nature of BH decay makes events with same/opposite flavor leptons roughly equally probable, whereas BRs of SUSY events favor same-flavor dileptons. For example, OS dileptons from the decay of $\widetilde{X}_2^0$ are of the same flavor. On the contrary, leptonic decays of charginos or third generation quarks and squarks would produce equal amounts of opposite-sign, opposite-flavor (OSOF) dilepton pairs compared to same flavor lepton pairs (see, for example decay chains at point LM1).

Table 2 shows the BRs of isolated dilepton events for SUSY and BH processes. The dominant channel is the OSSF channel for both models. However, SUSY and BH events can be easily discriminated by comparing the rate of same-flavor events to the rate of opposite-flavor events.

Table 2. BRs of high-P_T isolated dileptons for SUSY and BH models. 21,000 and 100,000 events were simulated in the two cases, respectively, yielding approximately 1000 dilepton events. OS(SS) stands for opposite(same) sign and OF(SF) denotes opposite(same) flavor.

High P_T isolated dileptons	*SUSY*	%	*BH*	%
OSSF	768	73	523	50
SSSF	65	6	103	10
OSOF	169	16	341	32
SSOF	52	5	87	8

The number of isolated high-P_T leptons is also an effective discriminator between BH and SUSY events and can be used to complement the above analysis [26]. SUSY decays could produce up to five isolated leptons [59]. On the contrary, events with three or more isolated leptons are suppressed in BH decays.

We now discuss the dilepton invariant mass distributions for the other SUSY points. Figure 9 shows the dilepton invariant mass distributions for points B and C. The background cuts remove the SM contribution. Unless otherwise stated all dilepton histograms show 2000 OSSF events, 900 SS events and 1100 OSOF events for both SUSY and BH. The SUSY dilepton invariant mass distribution is shown as the shaded (pink) histogram. The tail at large values M_{ll} for BHs is absent for SUSY events.

Table 3. BRs of high-P_T isolated dileptons for SUSY points B and C for 50,000 events. The high value of $\tan\beta$ =10 for point C manifests itself as a greater production of OSSF dileptons.

High P_T isolated dileptons	*OSSF*	*OSOF*
Point B	6017	5778
Point C	8083	5820

The dilepton invariant mass distribution for Point D is shown if Fig. 10. The difference in mass between the two lightest neutralinos ($M_{\tilde{X}_2^0} - M_{\tilde{X}_1^0}$ ~50 GeV) can be obtained from the edge in the OSSF invariant mass distribution (left panel of Fig. 10). P_T calculations are not carried out at this point [3] because lepton isolation requirements and the presence of b jets are effective in reducing the SM background [5]. The right panel of Fig. 10 shows the OSOF dilepton distribution for ~2600 SUSY and BH events. More than half of the background contribution is from $\bar{t}t$ events [38]. Simulations show that this background is negligible due to the low BR of W bosons into leptons.

Fig. 11 shows the dilepton invariant mass distribution for Point E. The OSSF invariant mass distribution is characterized by Z^0 peak. The invariant mass edge is not as distinct as at point A (see inset, Left Panel of Fig 11).

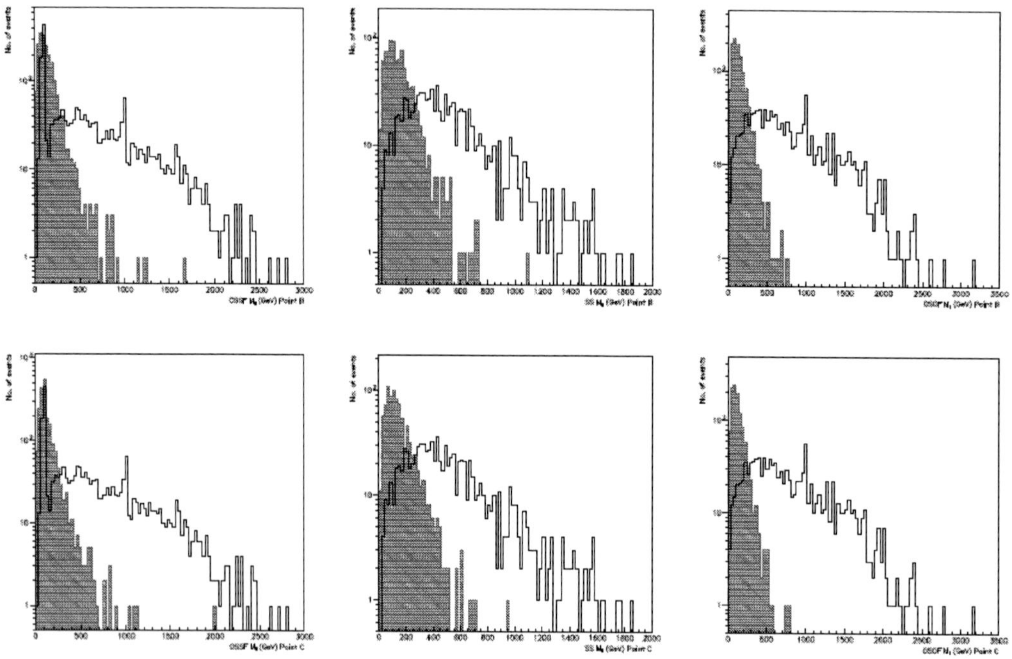

Figure 9. Invariant mass distributions at Point B (top panels) and Point C (bottom panels). OSSF (left panel), SS (middle panel) and OSOF (right panel) distributions clearly discriminate SUSY and BH events.

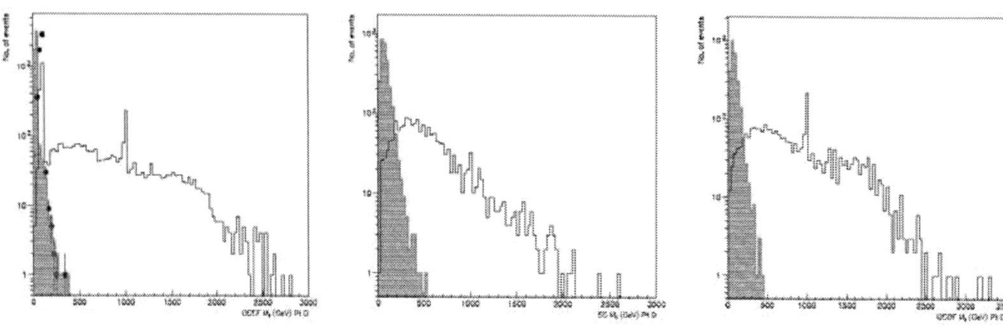

Figure 10. Left Panel: Distribution of 2000 SUSY and BH OSSF dilepton events for point D. P_T cut analysis is not considered at this point as explained in the text. The (blue) dots represent the SM background. Middle Panel: Event distribution for ~1700 SUSY and BH SS dilepton events. The SM background is negligible. Right Panel: SUSY and BH distribution for ~2600 OSOF events. The major SM background due to $\bar{t}t$ is negligible.

The dilepton distribution for point LM1 is shown in Fig. 12. The edge in the OSSF dilepton invariant mass distribution is not as evident as in point A because of the large amount of missing momentum which is carried away by the neutrinos in the τ decay. [38].

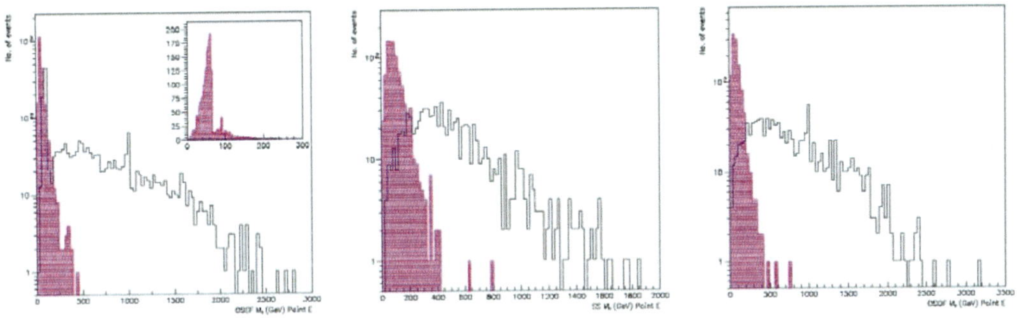

Figure 11. Left Panel: Invariant mass distribution for SUSY and BH OSSF dilepton events for point E. There is enhanced Z^0 production because of a high value of $\tan\beta$. Inset shows the invariant mass distribution OSSF leptons highlighting the Z^0 mass peak. The number of events is along the y axis and the OSSF invariant mass (in GeV) is along the x axis. Middle Panel: invariant mass distribution for same sign dileptons. Right Panel: OSOF invariant mass distribution

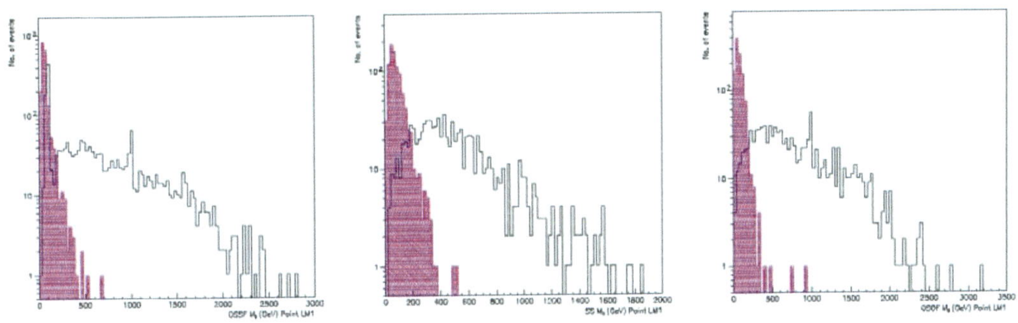

Figure 12. Left Panel: Invariant mass distribution for SUSY and BH OSSF dilepton events for the CMS benchmark point LM1. Likewise to point A, the SUSY distribution shows the endpoint due to the presence of the LSP. Middle Panel: invariant mass distribution for same sign dileptons. Right Panel: OSOF invariant mass distribution. A high value of $\tan\beta$ enhances the production of OSOF dileptons as explained in the text.

7. String and SM Event Analysis

String events are characterized by high values of visible energy and missing transverse momentum as the photon and the jet originate directly from the 2×2 interaction. Isolated photons provide a further means to extract string signals; isolated photons from string interactions, being directly produced from the string resonance, are harder than SM photons. Analysis of event-shape variables enhances the effectiveness of the above method.

Figure 13 shows the visible energy (left panel) and the transverse momentum of hadrons+photons (right panel) for 10 million string+SM and SM-only events. The visible energy and the transverse momentum are produced by the hard photons and the jets of the string decay. Their distributions are characterized by a long tail. The observation of events with transverse momentum (visible energy) in excess of 3 (6) TeV would provide strong evidence of the existence of a string resonance.

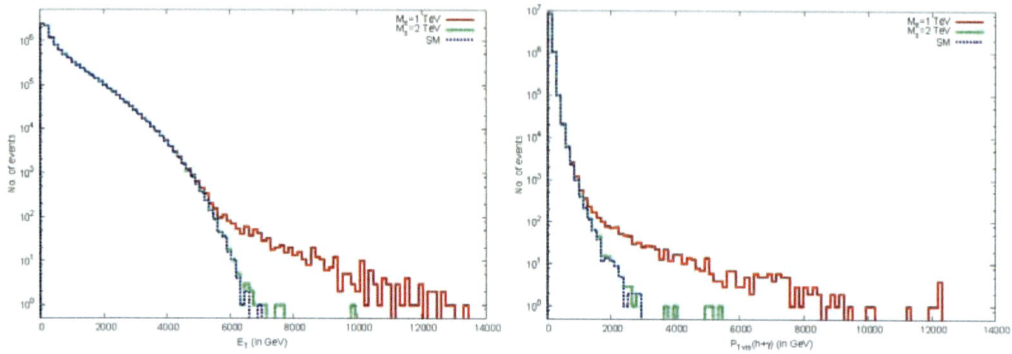

Figure 13. Left Panel: Visible energy distribution for string+SM and SM only events. The result for string events is shown by the solid red histogram (M_s =1 TeV) and the dashed green histogram (M_s =2 TeV). String events can be identified from the high-E_T tail for M_s =1 TeV. Right Panel: Distribution of visible P_T for γ +hadrons. The high-P_T tail is a strong indicator of the presence of string resonances.

In the analysis of the dijet invariant mass, the hadronic jets are selected according to the following criteria. The detector is assumed to have an absolute value of pseudorapidity $\eta = -\ln\{\tan(\frac{\theta}{2})\}$ =2.6. This ensures that the jets are originated in the hard 2×2 scattering rather than in multiple interactions or from the beam remnants. The contribution of jets that do not originate in the hard scattering are minimized by fixing the the minimum transverse energy of all particles comprising the jet ($\Sigma_i E_{T_i}$) to 40 GeV [60]. The particles of the jet must be within a cone of $R=\sqrt{(\Delta\eta^2+\Delta\phi^2)}$ = 0.2 from the jet initiator, where θ and ϕ are the azimuthal and polar angles of the particle $w.r.t$ the beam axis, respectively.

The invariant mass plot of the two highest P_T jets for each event is shown in the left panel of Fig. 14. Due to the nature of the interaction, the bulk of the events for both string+SM and SM-only events are comprised of dijets. These were selected using the same cuts as before. The jet invariant mass is

$$M_{j_1 j_2} = \sqrt{m_{j_1}^2 + m_{j_2}^2 + 2(E_{j_1}E_{j_2} - \overline{p}_{j_1}.\overline{p}_{j_2})}, \tag{17}$$

where $m_{j_{1(2)}}, E_{j_{1(2)}}$ *and* $\overline{p}_{j_{1(2)}}$ are the mass, energy and momenta of jet 1 (2), respectively. As expected, the SM invariant mass distribution is negligible beyond a certain threshold (~ 4 TeV). This is due to the production of direct soft photons and jets from the SM-only interaction. The string+SM distribution is characterized by a long tail of energy up to several

TeV (three times more than the SM). This tail is originated from the decay of string resonances into hard jets and photons. Therefore, the measure of a large invariant mass may provide strong evidence of a string mediated interaction.

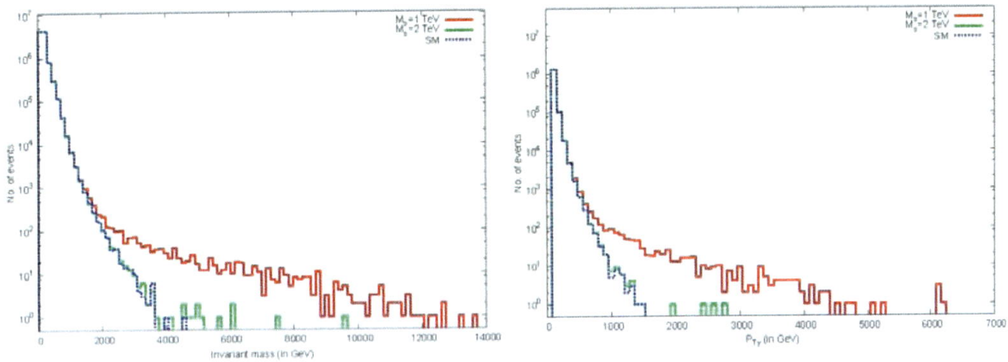

Figure 14. Left Panel: Dijet invariant mass distribution. String decays may result is large invariant mass. Right Panel: Distribution of the highest P_{T_γ} for each event. γ's with high P_T created in the string decay are the source of the long tail.

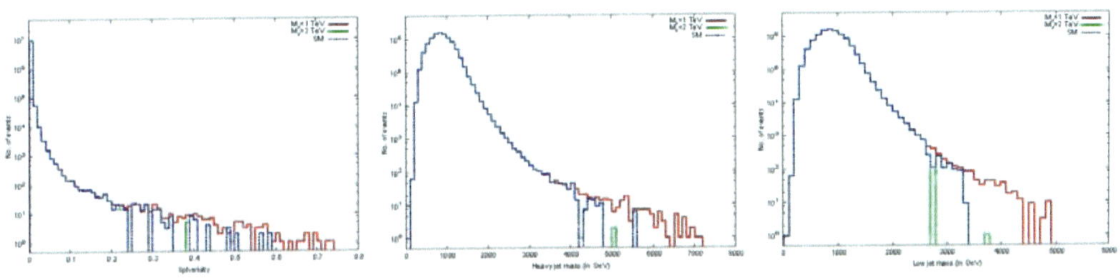

Figure 15. Histograms of event shape variables for 10 million string+SM and SM only events. String events are shown in solid red (M_s =1 TeV) and dashed green (M_s =2 TeV). SM events are shown in dotted blue. String events have on the average a slightly higher sphericity than SM events due to the slight increase in the number of jets (left panel). Similar conclusions are reached from the heavy and low jet mass distributions (middle and right panel, respectively).

The right panel of Fig. 14 shows the distribution of the highest P_{T_γ} of isolated photons for string+SM and SM-only events. Following Ref. [60], the cuts on the photon jets are $P_{T_\gamma} \geq 80$ GeV, $\eta < 2.6$ and an isolation cut $\Sigma_n P_T < 7$ GeV in a cone of $R = 0.4$. Single photons from string resonances are expected to have higher P_{T_γ} than SM photons because they are the direct products of the string decay. The main source of background for direct photons are jet fluctuations and photons originating from the initial and final state radiation [60]. In the former case, a jet consists of a few particles including high-P_T π_0's that decay into a pair of photons with a ~99% BR [60]. Due to the high boost, the photons have a relatively small angular separation and therefore "fake" a single photon in the electromagnetic calorimeter. The rate of this process is 1 out of $\approx 10^3$ *to* 10^4 events [60]. Simulations show that the maximum P_T of isolated π_0's is ~ 300 GeV. Other sources of fake photons are $H \rightarrow \gamma\gamma$ [61] or processes from other exotic phenomena, e.g. SUSY [62] or large extra dimensions [7]. Isolation cuts on the photon can effectively reduce the number of fake photons.

Histograms for different event shape variables are shown in Fig. 15. String events tend to be more spherical than the SM background: string+SM decays generally produce a distribution of high P_T jets at slightly higher values than the background. The jets originate from the decay of string resonances into photons and hadrons. The SM generates jets that are less heavier than their string counterparts (middle and right panels of Fig. 15).

8. String and BH Event Analysis

Searching for a Z^0 mass peak in the invariant mass of high-P_T leptons could be used to discriminate between BH and string events at the LHC. Z^0 production is highly suppressed in case of string events [19]. On the contrary BH decay is characterized by the production of a variety of particles with high transverse momentum. The rate of Z^0 and γ production is comparable (~2% to 3%). The Z^0 bosons decay into leptons with a 3.4% BR. The invariant mass distribution of BH events peaks at ~92 GeV, confirming the production of a Z^0 boson [26]. Therefore, the presence of a peak at ~92 GeV in the invariant mass of leptons would effectively rule out formation of string resonances. The above analysis is complemented by studying the particles produced in the BH decay. Since BH events produce uncorrelated particles, BH events are expected to have on average higher sphericity and a larger number of jets than string events; the $pp \rightarrow \gamma + jet$ channel of string events is dominated by dijets, whereas BH decay produces multijets.

9. Conclusions

We have discussed and compared SUSY and BH signatures at the LHC. A thorough analysis of event-shape variables and dilepton events has shown that it is possible to distinguish these two models at the ATLAS and CMS detectors. BH events are characterized by higher sphericity than SUSY processes. If a BH remnant is formed at the end of the evaporation phase, missing P_T and heavy jet mass are effective signatures to discriminate BH from SUSY events. Event-shape variables alone cannot unequivocally discriminate between SUSY and BHs but their knowledge may prove useful when combined with the analysis of the leptonic channel. Isolated dileptons could provide the "smoking gun" for detecting BHs at the LHC. The BH dilepton invariant mass shows a tail at high energy which is absent in the SM or

MSSM [4]. This analysis can be further strengthened by looking at the number and flavor of isolated leptons.

In Sec. 7, we have compared string resonance signals to the SM background. Our analysis has proven that string resonances could be detected when $M_s \approx$ 1 TeV. With the LHC operating at a luminosity of $2\times10^{33}cm^{-2}s^{-1}$, string scale of ~ 1 TeV could be detected within a year of the start of the LHC (assuming a signal-to-background ratio of 5). Additionally with the LHC luminosity upgrade to $10^{35}cm^{-2}s^{-1}$, string scale of the order of ~ 2 TeV could be detected within two years of running the LHC. These results are in agreement with [19, 21].

String events show higher sphericity and higher visible energy than the SM background. These quantities allow discrimination of string events from SM background when combined with the measure of the P_T of isolated photons and the dijet invariant mass. Since the final products of the string resonances are directly produced from the string decay, the dijet invariant mass is characterized by a tail at high energies which is absent in the SM.

Acknowledgments

The authors thank Luis A. Anchordoqui, Lucien Cremaldi, Alakabha Datta, Romulus Godang, David Sanders and Don Summers for fruitful discussions, Peter Skands for help with SLHA input file and Xerxes Tata for help on ISAJET. This work was supported (in part) by U.S. DoE contract DE-FG05-91ER40622.

References

[1] The CERN website, *www.cern.ch*

[2] J. Wess and B. Zumino, *Phys. Lett. B* **49**, 52 (1974); S. Weinberg, ``*The quantum theory of fields.* Vol. 3: Supersymmetry," *Cambridge, UK: Univ. Pr. (2000)* and references therein.

[3] H. Baer, C. h. Chen, F. Paige and X. Tata, *Phys. Rev. D* **52**, 2746 (1995) [arXiv:hep-ph/9503271]; H. Baer, C. h. Chen, F. Paige and X. Tata, *Phys. Rev. D* **53**, 6241 (1996) [arXiv:hep-ph/9512383]; F. E. Paige, *arXiv:*hep-ph/9801254.

[4] P. W. Higgs, *Phys. Lett.* 12, 132 (1964); F. Englert and R. Brout, *Phys. Rev. Lett.* **13**, 321 (1964); G. S. Guralnik, C. R. Hagen and T. W. B. Kibble, *Phys. Rev. Lett.* **13**, 585 (1964); P. W. Higgs, *Phys. Rev.* **145**, 1156 (1966); A. Djouadi, arXiv:hep-ph/0503172.

[5] ``*ATLAS detector and physics performance.* Technical design report. Vol. 2," A. Drozdetskiy [CMS and ATLAS Collaborations], ``The standard model Higgs boson at LHC: Recent developments," CERN CMS Note, CERN-CMS-CR-2007-022 (2007); D. Kcira, arXiv:0710.1957 [hep-ex].

[6] S. P. Martin, arXiv:hep-ph/9709356; M. E. Peskin, *arXiv*:0801.1928 [hep-ph].

[4] An important source of background is fake leptons that are caused by jets. However the study of these potential backgrounds and of the actual sensitivity of our method for specific experiments (CMS or ATLAS) is beyond the scope of this paper and will not be discussed here.

[7] N. Arkani-Hamed, S. Dimopoulos and G. R. Dvali, *Phys. Lett. B* **429**, 263 (1998) [arXiv:hep-ph/9803315]; I. Antoniadis, N. Arkani-Hamed, S. Dimopoulos and G. R. Dvali, *Phys. Lett. B* **436**, 257 (1998) [arXiv:hep-ph/9804398]; N. Arkani-Hamed, S. Dimopoulos and G. R. Dvali, *Phys. Rev. D* **59**, 086004 (1999) [arXiv:hep-ph/9807344].

[8] L. Randall and R. Sundrum, *Phys. Rev. Lett.* **83**, 3370 (1999) [arXiv:hep-ph/9905221]; L. Randall and R. Sundrum, *Phys. Rev. Lett.* **83**, 4690 (1999) [arXiv:hep-th/9906064].

[9] T. Appelquist, H. C. Cheng and B. A. Dobrescu, *Phys. Rev. D* **64**, 035002 (2001) [arXiv:hep-ph/0012100]; J. A. R. Cembranos, J. L. Feng and L. E. Strigari, *Phys. Rev. D* **75**, 036004 (2007) [arXiv:hep-ph/0612157].

[10] P. C. Argyres, S. Dimopoulos and J. March-Russell, *Phys. Lett. B* **441**, 96 (1998) [arXiv:hep-th/9808138]; T. Banks and W. Fischler, arXiv:hep-th/9906038; S. Dimopoulos and G. Landsberg, *Phys. Rev. Lett.* **87**, 161602 (2001) [arXiv:hep-ph/0106295]; S. B. Giddings and S. D. Thomas, *Phys. Rev. D* **65**, 056010 (2002) [arXiv:hep-ph/0106219]; E. J. Ahn, M. Cavaglià and A. V. Olinto, *Phys. Lett. B* **551**, 1 (2003) [arXiv:hep-th/0201042].

[11] E. J. Ahn and M. Cavaglià, *Gen. Rel. Grav.* **34**, 2037 (2002) [arXiv:hep-ph/0205168]; V. P. Frolov and D. Stojkovic, *Phys. Rev. Lett.* **89**, 151302 (2002) [arXiv:hep-th/0208102]; V. P. Frolov and D. Stojkovic, *Phys. Rev. D* **66**, 084002 (2002) [arXiv:hep-th/0206046]; K. m. Cheung, *Phys. Rev. Lett.* **88**, 221602 (2002) [arXiv:hep-ph/0110163], M. Cavaglià, S. Das and R. Maartens, *Class. Quant. Grav.* **20**, L205 (2003) [arXiv:hep-ph/0305223]; A. Chamblin, F. Cooper and G. C. Nayak, *Phys. Rev. D* **69**, 065010 (2004) [arXiv:hep-ph/0301239]; M. Cavaglià and S. Das, Class. Quant. Grav. 21, 4511 (2004) [arXiv:hep-th/0404050]; A. Chamblin, F. Cooper and G. C. Nayak, *Phys. Rev. D* **70**, 075018 (2004) [arXiv:hep-ph/0405054]; C. M. Harris, M. J. Palmer, M. A. Parker, P. Richardson, A. Sabetfakhri and B. R. Webber, *JHEP* **0505**, 053 (2005) [arXiv:hep-ph/0411022]; J. Tanaka, T. Yamamura, S. Asai and J. Kanzaki, Eur. Phys. J. C 41, 19 (2005) [arXiv:hep-ph/0411095]; B. Webber, *In the Proceedings of 33rd SLAC Summer Institute on Particle Physics (SSI 2005): Gravity in the Quantum World and the Cosmos, Menlo Park, California, 25 Jul - 5 Aug 2005, pp T030* [arXiv:hep-ph/0511128]; L. Lonnblad, M. Sjodahl and T. Akesson, *JHEP* 0509, 019 (200*5) [arXiv:hep-ph/0505181]; J. L. Hewett, B. Lillie and T. G. Rizzo,* Phys. Rev. Lett. 95, 261603 (2005) [arXiv:hep-ph/0503178]; G. C. Nayak and J. Smith, *Phys. Rev. D* 74, 014007 (2006) [arXiv:hep-ph/0602129]; H. Stoecker, *Int. J. Mod. Phys.* D 16, 185 (2007) [arXiv:hep-ph/0605062]; G. L. Alberghi, R. Casadio, D. Galli, D. Gregori, A. Tronconi and V. Vagnoni, arXiv:hep-ph/0601243; B. Koch, M. Bleicher and H. Stoecker, *J. Phys. G* **34**, S535 (2007) [arXiv:hep-ph/0702187]; D. M. Gingrich, *JHEP* **0711**, 064 (2007) [arXiv:0706.0623 [hep-ph]].

[12] J. L. Feng and A. D. Shapere, *Phys. Rev. Lett.* **88**, 021303 (2002) [arXiv:hep-ph/0109106]; A. Ringwald and H. Tu, *Phys. Lett. B* **525**, 135 (2002) [arXiv:hep-ph/0111042]; L. A. Anchordoqui, J. L. Feng, H. Goldberg and A. D. Shapere, *Phys. Rev. D* **65**, 124027 (2002) [arXiv:hep-ph/0112247]; E. J. Ahn, M. Ave, M. Cavaglià and A. V. Olinto, *Phys. Rev. D* **68**, 043004 (2003) [arXiv:hep-ph/0306008]; J. I. Illana, M. Masip and D. Meloni, Phys. Rev. D 72, 024003 (2005) [arXiv:hep-ph/0504234]; V. Cardoso, M. C. Espirito Santo, M. Paulos, M. Pimenta and B. Tome, *Astropart. Phys.* **22**, 399 (2005) [arXiv:hep-ph/0405056]; E. J. Ahn, M. Cavaglià and A. V. Olinto, *Astropart. Phys.* **22**, 377 (2005) [arXiv:hep-ph/0312249]; A. Cafarella, C. Coriano and

T. N. Tomaras, *JHEP* **0506**, 065 (2005) [arXiv:hep-ph/0410358]; E. J. Ahn and M. Cavaglià, *Phys. Rev. D* **73**, 042002 (2006) [arXiv:hep-ph/0511159]; M. Cavaglià and A. Roy, *Phys. Rev. D* **76**, 044005 (2007) [arXiv:0707.0274 [hep-ph]].

[13] G. F. Giudice, R. Rattazzi and J. D. Wells, *Nucl. Phys. B* **544**, 3 (1999) [arXiv:hep-ph/9811291]; E. A. Mirabelli, M. Perelstein and M. E. Peskin, *Phys. Rev. Lett.* **82**, 2236 (1999) [arXiv:hep-ph/9811337]; S. Cullen and M. Perelstein, *Phys. Rev. Lett.* **83**, 268 (1999) [arXiv:hep-ph/9903422].

[14] M. Cavaglià, *Int. J. Mod. Phys. A* **18**, 1843 (2003) [arXiv:hep-ph/0210296]; R. Emparan, *Black hole production at a TeV*, arXiv:hep-ph/0302226; S. Hossenfelder, *What black holes can teach us*, arXiv:hep-ph/0412265; P. Kanti, *Int. J. Mod. Phys. A* **19**, 4899 (2004) [arXiv:hep-ph/0402168]; G. Landsberg, *J. Phys. G* **32**, R337 (2006) [arXiv:hep-ph/0607297].

[15] J. Polchinski, *Cambridge, UK: Univ. Pr. (1998) 402 p*; J. Polchinski, Cambridge, UK: Univ. Pr. (1998) 531 p

[16] B. Zwiebach, Cambridge, UK: Univ. Pr. (2004) 558 p

[17] K. R. Dienes, *Phys. Rept.* **287**, 447 (1997) [arXiv:hep-th/9602045].

[18] J. D. Lykken, *Phys. Rev. D* **54**, 3693 (1996) [arXiv:hep-th/9603133], D. Chialva, R. Iengo and J. G. Russo, *Phys. Rev. D* **71**, 106009 (2005) [arXiv:hep-ph/0503125].

[19] L. A. Anchordoqui, H. Goldberg, S. Nawata and T. R. Taylor, *Phys. Rev. D* **78**, 016005 (2008) [arXiv:0804.2013 [hep-ph]].

[20] P. Burikham, T. Figy and T. Han, *Phys. Rev. D 71*, 016005 (2005) [Erratum-ibid. D 71, 019905 (2005)] [arXiv:hep-ph/0411094].

[21] L. A. Anchordoqui, H. Goldberg, D. Lust, S. Nawata, S. Stieberger and T. R. Taylor, *arXiv*:0808.0497 [hep-ph].

[22] ATLAS Collaboration, W.W. Armstrong *et al.*, Technical Proposal, *CERN/LHCC* 94-43, LHCC/P2, 15, December 1994.

[23] CMS Collaboration, G.L. Bayatian *et al.*, Technical Proposal, *CERN/LHCC* 94-38, LHCC/P1, 15, December 1994.

[24] T. G. Rizzo, *Phys. Rev. D* **64**, 095010 (2001) [arXiv:hep-ph/0106336]; C. Macesanu, C. D. McMullen and S. Nandi, *Phys. Rev. D* **66**, 015009 (2002) [arXiv:hep-ph/0201300]; A. Datta, K. Kong and K. T. Matchev, *Phys. Rev. D* **72**, 096006 (2005) [Erratum-ibid. D 72, 119901 (2005)] [arXiv:hep-ph/0509246]; M. Battaglia, A. K. Datta, A. De Roeck, K. Kong and K. T. Matchev, *In the Proceedings of 2005 International Linear Collider Workshop (LCWS 2005), Stanford, California, 18-22 Mar 2005, pp 0302* [arXiv:hep-ph/0507284]; M. Battaglia, A. Datta, A. De Roeck, K. Kong and K. T. Matchev, *JHEP* **0507**, 033 (2005) [arXiv:hep-ph/0502041]; P. Konar and P. Roy, *Phys. Lett. B* **634**, 295 (2006) [arXiv:hep-ph/0509161].

[25] A. Roy and M. Cavaglià, *Mod. Phys. Lett. A* **23**, 2987 (2008) [arXiv:0710.5490 [hep-ph]].

[26] A. Roy and M. Cavaglià, *Phys. Rev. D* **77**, 064029 (2008) [arXiv:0801.3281 [hep-ph]].

[27] D. M. Gingrich and K. Martell, *Phys. Rev. D* **78**, 115009 (2008) [arXiv:0808.2512 [hep-ph]].

[28] A. Roy and M. Cavaglià, *arXiv*:0902.1473 [hep-ph].

[29] A. Bartl *et al.*, *In the Proceedings of 1996 DPF / DPB Summer Study on New Directions for High-Energy Physics (Snowmass 96), Snowmass, Colorado, 25 Jun - 12 Jul 1996, pp SUP112.*

[30] M. Dittmar, *arXiv:hep-ex*/9901004.

[31] L. S. Stark, P. Hafliger, A. Biland and F. Pauss, *JHEP* **0508**, 059 (2005) [arXiv:hep-ph/0502197].

[32] G. Polesello, L. Poggioli, E. Richter-Was, J. Soderqvist, Precision SUSY measurements with ATLAS for SUGRA point 5, ATLAS Internal Note, *PHYS-No-111*, October 1997.

[33] I. Hinchliffe, F. E. Paige, M. D. Shapiro, J. Soderqvist and W. Yao, *Phys. Rev.* D 55, 5520 (1997) [arXiv:hep-ph/9610544].

[34] B. C. Allanach *et al.*, in *Proc. of the APS/DPF/DPB Summer Study on the Future of Particle Physics (Snowmass 2001)* ed. N. Graf, *In the Proceedings of APS / DPF / DPB Summer Study on the Future of Particle Physics (Snowmass 2001), Snowmass, Colorado, 30 Jun - 21 Jul 2001, pp P125* [arXiv:hep-ph/0202233].

[35] G. L. Bayatian *et al.* [CMS Collaboration], *J. Phys. G* **34**, 995 (2007).

[36] H. Baer, C. h. Chen, M. Drees, F. Paige and X. Tata, *Phys. Rev. D* **59**, 055014 (1999) [arXiv:hep-ph/9809223].

[37] D. Denegri, W. Majerotto and L. Rurua, *Phys. Rev. D* **60**, 035008 (1999) [arXiv:hep-ph/9901231].

[38] Yu. Andreev, N. Krasnikov and A. Toropin, *arXiv*:0706.2578 [hep-ph].

[39] K.S. Thorne, in: *Magic without magic: John Archibald Wheeler*, edited by J. Klauder (Freeman, San Francisco, 1972).

[40] V. Cardoso, E. Berti and M. Cavaglià, *Class. Quant. Grav.* **22**, L61 (2005) [arXiv:hep-ph/0505125].

[41] P. C. Aichelburg and R. U. *Sexl, Gen. Rel. Grav.* **2**, 303 (1971).

[42] H. Yoshino and Y. Nambu, Phys. Rev. *D* **67**, 024009 (2003) [arXiv:gr-qc/0209003]; H. Yoshino and V. S. Rychkov, *Phys. Rev. D* **71**, 104028 (2005) [arXiv:hep-th/0503171].

[43] E. Berti, M. Cavaglià and L. Gualtieri, *Phys. Rev. D* **69**, 124011 (2004) [arXiv:hep-th/0309203].

[44] S. W. Hawking, *Commun. Math. Phys.* **43**, 199 (1975) [Erratum-ibid. 46, 206 (1976)].

[45] R. Emparan, G. T. Horowitz and R. C. Myers, *Phys. Rev. Lett.* **85**, 499 (2000) [arXiv:hep-th/0003118]; M. Cavaglià, *Phys. Lett. B* **569**, 7 (2003) [arXiv:hep-ph/0305256].

[46] P. Kanti and J. March-Russell, *Phys. Rev. D* **66**, 024023 (2002) [arXiv:hep-ph/0203223]; P. Kanti and J. March-Russell, *Phys. Rev. D* **67**, 104019 (2003) [arXiv:hep-ph/0212199]; C. M. Harris and P. Kanti, *JHEP* **0310**, 014 (2003) [arXiv:hep-ph/0309054]; D. Ida, K. y. Oda and S. C. Park, *Phys. Rev. D* **67**, 064025 (2003) [Erratum-ibid. D 69, 049901 (2004)] [arXiv:hep-th/0212108]; A. S. Cornell, W. Naylor and M. Sasaki, *JHEP* 0602, 012 (2006) [*arXiv:hep*-th/0510009].

[47] V. Cardoso, M. Cavaglià and L. Gualtieri, *JHEP* **0602**, 021 (2006) [arXiv:hep-th/0512116]; V. Cardoso, M. Cavaglià and L. Gualtieri, *Phys. Rev. Lett.* **96**, 071301 (2006) [Erratum-ibid. 96, 219902 (2006)] *[arXiv:hep*-th/0512002].

[48] M. Cavaglià, R. Godang, L. Cremaldi and D. Summers, *Comput. Phys. Commun.* **177**, 506 (2007) [arXiv:hep-ph/0609001]; M. Cavaglià, R. Godang, L. M. Cremaldi and D. J. Summers, *JHEP* **0706**, 055 (2007) *[arXiv*:0707.0317 [hep-ph]].

[49] J. Pumplin, D. R. Stump, J. Huston, H. L. Lai, P. M. Nadolsky and W. K. Tung, JHEP 0207, 012 (2002) *[arXiv:hep*-ph/0201195].

[50] F. E. Paige, S. D. Protopopescu, H. Baer and X. Tata, *arXiv:hep*-ph/0312045.

[51] T. Sjostrand, S. Mrenna and P. Skands, JHEP 0605, 026 (2006) [arXiv:hep-ph/0603175]; See also: *http://projects.hepforge.org/pythia6.*

[52] B. Koch, M. Bleicher and S. Hossenfelder, *JHEP* **0510**, 053 (2005) [arXiv:hep-ph/0507138].

[53] P. Meade and L. Randall, *arXiv*:0708.3017 [hep-ph].

[54] Abdullin and F. Charles, *Nucl. Phys. B* **547**, 60 (1999) [arXiv:hep-ph/9811402]; M. Chiorboli, M. Galanti and A. Tricomi, *Acta Phys. Polon. B* **38**, 559 (2007).

[55] B. Abbott *et al.* [D0 Collaboration], *Phys. Rev. Lett.* **80**, 2063 (1998) [arXiv:hep-ex/9706014].

[56] http://pdg.lbl.gov/

[57] H. Baer, X. Tata and J. Woodside, *Phys. Rev. D* **41**, 906 (1990).

[58] H. Baer, X. Tata and J. Woodside, *Phys. Rev. D* **45**, 142 (1992).

[59] H. Baer *et al.*, *arXiv:hep*-ph/9503479.

[60] P. Gupta, B. C. Choudhary, S. Chatterji, S. Bhattacharya and R. K. Shivpuri, *arXiv*:0705.2740 [hep-ex].

[61] M. Pieri, S. Bhattacharya, I. Fisk, J. Letts, V. Litvin and J. G. Branson, *CERN-CMS-NOTE*-2006-112.

[62] A. Abulencia *et al.* [CDF Collaboration], *Phys. Rev. Lett.* **99**, 121801 (2007) [arXiv:0704.0760 [hep-ex]].

In: Black Holes and Galaxy Formation
Editors: A.D. Wachter and R.J. Propst, pp. 149-162
ISBN: 978-1-60741-703-3

Chapter 5

PRIMORDIAL BLACK HOLES, FORMATION AND EVOLUTION

Shawqi Al Dallal
University of Bahrain, Department of Physics,
P. O. Box 32038, Kingdom of BAHRAIN

Abstract

Research work conducted in the early 1970s have shown that density fluctuation in the early universe leads inevitably, under certain circumstances, to the formation of primordial black holes (PBH). The key idea is that a black hole is formed in overdense regions when the primordial fluctuations cause the gravitational energy to considerably exceed the kinetic energy of the expanding region. Alternative theories of the formation of primordial black holes assume that this process occurs during the inflation epoch, when the universe experienced a very short period of very rapid expansion.

A review of primordial black hole formation, evolution, detection, and identification from gamma ray burst signature is discussed.

Introduction

Theoretical investigations of the conditions prevailing in the early universe have shown the possibility of formation of primordial black holes (PBH) with a wide spectrum of masses. This is a direct consequence of the extreme densities associated with the big bang. If a portion of the early universe is sufficiently dense and inhomogeneous then it could collapse to form a black hole (Hawking, 1971). It was demonstrated by Choptuik (1993) and more recently by Kim (2000) that PBHs could have been formed in an inflationary universe by near- critical collapse in the expanding universe. During this epoch, the energy density of the universe underwent a dramatic decrease leading to a cosmological phase transition.

Very small black holes can also be formed by softening the equation of state (Canuto 1978), phase transition (Hawking 1982), or from collapse of cosmic strings (Hawking 1989).

In the first part of this review paper we discuss the various models and processes giving rise to the formation of primordial black holes with emphasis on the PBH mass problem, the equation of state, and gravitationally bound states called Holeums.

In the second part of this paper we address the mechanism of evaporation of PBHs by Hawking radiation, the controversies associated with this model, and the possible identification of the thermal nature of this radiation for micro black holes created in the Large Hadron Collider (LHC). It is shown that the evaporation mechanism is an essential ingredient for identifying the signature of gamma ray bursts originating from PBHs having masses corresponding to the present age of the universe.

In the last part of this paper we address the observational data related to PBHs abundance as derived from Galactic gamma ray background and entropy arguments.

Formation of Primordial Black Holes

Black holes may have formed at the early universe because of initial inhomogenities as pointed out by Hawking (1971). He argues that PBHs are formed in a wide spectrum of masses ranging from 10^{-5}g, corresponding to the Planck mass, to 10^{17} solar masses. We introduce in the following the argument presented by Hawking (1974) to set the conditions for the formation of primordial black holes. He assumed that certain overdense regions in the early universe may collapse to a black hole if the gravitational attraction overcomes the pressure forces and the velocity of expansion. This condition is realized when the potential energy of self–gravitation

$$\Omega \sim -\mu^2 R^5 \tag{1}$$

exceeds the kinetic energy of expansion

$$T \sim \mu R^3 \dot{R}^2 \tag{2}$$

Here R is the radius of a region in the early universe, and μ is the energy density. The units are such that $G = c = 1$. In a $k = 0$ Friedmann universe the sum of these energies is zero. Hence

$$\left(\frac{\dot{R}}{R}\right)^2 \sim \mu \tag{3}$$

Hawking assumes further that in the radiation epoch, when the mass of the particles are relativistic, the equation of state relating the pressure p and the energy density μ is: $p = 1/3\,\mu$, and μ is proportional to R^{-4}, Thus

$$\mu \sim t^{-2}, \qquad R \propto t^{1/2} \tag{4}$$

A collapse happens when the pressure forces requires that the gravitational energy, $-\Omega$, exceeds the internal energy U. When $p = 1/3\,\mu$, $U \sim \mu R^3$. Hence, a necessary condition for collapse to occur is

$$\mu R^2 > \sim 1 \tag{5}$$

And when $p \sim \mu_o \log \mu/\mu_o$, $U \sim \mu_o R^3 \log \mu/\mu_o$ m the necessary condition for collapse becomes

$$\mu R^2 > \sim \frac{\mu}{\mu_o} \log \frac{\mu}{\mu_o} \tag{6}$$

Whenever a black hole is formed in the early universe one would expect that it will grow by accreting some of the nearby matter. An estimate of the rate of accretion was first made by Zeldovitch and Navikov (1967) as

$$\frac{dM}{dt} \sim \mu R_g^{\ 2} \sim \mu M^2 \tag{7}$$

Where μ is the density of the background universe. But $\mu \sim t^{-2}$, hence

$$M \sim \frac{t}{1 + \frac{t}{t_o}\left(\frac{t_o}{M_o} - 1\right)} \tag{8}$$

Where M_o is the initial mass of the black hole and t_o is the time of formation. Thus, if M_o were small compared to t_o, i.e. if the black hole was small compared to the particle horizon, the $M - M_o$ remains small and there would be no much accretion. When M_o were of the same order as t_o, the Zeldovitch- Navikov argument indicates $M \sim$ t. In this case, the accretion would cause the black hole to grow at the same rate as the particle horizon, producing black holes of the order of Hubble radius if the growth continued to the present time, or it would reach a mass of 10^{15} to 10^{17} solar masses if its growth is at the same rate as the particle horizon. The Zeldovitch- Navikov argument is based on the assumption of quasi-stationary accretion, and is probably valid when the black hole is small compared to the particle horizon (Hawking 1974). Nevertheless, this assumption is no more valid in the critical case where the size of the black hole is of the order of the particle horizon, since in this case the expansion of the universe has to be taken into account.

Carr and Rees (1977) argue that if the universe started off with a photon-to-baryon ratio much less than that presently observed, massive black holes would have formed at early times even if the initial density fluctuations were very small. Those holes could have generated the rest of the back-ground radiation through accretion. In this way the universe might automatically evolve to have the photon to baryon ratio observed today. It could also produce galaxies with black hole "halos". This idea is in good agreement with the analysis we provide in the section on gamma ray burst.

Mass Spectrum of PBH

The equation of state is a key element in determining the conditions leading to the formation of PBHs and their characteristics. At times $t < 10^{-4}$ s after the big bang, the temperature was extremely high $(T \sim 10^{12}\ \mathrm{K})$, implying uncertainty about the equation of state that could range from the extremely soft (Franutschi, 1971) to the maximally stiff (Lin et al. 1976). This situation renders cosmological models to be highly speculative. Furthermore, our ignorance of the exact spectrum of fluctuations provides additional uncertainty in guesses as to the nature of the mass spectrum of PBHs. This idea was further discussed by Carr (1975). He examined the mass spectrum of PBHs in the early universe dominated by small density fluctuations superposed on a Friedmann background. He showed that only a certain type of fluctuation favors the formation of PBHs and that, consequentlym their spectrum should always have a particular form. Some authors set a lower mass limit of PBHs to be about one solar mass (Meszaros, 1974; Carr , 1977). On the other hand there was an initial worry that PBHs might grow rapaciously during the radiation dominated era, producing black holes of about 10^{16} solar masses by the time of the recombination era when the temperature dropped to about 4000 K (Zel'dovich and Mavikov, 1967). The stiff equation of state leads to a rapid growth of the black holes, and thus keeping pace with the horizon (Lin et al. 1976). Carr and Hawking (1974), have shown, however, that a hole is only likely to double its mass after the time when it comes within its particle horizon, and thus producing a wide spectrum of hole masses. Navikov and Thorne (1973) have demonstrated that significant formation of PBHs in a hot universe characterized by a red shift $z_f >> 10^{10}$ is somewhat unlikely. This problem can be avoided by postulating a cold chaotic universe in which the background radiation is created at times when $z_f << 10^{10}$ (Blandford and Thorne, 1979) .

A careful analysis by Carr and Hawking (1974) shows that, for $p = 1/3\,\mu$, there is no solution to the Einstein's equations in which a black hole is formed by purely local processes that can grow as fast as the particle horizon. Even though Carr and Hawking (1974) considered the case where $p = 1/3\,\mu$, the same argument applies for any equation of state of the form $p = \alpha\mu$ with α positive and less than 1.

The stiff models of the early universe were investigated by Lin et al. (1976). This is essentially determined by the strong interaction physics. One could expect the universe to be stiff after 10^{-4} s. The density at this time is less than the nuclear density, so strong interaction is unimportant and $\alpha \leq 1/3$. Nevertheless, before 10^{-4} s nucleons might develop a strong repulsion and in this case α is near 1. A model of this type was first suggested by Zel'dovich (1962). He found an equation of state of the form

$$p = 2\pi g^2 n^2 \lambda^2 \tag{9}$$

where $g^2/\hbar c$ is the coupling constant, $\lambda = \hbar/m_\pi c$, m_π is the meson mass, and n is the nucleon number. Subsequently, Zel'dovich (1972) used an equation of state having the form $p = \mu$ to explain how the amplitude of the universe's initial density fluctuation could explain both the entropy of the universe and the clustering of galaxies. Waleka (1974) and Canuto (1975) predict also a stiff equation of state using different approaches. Carr (1975) has shown that the density fluctuation in regions of average mass m at some initial time t_o has the form $\Delta m/m = \varepsilon(m/m_o)^{-n}$, where m_o is the initial horizon mass and ε and n are constants. For the above form of density fluctuation he has shown that only $n = 2/3$ favors the formation of a primordial black hole over an extended mass range. For a Gaussian distribution for the density fluctuations, the probability that a given region evolves into a black hole is (Carr, 1975)

$$P \approx \varepsilon \exp\left(\frac{-\alpha^2}{2\varepsilon^2}\right) \tag{10}$$

Thus, for α of order 1, and ε is much less than 0.1, a black hole is unlikely to be formed. Lin et al. (1976) have shown that if the equation of state is "stiff" ($p = \mu$) the Einstein equations permit a spherically symmertric solution in which pressure gradients cause a black hole to grow as fast as the universe. Carr (1977) argue that only hot models of the early universe are capable of producing PBHs too prolifically to be consistent with observation. However, if the early universe was cold (photonless), one could expect black holes to form prolifically without contravening observation. Carter et al. (1976) have shown that if high temperature matter is described by a relatively hard equation of state with an adiabatic index Γ greater than 6/5, then interactions between particles can probably be ignored so that the rate of creation will continue to be describable by Hawking's method above as below the threshold energy at which hadron creation sets in. On the other hand for softer equations of state the created matter will almost certainly be highly opaque and a hydrodynamic model of the emission process will be more appropriate.

Raf Guedens et al. (2002) considered the population evolution and evaporation of PBHs in the simplest braneworld cosmology: Randall-Sundrum type II. They demonstrate that black holes forming during the high energy phase of this theory (where the expansion rate is proportional to the density) have modified evaporation law, resulting in a longer time and lower temperature at evaporation, while those forming in the standard regime behave essentially as in the standard cosmology.

Harada (2006) considered the growth of super-horizon PBHs. He assumed that PBHs to have formed with the mass scale $M_{h,f} \simeq G^{-1}c^3 t_f$ which was contained within the Hubble horizon at the formation epoch, where G, c and t_f are the gravitational constant, the light speed, and the formation time from the big bang, respectively. The typical mass scale of the PBHs is the horizon mass scale at the formation epoch,

$$M_{PBH,f} \simeq M_{h,f} \simeq \frac{c^3 t_f}{G} \simeq 1 M_\odot \left(\frac{T_f}{100\, MeV} \right)^{-2} \tag{11}$$

The mass accretion rate of the black hole was estimated as

$$\frac{dM}{dt} = 4\pi\, \alpha\, r_A\, \upsilon_s\, \rho_\infty \tag{12}$$

Where $r_A = GM / \upsilon_s^{\,2}$ is the accretion radius, υ_s is the sound speed, ρ_∞ is the density at infinity, and α is a constant of order unity. Harada (2006) assumed that ρ_∞ is given by the density of the background Friedmann universe and υ_s is the order of the speed of light. Then the above equation is integrated to give (Harada and Carr, 2005)

$$M = \frac{At}{1 + \frac{t}{t_f}\left(\frac{At_f}{M_f} - 1 \right)} \tag{13}$$

where $A \simeq c^3 / G$ is a constant and M_f is the PBH mass at t_f of formation. Three categories of solutions, sub-horizon, self-similar and super-horizon are given by the author. In this argument, the effects of cosmological expansion are neglected, since they are important only for the cosmological horizon scale, whereas the above analysis is expected to be valid only for PBHs much smaller than the cosmological horizon scale.

Hawking Evaporation of Black Holes

In classical theory black holes can only absorb and not emit particles. However, Hawking (1975) has shown that particles can be created from the vacuum by an external field. Quantum gravitation effects are usually neglected in calculations of the formation and evolution of black holes. This is justified by the fact that the radius of curvature of spacetime outside the event horizon is very large compared to the Planck length $(G\hbar / c^3)^{1/2} \approx 10^{-33}$ cm, the length scale on which quantum fluctuations of the metric are expected to be of order unity (Hawking 1974b). This implies that the energy density of particles created by the gravitational field is small compared to the space time curvature. Even though quantum effects may be small locally, they may still, however, add up to produce significant effect over the lifetime of the Universe (10^{17}s) which is very long as compared to the Planck time (10^{-43}s). Using the usual quantum mechanical wave equation for a collapsing object with a post collapse classical wave metric, Hawking derived an expression of the emission spectrum that mimics a Planck law. For a particle of energy Q per unit of time *t*, the spectrum for each degree of freedom is

$$\frac{d^2N}{dQ\,dt} = \frac{\Gamma_s}{h\left[\exp\left(\frac{Q}{h\kappa/4\pi^2 c}\right) - (-1)^{2s}\right]} \tag{14}$$

where κ is the surface gravity, s is the spin of the emitted species and Γ_s is the absorption probability. In the above expression the contribution of angular velocity and electric potential have been neglected, since the black hole discharges and finishes its rotation much faster than it evaporates (Page 1977). Zaumen (1974), Gibbon (1975), and Carter (1974) have shown that a small black hole will quickly lose most of its charge. However, Page (1976) noted that there should remain random charge fluctuations of order $(\hbar c)^{1/2} = 11.7\,e$, where e is the positron charge. In his derivation, Hawking (1975) assumes that the matter fields, such as scalar, electro-magnetic, or neutrino fields, obey the usual wave equations with the Minkowski metric replaced by a classical space-time metric g_{ab}. He shows that quantum mechanical effects cause black holes to create and emit particles as if they were hot bodies of temperature

$$T = \frac{hc^3}{16\pi k G M} \approx 10^{-6}\left(\frac{M_\odot}{M}\right) K \tag{15}$$

The Hawking temperature T is inversely proportional to the black hole mass M. Thus as the black black hole radiates its temperature increases. In addition, since the surface gravity is inversely proportional to the black hole mass M, and the emitting area is proportional to M^2, the luminosity or the total emitted power is proportional to AT^4 or M^{-2}. As M decreases at this rate, the black hole lifetime will be proportional to M^3. Dimensional arguments indicate that the lifetime will be less than the age of the universe only if $M \leq 10^{15}$ g (Page 1976). Hence, the thermal emission is insignificant for black holes formed by stellar collapse where the life time exceeds about 10^{66} years for a black hole of several solar masses. However it is crucially important for black holes formed by fluctuations in the early universe. Page (1976b) has shown that the power emitted by a black hole increases as a function of its angular momentum, for a given mass, by factors of up to 13.35 for neutrinos, 107.5 for photons and 26380 for gravitons. Angular momentum is emitted several times faster than energy, so rapidly that a black hole spins down to nearly a non rotating state before most of its mass has been given up. As the rotation decreases, the thermal emission becomes dominant, drawing heat out of the black hole and decreasing its area. He showed that a non rotating primordial black hole with initial mass 5×10^{14} g would have just decayed away within the present age of the universe, a black hole created maximally rotating would have just died if its initial mass is 7×10^{14}. Primordial black holes created with larger masses would have a maximum rotation rate determined uniquely by the present mass.

In recent years the case of Hawking radiation becomes a source of controversy. Some authors found that Hawking radiation is so compelling that we should grade the possibility of theories of quantum gravity according to whether they reproduce this radiation (Carlip 2001). Others found that there are problems that can probably or hopefully be fixed (Wald 2001,

Parentani 2001). Gibbons (1977) found that the prediction depends on speculations about ultr-high energy physics, whereas Visser (2001) found that this is not the case. Berkenstein and Mukhanov (1995) and Ashtekar (1998) argue that quantum gravitational effects could completely alter or even obviate the thermal radiation. A review of the black radiation problem is given by Hefler (2003) The controversy of the Hawking evaporation of black holes can most likely be solved by detecting and identifying signatures arising from this process, as reviewed in the next section.

Recently, Kanti (2004) reviewed the existing literature on the creation of black holes during high energy particle collisions, both in the absence and presence of extra, compact spacelike dimensions. He then discussed the produced higher-dimensional black holes, including the physics that governs their evaporation, through the emission of Hawking radiation. Dimopoulos et al (2001) argue that at the scale of quantum gravity near a TeV, the LHC will be producing one black hole (BH) about every second. The BH decays into prompt, hard photons and charged leptons. The decay of the BH is governed by the higher dimensional Hawking temperature T_H law, which is proportional to the inverse radius, and is given by (Myers and Perry 1986):

$$T_H = M_p \left(\frac{M_P}{M_{BH}} \frac{n+2}{8\Gamma\left(n+3/2\right)} \right)^{1/(n+1)} \frac{n+1}{4\sqrt{\pi}} \tag{16}$$

Where M_P is the Planck mass, M_{BH} is mass of the created black hole, and n is the number of extra dimensions. The correlation between the black hole mass and its temperature, deduced from the energy spectrum of the decay products, can test experimentally the higher dimensional Hawking evaporation law.

This analysis provides undoubtedly a clue as to the nature of micro black holes formed at the Large Hadron Collider (LHC) and the mechanism of their evaporation.

Detection of Primordial Black Holes

Detection of PBHs was the subject of intensive research since the postulate of their existence in the early 1970s. The Hawking evaporation is the key process that allows a potential detection of PBHs. The evaporation is accompanied by a burst of emitted particles and γ-ray photons that provide a signature of their existence. In this section we review the main ideas proposed to identify this signature, and its effect on the observed universe.

The mass of PBHs evaporating at our epoch is of the order of $M \leq 10^{15}$ g (Hawking 1975; Page 1976). The evaporation process ends up with a burst of high energy particles and gamma rays (Vainer and Nasel'skii 1977). They were the first to consider the influence of evaporating PBHs on the cosmological synthesis of He^4. Page and Hawking (1976) examined the possibilities of detecting hard γ-rays produced by the quantum mechanical decay of PBHs. They argue that observations of the isotropic γ-ray background around 100 MeV place an upper limit of 10^4 pc^{-3} on the average number density of PBHs with initial masses around 10^{15} g. black holes of a mass of 5×10^{14} g would have a temperature of about $2.5\times10^{11}\mathrm{K}=20\,\mathrm{MeV}$. Page (1976) estimated the energy of radiated particles as

$$E=\frac{hc}{\lambda}\sim kT=\frac{\hbar c^{3}}{8\pi GM} \tag{17}$$

He indicates that such a black hole would radiate energy at a rate of $2.5\times10^{17}\,\mathrm{ergs\,s^{-1}}$ of which 1% is in gravitons, 45% is in neutrinos, 45% is in electrons and positrons, and 9% is in photons, in addition to some emission of muons and pions which is not included in the energy rate above. The number spectrum of the emitted photons would be peaked at 120 MeV, and could reach us from anywhere in the universe.

Zel'dovich et al (1977) analyzed the effect upon cosmological nucleosynthesis of high energy particles that would be emitted when PBHs of 10^{9} – 10^{12} g mass evaporation leading to what they call the deuterium problem. Rees (1977) pointed out that if a small black hole of mass $m\leq10^{11}\,\mathrm{kg}$ explodes impulsively as a consequence of radiating by Hawking (1974) process, then up to $\sim 50\%$ of the rest mass energy may eventually be transformed into electron-positron pairs. Blandford (1977) argue that if the black hole is surrounded by a typical interstellar magnetic field, the kinetic energy of the pairs may be efficiently transformed into radio waves. Cosmic rays from PBHs were investigated by MacGibbon and Carr (1991). They assumed that fundamental quarks, gluons, and leptons rather than the composite particles are emitted. The quark and gluon jets give rise to the stable photons, leptons and hadrons. Barrau et al. (2003) argue that the Hawking evaporation process could eventually lead to the formation of antideuterium nuclei that can be used as a probe of PBHs. Afshordi et al. (2003) considered the possibility that massive PBHs are the dominant form of dark matter. They found that black hole formation adds a Poisson noise to the matter density fluctuations. Recently, Ukwatta et al.(2009) propose spectral lag, which is the temporal delay between the high energy photon pulse and the low energy photon pulse, as a possible method to detect PBH evaporation events with the Fermi Gamma-ray Space Telescope observatory. Certainly, the above ideas and investigations about the nature of the emitted radiation from evaporating black holes constitute a wealth of information for scientists interested in observing this crucial cosmological phenomenon.

Gamma Ray Burst and PBHs

Gamma ray burst is one of the most important products of a potentially evaporated black hole that can be observed. The association of gamma ray bursts (GRBs) and primordial black holes (PBH) date back to the 1970's when the first GRB was discovered (Klebesadel, et al. 1973). However, this association was not clear primarily due to the lack of observational data. In 1994 Osborne et al. have established the existence of a diffuse component of gamma ray flux without providing a firm explanation of its origin. Dixon et al. (1998) have presented what they consider a quantitative statistical evidence for gamma-ray emission surrounding the Galactic halo. The flux level of this component is similar to the extragalactic diffuse flux.

Estimation of the Diffuse Galactic gamma ray flux is a powerful tool to investigate the possibility of PBHs evaporation in the Galactic halo. This flux depends primarily on the density and distribution of PBHs in the halo. In the past few decades investigations have been focused on the properties of black holes of mass $M_H \sim 10^{15} g$, which should evaporate by Hawking radiation in a time comparable with the present age of the universe. The precise value of M_H depends on the spin of the hole. The limits were given by Page (1976) as $5\times10^{14} \le M_H \le 7\times10^{14} g$. He also found that a significant fraction (≥ 0.2) of the evaporated luminosity appears in the form of γ-rays with a spectral distribution peaking at $\sim 100 MeV$. This value is in good agreement with that obtained more recently by Dixon et al. (1998).

The clustering of PBH around galaxies is considered by Page and Hawking (1976). The upper limit of the PBH density was computed to be $\bar{n}_H \le 10^4\, pc^{-3}$ for black holes of mass $\sim M_H$. This corresponds to a density parameter $\Omega_{PBH}(M_H) \le 10^{-8}$, where Ω_{PBH} is the ratio of their mean mass density to the critical density $\rho_c = 3H_o^2/8\pi G$ (H_o is the Hubble constant). It was inferred from the above limit that the closest individual source lies at a distance

$$10^{15}\left(\bar{n}_H/10^4\right)^{-1/3}\left(C/10^6\right)^{-1/3} cm \qquad (18)$$

where the clumping factor $C \le 10^6$. Therefore, the closest γ-ray source is at $10^{15} cm$, and delivers a flux of $\le 10^{-9}$ photons $cm^{-2}s^{-1}$, which is beyond the observational limit, unless a considerable amount of events occur simultaneously. The upper limit of permitted local density of heavier holes is obtained by requiring $\Omega_{PBH} \le 1$. Hence

$$n(M) \le 10^{11} C\left(M/10^{15} g\right)^{-1} pc^{-3} \qquad (19)$$

For $M \sim 10^{15} g$, the maximum allowed space density of PBHs is $n(M) \le 10^{11} C\, pc^{-3}$. As the random velocity of holes is unlikely to exceed 100 Km s^{-1}, the chances of hitting the Earth in this century are utterly negligible (Jackson and Ryan, 1973).

Evaporation of an isolated PBH occurs only for well specified mass corresponding to the actual age of the universe. Another possible scenario of interest is to consider gravitationally bound systems of PBHs which evolve by radiating gravitational waves. One possible effect of PBH binary system is an intensive particle creation and production of a fire ball due to the energy being released during the merging (Abdurashitov et al. 1999). If a significant fraction of all PBHs evolves in such a system, and if each system proceeds by contraction and coalescence to a single hole of mass $\sim M$ at some redshift $\sim z$, this gives rise to an isotropic gravitational-wave background of energy desntiy $\sim 0.1\Omega_{PBH} p_c c^2/(1+z)$, centered around a frequency $\sim c^3/[6\pi GM(1+z)]$, (Blandford and Thorne, 1979). Eichler and Solinger (1976) pointed out that if a comparable power is radiated into the electromagnetic background by these events, then it will be detected far more easily than a gravitational-wave background provided it has not been redshifted to frequencies $\leq 100MHz$. PBHs are considered as free moving dark matter component and therefore it is expected that they cluster in galaxies. Halzen et al. (1991) proposed a clustering factor as large as 10^7, giving densities of up to 10^{11} PBH pc^{-3} in the Galactic halo. An upper bound of 10^6 of the Galactic diffuse gamma-ray flux is adopted (MacGibbon and Carr 1991, Cline 1998).

Cline (1998) attributed some of the observed gamma ray flux to the evaporation of PBHs strongly clustered in the galactic halo. He showed also that the Galactic halo glow exhibits anisotropy in the flux. The proposed model relies on the observation by EGRET detector abroad the Compton Gamma Ray Observatory. Chavda and Chavda (2002) postulated the existence of gravitationally stable bound states called Holeums. These states were created and evolved in the early stages of the big bang. They are formed by a low mass PBHs in binary systems. Gamma rays may be produced by the ionization and subsequent evaporation of the component of this system.

Diffuse Galactic gamma ray flux depends primarily on the density and distribution of PBHs in the halo. In the past few decades investigations have been focused on the properties of black holes of mass $M_H \sim 10^{15} g$, which should evaporate by Hawking radiation in a time comparable with the present age of the universe. The precise value of M_H depends on the spin of the hole. The limits were given by Page (1976) as $5\times10^{14} \leq M_H \leq 7\times10^{14} g$. He also found that a significant fraction (≥ 0.2) of the evaporated luminosity appears in the form of γ-rays with a spectral distribution peaking at $\sim 100MeV$. This value is in good agreement with that obtained more recently by Dixon et al. (1998).

The clustering of PBH around galaxies is considered by Page and Hawking (1976). The upper limit of the PBH density was computed to be $\overline{n}_H \leq 10^4\ pc^{-3}$ for black holes of mass $\sim M_H$. This corresponds to a density parameter $\Omega_{PBH}(M_H) \leq 10^{-8}$, where Ω_{PBH} is the ratio of their mean mass density to the critical density $\rho_c = 3H_o^2/8\pi G$ (H_o is the Hubble constant). It was inferred from the above limit that the closest individual source is at a distance

$$10^{15}\left(\overline{n}_H/10^4\right)^{-1/3}\left(C/10^6\right)^{-1/3} cm \tag{20}$$

where the clumping factor $C \leq 10^6$. Therefore, the closest γ-ray source is at $10^{15} cm$, and delivers a flux of $\leq 10^{-9}$ photons $cm^{-2}s^{-1}$, which is beyond the observational limit, unless a considerable amount of events occur simultaneously. The upper limit of permitted local density of heavier holes is obtained by requiring $\Omega_{PBH} \leq 1$. Hence

$$n(M) \leq 10^{11} C \left(M / 10^{15} g\right)^{-1} pc^{-3} \tag{21}$$

For $M \sim 10^{15} g$, the maximum allowed space density of PBHs is $n(M) \leq 10^{11} C\, pc^{-3}$. As the random velocity of holes is unlikely to exceed 100 Km s^{-1}, the chances of hitting the Earth in this century are utterly negligible (Jackson and Ryan, 1973).

Halzen et al.(1991) calculated the bound on the number of PBH at their critical mass, constrained by the observed diffused gamma-ray background for the 10-100 MeV energy region as $N \leq 10^5 pc^{-3}$. MacGibbon and Carr (1991) have used the standard model of particle physics to estimate the PBH explosive density of events in the Galactic halo and found a value as high as $10 pc^{-3} yr^{-1}$. However, some uncertainties in the PBH evaporation mechanisms persist. For the Hagedron model (Hagedron 1965) the PBH mass would convert in extremely short time to hadronic matter at $T_{PBH} \sim 140-160 MeV$, whereas for the quark-gluon deconfinment phase transition model, the emitted free quarks and gluons would hadronize at some distance from the PBH horizon at a temperature $T \sim 100-300 MeV$ (Cline et al. 1997). The later author assumed a clumping factor of 5×10^5, corresponding to a density of PBHs of 10^{10} pc^{-3} in the Galactic halo. In this case the number of PBHs is estimated to be $\sim 10^{22}$. Using a decay rate of $\sim 3 \times 10^4 s^{-1}$, he found a photon flux $\sim 10^{38}$ ergs s^{-1} released into the halo. Cline (1997) assumed further a Page-Hawking bound of $2 \times 10^4 pc^{-3}$ and obtained a Galactic diffuse gamma ray flux of ~ 0.12 photons $m^{-2} s^{-1} sr^{-1}$. This result is consistent with the value obtained by Osborne et al. (1994) and Dixon et al. (1998). An important remark concerning the above estimation of the Galactic flux is that the evaporation is restricted to PBHs with masses corresponding to the present epoch.

Conclusion

Since postulating their existence in the early 1970s, primordial black holes remain as one of the most important astronomical mysteries. However, their existence is required to solve a wide spectrum of cosmological riddles, including the formation of galaxies, dark matter, and Galactic halo diffuse gamma ray background. It is expected that Gamma ray space missions, such as Swift, Fermi, and future satellites will provide a definitive answer regarding the existence and eventual distribution of primordial black holes. LHC, on the other hand, will provide most likely unprecedented experimental argument regarding the nature of processes leading to the formation and eventual evaporation of micro black holes, and possibly a test of the dimensionality of our world.

References

[1] Abdurashitov, J.N., Yants, V.E., and Parfenov, C.V., Astro-ph/9911093, 1999.
[2] Afshordi, N., McDonald, P., and Spergel, (2003), *The Astrophy. J.*, **594**: L71-L74.
[3] Alexeyev, S. O., Barrau, A., Boudoul, G., Khovanskaya, O., & Sazhin, M. (2002), *Class. Quantum Grav.*, **19**, 4431-4443.
[4] Barrau, A., Boudoul, G., Donato, F., Maurin, D. (2002), *A&A,* **388**, 676-687.
[5] Barrau, A., Boudoul, G, Donato, F., Maurin, D., Salati, P., Stefanon, I., and Taillet, R., (2003), arXiv:astro-ph/0207395.
[6] Barrow, J. D., (1977), *Nature*, **267**, 117.
[7] Bekenstein, J. D., and Mukhanov, F. F., (1995), *Phys. Lett.*, **B360**, 7-12.
[8] Blandford, R. D., (1977), *Mon. Not. Astr. Soc.*, 181, 489-498.
[9] Blandford, R. D. and Thorne, K. S. (1979), *General Relativity (Black Holes in Astrophysics),* An Einstein Survey, Ed. S. W. Hawking and W. Israel, Cambridge Press, P. 494.
[10] Carlip, S., (2001), *Rept Prog. Phys*. **64**, 2001.
[11] Canuto, V., (1978), *MNRAS*, **184**, 721.
[12] Carr, B.J., and Hawking, S.W., (1974), *Mon. Not. R. Astron. Soc.*, **168**, 399.
[13] Carr, B.J.,(1975), *Astrophys-J.,* **201**,1.
[14] Carr, B.J.,(1977), *Mon. Not R. Astron. Soc.,* **181**, 293.
[15] Carr, B. J. and Rees, M. J., (1977), *Astron. Astrophys.*, **61**, 705.
[16] Carter, B., (1974), *Phys. Rev. Lett.*, 33, 558.
[17] Carter, B., Gibbons, G. W., Lin, D.N.C., and Perry, M., (1976), *Astron. & Astrophys.* **52**, 427.
[18] Chardonnet, P., Orloff, J. & Salati, P., (1997), *Phys. Lett. B,* **409**, 313
[19] Chavda, L.K., Chavda, Abhijit L., *Class and Quantum Gravity*, **19**, 2927, 2002.
[20] Cline, D.B. and Sanders, D.A., *The Astrophys. J.*, **486**, 169-178, 1997.
[21] Cline, D.B., *The Astrophys. J.*, **501**:L1-L3, 1998.
[22] Choptuik, M. W., (1993), *Phys. Rev. Lett.*, **70**, 9.
[23] Connaughton, V., (1998), *Astropart. Phys.*, **8**, 179.
[24] Dixon, D.D., Hartmann, D.H., Kolaczyk, E.D., Samimi, J., Diehl, R., Kanbach, G., Mayer-Hasselwander, H., and Strong, A.W., *Evidence for a Galactic gamma-ray halo*, *New Astronomy*, Vol. 3, No.7, p.539-561, 1998.
[25] Dimopoulos, S., and Landsberg, G., (2001), arXiv:hep-ph/0106295.
[26] Donato, F., Fornengo, N. & Salati, P., (2000), *Phys. Rev. D,* **62**, 043003.
[27] Eicher, D and Solinger, A., *Astrophys. J.,* **203**, 1, 1976.
[28] Frautschi, S., (1971), *Phys. Rev.* **01**, 2821.
[29] Gibbon, G. W., (1975), *Math. Phys.* **44**, 245.
[30] Gibbon, G. W., (1977), in Proc. First Marcel Grossman Meeting on General Relativity, ed. R. Ruffini, 449-458.
[31] Guedens. R., Clancy, D., and Liddle, A. R., (2002), *Phys. Rev.* **D66**, 045315.
[32] Hagedron, R., *Nuovo. Cimento, A* **64**, 811, 1965.
[33] Halzen, F., MacGibbon, J. H., and Weekes, T. C., (1991), *Nature*, **353**,807.
[34] Harada, T., (2006), *J. of Phys. Conf. Series*, **31**, 111-114.
[35] Hagedorn, R., (1965), *Nuovo Cimento Suppl.*, **3,** 147.

[36] Hawking, S. W., (1971), *Mon. Not. R. Aston. Soc.*, **152**,75.
[37] Hawking, S.W.,(1974), *Nature,* **248**, 30.
[38] Hawking, S.W.,(1974b), *Nature,* **248**, 30-31.
[39] Hawking, S.W., (1975), *Comm.Math.Phys.***43**, 199.
[40] Hawking, S.W., (1982), *Phys. Rev. D,* **26**, 2681.
[41] Hawking, S.W. (1989), *Phys. Lett. B,* **231***,* 237.
[42] Hefler, A. D., (2003), arXiv: gr-qc/0304042
[43] Jackson, A. A. and Ryan, M.P. Jr. *Nature*, **216**, 578, 1973.
[44] Kanti, P., (2004), *Int. J. Modern Phys*. A, arXiv:hep-ph/0402168.
[45] Kim, H. I., Lee, C. H., MacGibbon., J. H., (1999), *Phys. Rev. D,* **59**, 063004.
[46] Kim, H. I. (2000), *Phys. Rev. D*, **62**, 063504.
[47] Lin, D.N.C., Carr, B.J. and Fall, S.M., (1976), *Mon. Not. R. Astron. Soc.,* **177**, 51.
[48] MacGibbon, J. H., & Carr, B. J., (1991), *ApJ,* **371**, 447.
[49] Mesazaros, P., (1974), *Astron. Astrophys.,* **37**, 225.
[50] Myers, R. C., and Perry, M. J., (1986), *Ann. Phys*. **172**, 304.
[51] Novikov, I. D. and Thorne, K. S., (1973*), In Black Holes* (Black Holes in Astrophysics), ed. C. DeWitt and B. DeWitt, p. 343. Gordon and Breach: New York.
[52] Osborne, J.L., Wolfendale, A.W., and Zhang, L., *J. Phys. G*, **20**, 1089, 1994.
[53] Page, D. N., (1976), *Phys. Rev*. **D13**, 198-206.
[54] Page, D. N. and Hawking, S.W., (1976), *Astroph. J.*, **206**, 1-7
[55] Page, D. N., (1976b), *Phys. Rev*. **D14**, 3260-3273.
[56] Page, D. N., (1977), *Phys. Rev*. **D16**, 2402-2411.
[57] Parentani, R, (1996), *Nucl. Phys*., **B465**, 4526.
[58] Parikh, M. K., & Wilczek, F., (2000), *Phys. Rev. Lett*., 85, 24.
[59] Rees, M.J., (1972), *Phys. Rev. Lett.,* **28**, 1660.
[60] Rees, M. J., (1977), *Nature*,**266**,333.
[61] Ukwatta, T. N., MacGibbon, J. H., Parke, W.C, Dhuga, K. S., Eskandarian, N. Gehrels, Maximon, L., and Morris, D. C., (2009), arXiv:0901.0542 (astroph)
[62] Vainer, B.V. and Nasel'skii, P.D.,(1977), *Pisma Astron. Zh*, **3**, 147.
[63] Visser, M., (2001), hep-thlo 0106111, *Essential and inessential features of Hawking radiation.*
[64] Wald, R. M., (1994), *Quantum Field Theory in Curved Spacetime and Black Hole Thermodynamics*, Chicago University Press.
[65] Zaumen, W.T., (1974), *Nature*, **247**, 530.
[66] Zel'dovich, Ya, B., (1962), *Sov. Phys.: JETP*, **16**, 1163.
[67] Zel'dovich, Ya, B. and Navikov, I.D., (1967), *Relativistic Astrophysics. Nauka: Moscow. Eng. Translation (1971). Relativistic Astrophysics*, Vol. 1*, Stars and Relativity,* University of Chicago Press: Chicago.
[68] Zel'dovich, Ya, B., (1972), *Mon. Not. R. Astro.Soc.*.**160**, 1.
[69] Zel'dovich, Ya, B., Srobinski, A. A., Khlopov, M. Yu, Chechätkin, V.M, (1977), *Sov. Astron.Lett.,* **3**(3).

In: Black Holes and Galaxy Formation ISBN 978-1-60741-703-3
Editors: A.D. Wachter and R.J. Propst, pp. 163-173

Chapter 6

ON THE INFORMATION PARADOX IN BLACK HOLES

Pedro F. González-Díaz
Colina de los Chopos, Centro de Física "Miguel Catalán",
Instituto de Física Fundamental,
Consejo Superior de Investigaciones Científicas, Serrano 121,
28006 Madrid (SPAIN) and Estación Ecológica de Biocosmología,
C/Pedro de Alvarado, 14 06411 Medellín, SPAIN

Abstract

We use a simple heuristic argument based on the conception that Planck length measures the upper resolution limit for all objects and use the quantum interest conjecture to show that the information paradox cannot be solved by the Susskind principle of black hole complementarity, and the simplest presuppositions of multiversal cosmology and the conservation laws for dark energy accretion onto black and worm-holes to see that the recent Hawking conjecture to recover information conservation can no longer work. Our conclusion therefore is that the information paradox is not solved and that, much like a vacuum is able to create energy pairs satisfying the Heisenberg principle, quantum gravity may create a quantum vacuum which can be able to apparently annihilate information, asking who is afraid of that possibly expectable loss. Some comments are also added to the connection between the information paradox and the origin of life in the context of the quantum multiverse.

PACS: 04.20.Jb, 04.70.-s, 98.80.Hw

Keywords: Black Holes; Loss of Quantum Coherence

1. Introduction

The story of quantum coherence in black holes is a curious one actually. It started in 1975 with Hawking's realization that since a black hole thermally radiates all its energy the information stored in the star that went into the gravitational collapse is irretrievably lost [1]. The story became more and more complicated thereafter and involved new concepts like superscattering matrices [2] and Euclidean wormholes [3], and was punctuated by funny situations and attitudes as bets and rather weird opinions. Apparently, it just finished when Hawking recently acknowledged that the contender's opinion that quantum coherence is not lost in gravity is the final verdict [4]. Actually, Hawking's final solution somehow implies that, in at least one sense, black holes do not really exist.

In this report we shall consider heuristic simple arguments that make inescapable the final—so far—conclusion first reached by Hawking 33 years ago, this time without placing in any danger the very existence of black holes, and relate the information paradox in the quantum multiverse with the problem of the origin of life in cosmology.

2. The Failed Tale of the Two Whales

We shall use two quantum results in order to get our first conclusion. On the one hand, we resort to the interpretation that the Planck length is the measure of the maximum resolution which can be reached in any experiment [5] and, on the other hand, we use the so-called quantum interest conjecture according to which the integrated magnitude of pulses of negative and positive energy cannot be the same if their mutual time separation is nonzero [6]. It will be seen that these two results together with the interpretation of black hole emission as a consequence from the particle creation by moving mirrors unavoidably leads to the conclusion that after radiating all its energy a black hole can no longer preserve all the information of the collapsed star and that there is some nonzero information stored in a little remaining region of space which cannot radiate anything, but it is fully hidden.

The known interpretation of the black hole evaporation process is that in the neighborhood of the black hole horizon pairs are created in such a way that whereas the member of a pair with positive energy escapes to infinity the corresponding member with negative energy enters the black hole and makes the black hole energy smaller by an amount which is exactly the same as that of the radiated particle [7]. The event horizon would thus behave like a membrane which is only permeable to particles with negative energy. However, if particles in the pair are assumed to be equivalent to energy pulses and the position of both pulses of one pair cannot be discriminated with unbounded precision, such as the existence of the Planck length seems to indicate, then the above interpretation cannot be sustained since we had then a situation in which the two pulses would unavoidably be created with a mutual nonzero time separation of the order the Planck time. But if this is actually the case then the quantum interest conjecture would inexorably dictate that the pulse with positive energy should overcompensate that with negative energy by an amount which is given by the approximate expression [6]

$$E_p = E_n(1 + \xi E_n t_P), \tag{1}$$

where E_p and E_n are the energy of the positive and negative pulse, respectively, ξ is a constant of order unity and t_P is the Planck time. Now, since for each radiation particle $E_p > E_n$ and for the observer [*at the horizon*] $\sum E_p$ should [*exceed*] or equal the total black hole energy in order to preserve energy conservation at infinity, it follows that once the thermal process has finished, depending on the placement of the observer, there is [*a vanishing*] or a nonzero amount of black hole energy which cannot be radiated where all or a part of the information could be stored. It follows that the information paradox cannot be avoided the way Susskind suggested [8] because even though two identical "whales" can be simultaneously dead and alive, they are inexorably made two different whales due to the quantum-gravity effect discussed above, so solving the complementarity paradox.

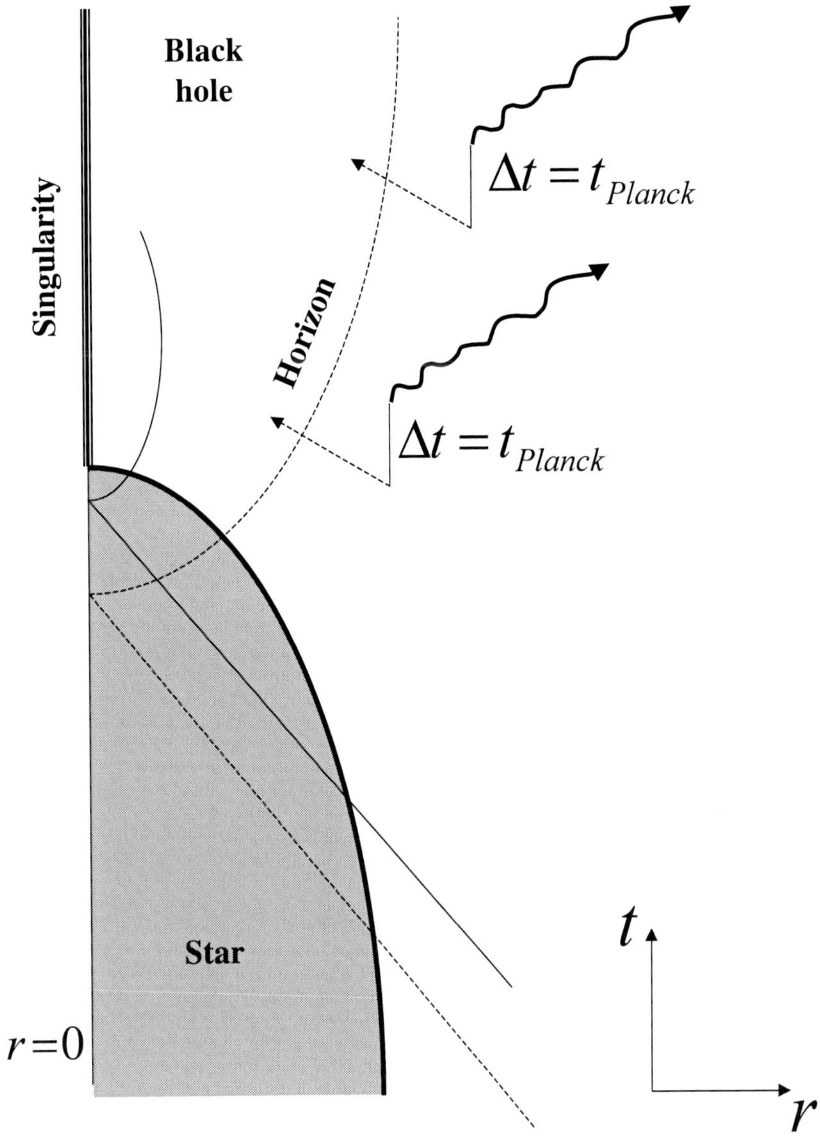

Figure 1. Diagram for black hole evaporation when no unbounded resolution is allowed.

The diagram for black hole radiation is then modified the way pointed out in Figure 1. In fact, from the standpoint of geometrical optics the situation outside the horizon in the evaporation diagram of Fig. 1 is conformally identical to that of a moving mirror [9] (Fig. 2) with acceleration a that emits a negative energy pulse at a given position x_i and time t_i and then at position $x_f = x_i + x_P$ (x_P being the Planck length) and time $t_f = t_i + t_P$ it stops accelerating emitting a pulse with positive energy. The horizon ceases to exist for the short while in which the pulse with positive energy is emitted for otherwise that pulse would be trapped by the black hole instead of escaping to infinity. This is precisely the situation considered by Ford and Roman [6] to show that the pulse with positive energy overcompensates that with negative energy by an amount which depends on the time separation between

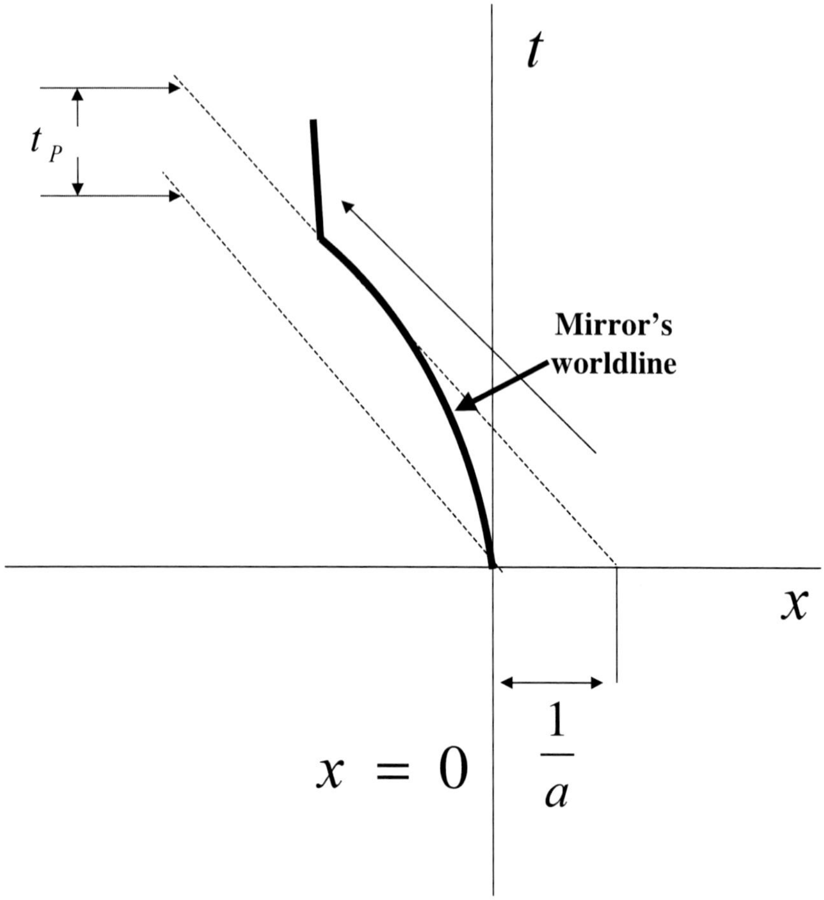

$$x = -t - A\exp(-2\kappa t) + B$$

Figure 2. A moving mirror which emits pulses of negative and positive energy which are time separated by a Planck time. The figure shows a picture that is equivalent to a mirror which accelerates away to the left for a short while along an asymptotically null trajectory after which it stops, emitting Hawking radiation. The broken line is the null asymptote $v = B$. The κ appearing in the trajectory equation would denote surface gravity.

the two pulses—the Planck time in black holes.

The conclusion of the above heuristic analysis is that a black hole can or cannot evaporate completely so that some remaining nonzero part of its energy is or is not left without being converted into thermal radiation. It follows that that non-radiated part of the black hole or the apparently smaller fully radiated out black hole can store or lose information which cannot be associated with what was contained in the original star that went collapsed into the black hole.

3. Black Holes in the Multiverse

Another attempt to keep information and quantum mechanics preserved is that which Hawking himself more recently uncovered in Dublin [4]. In this case black holes are placed in a set of universe histories (in the terminology preferred by Hawking) in such a way that some histories contain black holes and lose information by themselves and other histories do not contain black holes so preserving information. According to Hawking the latter histories which contain no black holes are more persistent and will dominate the future and therefore information is preserved at the end of the day. However, eliminating the cause it appears no surprise that no effect would occur; that is to say, if black holes are eliminated of course no loss of quantum coherence may take place. In order to check the viability of the Hawking argument, we shall consider in what follows a more general system formed by a multiverse [10] where at least one of the single universes contains a black hole. It is a multiversal general law that the probability that the whole set of independent space-times would contain a black hole is 100 percent and that then the probability that some nonzero information is lost in the multiverse becomes actual certainty.

However, at first sight it might be still possible that black holes belonging to distinct universes were connected to each other in such a way that the information lost in one of them would end up trapped in the other so that at the end of the day no information would be lost. Such a situation could be allowed if the universes involved would be filled with phantom energy and hence instead of into pinching-off Schwarzschild wormholes [11] the black holes finally became converted into Morris-Thorne wormholes [12]. Since no thermal radiation would be emitted by the latter wormholes, just like in the case of the above Hawking argumentation, no information paradox took place.

Nevertheless, in the case being considered the wormholes would undergo what is known as a big trip [13], a phenomenon by which the size of the wormhole throat would ultimately exceed that of the universe itself and there would be a disruption of causality which inexorably led to a loss of quantum coherence, also in this case. In fact, even before the advent of the future big rip singularity [14] at time t_{br} the involved universe would lose any causal order along its own time, just at a time

$$t_* = t_0 + \frac{t_{br} - t_0}{1 + CAm_0}, \tag{2}$$

where A is a constant, $C = \sqrt{8\pi\rho_0/3}$, ρ_0 is the original energy density of the phantom energy and m_0 is proportional to the initial throat radius of the wormhole.

But even in their black hole event horizon form black holes in a phantom universe lose all the information contained in them by means of an accretion process which supersedes the Hawking evaporation itself. In fact, just at the big rip time, t_{br}, all of the black hole has been annihilated by the accreted phantom energy leaving nothing instead. Babichev, Dokuchaev and Eroshenko have calculated [15] the rate of mass variation of a black hole with initial mass M_0 that accretes phantom energy. In the case where the scale factor is given by $a(t) = \left(a_0^{3(1+w)/2} + 3(1+w)(t-t_0)/2\right)^{2/[3(1+w)]}$, such a rate gives rise to the

the black hole integrated mass formula [16]

$$M(t) = \frac{M_0}{1 + \frac{(|w|-1)M_0}{\dot{M}_0}\left(\frac{t}{\bar{t}-\frac{3}{2}(|w|-1)t}\right)}, \tag{3}$$

in which $\bar{t} = a_0^{-3(|w|-1)/2}$ and $\dot{M}_= (4\pi A\rho_0)^{-1}$. It can be readily checked that the black hole mass vanishes at the big rip time, leaving nothing from the information originally contained in the black hole. Since the observational data appears to favor the phantom case where $w < -1$ we can conjecture that information paradox cannot still be solved.

4. Information and Life in the Multiverse

The idea that life and cosmology are intimately linked to one another is not new [17]. Neither is at all recent a set of assumed likely connections between the notion of life and the basic principles of quantum mechanics [18]. On the other hand, the current period of accelerating expansion of the universe could be related to a state where the universe would adopt a quantum mechanical behavior (see later) and therefore that period might also be most straightforwardly related with the emergence of life in the universe [18]. Moreover, life represents a way or another a gain of information and is therefore contradictory to the notion of information paradox. All of such questions are the subject of a new scientific discipline which is most usually dubbed "astrobiology", an activity which is flourishing in new research centers spread throughout the entire world. The second of the aims of this paper is nevertheless more akin to what can be denoted as biocosmology, which would be a branch of cosmology making use of the above biological and cosmological ideas together with the anthropic principles and the dark and phantom energy help, as that paper actually aims at extending all notions related with life and the participatory anthropic principle [19] into the realm of the multiverse through the concurrence of wormholes able to transfer information from one universe to other.

In fact, one can look at the problems of black holes in the multiverse dealt with in the previous session by considering that a couple of black holes, each placed in a different universe, might really form a Schwarzschild wormhole, that is an Einstein-Rosen bridge. Such objects are in fact highly unstable and pinch off through their throat. However, if the involved universes are filled with phantom energy, then what we have is a Lorentzian wormhole of the Morris-Thorne type. In what follows we shall comment on how all the information lost in a given universe through these formed wormholes does relate with the problem of the origin of life in the universe.

It appears quite a widespread accepted opinion that the origin of life is a cosmological problem [17]. In order for life to be an operative concept in this way two conditions are simultaneously required to hold: the formation of self-replicating long molecules and aminoacids and the synthesis of conveniently folded proteins made from such aminoacids. However, if we assume these two conditions to be fulfilled as a consequence from the evolution of the universe, then we are confronted with two big problems. On the one hand, such as it was many times stressed by Hoyle [20], the probability that self replicating molecules able to support life had been formed at any place of the universe is similar to that a tornado has of being able to build a Boeing 747 out of the materials from a junk-yard. On the other hand, the well-known Levinthal paradox [21] makes it sure that a supercomputer which is based on plausible physical-chemical and spectroscopic rules (such as internal hindered rotation, bending or wagging vibrations, etc.) would take 10^{127} years in finding the native configuration of a protein made of some 100 aminoacids, which is properly folded and has a suitable biological behavior. It follows that during its entire evolution until now the universe would not allow life to be created anywhere on it.

It has been believed for many years that only the smallest particles or objects show a quantum-mechanical behavior. However, recent years have seen that such a belief is no longer valid in certain realms of the accelerating current cosmology. In fact, if the present universe is filled with phantom energy (such as it appears to be most supported by astronomical data [22]), then the larger the universe size the larger its energy density and therefore a more quantum-mechanical behavior should be expected to be manifested for the current universe as far as it rapidly expands with time, tending to a true singularity when the size of the universe and its energy density are both simultaneously infinity, at the big rip [23]. Actually, the formation of molecules able to self replicate is by itself a quantum process. The billions of smaller molecules in the primordial soup collided and quantum-mechanically formed trillions of new molecules. The probability that these random collisions would produce a molecule able to self replicate is tiny actually, so tiny that such a molecule could not have been formed on Earth or on billions of planets alike [20]. Likewise, protein folding is also a process governed by quantum rules and describable by a wave equation that contains a power-law potential which can be expressed in terms of an order parameter expressible as a scalar field that jointly represents the set of all internal motions of the molecules along its normal coordinate modes [24]. A dependence of such a potential with temperature allows us to express the protein folding process as described by means of a mechanism of spontaneous symmetry breaking, the symmetry being the number of contacts among hydrophobic groups in the protein [24]. In any case the probability for the above whole process leading to the emergence of life to occur in a single universe is very small really.

It follows that the generation of life in a single universe like ours is an extremely unlikely process which requires feeding the living organisms with information. Anthropic principles correspond to a notion that is somehow against such a conclusion. In particular, the Wheeler notion of the participatory principle [19], according to which we exist in a universe which creates itself along a self-reference process. Of course, this idea has a quantum-mechanical origin and predicts the existence of observers who are by themselves able to create all the physical reality that they are able to observe, even the Big Bang and themselves. Thus, rather than intelligent observers being created by the universe, what matters here is a universe that is created and evolved by the observers, perhaps in order to satisfy the thermodynamic second law.

Another crucial notion is that of Boltzmann brains [25]. Perhaps ours is a typical civilization [25,26] that was created by some random fluctuation from vacuum and its condition of mediocrity [26] (typicalness) created the universe we are able to observe or imagine. Such a solution to the problem of unlikeness is actually not a solution because the Boltzmann spontaneous fluctuation is a process which is also extremely unlikely to occur along the life of the universe from the big bang until now.

It has been hitherto said that the probability that molecules able to self replicate be synthesized and that for even the smallest proteins to properly fold into their native structure, is actually tiny. Even so, strictly speaking such probabilities are not exactly zero. Therefore, since everything that may happen with whatever small but still nonzero probability actually happens with real certainty in the realm of the quantum multiverse [27], molecules able to self replicate and properly fold must have been immediately synthesized, and hence life must necessarily have emerged in the context of the quantum multiverse. That happened and it did rapidly. The solution of the problem of the origin of life in the context of the quantum multiverse must actually be based on an analysis of conditional probability rather than just probability. In fact, one can always be sure that however small can be the probability for protein folding to occur provided the corresponding biological molecules able to self replicate have already been synthesized, it will not be strictly zero and therefore the whole process for originating life in the quantum multiverse must be true certainty actually.

However, for that to become physically feasible in the cosmic realm we are dealing with, it ought to conform to both the Wheeler participatory anthropic principle and a civilization able to be typical with respect to the universe we know; that is to say, if we extend the notion of typicalness [25], and hence of mediocrity [26], to the whole multiverse, a civilization which is typical in a given universe, would be so also in the whole multiverse, that meaning that either the typical observers can in some way observe the universes and loss of information through the formation of black holes in the universes they are not living in or that such universes do not actually physically exist, at least from a participatory physical standpoint principle.

Physical space-time connections between two universes of the multiverse through which the observers can retrieve some relevant physical information from these two universes can only be achieved by means of Lorentzian wormholes with relative speeds between their mouths at all unspecified. The latter feature expresses the otherwise mutual independence between the space-times of the universes that form up the multiverse, and makes it impossible to establish any kind of simultaneity among distinct civilizations potentially living in different universes. Therefore, even though life is originated almost immediately in the whole context of the multiverse, it can only be realized in just one universe if we want observers to be typical, no matter whether they are able to perceive just one universe or many universes through connections by means of wormholes and loss of information through black holes, the second possibility being more probable certainly. This transfer of information through the wormholes is a process which is no doubt related to the information paradox and also to the emergence of life in a given universe, so the information paradox and the origin of life must be related to each other in such a way that life would not be possible without information loss in the formation of black hole systems.

The great scientist and scientific divulger Carl Sagan used to declare [28] that we all were somehow present in the primeval Big Bang and became later what has been dubbed as powder of stars. If life actually is an endeavor of the whole multiverse rather than a matter that concerns particular universes, then the Sagan's idea had to be actually extended to the context of the multiverse. Instead of his declared observations, one could well say that we all were somehow present at the moment in which the whole multiverse was created, that is to say eternity, quite likely.

Panspermia is an ever credit-gaining theory which shifts the origin of life on Earth from Earth itself to the single cosmological context [29]. Etymologically, it means seeds everywhere, so expressing the idea that the seeds of life are spread in a rather homogeneous form throughout the entire universe, and that such seeds once reached the Earth where they developed into the known living beings. If life is a matter concerning the whole extent of the multiverse, then one had to replace the notion of panspermia for that of what could be dubbed as holospermia. Etymologically, holospermia would mean *seeds in the wholeness* and would express instead the idea that the seeds of life were spread throughout the whole multiverse in our remote past, and that such seeds once reached our own universe, possibly through a wormhole.

It was Sir Fred Hoyle who coined the term panspermia for his cosmic theory for the origin of life on Earth [29]. It is possible- and actually claimed by the notion of holospermia- that, rather than being present at the Big Bang, we were all originated in the set of all universes making up the multiverse. Actually, if as stressed many times by Hoyle himself [20], the probability for life to have been spontaneously generated in our own universe is extremely tiny, and that on the contrary, it becomes full certainty in the set of all universes making up the multiverse, then the probability for holospermia to be responsible for the origin of life in our universe is by far much bigger than for panspermia making that job.

5. Conclusion

The results obtained in the two first paragraphs of this paper seem to indicate that information cannot be taken as a conserved quantity within the realm of quantum gravity, and that, much like quantum-mechanical vacuum is able to create energy satisfying the Heisenberg principle, apparently, albeit not actually, violating conservation of energy in some experiments and provoking mentally older physicists to become astonished , quantum gravity may create a quantum vacuum which can be able to apparently annihilate information and make most current physicists become unconvinced of this kind of quantum effect and also astonished from its possible implications. Thus, the present work can be symbolized by the following question: who is afraid of the information paradox? On the other hand, considering black holes in a quantum multiverse may help to solve the problem of the origin of life by means of the wormholes that might connect the universes of the multiverse, the Wheeler participatory anthropic principle and the Boltzmann brain idea.

Acknowledgements

The author thanks Carmen L. Sigüenza for useful discussions and the members of the theater group of Medellín, Spain, who allowed me to do this work in a nice scientific-artistic working atmosphere. This work was supported by MEC under Research Project No. FIS2005-01181.

References

[1] S.W. Hawking, *Commun. Math. Phys.* **43** (1975) 199; *Phys. Rev.* **D14** (1976) 2460; D13 (1976) 191.

[2] S.W. Hawking, *Commun. Math. Phys.* **87** (1982) 395.

[3] S.W. Hawking, *Phys. Rev.* **D37** (1987) 904.

[4] S.W. Hawking, Lecture at the *17th International Conference on General Relativity and Gravitation*, Dublin, 2004 (http://math.ucr.edu/home/baez/week207.html)

[5] T. Padmanabhan, *Gen. Relat. Grav.* **17** (2004) 215.

[6] L.H. Ford and A. Roman, *Phys. Rev.* **D60** (1999) 104018.

[7] S.W. Hawking, *On big bang and black holes*, World Scientific (Singapore, 1993).

[8] L. Susskind, *Phys. Rev. Lett.* **71**, 2367 (1993); *Tale of two whales*, NewScientist 5 July 2008 .

[9] P.C.W. Davies, *Rep. Prog. Phys.* **41** (1978) 1313.

[10] S. Robles-Pérez,, P. Mart'ın-Moruno, A. Rozas- Fernández, and P.F. González-Díaz, *Class. Quantum Grav.* **24** (2007) F41, and references therein.

[11] A. Einstein and N. Rosen, *Phys. Rev*. **48** (1935) 73.

[12] M.S. Morris, K.S. Thorne and Yursever, *Am. J. Phys.* **56** (1988) 395.

[13] González-Díaz, *Phys. Rev. Lett.* **93** (2004) 071301 .

[14] R.R. Caldwell, M. Kamionkowski and N.N. Weinberg, *Phys. Rev. Lett.* **91** (2003) 071301

[15] E. Babichev, V. Dokuchaev and Yu. Eroshenko, *Phys. Rev. Lett.* **93** (2004) 021102 .

[16] González-Díaz and C.L. Sigüenza, *Phys. Lett.* **B589** (2004) 78 .

[17] See, for example, L. Smolin, *The Life of the Cosmos* (Phoenix, London, UK, 1997), and references therein; Investigación y Ciencia, Temas 52: El Origen de la Vida.

[18] *Quantum aspects of life*, edited by D. Abbott, P.C.W. Davies and A.K. Pati (World Scientific Publishing, UK, 2008).

[19] J.A. Wheeler, Law without law. *In Quantum Theory and Measurement* edited by J.A. Wheeler and W.H. Zurek (Princeton University Press, uSA, 1983) pp. 182 -213 .

[20] F. Hoyle, *The Intelligent Universe* (Michael Joseph Limited, London, UK, 1983), pp. 18-19.

[21] C. Levinthal, Are there pathways for protein folding?, *J. Chim. Phys. et de Physico-Chemie Biologique* **65** (1961) 44

[22] D. J. Mortlock and R. L. Webster, The statistics of wide-separation lensed quasars, *Mon. Not. Roy. Astron. Soc.* **319** (2000) 872; A. G. Riess *et al.* [Supernova Search Team Collaboration], Observational Evidence from Supernovae for an Accelerating Universe and a Cosmological Constant, *Astron. J.* **116** (1998) 1009; S. Perlmutter et al. [Supernova Cosmology Project Collaboration], Measurements of Omega and

Lambda from 42 High-Redshift Supernovae, *Astrophys. J.* **517** (1999) 565; J. L. Tonry et al. [Supernova Search Team Collaboration], Cosmological Results from High-z Supernovae, *Astrophys. J.* **594** (2003) 1; D. N. Spergel et al. [WMAP Collaboration], First Year Wilkinson Microwave Anisotropy Probe (WMAP) Observations: Determination of Cosmological Parameters, *Astrophys. J. Suppl.* **148** (2003) 175; C. L. Bennett et al., First Year Wilkinson Microwave Anisotropy Probe (WMAP) Observations: Preliminary Maps and Basic Results, *Astrophys. J. Suppl.* **148** (2003) 1; M. Tegmark et al. [SDSS Collaboration], Cosmological parameters from SDSS and WMAP, *Phys. Rev. D* **69**, 103501 (2004).

[23] R.R. Caldwell, *Phys. Lett.* **B545** (2002) 23; R.R. Caldwell, M. Kamionkowski and N.N. Weinberg, *Phys. Rev. Lett.* **91** (2003) 071301; P.F. González-Díaz, *Phys. Lett.* **B586** (2004) 1 .

[24] P.F. González-Díaz and C.L. Sigüenza, *Protein folding and cosmology*, astro-ph/9706040 .

[25] L. Dyson, M. Kleban and L. Susskind, Disturbing implications of a cosmological constant, *JHEP210* (2002) 011 .

[26] A. Vilenkin, Unambiguous probability in an eternally inflating universe, *Phys. Rev. Lett.* **81** (1998) 5501.

[27] See the contributions to *Universe or Multiverse*, edited by B. Carr (Cambridge University Press, Cambridge, UK, 2007).

[28] C. Sagan, TV series programme "Cosmos", Episode 1

[29] The idea for an origin extraterrestrial of life can be traced back to the Greek philosopher Anaxagoras (V century b.c) and was continued by scientists such as H. von Helmholtz (1879) and S. Arrhenius (1903). The modern theory of panspermia was established by F. Hoyle and N.C. Wickramasinghe, *Astronomical Origins of Life: Steps towards Panspermia* (Kluwer Academic Press, USA, 2000).

In: Black Holes and Galaxy Formation ISBN 978-1-60741-703-3
Editors: A.D. Wachter and R.J. Propst, pp. 175-194 © 2010 Nova Science Publishers, Inc.

Chapter 7

GEOMETRICAL AND NUMERICAL ASPECTS OF BLACK HOLE EVOLUTION

J.A. González and F.S. Guzmán
Instituto de Física y Matemáticas,
Universidad Michoacana de San Nicolás de Hidalgo.
Edificio C-3, Cd. Universitaria, A. P. 2-82,
58040 Morelia, Michoacán, México

Abstract

In this chapter we present a detailed study about the black hole evolution. We focus on the dynamical time-dependent process which requires the use of numerical techniques. For the sake of clarity we restrict our analysis to a spherically symmetric space-time. We set the construction of the solutions as an initial value problem consisting of a 3+1 decomposition of Einstein's equations under the BSSN formulation. We also show the case of a growing black hole by accretion. The type of matter accreted by the black hole is a massless scalar field minimally coupled to gravity. Among the techniques needed to achieve a dynamical Black Hole evolution we present: the gauge election, initial data and boundary conditions. The analysis is restricted to the geometrical aspects of the process including the following diagnostics: apparent horizon and event horizon tracking, causal structure and gauge dependence. The contents of this manuscript allows the understanding of the tools currently applied to more general space-times in numerical relativity like the binary black hole case.

1. Introduction

In classical General Relativity the black holes are the most important space-time solutions of Einstein's equations together perhaps with the cosmological solutions, being both the most studied from an astrophysical point of view. The reason is that in the case of black holes, a considerable amount of astrophysical candidates have been found, like the supermassive black hole candidates localized at the center of some galaxies and dark stellar-size compact objects in binary systems.

Black holes are particularly important in the gravitational wave astronomy based on interferometers which nowadays are at the threshold of sensitivity to measure this phenomenon [1]. Specially relevant in the gravitational wave forms and templates used to filter the observations are the orbiting binary black hole and neutron star systems [2] and the collapse or perturbation of rotating single objects [3]. Until very recently the templates used were estimated using only post-Newtonian approximations that helped calculating wave forms because there were no confident numerical calculations based on the solution of the full Einstein's equations. By the time, the numerical relativity community was developing the codes necessary to solve Einstein's equations for black holes with no symmetries using three dimensional codes (see e.g. [4, 5]). Also hybrid approaches were essayed, in which post-Newtonian approximations were used to estimate the wave forms in the orbiting stage of the binary, full general numerical relativity was used for the stage of the plunge and finally a perturbative approach was used to study the resulting final black hole in its ring-down stage as the result of the collision and thus the gravitational wave signal form due to the whole process was constructed [6].

The fact that Einstein's equations could be solved using numerical methods for particular cases and under restrictive symmetries was not a guarantee that at some point the binary black hole problem could be solved numerically, from the orbiting stage to the ring-down of the final black hole, based only on the increase of the hardware power, so that a deeper understanding of the equations was required. Within numerical relativity the solution of Einstein's equations has been mainly tackled using a 3+1 decomposition of the space-time, being the ADM formulation of General Relativity the most popular one. That is, Einstein's equations are split up in such a way that they become an Initial Value Problem (IVP) with suitable initial data consistent with the Hamiltonian and Momentum Constraints, which are satisfied also during the evolution due to the Bianchi identities.

Being the most popular formulation of Einstein's equations, the ADM formulation did not constitute a well posed IVP because in general for most of the gauge conditions such formulation does not constitute a strongly hyperbolic system of evolution equations, which is a strong enough reason to forbid the numerical evolution of a single black hole for a long time [5]. Thus new formulations of the 3+1 split were proposed with hyperbolicity properties that allowed a well posed IVP. The most important one is the BSSN (stands for Baumgarte, Shapiro, Shibata, Nakamura) formulation [7, 5]. This formulation allowed us to obtain more stable numerical solutions of the IVP for the Einstein's equations that lasted a along enough time (to extract the physical end geometrical information) in the cases of single black holes and hydrodynamical systems using codes without symmetries. The goal turned to be the evolution of a binary black hole system. The first step was the construction of initial data for two black holes, which was solved using the concept of puncture and excision. The puncture is a wormhole solution with an apparent horizon at the throat and one of the two ends is compactified to one point, thus in order to start an evolution with two black holes there were needed two punctures with assigned linear momentum, spin and orbital angular momentum [8] (for a recent review on the construction of initial data for 3+1 numerical relativity see [9]).

The first stable evolution of orbiting binary black holes was performed using the harmonic formulation of General Relativity [10]. It was until 2006 that it was possible to keep the binary black hole system evolving for various orbital times using the BSSN formulation, and most important, the extraction of the corresponding gravitational wave signals that had been expected for a long time was possible [11, 12]. The nowadays numerical relativity studies the evolution of binary black hole systems related to various astrophysical scenarios, like the collision of black holes with different spin orientations [13], different mass ratios [14] and the after-collision effects like kicks [15].

Based on the importance of space-times with black holes, in this chapter we present the evolution of a single black hole and the accretion of a toy material into the black hole using the BSSN formulation and puncture data in order to illustrate the numerical and analysis tools used in current numerical relativity studies.

It is also of our interest to present a numerical and geometrical approach to black holes related to time dependent situations, like the growth of a black hole due to accretion, because it is possible to show differences between important concepts like apparent and event horizons which sometimes remain mixed up under very restrictive conditions like stationarity or staticity symmetries. We consider the spherically symmetric case because its evolution and growth can be done in detail and allows us to illustrate various significant geometrical aspects of the space-time, specially those related to the different concepts of mass of the system, which again, have the same values for stationary and static situations.

The chapter is organized as follows. In section 2 we describe the 3+1 decomposition of the space-time and the evolution equations we use to construct the space-time, in section 3 we illustrate the way an IVP is solved using the finite differences approximation, in section 4 we show the evolution of a black hole in isotropic coordinates whereas in section 5 we illustrate the accretion of a scalar field into a black hole. Finally in section 6 we present final comments related to the problems presented here.

2. 3+1 Formalism of General Relativity

The 3+1 decomposition of Einstein's equations assumes the 3+1 decomposition of the space-time itself that starts with a foliation of the space-time with space-like hypersurfaces Σ_t which label is a time-like coordinate t.

The key ingredients of the decomposition are: (1) the three metric of the hypersurfaces with components γ_{ij}, $i,j = 1,2,3$ that measures the proper distance on the hypersurface $dl^2 = \gamma_{ij}dx^i dx^j$, (2) the *lapse function* α that measures the proper time of Eulerian observers moving along the normal directions to the hypersurfaces $d\tau = \alpha(t,x^i)dt$, (3) the relative velocity between Eulerian observers and the lines of constant spatial coordinates $\beta^i = \beta^i(t,x^j)$, known as the *shift vector*, which is defined such that $x^i_{t+dt} = x^i_t - \beta^i(t,x^j)dt$, where t labels the hypersurface. In terms of these definitions the line element of the space-time is

$$ds^2 = g_{\mu\nu}dx^\mu dx^\nu = (-\alpha^2 + \gamma_{ij}\beta^i\beta^j)dt^2 + 2\beta_i dt dx^i + \gamma_{ij}dx^i dx^j, \quad (1)$$

where $\beta_i = \gamma_{ij}\beta^j$. In terms of these variables, Einstein's equations $G_{\mu\nu} = \kappa T_{\mu\nu}$ split into a set of six evolution equations for the extrinsic curvature components K_{ij}

$$\begin{aligned}(\partial_t - \mathcal{L}_\beta)\gamma_{ij} &= -2\alpha K_{ij},\\ (\partial_t - \mathcal{L}_\beta)K_{ij} &= -D_iD_j\alpha + \alpha(R_{ij} - 2K_{il}K^l_j + KK_{ij} - S_{ij} - \frac{1}{2}\gamma_{ij}(\rho - S)), \quad (2)\end{aligned}$$

where it is worth noticing that the first set of six equations defines the extrinsic curvature and the second set are the evolution equations, K is the trace of K_{ij}, ${}^{(3)}R_{ij}$ are the components of the 3-Ricci tensor of the hypersurface, D_i is the covariant derivative associated to the 3-metric γ_{ij}, $S_{ij} = g_{i\mu}g_{j\nu}T^{\mu\nu}$ and S its trace, and $\rho = n^\mu n^\nu T_{\mu\nu}$ is the energy density of the matter measure by an Eulerian observer, where $n_\mu = (-\alpha, 0, 0, 0)$ and thus $n^\mu = (1/\alpha, -\beta^i/\alpha)$ is the vector normal to the hypersurface. There are also four constraint equations

$$\begin{aligned} H &= {}^{(3)}R + K^2 - K_{ij}K^{ij} - 16\pi\rho = 0, \\ M^i &= D_j(K^{ij} - \gamma^{ij}K) - 8\pi S^i = 0, \end{aligned} \tag{3}$$

where H stands for the Hamiltonian constraint and M^i for three momentum constraints, ${}^{(3)}R$ is the Ricci scalar of the hypersurface. The source terms are again the energy density ρ and the fluxes along the spatial directions $S_i = -g_{i\mu}n_\nu T^{\mu\nu}$. These equations are obtained through appropriate projections of Einstein's equations, both along the normal directions and along the space-like hypersurfaces themselves [16]. Notice that this description does not provide any information about the gauge.

2.1. The BSSN Formulation

The BSSN formulation consists in choosing evolution variables different from those used in the ADM formulation (2). First, a new 3-metric is defined in terms of γ_{ij} through a conformal factor: $\tilde{\gamma}_{ij} = \psi^{-4}\gamma_{ij}$ with the condition $\det(\tilde{\gamma}_{ij}) = 1$; the trace-free part of the extrinsic curvature is promoted to evolution variable $A_{ij} = K_{ij} - \frac{1}{3}\gamma_{ij}K$, which in the conformal frame becomes $\tilde{A}_{ij} = \psi^{-4}A_{ij}$; also three variables are promoted to evolution variables, these are the conformal connection functions $\tilde{\Gamma}^i = \tilde{\gamma}^{jk}\tilde{\Gamma}^i_{jk} = -\partial_j\tilde{\gamma}^{ij}$, where $\tilde{\Gamma}^i_{jk}$ are the Christoffel symbols associated with the conformal metric $\tilde{\gamma}_{ij}$. Thus, instead of evolving the components of the 3-metric and the extrinsic curvature, the variables to be evolved are: the components of the conformal 3-metric, the conformal factor, the trace-free part of the extrinsic curvature, the trace of the extrinsic curvature and the contracted Christoffel symbols. Thus the system of equations becomes [16]

$$\begin{aligned} (\partial_t - \mathcal{L}_\beta)\tilde{\gamma}_{ij} &= -2\alpha\tilde{A}_{ij}, \\ (\partial_t - \mathcal{L}_\beta)\phi &= -\frac{1}{6}\alpha K, \\ (\partial_t - \mathcal{L}_\beta)\tilde{A}_{ij} &= e^{-4\phi}\left[-D_iD_j\alpha + \alpha R_{ij} + 4\pi\alpha(\gamma_{ij}(S-\rho) - 2S_{ij})\right]^{TF} + \alpha(K\tilde{A}_{ij} - 2\tilde{A}_{ik}\tilde{A}^k_j \\ (\partial_t - \mathcal{L}_\beta)K &= -D_iD^i\alpha + \alpha(\tilde{A}_{ij}\tilde{A}^{ij} + \frac{1}{3}K^2) + 4\pi\alpha(\rho + S), \\ \partial_t\tilde{\Gamma}^i &= -\partial_j(\mathcal{L}_\beta\tilde{\gamma}^{ij}) - 2(\alpha\partial_j\tilde{A}^{ij} + \tilde{A}^{ij}\partial_j\alpha) \end{aligned} \quad ($$

where ϕ is defined by $\psi = e^\phi$ and TF stands for trace-free part.

2.2. Gauge

In order to complete the system of evolution equations it is necessary to specify the slicing used during the evolution of the spatial hypersurfaces. This is achieved by choosing appropriate lapse and shift. In the present chapter we use a "1 + log" condition for the lapse and a $\Gamma-$driver condition for the shift, which can be written as [17]

$$\begin{aligned}
\partial_t\alpha &= \beta^a\partial_a\alpha - 2\alpha K, \\
\partial_t\beta^a &= \frac{3}{4}B^a + \beta^c\partial_c\beta^a, \\
\partial_t B^a &= \partial_t\Gamma^a + \beta^c\partial_c B^a - \beta^c\partial_c\Gamma^a - \eta B^a,
\end{aligned} \quad (5)$$

which together with the evolution equations (4) constitute the full set of equations to be solved.

2.3. The Spherically Symmetric Case

We write the metric in the general form for the coordinate system (t, r, θ, φ) in spherical coordinates:

$$ds^2 = -(\alpha^2 - \beta^r\beta^r\frac{g_{rr}}{\chi})dt^2 + 2\frac{\beta^r g_{rr}}{\chi}dtdr + \frac{1}{\chi}[g_{rr}dr^2 + g_{\theta\theta}(d\theta^2 + \sin^2\theta d\varphi^2)], \quad (6)$$

where χ acts as a conformal factor relating this metric to a space-like flat metric, β^r is the only non-zero component of the shift vector and α is the lapse function as in [17].

Based on the line element (6) the trace-free part of the extrinsic curvature A_{ij} and Γ^i can be written

$$A_{ij} = A_{rr}Diag(1, -g_{\theta\theta}/(2g_{rr}), -g_{\theta\theta}\sin^2\theta/(2g_{rr})), \quad \Gamma^i = (\Gamma^r, -\cos\theta/(g_{\theta\theta}\sin\theta), 0)^T$$

so that the system of evolution equations is reduced to the variables χ, g_{rr}, $g_{\theta\theta}$, A_{rr}, K and Γ^r, all of which depend in general on r and t. In spherical symmetry the determinant of the conformal metric is not equal to one, and it is necessary to adapt the BSSN system of equations to this case. It is standard to use the Generalized BSSN equations (GBSSN) where the determinant of the conformal metric is not assumed to be equal to one. According to the parameterization (6) of the metric, the evolution equations in the GBSSN evolution system, under the Eulerian condition $(\partial_t - \mathcal{L}_\beta)(det(g_{ij})) = 0$, is [17]:

$$
\begin{aligned}
\partial_t \chi &= \frac{2}{3} K \alpha \chi + \beta^r \chi' \\
\partial_t g_{rr} &= -2 A_{rr} \alpha + \beta^r g'_{rr} + 2 g_{rr} \beta^{r\prime} \\
\partial_t g_{\theta\theta} &= \frac{A_{rr} g_{\theta\theta} \alpha}{g_{rr}} + \beta^r g'_{\theta\theta} \\
\partial_t A_{rr} &= -\frac{2\alpha A_{rr}^2}{g_{rr}} + K \alpha A_{rr} + 2\beta^{r\prime} A_{rr} \\
&+ \frac{2\alpha\chi (g'_{rr})^2}{3 g_{rr}^2} - \frac{\alpha\chi (g'_{\theta\theta})^2}{3 g_{\theta\theta}^2} - \frac{\alpha (\chi')^2}{6\chi} - \frac{2 g_{rr} \alpha\chi}{3 g_{\theta\theta}} + \beta^r A'_{rr} + \frac{2}{3} g_{rr} \alpha \chi \Gamma^{r\prime} \\
&- \frac{\alpha\chi g'_{rr} g'_{\theta\theta}}{2 g_{rr} g_{\theta\theta}} + \frac{\chi g'_{rr} \alpha'}{3 g_{rr}} + \frac{\chi g'_{\theta\theta} \alpha'}{3 g_{\theta\theta}} - \frac{\alpha g'_{rr} \chi'}{6 g_{rr}} - \frac{\alpha g'_{\theta\theta} \chi'}{6 g_{\theta\theta}} - \frac{2\alpha' \chi'}{3} - \frac{\alpha\chi g''_{rr}}{3 g_{rr}} \\
&+ \frac{\alpha\chi g''_{\theta\theta}}{3 g_{\theta\theta}} - \frac{2\chi\alpha''}{3} + \frac{\alpha\chi''}{3} - [\alpha\chi S_{rr}]^{TF} \\
\partial_t K &= \frac{3\alpha A_{rr}^2}{2 g_{rr}^2} + \frac{K^2 \alpha}{3} + \beta^r K' + \frac{\chi g'_{rr} \alpha'}{2 g_{rr}^2} - \frac{\chi g'_{\theta\theta} \alpha'}{g_{rr} g_{\theta\theta}} + \frac{\alpha' \chi'}{2 g_{rr}} - \frac{\chi \alpha''}{g_{rr}} + \frac{1}{2}(S + \rho)\alpha \\
\partial_t \Gamma^r &= +\frac{A_{rr} \alpha g'_{\theta\theta}}{g_{rr}^2 g_{\theta\theta}} + \frac{\beta^{r\prime} g'_{\theta\theta}}{g_{rr} g_{\theta\theta}} + \beta^r \Gamma^{r\prime} + \frac{A_{rr} \alpha g'_{rr}}{g_{rr}^3} - \frac{4\alpha K'}{3 g_{rr}} - \frac{2 A_{rr} \alpha'}{g_{rr}^2} - \frac{g'_{rr} \beta^{r\prime}}{2 g_{rr}^2} \\
&- \frac{3 A_{rr} \alpha \chi'}{g_{rr}^2 \chi} + \frac{\beta^{r\prime\prime}}{g_{rr}} - 2\alpha g^{rr} S_r
\end{aligned}
\tag{7}
$$

where a prime denotes derivative with respect to r. The constraint equations are written as

$$
\begin{aligned}
H &= -\frac{3 A_{rr}^2}{2 g_{rr}^2} + \frac{2K^2}{3} - \frac{-5\chi'^2}{2\chi g_{rr}} + \frac{2\chi''}{g_{rr}} + \frac{2\chi}{g_{\theta\theta}} - \frac{2\chi g''_{\theta\theta}}{g_{rr} g_{\theta\theta}} \\
&+ \frac{2\chi' g'_{\theta\theta}}{g_{rr} g_{\theta\theta}} + \frac{\chi g'_{rr} g'_{\theta\theta}}{g_{rr}^2 g_{\theta\theta}} - \frac{\chi' g'_{rr}}{g_{rr}^2} + \frac{\chi g'^2_{\theta\theta}}{2 g_{rr} g_{\theta\theta}^2} - 2\rho, \\
M_r &= \frac{A'_{rr}}{g_{rr}} - \frac{2K'}{3} - \frac{3 A_{rr} \chi'}{2\chi g_{rr}} + \frac{3 A_{rr} g'_{\theta\theta}}{2 g_{rr} g_{\theta\theta}} - \frac{A_{rr} g'_{rr}}{g_{rr}^2} - S_r, \\
G^r &= -\frac{g'_{rr}}{2 g_{rr}^2} + \Gamma^r + \frac{g'_{\theta\theta}}{g_{rr} g_{\theta\theta}},
\end{aligned}
\tag{8}
$$

where the third constraint appears since Γ^i are promoted to dynamical variables. What has to be done next is to set up initial data for the metric functions that satisfy the constraints and then solve the evolution equations, but first we show how an IVP is solved using the finite differences approximation of partial differential equations (PDEs).

3. The Numerical Implementation of an Initial Value Problem

The item that has to be kept in mind from the beginning, is that when using finite differences, one is assuming that the domain of the PDEs is discrete. That is, the functions involved in a PDE are evaluated only in a finite number of points. The implication of such restriction is that it is not possible to calculate the solutions with infinite high precision, and there will be always a contribution of a certain error (called discretization error) that contaminates off the continuum problem. Of course, if one thinks on the variety of problems that can be solved numerically that are still unsolved through analytic methods, it seems that the discretization error is a good price to pay.

There are different approximations to the system of equations to be solved, for example, spectral methods assume that the functions involved in the system of differential equations can be expanded as a series of orthogonal functions on a given domain; then orthogonality conditions and recurrence relations are useful to reduce the system to a simpler system of equations for the coefficients of the expansion [21].

The approximation using finite differencing works in a different way. In order to illustrate how discretization works assume the case of a hypothetical finite domain with a time coordinate t and a spatial coordinate x. Spatial coordinates are defined as a discrete set of points given by $x_j = j\Delta x$, and the boundaries correspond to the points x_0 (on the left) and x_N (on the right). Time $t^n = n\Delta t$ is also defined only for certain values of the continuum time. Thus a function is defined only for the values of x and t that correspond to points in the mesh in such a way that for a given continuous function f there are available values of it at (t^n, x_j), denoted here by f^n_j. For a uniformly discretized domain, $\Delta x = x_{i+1} - x_i$ and $\Delta t = t^{n+1} - t^n$ indicate the resolution in the spatial and time coordinates respectively.

The finite differencing approximation assumes that the functions involved can be expanded in a Taylor series around every point of the mesh up to a desired order approximation. Therefore, considering the function is defined at the spatial point x_i the value of the function in the nearest neighbors can be calculated as follows:

$$\begin{aligned} f(x_{j-1}) &= f(x_j) - \Delta x f'(x_j) + \frac{\Delta x^2}{2} f''(x_j) + O(\Delta x^3) \\ f(x_j) &= f(x_j) \\ f(x_{j+1}) &= f(x_j) + \Delta x f'(x_j) + \frac{\Delta x^2}{2} f''(x_j) + O(\Delta x^3) \end{aligned} \tag{9}$$

where the prime denotes derivative with respect to x. Starting from these approximations it is possible to construct difference operators for the derivatives of f^n_i. For instance, by adding up the first and the third expressions one obtains an expression for the first derivative at the point x_j with a second order error $f'(x_j) = \frac{f(x_{j+1}) - f(x_{j-1})}{2\Delta x} + O(\Delta x^2)$; notice that the value of the function at the left and right nearest neighbors is needed in order to calculate this derivative, that is why it is called a centered finite differencing approximation.

In order to obtain the second derivative of f it suffices to write the combination $\frac{f(x_{j+1}) - 2f(x_j) + f(x_{j-1})}{\Delta x^2} = f''(x_j) + O(\Delta x^2)$, which implies the desired expression for the second derivative with second order accuracy as well. As in the previous case, this is also a centered approximation of the second derivative.

In the case of the spatial boundaries, one of the nearest neighbors would be missing, and a centered approximation would require the addition of an extra point (a ghost point) to the domain; in most problems one prefers to live without ghost points for the purpose of imposing boundary conditions. Therefore one considers that only points to the right in the spatial domain are available (the case of the left boundary), thus one proceeds by writing the approximations of the function considering only points on the right as follows:

$$\begin{aligned} f(x_j) &= f(x_j) \\ f(x_{j+1}) &= f(x_j) + \Delta x f' + \frac{\Delta x^2}{2} f'' + O(\Delta x^3) \\ f(x_{j+2}) &= f(x_j) + 2\Delta x f' + 2\Delta x^2 f'' + O(\Delta x^3). \end{aligned} \tag{10}$$

The combination $f(x_{j+2}) - 4f(x_{j+1}) + 3f(x_j) = 2\Delta x f'$ implies the desired expression for the first spatial derivative of the function with second order accuracy.

If one considers now the discretization in time, the operations above apply in a similar way. It is of course possible to construct more accurate approximations to derivatives which would result in better approximations to the continuous equations (see e.g. [23]). However more accurate approximations involve points beyond $x_{j\pm 1}$ and are unpractical for the purpose of this chapter.

On the other hand, when the points in time are considered, it is possible to use the expressions constructed for the calculation of spatial derivatives, for instance, the first time derivative at time t^n is approximated to second order by $\frac{\partial f}{\partial t} = \frac{f_j^{n+1} - f_j^{n-1}}{\Delta t} + O(\Delta t^2)$. The second derivative calculated to second order centered at time t^n reads $\frac{\partial^2 f}{\partial t^2} = \frac{f_i^{n+1} - 2f_i^n + f_i^{n-1}}{(\Delta t)^2} + O(\Delta t^2)$. With the operators for the second derivatives it is easy to construct the discretized version of the wave equation $\frac{\partial^2 f}{\partial t^2} - \frac{\partial^2 f}{\partial x^2} = 0$ centered at the point (t^n, x_i):

$$\frac{f_i^{n+1} - 2f_i^n + f_i^{n-1}}{(\Delta t)^2} = \frac{f_{i+1}^n - 2f_i^n + f_{i-1}^n}{(\Delta x)^2} + O(\Delta x^2, \Delta t^2) \tag{11}$$

for every i and n. From this expression it is possible to solve for f_i^{n+1}, and provided values of the function f at the time labels n and $n-1$ it is possible to calculate the values of f at the label $n+1$. This is an example of how, given initial conditions, the solution for further times can be constructed. An important note about finite differences is that because the approximations of the differential operators involved in the PDEs are obtained through a Taylor series expansion, an error in the definition of the differential equations is introduced due to the truncation of the Taylor series up to a given order. In fact, it is expected that the error term in (11) decreases when decreasing the resolution in time and space and the solution is expected to converge to the continuous solution.

In what follows the equations one deals with are discretized in a similar manner in order to construct a global solution for a given set of PDEs in terms of initial conditions.

3.1. Evolution Using Finite Differences

The evolution of data consists in calculating the function f_j^{n+1} from data in the previous time slice. As we have seen in the previous section, the evolution equations of the spacetime are first order in time, thus we illustrate our evolution algorithm with the advection equation $\partial_t u = \partial_x u$, again in a 1+1 mesh with points at (t^n, x_i). We used a semi-discretized version of this equation which we write, discretizing the spatial derivative

$$\partial_t u = \frac{u_{i+1} - u_{i-1}}{2\Delta x} + \mathcal{O}(\Delta x^2). \tag{12}$$

Next, it is assumed that for each i the PDE satisfies an ordinary differential equations (ODE) along the lines of $x_i = constant$. With this in mind, it suffices to have the discretization provided by (12) and integrate in time the resulting differential equation for $\partial_t u$. The terms other that the time derivative of the PDE are considered to belong to the right hand side of the ODE in time. This approach is named Method of Lines (MoL) and is the standard tool used within numerical relativity. Thus only an ODE integrator is required to evolve the data from one time slice to the next one. One only needs to choose the integrator, which is selected according to the accuracy, the dissipation and the stability properties that depend on the restrictions on the Courant factor $\Delta t/\Delta x$ (to learn about the properties of evolution algorithms we refer the reader to references [18, 19, 20]). In the present chapter the fourth order Runge-Kutta algorithm is the one used for the evolution of the different systems.

Causal disconnection. A particular important aspect of relativistic numerical solutions is related to the advection terms in the right hand sides of the equations. These terms are of the form $\partial_t p = \beta \partial_x q$, with β a component of the shift vector, ∂_x the spatial derivative along x and p, q are arbitrary functions, and the subtlety is due to the fact that β is a velocity (with any magnitude) of the coordinates between one spatial hypersurface and the next one across the discrete domain, thus defining a causal restriction on the time integration with MoL. In words, assuming that we want to calculate p_k^{n+1} for which we may use the centered approximation $RHS(p_k) = \beta_k(q_{k+1} - q_{k-1})/(x_{k+1} - x_{k-1})$, (where RHS stand for right hand side) however the value of β_k may positive and big enough such that the point x_{k-1} at time t^n may be causally disconnected from x_x at time t^{n+1}. A solution to this case consists in approximating the spatial derivative using one-sided operators so that only points to the right of x_k are used. These operators can be calculated using a Taylor expansion either to the right or to the left of x_k which are used respectively when β_k is positive or negative. To second order the approximations read:

$$(\partial_r u)_k = \frac{-3u_k + 4u_{k+1} - u_{k+2}}{2\Delta x}, \quad \beta > 0 \tag{13}$$

$$(\partial_r u)_k = \frac{3u_k - 4u_{k-1} + u_{k-2}}{2\Delta x}, \quad \beta < 0, \tag{14}$$

In Fig. 1 we show the schematic version of the stencils used to approximate derivatives for the centers and one-sided approximations.

Convergence. A criterion of crucial importance when a system of PDEs is being solved is the convergence of the solution in the continuum limit. Convergence is a test that validates any numerical results presented. In order to illustrate how a convergence test works, consider a function f_j that is second order accurate (like the operators described above) such that $f(x) = f_0(x) + E(x)(\Delta x^2) + O(\Delta x^3)$, where $f_0(x)$ is the exact solution at the continuum, E denotes the unknown error term. If $f_0(x)$ is known, it is enough to compare

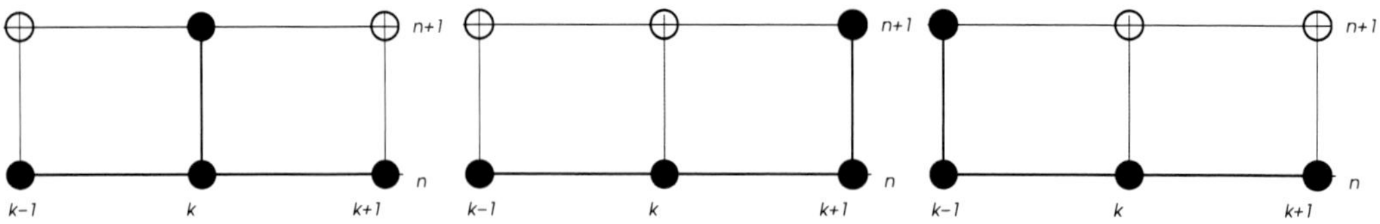

Figure 1. We show the stencil used for the centered Finite Differences approximation to second order, the one-sided used for a positive shift and the one-sided used for the negative shit case.

the results obtained with two different resolutions f_1 obtained using Δx and f_2 obtained using $\Delta x/2$ [22]. Then the theory predicts that

$$\frac{f_1 - f_0}{f_2 - f_0} = \frac{\Delta x^2 + O(\Delta x^3)}{\frac{1}{4}\Delta x^2 + O(\Delta x^3)} = 4 + O(\Delta x^3). \tag{15}$$

The number 4 is the ratio between the error obtained using one resolution and the double resolution, and such factor should be verified. In the case of Einstein's equations the system is over-determined and aside of the evolution equations there are constraints which solution we know $H = M_r = G^r = 0$, and thus we can apply this simple criterion to verify whether or not our calculations converge to the correct solution in the continuum limit. So, in order to backup our results we use this convergence test using the fact that $f_0 = H = M_r = G^r = 0$. Similar Cauchy convergence criteria are used when the solution is unknown.

In practice, instead of calculating the convergence factor for a function at every time step, it is common to estimate a norm of the error, that is, the convergence of a scalar quantity representing a function in the whole spatial domain at every time step. For instance, here we use the L_2 norm of a function, say H which is the violation of the Hamiltonian constraint. Then the L_2 norm is calculated using $||H||_2 = \sqrt{\Sigma_k \frac{H_k^2}{N_x}}$ where k labels each grid point, the summation is calculated over the whole grid and N_x is the total number of points of the grid.

4. Evolution of a Spherically Symmetric Black Hole

The evolution of a single black hole space-time will be driven by the system of evolution equations (7). Among the subtleties to be considered before implementing the evolution of initial data, specially important in GR is the election of coordinates at initial time and possibly during the evolution.

Being the event horizon a fundamental ingredient of a black hole, it is specially important to start-up an evolution using coordinates that are regular across the horizon. In this sense, the Schwarzschild solution in Scwarzschild coordinates is inappropriate, because the 3-metric diverges at the horizon. Ingoing Eddington-Finkelstein (EF) coordinates on the other hand are useful because they are regular at the horizon, however they offer a new problem: the singularity at the origin of coordinates belongs to the domain of evolution and in order to track the evolution of this solution it is required a surgery called excision. The excision is a process that consists in removing a bounded chunk of the domain of the IVP inside the horizon. That is, if $r_h = 2M$ is the radius of the event horizon in EF coordinates then removing the domain $S^2 \times R$ from the domain at $r = r_{ex} < 2M$ and imposing boundary conditions at $r = r_{ex}$ allows a successful evolution [23]. The fact that the excision region remains inside the event horizon in principle guarantees that the light cones point toward the singularity and no signals (including numerical errors) come out from that region to the exterior of the event horizon except for superluminal gauge modes.

Another useful and more modern approach is to consider isotropic coordinates and the BSSN formulation together. In this case there is no need for excision and as well as in the EF coordinates case, there are also initial data satisfying the constraints at initial time for a time-symmetric slice, and is the case presented in this chapter.

4.1. Initial Data

The Schwarzschild solution in isotropic coordinates is obtained by setting the values:

$$g_{rr} = 1, \quad g_{\theta\theta} = r^2, \quad \chi = \left(1 + \frac{M}{2r}\right) \tag{16}$$

where M is the mass of the apparent horizon. This mass is approximately the same as the ADM mass of the puncture. For the gauge we choose an initial zero shift $\beta^r = B^r = 0$ and a pre-collapsed lapse $\alpha = (1 + M/2r)^{-2}$. Because we choose the initial slice to be time-symmetric the extrinsic curvature components are all zero initially. It can be easily verified that the the constraints (8) are satisfied automatically with these initial data.

4.2. Diagnostics

Apparent horizon location. In order to track the radius, area and mass of the apparent horizon during the evolution, we define a marginally trapped surface (MTS) through the condition $\Theta = \nabla_i n^i + K_{ij} n^i n^j - K = 0$, where n^i is an outward pointing unit vector normal to the horizon, K_{ij} are the components of the extrinsic curvature and K its trace of a space-like hypersurface on which one calculates the MTSs. The apparent horizon is the outermost MTS. For the metric (6) this equation is written as

$$\Theta = \frac{\partial_r(g_{\theta\theta})}{\sqrt{g_{rr}} g_{\theta\theta}} + \frac{2}{\chi}\left(\frac{A_{rr}}{2g_{rr}} - \frac{1}{3}K\right). \tag{17}$$

In order to track the evolution of the apparent horizon we calculate Θ at every time step and locate the outermost zero of it at the value of the areal radius $R_{AH} = (g_{\theta\theta}/\chi)_{AH}$. Then calculate the area of the corresponding 2-sphere $A_{AH} = 4\pi R_{AH}^2$ and its mass $M_{AH} = R_{AH}/2$.

Geometric quantities. We want to know whether the black hole we are evolving approximates a Schwarzschild black hole. In order to monitor this property we calculate the following scalar

$$M_{MS} = \frac{R}{2}\left(1 + \frac{1}{\alpha^2}(\partial_t R)^2 - \frac{\chi}{g_{rr}}(\partial_r R)^2\right), \tag{18}$$

which is the Misner-Sharp mass, where $R = \sqrt{g_{\theta\theta}/\chi}$ is the areal radius. This quantity is useful since for the Schwarzschild black hole it approaches the value $2M/r = 1$ at the event horizon. We calculate this scalar at the apparent horizon location in order to monitor whether or not the apparent horizon shows the properties of the event horizon of Schwarzschild solution. We also notice that the ADM mass can be computed taking the limit of the Misner-Sharp mass when $r \to \infty$.

Causal structure. We track a bundle of radial outgoing null rays in order to i) locate an approximation to the event horizon trough the evolution as the 2-sphere from which outgoing null rays diverge when launched toward the future and ii) certify that the cones point in the inward direction at the event horizon.

4.3. Boundary Conditions

Even though it is possible to implement radiative boundary conditions on the evolution variables, that is, assuming that the fields behave as outgoing waves at the boundary of the numerical domain -in a far region- we do not apply any boundary condition. Instead, we define two domains: i) the numerical domain where we compute the evolution of the fields, ii) the physical domain, contained in the numerical domain, where we extract the physical properties of the solution. We make sure that the boundary of the physical domain is causally disconnected from the boundary of the numerical domain during the time evolution of our simulations. This approach is feasible because the spherically symmetric problems reduce to a 1+1 IVP and it is possible to afford the computational resources.

4.4. Results

In all our results in this chapter we use a physical spatial domain $r \in [0, 50]$ and various spatial resolutions $\Delta r = M/25, M/50, M/100$, where the scale M is the mass of the apparent horizon at initial time and is the same as the ADM mass of the system. We also apply the second order approximation of the spatial derivatives and use the Method of Lines, just as described in section 3.

Specially important before showing any physical results is to show the satisfaction of the constraints of the solution. In the present case, since there are three constraints that need to be satisfied, it is necessary to verify that they are satisfied in the continuum limit. In Fig. 2 we show the log of the L_2 norm of the violation of the three constraints. This is the type of convergence tests that validate numerical results as appropriate approximations to the solution in the continuum limit.

Another illustrative results has to do with the use of the shift. In Fig. 3 we show a snapshot of the g_{rr} metric function at $t = 50M$. In the first case a zero shift is used and the resulting metric function shows a quick growth at a location near the horizon; this effect is called the slice stretching problem. This problem appears because there is no shift that could pull the coordinates out from the horizon, which causes the development of steep gradients in the evolution variables and eventually brings the code to a non-convergent stage and possible blow-up. In the second case we use a none-zero shift that evolves according to the Γ-driver condition (5), which shows a more modest growth of the metric functions, showing in this way that the use of a non-trivial shift helps at keeping the space-time under control.

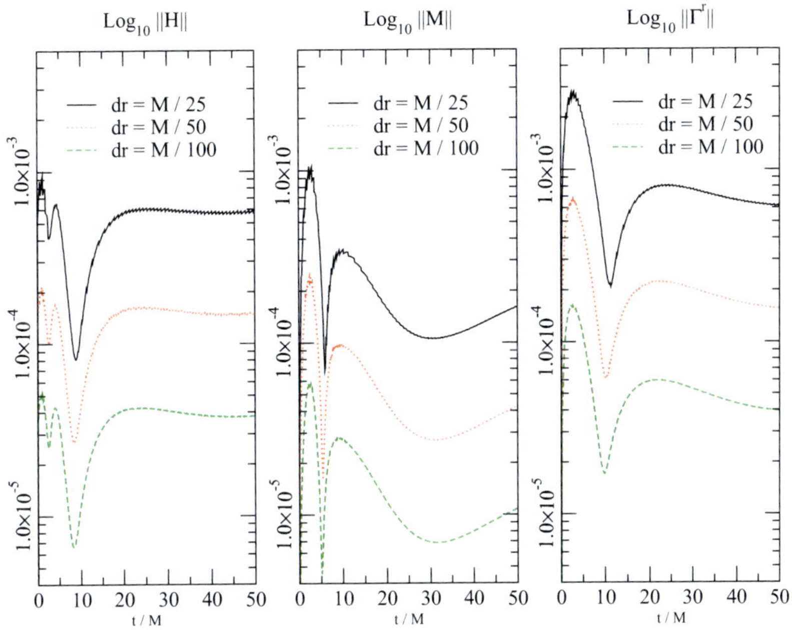

Figure 2. We show the log of the L_2 norm of the violation of the three constraints of the evolution described below. These plots show that the violation of the error tends to zero when the spatial resolution is increased. In fact, as the finite differences approximation we are using is second order accurate the violation should be four times smaller when the resolution is doubled.

In Fig. 4 we show the apparent horizon and a bundle of outgoing null rays. Two important notes are in turn: i) since the initial data have been evolved during a finite coordinate time t we can now explore the location of the event horizon, which is approximately located at the surface from which the outgoing null rays diverge, ii) the apparent horizon location is estimated at every time slice and tells nothing about the global structure of the space-time. In practice, what is done in full 3D numerical calculations is that the event horizon is located after the initial data have been evolved during a finite time, and then a set of null 2-surfaces are launched from the future to the past and the event horizon is estimated as the surface at which those null surfaces launched backwards converge to [24]. In our spherically symmetric case we know the surface is a sphere, and moreover, since no important computational restrictions are in turn, we fine-tune a bundle of outgoing null rays and integrate the null geodesic equation toward the future on the fly.

5. Accretion of a Scalar Field into a Black Hole

5.1. The Scalar Field

As an example of a dynamical black hole we present the full non-linear evolution of a scalar field minimally coupled to gravity, that is we introduce the scalar field into Einstein's

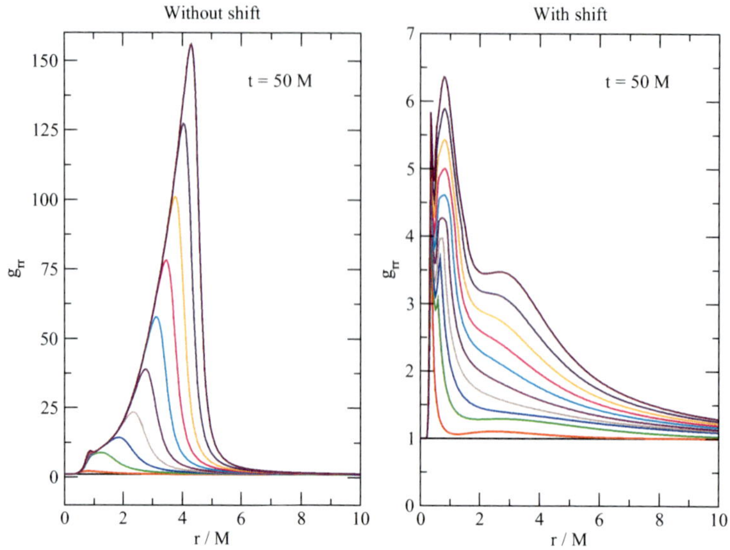

Figure 3. We show snapshots of the metric function g_{rr} with and without shift. The result in the first case is the slice stretching effect whereas a non-trivial shift allows the control of the evolution for longer times since it prevents the involved metric functions from developing important gradients.

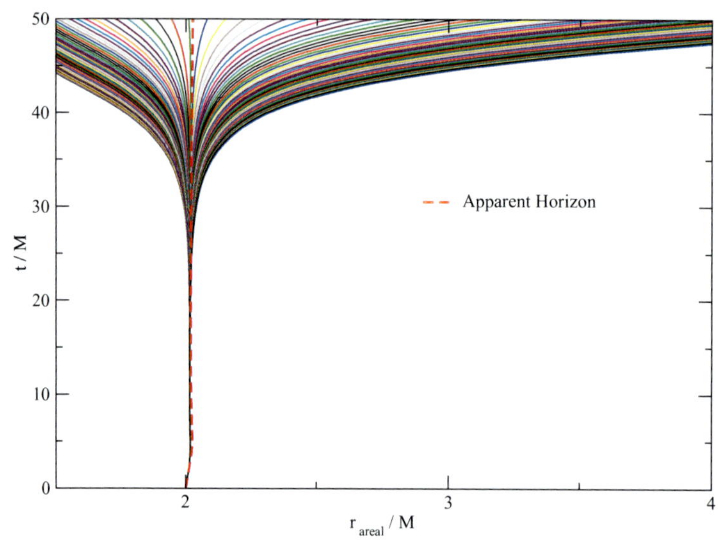

Figure 4. We show the location of the apparent horizon in time and a bundle of outgoing null rays that diverge from a surface that is expected to be the event horizon. This plot is particularly illustrative of the difference of an apparent horizon (a 2-surface, a local property) estimated at every slice, and an event horizon (a 3-surface, a global property) estimated only after a chunk of space-time is available.

equations through the Lagrangian density $\mathcal{L} = -R + \frac{1}{2}g^{\mu\nu}\partial_\mu\Phi\partial_\nu\Phi$, where Φ is the scalar field and R is the Ricci scalar of the space-time.

This introduces a new evolution equation for the scalar field, namely the wave equation

$$\Box\Phi = \frac{1}{\sqrt{-g}}\partial_\mu[\sqrt{-g}g^{\mu\nu}\partial_\nu\Phi] = 0, \tag{19}$$

which using the variables $\pi = \partial_t\Phi$ and $\xi = \partial_r\Phi$ can be cast in the following system of first order evolution equations

$$\begin{aligned}
\partial_t\pi &= \partial_t\beta^r + (-\pi + \beta^r)\partial_t\left[\ln\left(\frac{\sqrt{g_{rr}}g_{\theta\theta}}{\alpha\chi^{3/2}}\right)\right] \\
&+ \partial_r\left[\ln\left(\frac{\sqrt{g_{rr}}g_{\theta\theta}}{\alpha\chi^{3/2}}\right)\right]\left[\left(\frac{\alpha^2\chi}{g_{rr}} - \beta^r\right)\psi + \beta^r\pi\right] \\
&+ \partial_r\left[\left(\frac{\alpha^2\chi}{g_{rr}} - \beta^r\right)\psi + \beta^2\pi\right], \\
\partial_t\xi &= \partial_r\pi,
\end{aligned} \tag{20}$$

that has to be solved at once with the GBSSN evolution equations and those for the lapse and the shift.

5.2. Initial Data

In order to add the contribution of the scalar field it is necessary to solve the constraints at initial time slice which we assume for simplicity to be time-symmetric. For this we use a pulse of scalar field with a Gaussian profile located outside the horizon and solve the Hamiltonian constraint. In order to achieve a smooth coordinate system we use the ansatz that the space-time metric has a form similar to that of a Schwarzschild black hole in isotropic coordinates. Thus we assume that the metric at initial time has the form

$$g_{rr} = 1, \quad g_{\theta\theta} = r^2, \quad \chi = \left(1 + \frac{M}{2r} + u\right), \tag{21}$$

where M is the mass of the apparent horizon and $u = u(r, t = 0)$ is the function to be determined through the solution of the Hamiltonian constraint. The Hamiltonian constraint is thus reduced to the following ODE for u:

$$\partial^2_{rr}u = -\frac{(\partial_r\Phi)^2}{8}\left(1 + \frac{M}{2r} + u\right) - \frac{2\partial_r u}{r}, \tag{22}$$

where we have used the appropriate ρ in (3) with $T_{\mu\nu} = \partial_\mu\Phi\partial_\nu\Phi - \frac{1}{2}g_{\mu\nu}\partial^\alpha\Phi\partial_\alpha\Phi$. We solve (22) first by providing a profile for Φ and consequently for ξ, we also assume that at initial time $\pi = 0$. Then we solve numerically for u using a fourth order Runge-Kutta integrator. Finally the coordinate r is rescaled in order to recover the asymptotic behavior $\chi \to 1$ when $r \to \infty$. The initial scalar field profile used to obtain the following results is $\Phi = 0.25e^{-(r-5)^2}$.

As in the black hole case, we initialize the gauge using a pre-collapsed lapse $\alpha = (1 + M/2r)^{-2}$, a zero shift $\beta^r = 0$ and so its initial time derivative, and also choose the extrinsic curvature components and its trace to vanish so that the time-symmetric condition is satisfied.

These initial data are used to start up a Cauchy evolution using the GBSSN evolution equations mentioned, the gauge conditions and the scalar field equation (20).

6. Results

The results presented in this section are rescaled with the initial apparent horizon mass M^0_{AH}. Due to the accretion of the scalar field into the black hole, what happens is that the area of the hole increases. In Fig. 5 we show the apparent horizon location and also a bundle of outgoing null rays so that the event horizon surface is approximately tracked down.

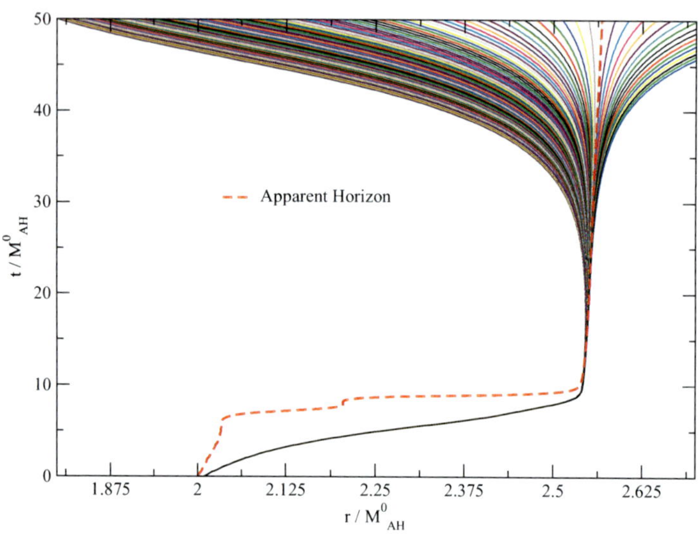

Figure 5. We show the apparent horizon radius in coordinate time and a bundle of outgoing null rays that diverge from the event horizon. Notice that the apparent horizon lives inside the event horizon.

In Fig. 6 we show the ADM, Misner-Sharp and apparent horizon masses. The ADM mass is estimated as the limit for $r \to \infty$ of the Misner-Sharp mass (18). The Misner-Sharp mass has been calculated at $r = 15M$. In Fig. 7 we show a few snapshots of the density of the scalar field measured by an Eulerian observer, so that it is possible to track its location during the evolution. It is natural that because the initial data are time-symmetric, the initial Gaussian splits into two pulses, one traveling toward the black hole and the other one departing from it. Finally in Fig. 8 we show the second order convergence of the runs.

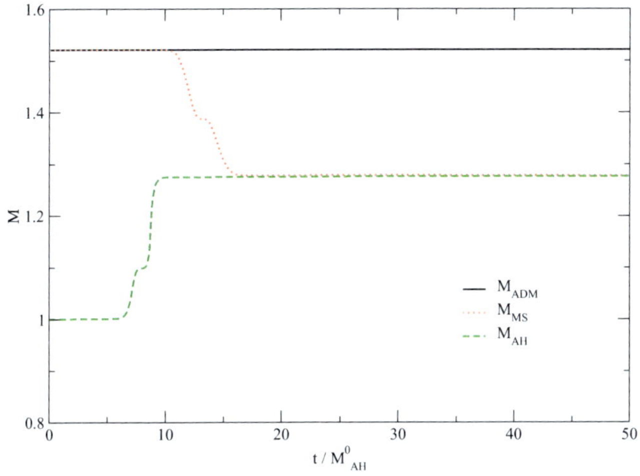

Figure 6. We show the ADM, Misner-Sharp measured at $r = 15M$ and the apparent horizon masses during the evolution. It can be noticed that once the pulse of scalar field has been accreted, the mass of the apparent horizon and the Misner-Sharp mass coincide, which is an indication that the final black hole has stabilized. The difference between the Misner-Sharp mass and the ADM mass at late stages stands for the pulse of the scalar field travelling outwards.

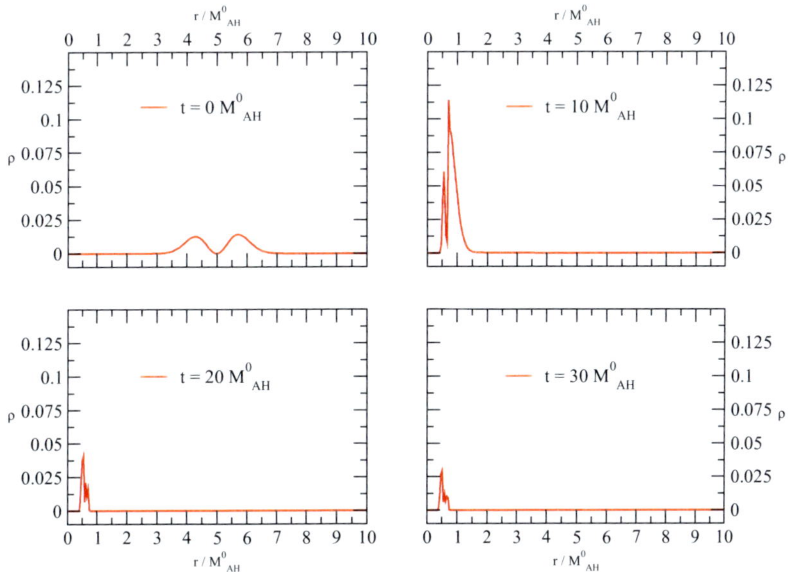

Figure 7. Snapshots of the density ρ of the scalar field is shown. As expected, part of the initial time-symmetric pulse is accreted by the hole, whereas there is another pulse that runs away from the hole. Notice that r is the coordinate radius. The apparent horizon in the coordinates used grows in such a way that at late times (say the last two snapshots) the pulse of scalar field is well inside the apparent horizon.

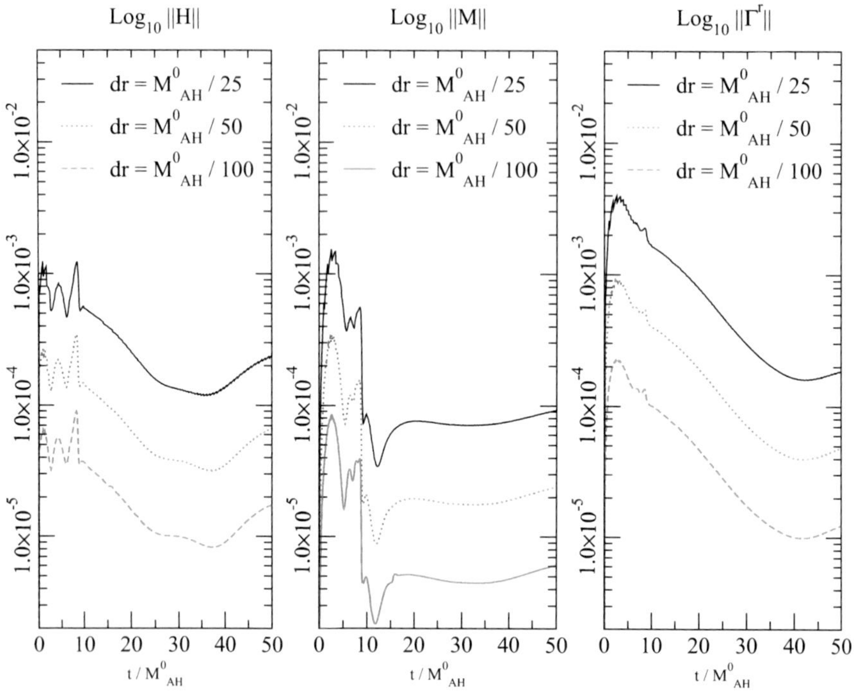

Figure 8. We show the second order convergence of the L_2 norm of the violation of the constraints in time.

7. Final Comments

In this chapter we have presented the numerical and mathematical methods used to evolve a black hole space-time. We have pointed out the obstacles found when using inappropriate coordinates that might be faced when dealing with dynamical space-times, and the use of a non-trivial shift in order to avoid the slice stretching effect.

The use of a non-trivial shift is manifest in the case of the black hole evolution already and becomes crucial in processes involving the black hole accretion and formation.

We have also discussed different techniques used to extract physical information from the numerical simulations. We have emphazised the importance of the apparent and event horizons and the estimation of global and local quantities related to the mass of the system.

We expect that these recipes inspire the reader to get in contact with other general relativistic problems in spherical symmetry and perhaps full 3D systems without symmetries that can be closer to current astrophysical scenarios.

Acknowledgments

This work was supported in part by grants CIC-UMSNH 4.9 and 4.23, PROMEP UMICH-PTC-121, UMICH-PTC-210, UMICH-CA-22 from SEP Mexico, COECyT Michoacan S08-02-28 and CONACyT grant numbers 79601 and 79995.

References

[1] B. Abbott et al. (LSC), *Phys. Rev. Lett.* **94**, 181103 (2005).

[2] B. Abbott et al. (LSC), *Phys. Rev. D* **72**, 082001 (2005). B. Abbott et al. (LSC), *Phys. Rev. D* **72**, 082002 (2005).

[3] B. Abbott et al. (LSC), *Phys. Rev. D* **72**, 102004 (2005).

[4] M. Alcubierre et al. *Phys. Rev. D* **62**, 044034 (2000). M. Alcubierre et al. *Phys. Rev. D* **67**, 084023 (2003). M. Alcubierre et al. *Phys. Rev. D* **72**, 044004 (2005).

[5] T. W. Baumgarte and S. L. Shapiro, *Phys. Rev. D* **59**, 024007 (1999). H-J Yo, T. W. Baungarte, S. L. Shapiro, *Phys. Rev. D* **66**, 084026 (2002).

[6] J. Baker, B. Bruegmann, M. Campanelli, C. Lousto, R Takahashi, *Phys. Rev. Lett.* **87**, 121103 (2001). J. Baker, M. Campanelli, C. Lousto, *Phys. Rev. D* **65**, 044001 (2002).

[7] M. Shibata, T. Nakamura, *Phys. Rev. D* **52**, 5428 (1995).

[8] S. Brandt and B, Bruegmann, *Phys. Rev. Lett.* **78**, 3606 (1997).

[9] E. Gourgoulhon, *J. Phys. Conf. Series* **91**, 012001 (2007).

[10] F. Pretorious, *Phys. Rev. Lett.* **95**, 121101 (2005).

[11] M. Campanelli, C. Lousto, Y. Zlochower, *Phys. Rev. D* **73**, 061501(R), (2006). M. Campanelli, C. Lousto, Y. Zlochower, *Phys. Rev. Lett* **96**, 111101 (2006).

[12] J. Baker, et al. *Phys. Rev. D* **73**, 104002 (2006). J. Baker, et al. *Phys. Rev. Lett.* **96**, 111102 (2006).

[13] M. Campanelli, C. Lousto, Y. Zlochower, B. Krishnan, D. Merrit, Phys. Rev. D 75, 064030, (2007). D. Shoemaker, B. Vaishnav, I. Hinder, F. Herrmann, *Class.Quant.Grav.* **25**, 114047 (2008).

[14] J. A. González, U Sperhake, B. Bruegmann, arXiv:0811.3952 [gr-qc]. A. Nagar, T. Damour, A. Tartaglia, *Class. Quantum Grav.* **24**, S109 (2007).

[15] J. A. González, U. Sperhake, B. Bruegmann, M. Hannam, S. Husa, *Phys. Rev. Lett.***98**, 091101 (2007). J. G. Baker et al. ApJ 653 L93 (2006).

[16] M. Alcubierre, *Introduction to 3+1 Numerical Relativity.* Cambridge University press 2008.

[17] Brown J.D., 2008, *Class. Quantum Grav.,* **25**, 205004.

[18] B. Gustafsson, H-O. Kreiss, J. Oliger, *Time Dependent Problems and Difference Methods*. Wiley-Interscience, 1996.

[19] R. J. LeVeque, in *Numerical methods for conservation laws*. Birkhauser, Basel (1992).

[20] W. H. Press, S. A. Teukolsky, W. T. Watterling and B. P. Flannery, *Numerical Recipes in Fortran*. Cambridge University Press, 1992.

[21] J. P. Boyd, 1989. *Chevyshev and Fourier Spectral Methods,* New York: Springer-Verlag.

[22] I. Hawke, S. Husa and B. Szilagyi, http://numrel.aei.mpg.de/Education/Tutorials/School04/tov

[23] Thornburg J., 1999, *Phys. Rev. D* **59**, 104007, Ibid. arXiv:gr-qc/9906022.

[24] P. Diener, *Class. Quantum Grav.* **20**, 4901 (2003).

In: Black Holes and Galaxy Formation ISBN 978-1-60741-703-3
Editors: A.D. Wachter and R.J. Propst, pp. 195-220 © 2010 Nova Science Publishers, Inc.

Chapter 8

Quantum Loss of Charge by Non-rotating Black Holes with Cosmological Constant

F. Belgiorno[1]****and S.L. Cacciatori***[2†]
[1]Dipartimento di Fisica, Università di Milano, 20133 Milano, Italy, and I.N.F.N., sezione di Milano, Italy
[2]Dipartimento di Fisica, Università dell'Insubria, 22100 Como, Italy, and I.N.F.N., sezione di Milano, Italy

Abstract

We review the problem of the spontaneous loss of electric charge by non-rotating black holes with a cosmological constant. The pair-creation process which is allowed in the presence of a charged black hole and causes its discharge is a long-standing topic in the framework of quantum effects on a black-hole background, at least as old as the Hawking effect but still different in its origin. In order to study this process, a very effective method is represented by the Damour-Deruelle-Ruffini approach, which identifies the existence of level-crossing between positive and negative continuous energy states for charged matter fields on the given background as a signal of the quantum instability leading to the discharge of the black hole. After a review of this approach, we apply it to the case of charged non-rotating black holes in the presence of a non-vanishing cosmological constant Λ, and we take into account both the case of Λ negative (Anti-de-Sitter background) and the case of Λ positive (de-Sitter background). The analysis for $\Lambda < 0$ is more difficult, because of the boundary-like behavior of infinity in a AdS background. Level-crossing is shown to be allowed but not ensured, as happens also when $\Lambda = 0$. When Λ is positive, a novel feature appears, indeed level-crossing always occurs, due to the relevant and peculiar presence of two event horizons; nevertheless a significant effect of pair creation is expected only when the potential barrier between the positive and negative energy states is as narrow as possible. For $\Lambda > 0$ also the Nariai geometry and the ultracold cases are considered.

*E-mail address: belgiorno@mi.infn.it
†E-mail address: sergio.cacciatori@uninsubria.it

1. Introduction

The loss of charge of a charged black hole by quantum effects is a phenomenon as old as the well-known phenomenon of black hole evaporation, and the first analysis is due to Gibbons [1] for the case of Reissner-Nordström black holes. The occurrence of quantum discharge of black holes can be related to the presence of a quantum instability which takes place for charged quantum fields in the background of a charged black hole, and makes unstable even extremal black holes, which are not affected by quantum evaporation: spontaneous pair creation of charged particles induces a loss of charge which is associated with a spontaneous decay from extremality to non-extremality of extremal charged black holes. A relation with the phenomenon of pair-creation in external homogeneous electrostatic field, which was discovered by Schwinger in his seminal paper [2], has been soon realized in the original studies upon this topic. See also [3]. The approximation of almost homogeneous field has been also exploited. The main difference between the analysis in the black hole case and Schwinger's calculation is in the fact that in the latter the calculation of the probability of pair creation for unit time and unit volume is exact, whereas in the case of black holes in general only WKB approximation is available, due to computational difficulties. Notwithstanding, at least qualitative tools for pinpointing quantum instability related to the quantum discharge are available, and consist in an interesting mixing of relativistic quantum field theory in curved spacetime and solid state physics: an analysis of relativistic wave equations for charged scalar and Dirac particles leads to the discovery of level-crossing between positive energy states and negative energy ones on the background of a charged black hole, and this is to be associated with the aforementioned instability. This kind of analysis is mainly due to Damour-Deruelle-Ruffini in the middle of the '70th, in a series of papers [6, 7, 8, 9, 10] which has been recently kept into account for the case of Dirac particles in [11, 12]. We also mention in passing the instanton methods introduced by Kim and Page for pair production in external fields [13], also applied to the black hole case [14]. In the present review, we mean to extend our analysis to the case of scalar fields in the background of charged spherosymmetric black holes with cosmological constant. We consider both the case of negative cosmological constant (AdS) and the case of positive cosmological constant, which are extensions of the well-known Reissner-Nordström case.

1.1. AdS Case

AdS background geometry has been considering a challenging field for quantum field theory in different frameworks, including supersymmetry and string theory, both for the well-known relevance of the AdS backgrounds in the AdS-CFT conjecture and in supergravity, and for the peculiar thermodynamical properties of AdS black holes, for which the canonical ensemble is well-defined [15]. We take into account the problem of the discharge of a Reissner-Nordström-AdS black hole by quantum effects, by extending the analysis carried out in [1, 6, 7, 8, 9, 10, 3] to the case of a charged black hole which is asymptotically AdS. This kind of analysis, which is addressed to the case of massive charged scalar fields, represents also a completion on the quantum side of the analysis carried out in Ref. [4], where classical superradiance effects in a Kerr-Newman-AdS black hole background are considered for the case of an uncharged scalar field, both in the case of reflecting boundary conditions and in the case of transparent boundary conditions.

1.2. dS Case

Black holes in de Sitter space are an interesting subject of investigation, both on the theoretical side, because of the contextual presence of a black hole event horizon and of a cosmological event horizon, to be associated with the corresponding quantum emission of thermal radiation [5], and also from the experimental side, due to the actual measurements confirming the presence of a small cosmological constant. The presence of level-crossing according the aforementioned criterion introduced in [6, 7, 8, 9, 10] is a still valid tool for investigating pair-creation of charged scalar particles even in the presence of a positive cosmological constant. Attention is also focused on special cases, like the Nariai and the ultracold ones.

The present analysis represents the extension to the case of a charged scalar field of the analysis carried out in [11, 12] for the case of a massive Dirac field on the same manifolds. As such, parameters in the Figures are purposefully the same as in [11, 12], except for the angular momentum eigenvalue which is of course different and is chosen to be the most favorable for the pair-creation process.
The plan of this chapter is the following: in section 2. we describe the black hole backgrounds of interest for our study; in section 3. the Klein-Gordon equation is written in its (two-components) Hamiltonian form in each background; in section 4. potentials which are relevant to the physical behavior of the scalar particles are discussed and then the fundamental feature of level-crossing is considered. In section 5. the relation between level-crossing and spontaneous discharge of charged black holes is addressed, and further pinpoint some relevant problems for quantum field theory on the considered manifolds. In section 6. we summarize the main results of our analysis.

2. Geometrical Backgrounds

We first consider manifolds where the metric has the following form:

$$ds^2 = -f(r)dt^2 + \frac{1}{f(r)}dr^2 + r^2 d\Omega^2; \tag{1}$$

the function $f(r)$ has to be specified, as well as the manifold. Geometrical parameters of interest are the black hole mass M and charge Q, as well as the cosmological constant Λ. The vector potential associated with the solutions we are interested in is $A_\mu = (-Q/r, 0, 0, 0)$.
Then we take into account some special cases occurring for positive cosmological constant. We use natural units $\hbar = c = G = 1$ and unrationalized electric units.

2.1. RN-AdS Background

As in the case of the Dirac equation [11], the universal covering space of the Reissner-Nordström-AdS black hole geometry is considered (in such a way that closed timelike curves do not occur; see e.g. [15]); we refer to it for simplicity as RN-AdS in the following. The function $f(r)$ for the metric (1) of the RN-AdS manifold ($t \in R$; $r \in (r_+, +\infty)$; $\Omega \in S^2$) is

$$f(r) = 1 + \frac{r^2}{L^2} - \frac{2M}{r} + \frac{Q^2}{r^2}; \tag{2}$$

r_+ is the radius of the black hole event horizon (see also the Appendix), L is a length associated with the cosmological constant Λ by $\Lambda = -3/L^2 \Leftrightarrow L = \sqrt{\frac{3}{|\Lambda|}}$. The following re-parameterization of $f(r)$ in terms of L^2, r_+, r_- is also interesting:

$$f(r) = \frac{1}{L^2 r^2}(r - r_+)(r - r_-)[r^2 + (r_+ + r_-)r + L^2 + r_+^2 + r_-^2 + r_+ r_-], \tag{3}$$

with

$$M = \frac{(r_+ + r_-)(\eta - r_+ r_-)}{2L^2} \tag{4}$$

$$Q^2 = \frac{\eta r_+ r_-}{L^2} \tag{5}$$

$$\eta = L^2 + r_+^2 + r_-^2 + r_+ r_-. \tag{6}$$

2.2. RN-dS Background

For the Reissner-Nordström-dS black hole manifold (RN-dS black hole in the following) ($t \in \mathbb{R}$; $r \in (r_+, r_c)$; $\Omega \in S^2$) the function $f(r)$ for the metric (1) is

$$f(r) = 1 - \frac{2M}{r} + \frac{Q^2}{r^2} - \frac{\Lambda}{3} r^2; \tag{7}$$

We also define $L = \sqrt{\frac{3}{\Lambda}}$.
The equation $f(r) = 0$ is assumed to admit solutions $r_c > r_+ \geq r_- > r_0$; then one obtains

$$f(r) = \frac{1}{L^2 r^2}(r_c - r)(r - r_+)(r - r_-)(r - r_0). \tag{8}$$

r_c is the radius of the cosmological horizon, r_+ is the radius of the black hole event horizon, r_- is the radius of the Cauchy horizon. Moreover, due to the actual lack of a term proportional to r^3, one has $r_0 = -(r_c + r_+ + r_-)$. The above reparameterization of the metric amounts to implementing the following relations between r_c, r_+, r_- and M, Q, L:

$$\begin{aligned} L^2 &= r_c(r_+ + r_- + r_c) + r_+^2 + r_-^2 + r_+ r_- \\ 2L^2 M &= r_c^2 r_+ + r_c r_+^2 + 2 r_c r_+ r_- + r_c^2 r_- + r_c r_-^2 + r_+^2 r_- + r_+ r_-^2 \\ L^2 Q^2 &= r_c r_+ r_- (r_c + r_+ + r_-). \end{aligned} \tag{9}$$

It is not difficult to show that four real zeroes of $f(r) = 0$ exist (and three are positive) if and only if the following conditions are implemented [12]:

$$0 < Q^2 < \frac{L^2}{12} \tag{10}$$

$$M_{extr} \leq M < M_{max}, \tag{11}$$

where

$$M_{extr} = \frac{L}{3\sqrt{6}}\sqrt{1-\sqrt{1-12\frac{Q^2}{L^2}}}\left(2+\sqrt{1-12\frac{Q^2}{L^2}}\right) \tag{12}$$

is the mass of the extremal black hole with $r_- = r_+$, and

$$M_{max} = \frac{L}{3\sqrt{6}}\sqrt{1+\sqrt{1-12\frac{Q^2}{L^2}}}\left(2-\sqrt{1-12\frac{Q^2}{L^2}}\right) \tag{13}$$

is the mass of the black hole with $r_+ = r_c$ (see sect. 2.3.). The analysis of the more general Kerr-Newman-dS case is carried out in [16].

2.3. Nariai Solution

We take also into consideration the special case of the so-called charged Nariai solution [20, 21], which occurs for positive cosmological constant and corresponds to a black hole solution with $r_- < r_+ = r_c$. As known, the metric (2) is no more valid and a suitable transformation is necessary. It can be shown that the manifold can be described by

$$ds^2 = \frac{1}{A}(-\sin^2(\chi)d\psi^2 + d\chi^2) + \frac{1}{B}(d\theta^2 + \sin^2(\theta)d\phi^2) \tag{14}$$

with $\psi \in \mathbb{R}, \chi \in (0,\pi)$ and where $B = \frac{1}{2Q^2}\left(1-\sqrt{1-12\frac{Q^2}{L^2}}\right)$ and $A = \frac{6}{L^2} - B$ are constants such that $\frac{A}{B} < 1$ [20, 21]. We note that there is no warping factor in the metric between the "radial" part and the S^2 part. For an electrically charged black hole we can choose $A_i = -Q\frac{B}{A}\cos(\chi)\delta_i^0$.

2.4. Ultracold Solutions

A further special case of static charged black hole solutions in presence of a positive cosmological constant is the so-called ultracold case [20], where the three horizons coincide: $r_- = r_+ = r_c$. Also in this case the metric (2) is no more valid, and a suitable limit has to be constructed [20]. Actually, one can introduce two different metrics for the ultracold case, and then two possible backgrounds for our analysis.

2.4.1. Ultracold I

A first metric [20] is

$$ds^2 = -\chi^2 d\psi^2 + d\chi^2 + \frac{1}{2\Lambda}(d\theta^2 + \sin^2(\theta)d\phi^2), \tag{15}$$

with $\chi \in (0,\infty)$ and $\psi \in \mathbb{R}$. One gets $\Gamma^0_{01} = \frac{1}{\chi}, \Gamma^1_{00} = \chi, \Gamma^2_{33} = -\sin(\theta)\cos(\theta), \Gamma^3_{23} = \cot(\theta)$. The electromagnetic field strength is $F = \sqrt{\Lambda}\chi d\chi \wedge d\psi$, and we can choose $A_0 = \frac{\sqrt{\Lambda}}{2}\chi^2$ and $A_j = 0, j = 1,2,3$.

2.4.2. Ultracold II

The second allowed metric [20] for the ultracold case is

$$ds^2 = -d\psi^2 + dx^2 + \frac{1}{2\Lambda}(d\theta^2 + \sin^2(\theta)d\phi^2), \tag{16}$$

with $x \in \mathbb{R}$ and $\psi \in \mathbb{R}$. The electromagnetic field strength is $F = -\sqrt{\Lambda}d\psi \wedge dx$, and we can choose $A_0 = \sqrt{\Lambda}x$ and $A_j = 0, j = 1, 2, 3$.

3. Klein-Gordon Hamiltonian

We associate to the Klein-Gordon equation an appropriate Hamiltonian, such that the equation becomes equivalent to a first-order system. We consider scalar massive charged particles on the backgrounds described in the previous section. Coupling is both with the gravitational field and with the electrostatic field of the background solution.

3.1. RN-AdS and RN-dS Cases

We define for both the metrics defined above the following operator:

$$B_0 \equiv \left[\begin{array}{cc} V & I \\ H_0 & V^+ \end{array} \right], \tag{17}$$

where

$$H_0 = -\frac{1}{r^2} f \partial_r (f r^2 \partial_r) + f \left(\mu^2 + \frac{-\nabla^2_\omega}{r^2} \right) \tag{18}$$

(∇^2_ω is the laplacian on the unit sphere) and

$$V = \left(\frac{eQ}{r} \right) = V^+. \tag{19}$$

μ and e correspond to particle parameters mass and charge respectively.
The Klein-Gordon equation can be obtained as follows:

$$i\frac{d}{dt}\vec{g} = B_0 \vec{g}, \tag{20}$$

where $\vec{g} \equiv \left(\begin{array}{c} u \\ v \end{array} \right)$.
It is easy to show that, for $u = u_0(x)\exp(-i\omega t)$ one obtains the following equation

$$\left[H_0 - (V - \omega)^2 \right] u_0 = 0. \tag{21}$$

We mean to perform variable separation for both cases. By defining

$$S_{l,m} \equiv h(r) \otimes Y_{lm} \tag{22}$$

one obtains by variable separation

$$H_0 = \sum_{l,m}^{\oplus} (h_l^{(0)} \otimes I_{lm}), \tag{23}$$

where I_{lm} stays for the identity on the subspace $S_{l,m}$ and

$$h_l^{(0)} = -\frac{1}{r^2} f \partial_r (f r^2 \partial_r) + f \left(\mu^2 + \frac{l(l+1)}{r^2} \right). \tag{24}$$

The Hamiltonian B_l obtained by restricting B_0 to the eigenspaces for the angular operator Δ_Ω is:

$$B_l \equiv \begin{bmatrix} \frac{eQ}{r} & I \\ h_l^{(0)} & \frac{eQ}{r} \end{bmatrix}, \tag{25}$$

We do not discuss herein mathematical properties of B_0. We limit ourselves to point out that both H_0 and $h_l^{(0)}$ play an important role in the constructions; in particular, for each l, $h_l^{(0)}$ is a non-negative operator in the Hilbert space $L^2((r_+, +\infty), \frac{r^2 dr}{f(r)})$ in the RN-AdS case and $L^2((r_+, r_c), \frac{r^2 dr}{f(r)})$ in the RN-dS case. It is not difficult to show that $h_l^{(0)}$ is essentially self-adjoint on $C_0^\infty(r_+, +\infty)$ in the RN-AdS case for

$$\mu^2 L^2 \geq -\frac{5}{4}, \tag{26}$$

whereas it is essentially self-adjoint on $C_0^\infty(r_+, r_c)$ in the RN-dS case for any value of the mass. Details are deferred to [17].

3.2. Nariai Case

It is useful to recall that the non-vanishing Christoffel symbols are the following: $\Gamma^0_{01} = \cot(\chi), \Gamma^1_{00} = \sin(\chi)\cos(\chi), \Gamma^2_{33} = -\sin(\theta)\cos(\theta), \Gamma^3_{23} = \cot(\theta)$.
The Klein-Gordon equation on this manifold is

$$(\partial_t - ieA_0)^2 \phi - \sin(\chi)\partial_\chi(\sin(\chi)\partial_\chi \phi) + \frac{\sin^2(\chi)}{A}(-B\Delta_\Omega + \mu^2)\phi = 0. \tag{27}$$

A first order equation can be obtained as in the previous sections, with $V = eA_0$ and $H_0 = -\sin(\chi)\partial_\chi(\sin(\chi)\partial_\chi) + \frac{\sin^2(\chi)}{A}(-B\Delta_\Omega + \mu^2)$. The Hamiltonian B_l obtained by restricting B_0 to the eigenspaces for the angular operator Δ_Ω is:

$$B_l \equiv \begin{bmatrix} eA_0 & I \\ h_l^{(0)} & eA_0 \end{bmatrix}, \tag{28}$$

with

$$h_l^{(0)} = -\sin(\chi)\partial_\chi(\sin(\chi)\partial_\chi) + \frac{\sin^2(\chi)}{A}(Bl(l+1) + \mu^2). \tag{29}$$

The coordinate transformation

$$x = \log(\tan(\frac{\chi}{2})) \longleftrightarrow \chi = 2\arctan(\exp(x)) \tag{30}$$

is such that $x \in \mathbb{R}$ and, furthermore, $h_l^{(0)}$ becomes

$$h_l^{(0)} = -\partial_x^2 + \left(\frac{B}{A} l(l+1) + \frac{\mu^2}{A}\right) \frac{1}{\cosh^2(x)}. \tag{31}$$

3.3. Ultracold Cases

We have to split in two parts our analysis, because two metrics are available in the ultracold case.

3.3.1. Ultracold I

One gets the $\Gamma^0_{01} = \frac{1}{\chi}, \Gamma^1_{00} = \chi, \Gamma^2_{33} = -\sin(\theta)\cos(\theta), \Gamma^3_{23} = \cot(\theta)$.
Calculations which are strictly analogous to the ones performed in the Nariai case and the variable change $\chi = \exp(x)$, $x \in \mathbb{R}$, lead to the following reduced Hamiltonian:

$$B_l \equiv \begin{bmatrix} eA_0 & I \\ h_l^{(0)} & eA_0 \end{bmatrix}, \tag{32}$$

with $A_0 = \frac{\sqrt{\Lambda}}{2} \exp(2x)$ and with

$$h_l^{(0)} = -\partial_x^2 + (2\Lambda l(l+1) + \mu^2) \exp(2x). \tag{33}$$

3.3.2. Ultracold II

One gets $\Gamma^2_{33} = -\sin(\theta)\cos(\theta), \Gamma^3_{23} = \cot(\theta)$.
Again calculations as above lead to the following reduced Hamiltonian:

$$B_l \equiv \begin{bmatrix} e\sqrt{\Lambda}x & I \\ -\partial_x^2 + (2\Lambda l(l+1) + \mu^2) & e\sqrt{\Lambda}x \end{bmatrix}, \tag{34}$$

and then

$$h_l^{(0)} = -\partial_x^2 + (2\Lambda l(l+1) + \mu^2). \tag{35}$$

4. Level-Crossing

In this section, as in the Dirac case, we limit ourselves to the study of the case where essential self-adjointness is implemented without imposing boundary conditions.
In the RN black hole background, the presence of an effect of particle creation can be related to the Klein paradox, i.e. to the possibility to find regions where negative energy states overlap with positive energy states. Our initial discussion is limited to the RN-AdS and to the RN-dS cases, because this allows a straightforward comparison with previous literature. The remaining cases can be treated analogously.
We follow Ruffini-Damour-Deruelle approach [6, 7, 8, 9], in which one identifies the effective potentials $E_0^{\pm}(r)$ in the RN case (and also in the Kerr case) for the positive and negative energy states respectively (cf. eq. (40) below); they represent the classical turning

points for the ptc. motion and lead to the definition of the so-called effective ergosphere. These potentials enter the Hamilton-Jacobi (HJ) equation for a classical particle. They can be interpreted also at the quantum level, as in [9]. In particular, they indicate the regions of level-crossing between positive and negative energy states [7, 8]. See also [10]. Level-crossing amounts to the presence of overlap between the range of E_0^+ and the range of E_0^-, signalling the possibility of a tunneling between positive energy states and negative energy ones; the latter phenomenon is in turn interpreted as pair-creation at the barrier potential, and is strictly related to the Klein paradox (incidentally, it could be called for this reason also Klein effect [18]). We give some details below.

We first recall the HJ equation and find $E_0^{\pm}(r)$, then we show that it is possible to find a level-crossing even in the RN-AdS case. The HJ equation for a classical ptc. is

$$g^{\mu\nu}(\partial_\mu S - eA_\mu)(\partial_\nu S - eA_\nu) + \mu^2 = 0. \tag{36}$$

Variable separation leads to $S = -\omega t + \Theta(\theta, \phi) + R(z)$, where $dz = -dr/\Delta$, with $\Delta = r^2 f(r)$, and

$$\left(\frac{dR}{dz}\right)^2 = -Z, \tag{37}$$

where

$$Z \equiv \Delta(\mu^2 r^2 + K) - [\omega r^2 - eQr]^2, \tag{38}$$

and, in the RN case, $\Delta = r^2 - 2Mr + Q^2$ (K is square of the angular momentum vector) [7]. The classical region of accessibility is defined by $Z \leq 0$. One has

$$\Delta^2 \left(\frac{dR}{dr}\right)^2 = r^4(E - E_0^+(r))(E - E_0^-(r)), \tag{39}$$

with

$$E_0^{\pm}(r) = \frac{eQ}{r} \pm \frac{1}{r^2}\sqrt{\Delta(\mu^2 r^2 + K)}. \tag{40}$$

From a classical point of view, $E_0^{\pm}(r)$ represent effective potentials for the solutions having positive and negative energy, and particles states are defined for $E > E_0^+(r)$. Classical bound states correspond to circular or elliptical orbits and require the presence of a couple of turning points. States such that $E_0^-(r) < E < E_0^+(r)$ are forbidden (they correspond to particles having imaginary momentum). States with $E < E_0^-(r)$ are also classically forbidden, because they correspond to particles with negative mass (and negative energy). The latter states are meaningful at the quantum level (anti-particles)[7, 8].
Variable separation in the quantum case allows to obtain an obvious improvement of the above classical formulas, amounting in replacing the classical value of K with the quantum eigenvalues of the corresponding quantum operator $\hat{K}$. The analogy with the replacement of the classical centrifugal term $\frac{\vec{L}^2}{2mr^2}$ with the quantum term $\frac{\hbar^2 l(l+1)}{2mr^2}$ in the spherosymmetric case for the Schrödinger equation is self-evident. In subsection 4.1., the meaning of the aforementioned improved potentials will be discussed.

It can be shown that it is possible to find level-crossing both in the RN-AdS case, where $\Delta = r^2 - 2Mr + Q^2 + r^4/L^2$, and in the RN-dS case, where $\Delta = r^2 - 2Mr + Q^2 - r^4/L^2$.

4.1. The Meaning of $E_0^{\pm}(r)$

Let us consider the Hamiltonian B_l (25), and let us heuristically separate the part containing potentials from the one containing the differential operator:

$$B_l = \begin{bmatrix} 0 & 0 \\ -\frac{1}{r^2} f\partial_r (f r^2 \partial_r) & 0 \end{bmatrix} + P_l(r), \tag{41}$$

where

$$P_l(r) = \begin{bmatrix} \frac{eQ}{r} & I \\ f\left(\mu^2 + \frac{l(l+1)}{r^2}\right) & \frac{eQ}{r} \end{bmatrix}. \tag{42}$$

Let us also calculate the eigenvalue $\lambda^{\pm}(r)$ (we omit for simplicity of notation the obvious dependence on l) of the matrix $P_l(r)$; one easily obtains

$$\lambda^{\pm}(r) = E_0^{\pm}(r), \tag{43}$$

i.e., the potentials coincide with the eigenvalues of the matrix potential term in the reduced Hamiltonian. Moreover, one has

$$E_0^{\pm}(r) = \frac{eQ}{r} \pm \frac{1}{r^2}\sqrt{\Delta(\mu^2 r^2 + K)} = \frac{eQ}{r} \pm \sqrt{f(r)(\mu^2 + \frac{k^2}{r^2})}. \tag{44}$$

Note again that the square of the classical angular momentum term is replaced by the eigenvalues $l(l+1)$ for the quantum angular momentum. Notice also that E_0^+, as a function of μ^2, is increasing, and the same is true for its dependence on l; both these properties are reversed in the case of E_0^-, which is in fact decreasing. As a consequence, one expects a more difficult level-crossing for increasing values of μ^2 or l. Is it also to be noted that they appear in the expression $\mu^2 r^2 + l(l+1)$, and then the angular momentum contribution is suppressed by a factor $\frac{1}{r^2}$ with respect to the μ^2 contribution, and the angular momentum contribution is maximal near r_+.

We point out that also in the Nariai case and the ultracold ones the potentials $E^{\pm}$ coincide with the eigenvalues $\lambda^{\pm}$ of the potential part of the operator B_l, and are displayed in the following subsections.

4.2. Level-Crossing in the RN-AdS Case

It is useful to define for $eQ > 0$

$$P(r) \equiv E_0^+(r) - \frac{eQ}{r_+} = -\frac{eQ}{rr_+}(r - r_+) + \frac{1}{r^2}\sqrt{\Delta(\mu^2 r^2 + l(l+1))}, \tag{45}$$

and

$$N(r) \equiv E_0^-(r) - \frac{eQ}{r_+} = -\frac{eQ}{rr_+}(r - r_+) - \frac{1}{r^2}\sqrt{\Delta(\mu^2 r^2 + l(l+1))}. \tag{46}$$

It is evident that, for any $r > r_+$, one has $N(r) < 0$ (for $eQ < 0$ an obvious change in the definitions above is required). This means that $P(r) < 0$ is the condition which has to be satisfied in order to find a level-crossing ($P(r) \geq 0$ for all $r > r_+$ means $E_0^+ > E_0^-$ in the same region, i.e., no level-crossing). One has to study the roots of the equation $P(r) = 0$ for $r \geq r_+$ and then find out the sign of $P(r)$ between two consecutive roots. Of course $r = r_+$ is a root of $P(r)$. It is necessary to find another root for $r > r_+$ in order to find level-crossing. It is useful to write

$$\Delta = (r - r_+)(r - r_-)G(r), \tag{47}$$

where $G(r) \equiv (1/L^2)[r^2 + (r_+ + r_-)r + L^2 + r_+^2 + r_-^2 + r_+r_-]$. r_+ and r_- are the event horizon radius and the Cauchy horizon radius respectively. One has to implement $P(r) < 0$, which amounts to

$$\frac{1}{r^2L^2}(r-r_+)(r-r_-)(r^2+(r_+ + r_-)r+\eta)(\mu^2r^2+l(l+1)) < \left(\frac{eQ}{r_+}\right)^2 (r-r_+)^2; \tag{48}$$

then, for the extremal case one obtains

$$\frac{1}{r^2L^2}(r^2 + 2r_+r + 3r_+^2 + L^2)(\mu^2r^2 + l(l+1)) < \left(\frac{eQ}{r_+}\right)^2, \tag{49}$$

i.e.

$$\mu^2r^4 + 2r_+\mu^2r^3 + \left((3r_+^2 + L^2)\mu^2 + l(l+1) - \left(\frac{eQ}{r_+}\right)^2 L^2\right) r^2$$
$$+2l(l+1)r_+r + l(l+1)(3r_+^2 + L^2) < 0.$$

From a naive inspection, one sees that the following necessary condition has to be implemented:

$$(3r_+^2 + L^2)\mu^2 + l(l+1) - \left(\frac{eQ}{r_+}\right)^2 L^2 < 0. \tag{50}$$

A more general criterion for the extremal case, which holds true whichever sign of eQ is given, is the following:

$$\frac{dE_0^+}{dr}(r_+)\frac{dE_0^-}{dr}(r_+) > 0, \tag{51}$$

which means that both potentials are increasing or decreasing near $r = r_+$. One has to impose

$$\frac{dE_0^+}{dr}(r_+)\,\frac{dE_0^-}{dr}(r_+) = \frac{1}{r_+^2}\left[\Phi_+^2 - \frac{1}{L^2}\left(6r_+^2 + L^2\right)\left(\mu^2 + \frac{l(l+1)}{r_+^2}\right)\right] > 0, \tag{52}$$

and then, under the hypotheses $\frac{l(l+1)}{r_+^2} \ll 1$ and $r_+ \ll L$, one obtains

$$\frac{dE_0^+}{dr}(r_+)\,\frac{dE_0^-}{dr}(r_+) \simeq \frac{1}{r_+^2}\left(\Phi_+^2 - \mu^2\right) > 0, \tag{53}$$

where we define

$$\Phi_+ = \frac{eQ}{r_+}, \tag{54}$$

and then also the standard condition

$$\left(\frac{eQ}{r_+}\right)^2 > \mu^2, \tag{55}$$

for the standard Reissner-Nordström case is implemented.

In the general case, it is difficult to find out analytically the overlap region between positive and negative energy states. It is still possible to find out level-crossing by means of graphical tools. We show below two examples. They concern the case of an extremal RN-AdS black hole and of a non-extremal one, respectively. In Figure 1 the extremal case is displayed in the case that the particle and the black hole have charge with the same sign; for the same parameters, in Figure 2 the opposite sign case is shown. Analogously, in Figure 3 and Figure 4 the non-extremal case is displayed. Note also that, for fixed values of the mass μ for which level-crossing occurs at $l = 0$, one expects to find out values of $l > 0$ such that level-crossing disappear. We do not show explicitly this phenomenon, we limit ourselves to indicate in the captions the values of l such that level-crossing disappear in the following Figures.

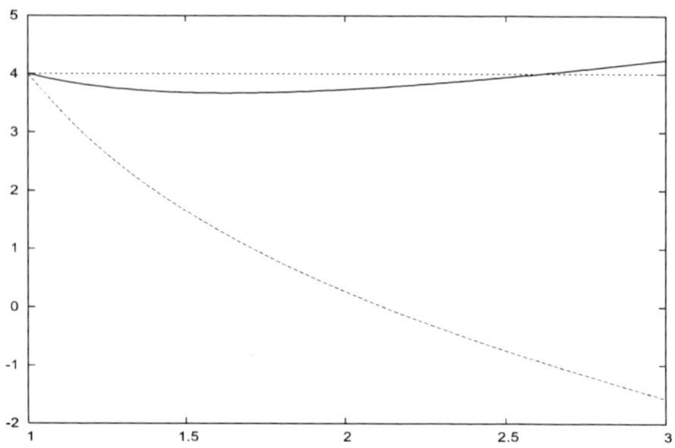

Figure 1. Level-crossing in the case of an extremal RN-AdS black hole, with $L = 1, M = 3, Q = 2, \mu = 1, e = 2, l = 0$. The particle and the black hole have charges with the same sign. Black hole horizon occurs for $r = 1$. The straight line represents the value eQ/r_+. The upper potential is $E_0^+(r)$, the lower one is $E_0^-(r)$. Level-crossing occurs where $E_0^+(r) < eQ/r_+$, and disappears for $l \geq 1$.

4.3. Level-Crossing in the RN-dS Case

It is useful to define, analogously to (54),

$$\Phi_c = \frac{eQ}{r_c}. \tag{56}$$

There is a peculiar property in the case of RN-dS black holes: due to the inequality $\Phi_c \neq \Phi_+$, an overlap is always present, as it was shown for the first time for Dirac particles in [12], indeed one gets also for scalar particles

$$\lim_{r \to r_+} E_0^{\pm}(r) = \Phi_+, \tag{57}$$

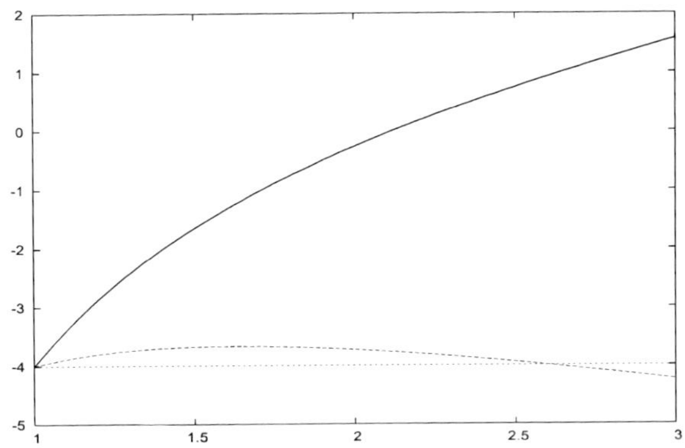

Figure 2. Level-crossing in the case of an extremal RN-AdS black hole, with $L = 1, M = 3, Q = 2, \mu = 1, e = -2, l = 0$. The particle and the black hole have charges with opposite sign. Black hole horizon occurs for $r = 1$. The straight line represents the value eQ/r_+. The upper potential is $E_0^+(r)$, the lower one is $E_0^-(r)$.

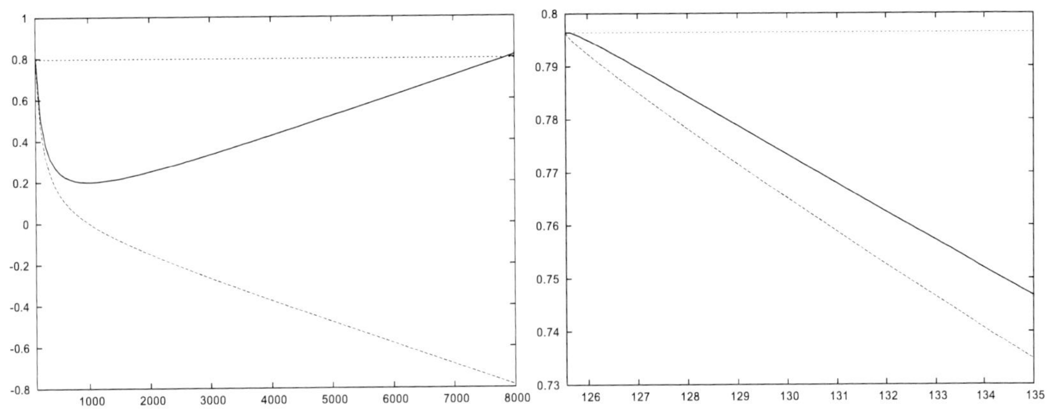

Figure 3. Level-crossing in the case of a non-extremal RN-AdS black hole, with $L = 10$,$M = 10000$,$Q = 100$, $\mu = 10^{-3}$,$e = 1$, $l = 0$. The particle and the black hole have charges with the same sign. Black hole horizon occurs for $r \sim 125.56$. The straight line represents the value eQ/r_+. The upper potential is $E_0^+(r)$, the lower one is $E_0^-(r)$. Level-crossing occurs where $E_0^+(r) < eQ/r_+$, and disappears for $l \geq 7$. The figure on the right displays the potentials near $r = r_+$.

and

$$\lim_{r \to r_c} E_0^{\pm}(r) = \Phi_c. \tag{58}$$

For $eQ > 0$, one finds

$$E_0^-(r_+) = \Phi_+ > \Phi_c = E_0^+(r_c), \tag{59}$$

and level-crossing occurs for energy ω such that

$$\Phi_c = \min E_0^+(r) \leq \omega \leq \max E_0^-(r) = \Phi_+. \tag{60}$$

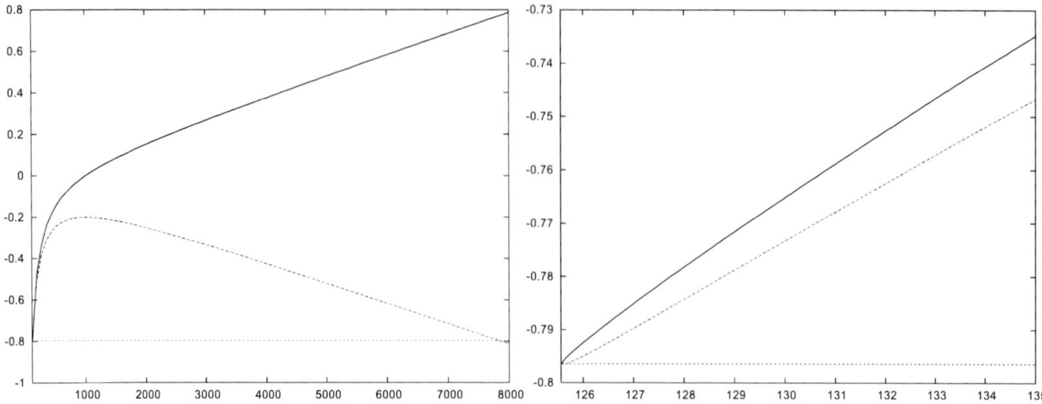

Figure 4. Level-crossing in the case of a non-extremal RN-AdS black hole, with $L = 10$,$M = 10000$,$Q = 100$, $\mu = 10^{-3}$,$e = -1$, $l = 0$. The particle and the black hole have charges with opposite sign. Black hole horizon occurs for $r \sim 125.56$. The straight line represents the value eQ/r_+. The upper potential is $E_0^+(r)$, the lower one is $E_0^-(r)$. The figure on the right displays the potentials near $r = r_+$.

For $eQ < 0$ the overlap still exists:

$$E_0^-(r_c) = \Phi_c > \Phi_+ = E_0^+(r_+), \tag{61}$$

and then level-crossing takes place for

$$\Phi_+ \leq \omega \leq \Phi_c. \tag{62}$$

We stress that this feature, which characterizes RN-dS black holes and arises because of the occurrence of two event horizons, is novel in the framework of this kind of phenomenon, and represents a further signature distinguishing black hole manifolds in presence of a positive cosmological constant with respect to previously analyzed cases.
As to the instability implied by this always present level-crossing, we can easily conclude that it is a necessary condition for the discharge process, but *per se* it is not yet enough for concluding that an efficient pair-creation process takes place on the given manifold. The point is that also the largeness of the potential barrier to be overcome has to be taken into account, and a priori it can still be as large as the whole manifold, thus making negligible the aforementioned process. This is made clear by the following Figures 5-11, where we keep fixed the geometric background parameters L, M, Q and the charge e of the scalar particle and consider two sample values of the particle mass μ. On the left, for the larger value of μ, a level-crossing with a barrier as large as $\sim r_c - r_+$ is found, and then a negligible pair-creation process is expected. On the right, for a smaller mass μ, we obtain instead a level-crossing associated with a much smaller extent of the barrier. The latter case is expected to be involved in a effective phenomenon of pair creation at the barrier. In Figures 5, 7 and 8 non-extremal cases are displayed. The dependence on l for fixed μ is shown in Figure 6. With respect to the AdS case, one cannot find the disappearing of level-crossing for sufficiently high values of the angular momentum, because the above discussion implies

that level-crossing always occurs in the dS case, whichever value of l (and also of μ) is considered. In Figure 8 we show the details of the potentials near the horizons. In Figure 9, an extremal case is also shown. In Figure 10 the so-called lukewarm case [22, 20], which is such that the same temperature occurs in the case of the cosmological horizon and of the black hole event horizon, still with $r_+ < r_c$ (this happens for $Q = M$). Even if the latter case is special, no peculiar behavior appears, as to the pair-creation phenomenon. The agreement with the case of Dirac particles analyzed in [12] is evident. A narrower barrier occurs because a smaller minimal angular momentum eigenvalue is allowed in the scalar case ($l = 0$ to be compared with $|k| = 1$ in the Dirac case).

As for the AdS case, we look for the conditions allowing to obtain a sensible level-crossing with a narrow potential barrier, and also try to determine the relevance of the standard condition (55) for RN-dS black holes. As for the Dirac case [12], we can show that (55) is sufficient in the extremal black hole case under the hypothesis $\frac{l(l+1)}{r_+^2} \ll 1$ (which can be implemented in the case of scalar field by imposing e.g. $l = 0$), which allows to neglect the angular momentum contribution in the potentials, and also for $r_+ \ll r_c$. In the extremal case, condition (51) holds. One has to impose

$$\frac{dE_0^+}{dr}(r_+)\,\frac{dE_0^-}{dr}(r_+) = \frac{1}{r_+^2}\left[\Phi_+^2 - \frac{r_c^2}{L^2}\left(1-\frac{r_+}{r_c}\right)\left(1+3\frac{r_+}{r_c}\right)\left(\mu^2+\frac{l(l+1)}{r_+^2}\right)\right] > 0, \tag{63}$$

and then, under the above hypotheses $\frac{l(l+1)}{r_+^2} \ll 1$ and $r_+ \ll r_c$, one obtains

$$\frac{dE_0^+}{dr}(r_+)\,\frac{dE_0^-}{dr}(r_+) \simeq \frac{1}{r_+^2}\left(\Phi_+^2 - \mu^2\frac{r_c^2}{L^2}\right) > 0, \tag{64}$$

for which, being $\frac{r_c}{L} < 1$, the aforementioned condition (55) is sufficient. In Figure 11 we show an example of extremal black hole where a significant level-crossing occurs but condition (55) is not satisfied. It has the same geometric parameters as in [12].

4.4. Level-Crossing in the Nariai Case

The potentials $E_0^{\pm}(\chi)$ in the Nariai case are

$$E_0^{\pm}(\chi) = eQ\frac{B}{A}\cos(\chi) \pm \sqrt{\frac{\mu^2}{A} + l(l+1)\frac{B}{A}}\sin(\chi). \tag{65}$$

Also in this case level-crossing is always present, being $E_0^+(\pi) < E_0^-(0)$ for $eQ > 0$ and $E_0^+(0) < E_0^-(\pi)$ for $eQ < 0$. Level-crossing occurs for energies ω such that $E_0^+(\pi) \le \omega \le E_0^-(0)$ in the former case and for $E_0^+(0) \le \omega \le E_0^-(\pi)$ in the latter one. Again, the largeness of the potential barrier depends on the choice of the parameters. See Figure 12.

4.5. Level-Crossing in the Ultracold Cases

We obtain

$$E_I^{\pm}(\chi) = -\frac{e\sqrt{\Lambda}}{2}\chi^2 \pm \sqrt{2\Lambda l(l+1) + \mu^2}\,\chi \tag{66}$$

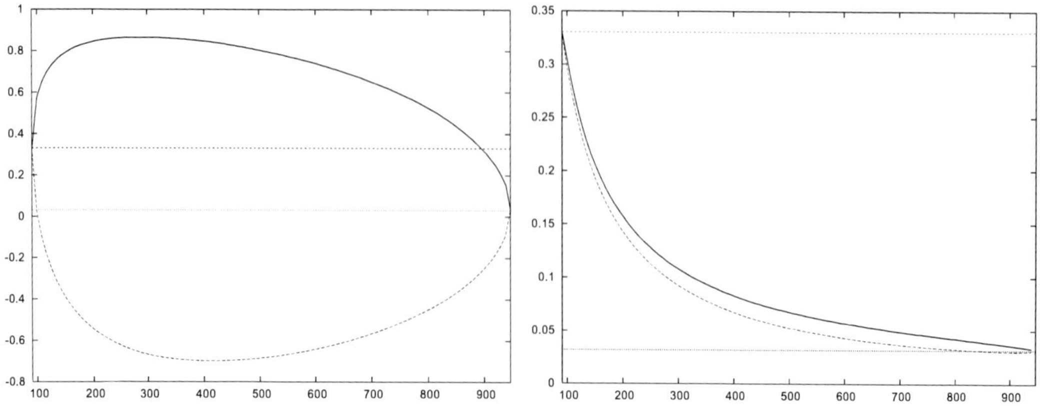

Figure 5. On the left, we display the level-crossing in the case of a non-extremal RN-dS black hole, with $L = 1000, M = 50, Q = 30, \mu = 1, e = 1, l = 0$. The particle and the black hole have charges with the same sign. One finds $r_+ \sim 90.842$, $r_c \sim 946.214$, $r_- \sim 9.999$ and $r_0 \sim -1047.056$. The upper straight line represents eQ/r_+, the lower is eQ/r_c. The upper potential is $E_0^+(r)$, the lower one is $E_0^-(r)$. Level-crossing occurs, but the potential barrier is as large as the whole spacetime region at hand. On the right, the only change with respect to the previous figure stays in the smaller value $\mu = 0.01$ of the scalar particle mass. The upper straight line represents eQ/r_+, the lower is eQ/r_c. The upper potential is $E_0^+(r)$, the lower one is $E_0^-(r)$. Level-crossing occurs in this case with a much smaller extent of the potential barrier with respect to $r_c - r_+$.

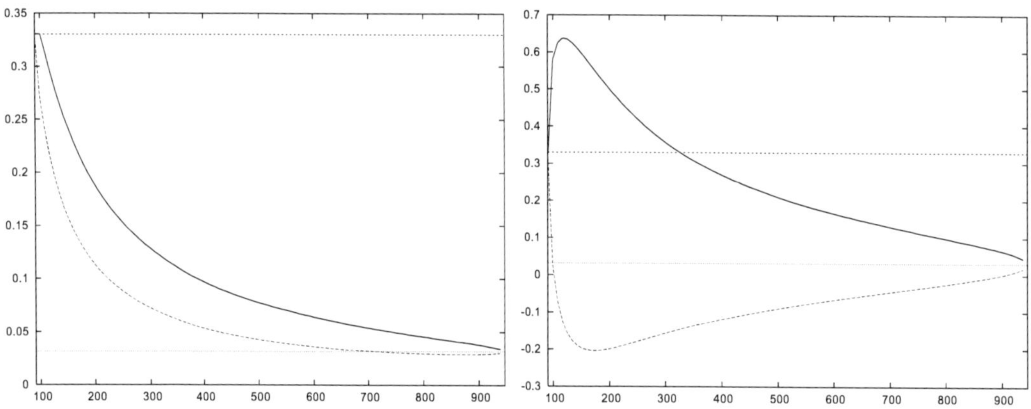

Figure 6. In the previous case of a non-extremal RN-dS black hole, with $L = 1000, M = 50, Q = 30, \mu = 0.01$, we consider two different non-zero values for the angular momentum. On the left, we display the case of $l = 10$. The barrier begins being large. On the right, $l = 100$ is considered, and the barrier becomes very large.

in the case of the first kind of ultracold solution (15), and level-crossing requires that $\omega \leq 0$ for $e > 0$ and $\omega \geq 0$ for $e < 0$. In the case (16) one finds

$$E_{II}^{\pm}(x) = -e\sqrt{\Lambda}x \pm \sqrt{2\Lambda l(l+1) + \mu^2}, \tag{67}$$

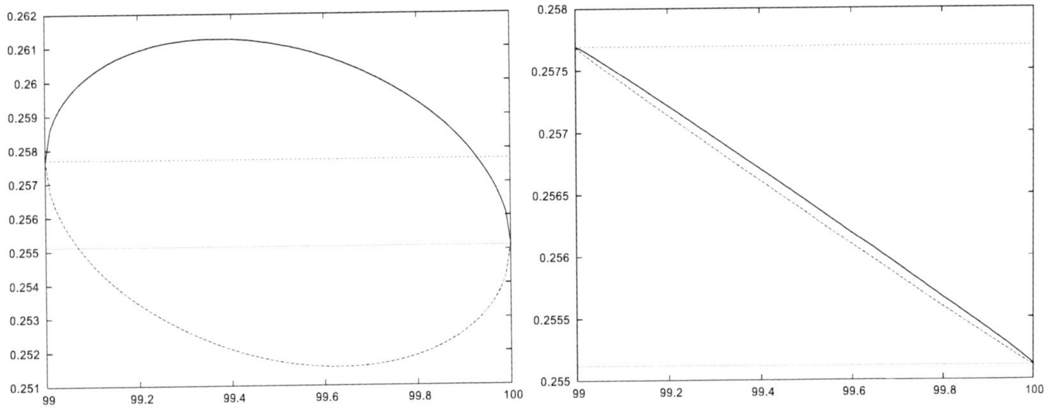

Figure 7. Level-crossing in the case of a non-extremal RN-dS black hole having $L = \sqrt{31791}$, $M = 1193005/31791$ and $Q = 330\sqrt{190}/\sqrt{31791}$, which are such that $r_c = 100$, $r_+ = 99$ and $r_- = 10$. Moreover, we choose $e = 1$, $l = 0$, and the same sign for the black hole charge and for the particle charge. The upper straight line represents eQ/r_+, the lower is eQ/r_c. The figure on the right displays the potentials for $\mu = 1$, and shows that a very large potential barrier occurs. In the figure on the right one has $\mu = 0.01$ and a very narrower potential barrier.

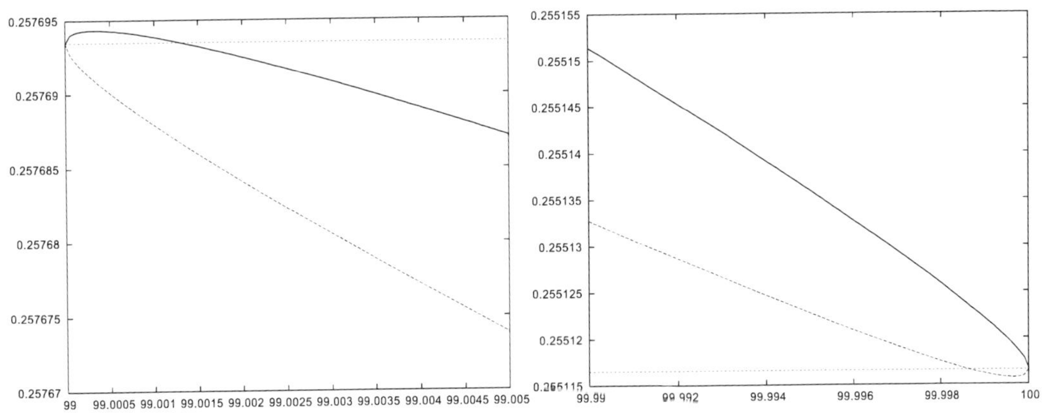

Figure 8. Level-crossing in the case of a non-extremal RN-dS black hole having $L = \sqrt{31791}$, $M = 1193005/31791$ and $Q = 330\sqrt{190}/\sqrt{31791}$, as in the previous figure, with $e = 1$, $l = 0$. We display on the left the presence of a bump near the black hole horizon in the case of E_0^+. Analogously, on the right we show the behavior of the potentials very near the cosmological horizon.

It is evident that for any $\omega \in \mathbb{R}$ one obtains level-crossing. In Figures 13 and 14 level-crossing is displayed for both the ultracold I and the ultracold II cases.

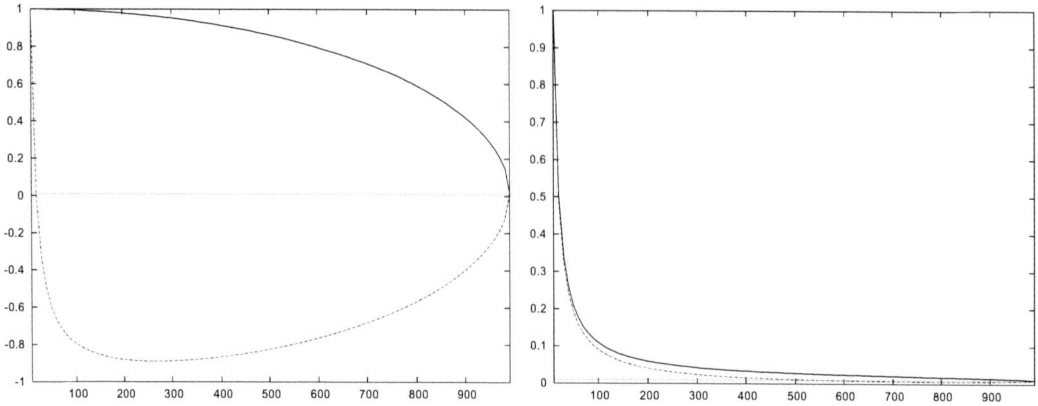

Figure 9. Level-crossing in the case of an extremal RN-dS black hole, with $L = 1000, Q = 10$ and then $M \simeq 9.999499$ (see eq. (12)). Particle parameters $e = 1, l = 0$ are kept fixed, whereas it holds $\mu = 1$ on the left and $\mu = 0.01$ on the right. Level-crossing is more effective in the latter case, and it occurs without any bump near the black hole event horizon.

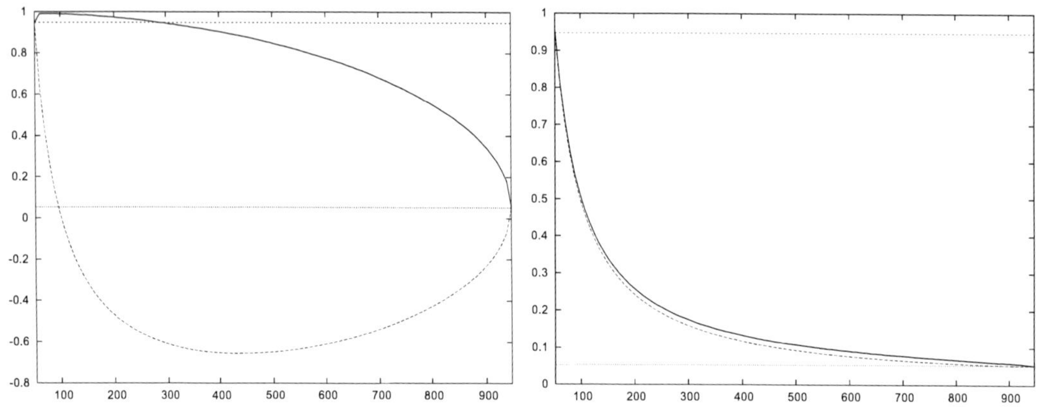

Figure 10. Level-crossing in the case of a lukewarm RN-dS black hole, with $L = 1000, M = 50, Q = 50, e = 1, l = 0$, and with $\mu = 1$ on the left and $\mu = 0.01$ on the right. One finds $r_+ \sim 52.786$ and $r_c \sim 947.214$. Level-crossing is qualitatively similar to the one displayed in Figure 5.

5. Pair Creation and Discharge

Quantum field theory in an external field (like e.g. an external electrostatic field and a static non-flat spacetime metric) displays features which cannot in general be related in a straightforward way to the spectral decomposition of the one-particle Hamiltonian H into positive and negative energy states. This of course happens in the free case in flat spacetime, but, in the presence of external fields, an intrinsic relevance of the classical external potentials in determining the structure of the quantum field theory associated with the given one-particle Hamiltonian has to be stressed. It is evident also from the previous

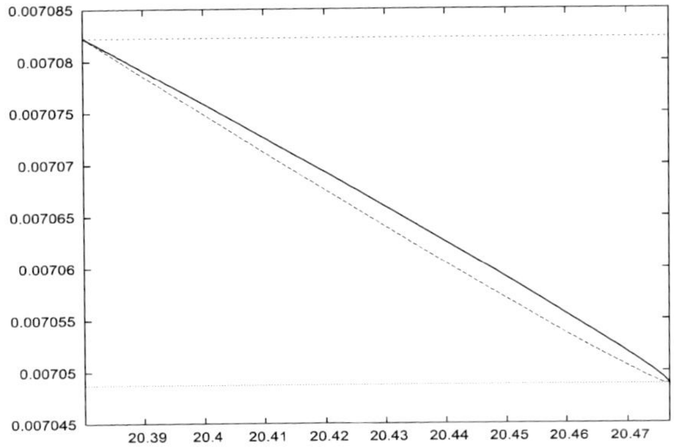

Figure 11. Level-crossing in the case of an extremal RN-dS black hole, with $L = 50, Q = 14.4336845607765725$, which are such that $M = 13.6082252638718051 21$, $r_+ \simeq 20.380115$ and $r_c \simeq 20.476977$. With $\mu = 0.01, e = 0.01, l = 0$ one gets $\Phi_+ \simeq 0.708\mu$, which violates (55).

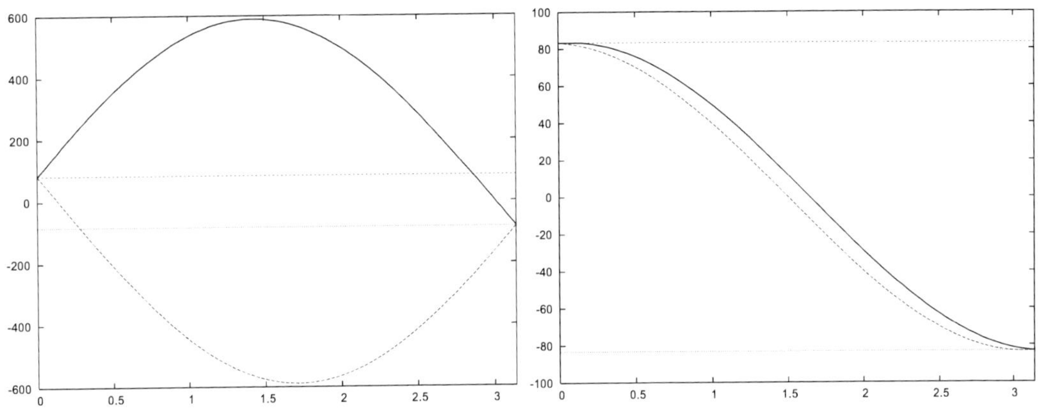

Figure 12. Level-crossing in the case of a Nariai solution. The potentials $E_0^{\pm}(\chi)$ are plotted, with $L = 1000, Q = 80$. Particle parameters $e = 1, l = 0$ are kept fixed; particle mass is chosen to be $\mu = 1$ on the left and $\mu = 0.01$ on the right. Also in the latter case, a bump is present for E_0^+ near $\chi = 0$ and a hollow occurs for E_0^- near $\chi = \pi$.

discussion that the potentials λ^+ and λ^- not only delimit the allowed regions of the classical motion, but also give physical insights about the characteristics of the quantum field theory one can build up by starting from the given H. We have pointed out for the Dirac particle case in [11] that their role is analogous to the one of the potential in non-relativistic quantum mechanics for a single particle: the potential defines the allowed regions of the classical motion, gives rises to quantum resonances and to quantum tunnelling phenomena and its behavior at infinity fixes also the limits of the essential spectrum, and this holds true also for scalar particles. It may also happen that, as in the cases of scalar and Dirac particles on the previously considered black hole backgrounds, a phenomenon of level-crossing occurs, i.e. negative energy states can overlap with positive energy ones. This lack of a clear separation

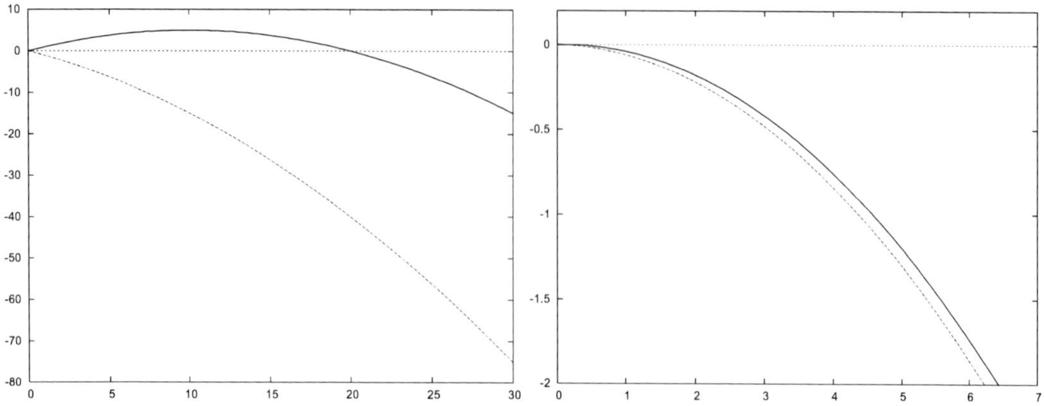

Figure 13. Level-crossing in the case of the ultracold I metric. We choose $\Lambda = 0.01, l = 0, e = 1$ in both cases and $\mu = 1$ on the left and $\mu = 0.01$ on the right. We display only a part of the full plot (but qualitatively all relevant information is given). A different scale is chosen on the left in order to make visible the non-coincidence of the two potentials, which converge both to zero as $\chi \to 0$ and to $-\infty$ as $\chi \to +\infty$.

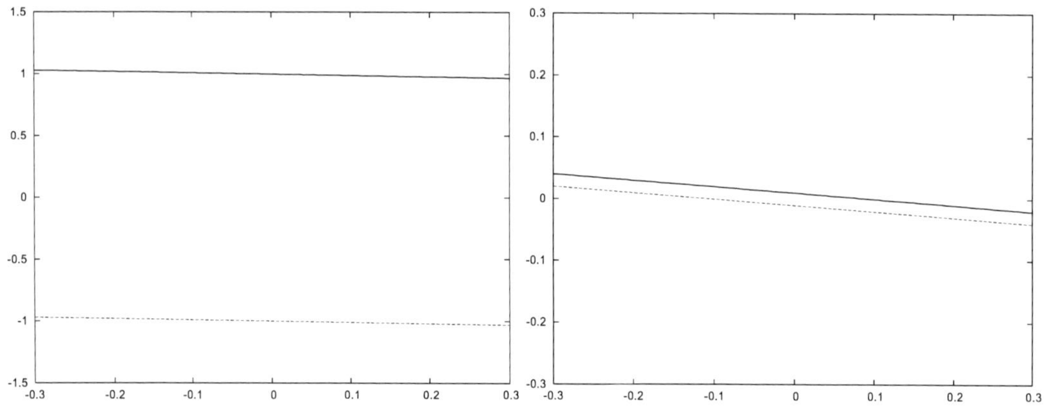

Figure 14. Level-crossing in the case of the ultracold II metric. We choose $\Lambda = 0.01, l = 0, e = 1$ in both cases and $\mu = 1$ on the left and $\mu = 0.01$ on the right. We display only a part of the full plot, which shows of course a linear behavior.

between the positive and the negative energy solutions can be physically interpreted as the occurrence of an instability phenomenon which leads to pair creation [7]. In particular, defining as in Ref. [7] a complete set of positive and negative 'in' modes p_i^{in} and n_i^{in} and a complete set of 'out' modes p_i^{out} and n_i^{out}, with the channel for the transition $n_k^{in} \to p_i^{out}$ a non-vanishing transition amplitude

$$T_{ik} = (p_i^{out}, n_k^{in}) \tag{68}$$

can be associated. The mean number of pairs which are created is [7]

$$< N >= \sum_{\text{channels } i,k} |T_{ik}|^2. \tag{69}$$

This should be compared with the transmission coefficient in the Klein paradox situation, where the transmission coefficient is

$$|T|^2 = \frac{|\text{transmitted flux}|}{|\text{incident flux}|}. \tag{70}$$

The absence of overlapping between the ranges of λ^+ and λ^- avoids the occurrence of level-crossing. When an overlap occurs, the possibility that a positron state can emerge in a suitable range as a positive energy scattering state is at the root of the Klein paradox and also of the pair-creation process. Semi-quantitative estimates for this phenomenon can be given, e.g. in the WKB approximation, as in [7]:

$$|T|^2 = \exp(-2 \int_{\text{barrier}} dz \sqrt{Z}). \tag{71}$$

5.1. A Picture for Spontaneous Quantum Discharge

We start by discussing the AdS case for a scalar field with negatively charged particles. According to the classical potentials described above, the negative energy state with suitable energy can tunnel beyond the barrier and reach the external region, i.e. in the classically allowed region where an observer can measure it and also determine that, by charge conservation, the charge of the black hole is diminished. This transition, in a WKB approximation, is exponentially suppressed. Near infinity, it meets a classical turning point where it is reflected back towards the barrier. The possibility of a further passage of the barrier is exponentially suppressed, and then the whole process of emission-re-absorption of the given state by the black hole is suppressed with respect to the simple emission process. As a consequence, the phenomenon of pair-creation and emission of positively charged anti-particles by a positively charged black hole is allowed also in the RN-AdS black hole manifold. See also Figure 15.

A special feature of the behavior of these emitted particles on RN-AdS black hole background is that they are are mostly confined in the classically allowed region, and they classically behave as bound states with a minimal and a maximal distance from the black hole. For Dirac particles, we have shown that they do not correspond to stable quantum mechanical bound states, because there is no discrete spectrum for the Hamiltonian [11], and, as a consequence, they could at most correspond to resonances (cf. also [7]).

In the dS case, the picture is similar, with the difference that positively charged anti-particles emerge from near the black hole horizon in the classically allowed region (and then they produce the discharge of the black hole) without meeting any more the potential barrier. Analogous considerations hold true for the special Nariai and ultracold cases.

5.2. Remarks on Quantum Field Theory

As in the discussion for the Dirac case displayed in [11, 12], we need to address for completeness some considerations about the problem of the choice of the quantum state playing the role of vacuum. If one were to assume that the positive and negative frequencies associated with the Hamiltonian define the vacuum, one would end up with the so-called

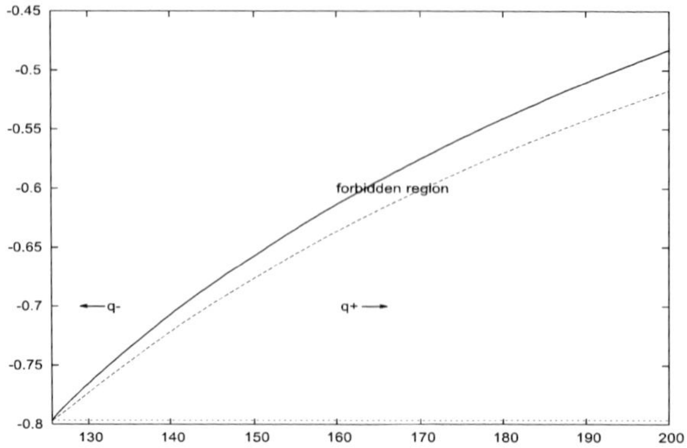

Figure 15. Pair creation in the case of a non-extremal RN-AdS black hole, with $L = 10$,$M = 10000$,$Q = 100$, $\mu = 10^{-3}$,$e = -1$, $l = 0$. The particle and the black hole have charges with opposite sign. Black hole horizon occurs for $r \sim 125.56$. The straight line represents the value eQ/r_+. The upper potential is $E_0^+(r)$, the lower one is $E_0^-(r)$. A pair is also sketched ($q+$ stays for a particle with positive charge, $q-$ for a particle with negative charge).

Boulware vacuum, which is viable as the real vacuum only in the case where the background temperature is zero. This is the case for an extremal RN-AdS black hole ($\Lambda < 0$) and for the ultracold cases ($\Lambda > 0$).

In the non-extremal case RN-AdS case, one has to take into account the fact that the regular quantum state corresponds to the so-called Hartle-Hawking state, which is the thermal state at the Hawking temperature, whereas the Boulware vacuum is singular on the horizon. Suitable analyticity requirements for the fields on the extended manifold lead to the thermal state [1]. Notwithstanding, the thermal state can be obtained by "heating up" the Boulware one at least in the Schwarzschild case, where the construction can be implemented rigorously for the scalar field [23]. In presence of electric charge, like in the present case, there is a further difficulty which is associated with the presence of level-crossing (it has manifestly the characteristics of the Klein paradox in the Reissner-Nordström case) which signals the presence of a further source of particle creation which is different from the one associated with the Hawking effect.
Analogous considerations hold true for the extremal RN-dS case, as well as for the lukewarm RN-dS black hole, and also for the Nariai case, because only a single temperature is involved. Instead, for a generic non-extremal Reissner-Nordström-de-Sitter black hole background, a further difficulty arises due the presence of two different temperatures (a cosmological temperature and of a black hole temperature), involving a true non-equilibrium situation which is more tricky and challenging to be dealt with.

6. Conclusion

In agreement with the analysis for Dirac fields carried out in [11, 12], we have shown that the discharge phenomenon of charged AdS and dS black holes through the quantum

emission of pairs of charged scalar particles can be analyzed by means of the study of the potentials $\lambda^{\pm}$, which were introduced in a semi-classical approximation [7]. In particular, we have heuristically shown how to relate $\lambda^{\pm}$ to the eigenvalues of the potential part of the Hamiltonian B_l. The agreement with the analysis of the Dirac case is remarkable. The main conceptual tool for determining the possibility to get a sensible pair-creation phenomenon has been identified with level-crossing, i.e. with the presence of an overlapping between states with energy $E > \lambda^+$, which correspond to classically allowed states, and states with energy $E < \lambda^-$, which correspond to the classical counterpart of quantum negative energy states. This overlapping, as well-known from previous literature on this topic, indicates the possibility to find out a phenomenon of pair-creation by means of tunneling from the negative energy states to the classically allowed states. WKB approximation can be implemented in order to describe the corresponding tunneling amplitude (see e.g. [12]). Difference between AdS and dS backgrounds have been pointed out, in particular in the dS case there is no condition to be imposed in order to obtain level-crossing; notwithstanding, one has to further ensure that this necessary condition for the discharge becomes also sufficient, by allowing for a suitably narrow potential barrier between positive energy states and negative energy ones.
Quantitative calculations for the pair creation process will be the subject of future investigations.

A. Horizon Radius r_+ for an RN-AdS Black Hole

The radius r_+ of an RN-AdS black hole is determined by the higher real root of

$$1 + \frac{r^2}{L^2} - 2\frac{M}{r} + \frac{Q^2}{r^2} = 0 , \tag{72}$$

or

$$x^4 + x^2 - 2\mu x + q^2 = 0 , \tag{73}$$

where

$$x = \frac{r}{L}, \quad \mu = \frac{M}{L}, \quad q = \frac{Q}{L} . \tag{74}$$

The standard way to solve this equation is to introduce a free parameter y and rewrite the equation in the form

$$(x^2 - y)^2 + 2yx^2 - y^2 + x^2 - 2\mu x + q^2 = 0 , \tag{75}$$

(which is true for any y) and then

$$(x^2 - y)^2 = -(2y+1)\left[x^2 - \frac{2\mu}{2y+1}x + \frac{q^2 - y^2}{2y+1}\right] . \tag{76}$$

The strategy is to choose y so that the r.h.s. of the equation is a square too. This happen if

$$\left(\frac{\mu}{2y+1}\right)^2 = \frac{q^2 - y^2}{2y+1} . \tag{77}$$

Then

$$y^3+\frac{y^2}{2}-q^2y+\frac{\mu^2-q^2}{2}=0\,, \tag{78}$$

or, using $z=y+1/6$

$$z^3-z\left(q^2+\frac{1}{12}\right)-\left(\frac{q^2}{3}-\frac{\mu^2}{2}+\frac{1}{2^23^3}\right)=0\,, \tag{79}$$

This has the form

$$z^3=az+b\,, \tag{80}$$

which can be solved by the Cardano formula

$$z=\left(\frac{b}{2}+\sqrt{\frac{b^2}{4}-\frac{a^3}{27}}\right)^{\frac{1}{3}}+\left(\frac{b}{2}-\sqrt{\frac{b^2}{4}-\frac{a^3}{27}}\right)^{\frac{1}{3}}\,. \tag{81}$$

We get

$$\begin{aligned} y(\mu,q)= & -\frac{1}{6}+\left(\frac{q^2}{6}-\frac{\mu^2}{4}-\frac{1}{6^3}+\sqrt{\sigma(\mu,q)}\right)^{\frac{1}{3}} \\ & +\left(\frac{q^2}{6}-\frac{\mu^2}{4}-\frac{1}{6^3}-\sqrt{\sigma(\mu,q)}\right)^{\frac{1}{3}}\,, \end{aligned} \tag{82}$$

with

$$\sigma(\mu,q)=\frac{\mu^4}{2^4}-\frac{q^6}{3^3}-\frac{\mu^2q^2}{12}+\frac{q^4}{54}+\frac{\mu^2-q^2}{2^43^3}\,. \tag{83}$$

If q is small (take it very small) enough then $\sigma(\mu,q)$ is positive and y is real and from (77) we see that then $2y+1$ is negative and (76) takes the form

$$x^2-y=\pm\left(x\sqrt{|2y+1|}+\frac{\mu}{\sqrt{|2y+1|}}\right)\,. \tag{84}$$

The discriminant is

$$\Delta=-(2y+1)+4y\pm4\sqrt{y^2-q^2}\,, \tag{85}$$

and is positive only if we take the plus sign. Then the real roots satisfy

$$x^2-x\sqrt{|2y+1|}-y-\frac{\mu}{\sqrt{|2y+1|}}=0\,. \tag{86}$$

Note that

$$-y-\frac{\mu}{\sqrt{|2y+1|}}=|y|-\sqrt{y^2-q^2}>0 \tag{87}$$

so that both the roots are positive and

$$r_\pm=\frac{L}{2}\left[\sqrt{|2y+1|}\pm\sqrt{|2y+1|+4\left(y+\sqrt{y^2-\frac{Q^2}{L^2}}\right)}\right]\,, \tag{88}$$

with $y=y(M/L,Q/L)$ given by (82).

References

[1] Gibbons, G. *Comm. Math. Phys.* (1975), vol. 44, 245.

[2] Schwinger, J. *Phys. Rev.* (1951), vol. 82, 664.

[3] Khriplovich, I.B. *Phys. Rep.* (1999), vol. 320, 37.

[4] Winstanley, E. *Phys. Rev. D* (2001) vol. 64, 104010.

[5] Gibbons, G.W.; Hawking, S.W. *Phys. Rev. D* (1977) vol. 15, 2738.

[6] Christodoulou, D.; Ruffini, R. *Phys. Rev. D* (1971) vol. 12, 3552.

[7] Damour, T., Klein paradox and vacuum polarization. In: *Proc. first Marcel Grossmann Meeting on General Relativity*; Ruffini R.; Ed.; North–Holland: Amsterdam, 1977; p. 459.

[8] Deruelle, N., Classical and quantum states in black hole physics. In: *Proc. first Marcel Grossmann Meeting on General Relativity*; Ruffini R.; Ed.; North–Holland: Amsterdam, 1977; p. 483.

[9] Ruffini R.; Deruelle, N. *Phys. Lett. B* (1974) vol. 52, 437.

[10] Ruffini R.; Deruelle, N. *Phys. Lett. B* (1975) vol. 57, 248. Damour, T.; Deruelle, N.; Ruffini R. *Lett. Nuovo Cimento* (1976) vol. 15 (1976), 257. Damour, T. *Lett. Nuovo Cimento* (1975) vol. 12, 315.

[11] Belgiorno, F.; Cacciatori, S.L. *Class. Quant. Grav.* (2008) vol. 25, 105013.

[12] Belgiorno, F.; Cacciatori, S.L. (2008). *Massive Dirac particles on the background of charged de-Sitter black hole manifolds*. ArXiv:0810.1642

[13] Kim, Sang Pyo; Page, Don N. *Phys. Rev. D* (2002) vol. 65, 105002; ibid. (2006) vol 73, 065020; ibid. (2007) vol. 75, 045013.

[14] Kim, Sang Pyo; Page, Don N. *Nuovo Cim.* (2005) vol. 120, 1193.

[15] Hawking, S.W.; Page, Don N. *Commun. Math. Phys.* (1983) vol. 87, 577.

[16] Belgiorno, F.; Cacciatori, S.L. (2008). *Absence of Normalizable Time-periodic Solutions for the Dirac Equation in Kerr-Newman-dS Black Hole Background.* ArXiv:0807.4310

[17] Belgiorno, F.; Cacciatori, S.L. In preparation.

[18] Belgiorno, F.; Martellini, M. *Phys. Lett. B* (1999) vol. 453, 17.

[19] Avis, S.J.; Isham, C.J.; Storey, D. *Phys. Rev. D* (1978) vol. 18, 3565.

[20] Mann, R.B.; Ross, S.F. *Phys. Rev. D* (1995) vol. 52, 2254.

[21] Bousso, R. *Phys. Rev. D* (1999) vol. 60, 063503.

[22] Romans, L.J. *Nucl. Phys. B* (1992) vol. 383, 395.

[23] Kay, B.S.; Wald, R.M. *Phys. Rep.* (1991) vol. 207, 49.

In: Black Holes and Galaxy Formation ISBN 978-1-60741-703-3
Editors: A.D. Wachter and R.J. Propst, pp. 221-247

Chapter 9

BLACK HOLES IN HIGHER ORDER CURVATURE GRAVITY

*S.O. Alexeyev**
Sternberg Astronomical Institute, Lomonosov Moscow State University
Universitetskii Prospekt, 13, Moscow, 119991, RUSSIA

Abstract

Gauss-Bonnet gravity provides one of the most promising frameworks to study curvature corrections to the Einstein action, for example, in supersymmetric string theories, because of avoiding ghosts and keeping second order field equations. Gauss-Bonnet black holes with there original properties: presence of dilatonic hair, down limit on minimal black hole mass, new types of space-time singularities are reviewed in $N = 4$ dimensions. As the endpoint of black-hole evaporation is a very intriguing problem of modem physics, we show (based on the Einstein-dilaton-Gauss-Bonnet four-dimensional string gravity model) that black holes do not disappear completely and should become relics at the end of the evaporation process, so that the standard Bekenstein-Hawking evaporation formula must be modified near the last stages. This solution could represent a model for relic remnants of primordial black holes. Higher order curvature corrections also play an important role in braneworld gravity models, therefore, we discuss the forms of black hole solutions in $N > 4$ dimension gravity models. The possibility of direct and indirect experimental detection of outlined black holes in astrophysics, cosmology and new generation of colliders is discussed as well.

1. Introduction

At the present time the physics of black holes contains a lot of unsolved (and even non-understanded) problems. One of them is the question on the nature of the black hole inner singularities. Studying them we hope to clarify some important aspects of the Cosmic Censorship applications [1, 2, 3, 4, 5, 6]. Moreover we can also examine the boundaries of the applicability of the General Relativity. Another interesting and completely unsolved question is: what is the endpoint of the black hole evaporation? [2, 7, 8]. In order to find

*E-mail address: alexeyev@sai.msu.ru

the more comprehensive solutions of these problems it would be desirable to use the non-minimal gravity model which is the effective low energy limit of some great unification theory. That is why during last years the four-dimensional dilatonic black holes attractes the great attention of scientific community because this type of black holes represents the solution of the string theory at its low energy limit [9, 10, 11, 12, 13, 14, 15, 16, 17].

It is important to note that the string theory predicts the Einstein equations to be modified by higher order curvature corrections in the range where the curvature of space-time has the near-Planckian values. At the present time the form of the higher order curvature corrections in the string effective action is not investigated completely [15]. We do not know the general structure of the expansion, and, hence, the direct summing up is impossible. But as we deal with the expansion, the most important correction is the second order curvature one which is the product of the Gauss-Bonnet and dilatonic terms. It increases the order of the differential equations till the second one and the existence of the dilatonic term makes the contribution of the Gauss-Bonnet term to be dynamic. So, the action is (for simplicity, only bosonian part is taken into account)

$$S = \frac{1}{16\pi}\int d^4x\sqrt{-g}\Big[m_{Pl}^2\left(-R + 2\partial_\mu\phi\partial^\mu\phi\right) + \lambda e^{-2\phi}S_{GB}\Big]. \tag{1}$$

Here ϕ is the dilatonic field and λ is the string coupling constant showing the influence of the Gauss-Bonnet term $S_{GB} = R_{ijkl}R^{ijkl} - 4R_{ij}R^{ij} + R^2$.

The most careful analysis of the discussed model was proceed in Refs [9, 10, 11, 12, 13, 14, 15, 16, 17] but the model predictions are not investigated completely till now. It predicts a change of the solution behavior near singularities or in the regions where the influence of the higher order curvature corrections becomes strong. As the differential equations have a very complicated form, the solutions were obtained by the perturbative [14, 17] or numerical [9, 10, 11, 12, 13] methods. For example, using these methods a new solution called the "neutral Gauss-Bonnet black hole" was found [9, 10, 11].

The main purpose of our work is to discuss external and internal black hole solutions with dilatonic hair and their main properties. This means that we are interesting in static, spherically symmetric, asymptotically flat solutions providing a regular horizon. Therefore, the most convenient choice of metric (which is usually called as the "curvature gauge") is

$$ds^2 = \Delta dt^2 - \frac{\sigma^2}{\Delta}dr^2 - r^2(d\theta^2 + \sin^2\theta d\varphi^2), \tag{2}$$

where $\Delta = \Delta(r), \sigma = \sigma(r)$. We use this curvature gauge (and the Einstein frame) for more convenient comparison with the Schwarzschild solution.

2. Numerical Results

For searching the solution in the maximal widest range of the radial coordinate it was necessary to use the most "strong" method for a numerical integration of the systems of the differential equations with particular points. The problem was solved by a modernization of the methods of integration over the additional parameter [11]. This allowed us to investigate the internal structure and the particular points of the black hole using the analysis of the main determinant zeros of the linear system of the differential equations in the non-evident form. So, the system has the following matrix form

$$a_{i1}\Delta'' + a_{i2}\sigma' + a_{i3}\phi'' = b_i, \tag{3}$$

where $i = 1, 2, 3$, matrices a_{ij} b_i are:

$$\begin{aligned}
a_{11} &= 0,\\
a_{12} &= -m_{Pl}^2\sigma^2 r + 4\mathrm{e}^{-2\phi}\lambda\phi'(\sigma^2 - 3\Delta),\\
a_{13} &= 4\mathrm{e}^{-2\phi}\lambda\sigma(\Delta - \sigma^2),\\
a_{21} &= m_{Pl}^2\sigma^3 r + 4\mathrm{e}^{-2\phi}\lambda\phi' 2\Delta\sigma,\\
a_{22} &= -m_{Pl}^2\sigma^2(\Delta' r + 2\Delta) - 4\mathrm{e}^{-2\phi}\lambda\phi' 6\Delta\Delta',\\
a_{23} &= 4\mathrm{e}^{-2\phi}\lambda 2\Delta\Delta'\sigma,\\
a_{31} &= 4\mathrm{e}^{-2\phi}\lambda\sigma(\Delta - \sigma^2),\\
a_{32} &= 2m_{Pl}^2\sigma^2\Delta r^2\phi' + 4\mathrm{e}^{-2\phi}\lambda\Delta'(-3\Delta + \sigma^2),\\
a_{33} &= -2m_{Pl}^2\sigma^2\Delta r^2\sigma,
\end{aligned}$$

$$\begin{aligned}
b_1 &= -m_{Pl}^2\sigma^3 r^2(\phi')^2 + 4\mathrm{e}^{-2\phi}\lambda\sigma(\Delta - \sigma^2)2(\phi')^2,\\
b_2 &= -m_{Pl}^2\sigma^3(2\Delta' + 2\Delta r(\phi')^2) + 4\mathrm{e}^{-2\phi}\lambda 4\sigma\Delta\Delta'(\phi')^2\\
&\quad -4\mathrm{e}^{-2\phi}\lambda\phi' 2(\Delta')^2\sigma,\\
b_3 &= 2m_{Pl}^2\sigma^3 r\phi'(\Delta' r + 2\Delta) - 4\mathrm{e}^{-2\phi}\lambda(\Delta')^2\sigma.
\end{aligned}$$

Matrix (3) represents the linear system of ordinary differential equations (relatively the oldest derivatives) given in a non-evident form. Hence, according to the existence theorem the system (3) has a single solution only in the case of its main determinant to be not equal to zero. In the case of zero main determinant at some point of the solution trajectory, the uniqueness of the solution (3) will be violated. Different types of zeros of the main system determinant correspond to the different types of the particular points of the solution and only three types of zeros present in the asymptotically flat solution of Eqs. (3). So, as the main determinant has the following structure

$$D_{main} = \Delta\Big[A\Delta^2 + B\Delta + C\Big], \tag{4}$$

where

$$\begin{aligned}
A &= (-32)e^{-4\phi}\sigma^2\lambda^2\Big[4\sigma^2\phi'^2 m_{Pl}^2 r^2 - \sigma^2 m_{Pl}^2 + 12e^{-2\phi}\Delta'\phi'\lambda\Big],\\
B &= (-32)e^{-2\phi}\sigma^4\lambda\Big[\sigma^2\phi' m_{Pl}^4 r^3 + 2e^{-2\phi}\sigma^2\lambda m_{Pl}^2 - 8e^{-4\phi}\Delta'\phi'\lambda^2\Big],\\
C &= 32e^{-4\phi}\sigma^8\lambda^2 m_{Pl}^2 - 2\sigma^8 m_{Pl}^6 r^4 + 64e^{-4\phi}\sigma^6\Delta'\lambda^2 m_{Pl}^2 r\\
&\quad +128e^{-6\phi}\sigma^6\Delta'\phi'\lambda^3,
\end{aligned}$$

zeros of the main determinant are

$$(a)\quad \Delta = 0,\ \ C \neq 0$$

$$(b) \quad A\Delta^2 + B\Delta + C = 0, \;\; \Delta \neq 0, \; C \neq 0 \tag{5}$$
$$(c) \quad \Delta = 0, \; C = 0.$$

The results of numerical integration are shown in the Fig. 1. It presents a dependence of the metric functions Δ (a), σ (b) and the dilaton function $e^{-2\phi}$ (c) against the radial coordinate r. The behavior of the solution outside the horizon has the usual Schwarzchild form. Under the influence of the Gauss-Bonnet term, the structure of the solution inside the horizon changes: new singular point, so-called r_s, appears (see Fig. 1). The type (b) of zero of the main system determinant is realized in this point. The solution turns to the other branch when reaching this point. Only two branches exist near the position r_s. There are no any branches between r_s and the origin. Therefore, a new singular surface with the topology $S^2 \times R^1$ appears (it is an infinite "tube" in the t direction). Such a singularity is absent in the first order curvature gravity. This singularity does exist in the different kinds of metric choice as we tested.

3. Minimal Black Hole

3.1. Numerical Investigations

Fig. 2 shows the graph of the metric function Δ versus the radial coordinate r at the different values of the event horizon r_h when $q = 0$. The curve (a) represents the case where r_h is rather large and is equal to 30.0 Planck unit values (P.u.v.). The curve (b) shows the changes in the behavior of $\Delta(r)$ when r_h is equal to 7.5 P.u.v. The curve (c) represents the boundary case with $r_h = r_{hmin}$ where all the particular points merge and the internal structure disappears. The curve (d) shows the case where $2M \ll r_{hmin}$ and any horizon is absent. Here it is necessary to note that for minimal and near-minimal values of r_h the metric functions can be approximated by the following formulas:

$$\begin{aligned} \Delta &= 1 - \frac{r_h}{r}, \\ \sigma &= 1 - \frac{s_h}{r^8}, \end{aligned} \tag{6}$$

where $s_h = s_h(r_h)$.

3.2. Analytical Investigations

One can also obtain the existence of a minimal dilatonic black hole from the analytical manipulations. This type of a system singular point corresponds to the type (5a) of zeros of the main system determinant. The asymptotic expansions near the position r_h have the form

$$\begin{aligned} \Delta &= d_1(r - r_h) + d_2(r - r_h)^2 + \ldots, \\ \sigma &= s_0 + s_1(r - r_h) + \ldots, \\ \phi &= \phi_{00} + \phi_1(r - r_h) + \phi_2(r - r_h)^2 + \ldots, \end{aligned} \tag{7}$$

where $(r - r_h) \ll 1$.

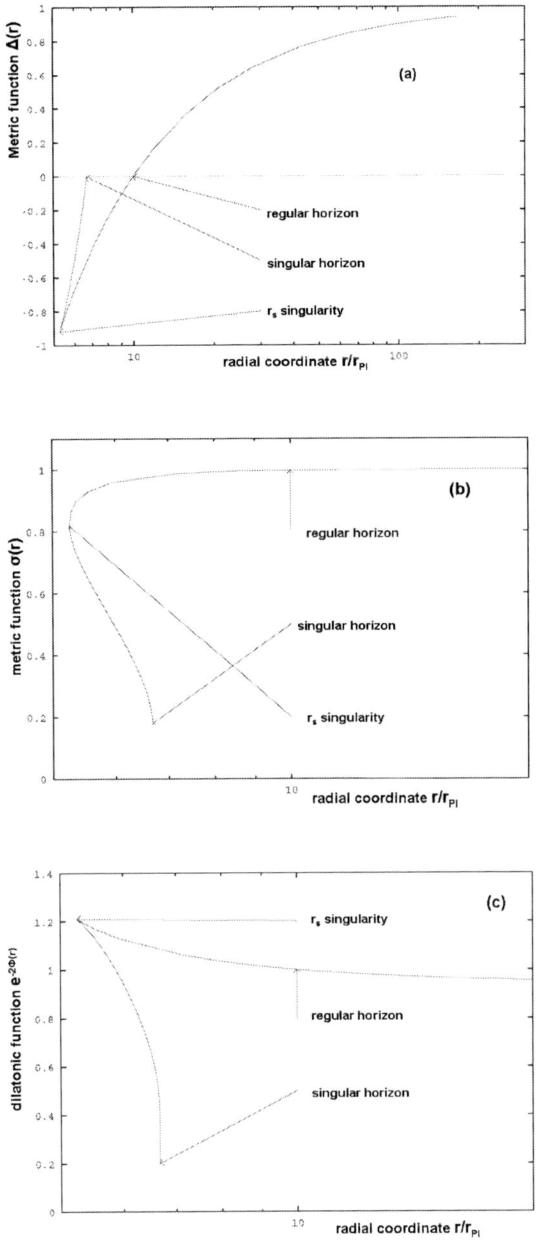

Figure 1. Dependence of the metric functions Δ (a), σ (b) and the dilaton function $e^{-2\phi}$ (c) against the radial coordinate r.

Substituting the formulas (7) into the system (3), one obtains the following relations between the expansion coefficients (s_0, $\phi_0 = e^{-2\phi_{00}}$ and r_h are free independent parameters):

$$d_1(z_1 d_1^2 + z_2 d_1 + z_3) = 0, \tag{8}$$

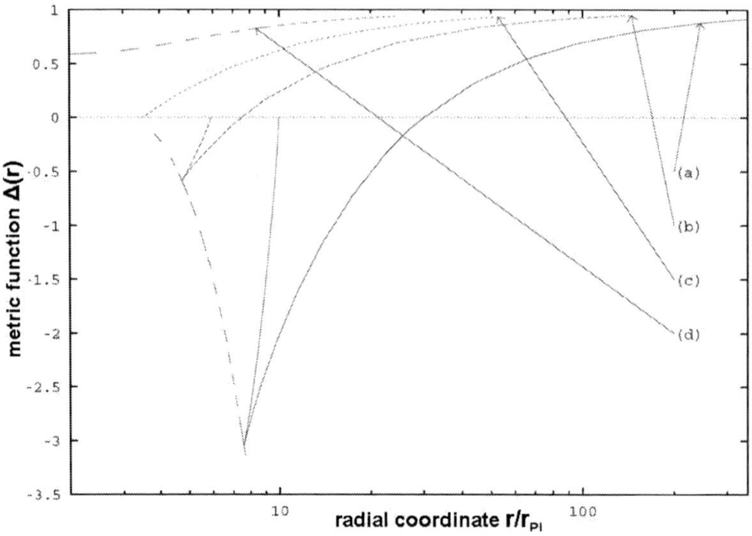

Figure 2. The dependence of the metric function Δ versus the radial coordinate r at the different values of the event horizon r_h when $q = 0$. The curve (a) represents the case where r_h is rather large and is equal to 30.0 Planck unit values (P.u.v.). The curve (b) shows the changes in the behavior of $\Delta(r)$ when r_h is equal to 7.5 P.u.v. The curve (c) represents the boundary case with $r_h = r_{hmin}$ where all the particular points merge and the internal structure disappears. The curve (d) shows the case where $2M \ll r_{hmin}$ and any horizon is absent.

where:

$$
\begin{aligned}
z_1 &= 24\lambda^2\phi_0^2, \\
z_2 &= -m_{Pl}^4 r_h^3 s_0^2, \\
z_3 &= m_{Pl}^4 r_h^2 s_0^4,
\end{aligned}
$$

and the parameter ϕ_1 for $d_1 \neq 0$ is equal to: $\phi_1 = \left[(m_{Pl}^2)/(4\lambda d_1 \phi_0)\right] * \left[r_h d_1 - s_0^2\right]$.

When $d_1 = 0$, the metric function Δ has the double (or higher order) zero. In such a situation the equation for d_2 (d_3, d_4, ...) is a linear algebraic one and there are no asymptotically flat branches.

When $d_1 \neq 0$ the solution of the black hole type takes place only if the discriminant of the equation (8) is greater or equal to zero and, hence, in this case $r_h^2 \geq 4\lambda\phi_0\sqrt{6}$ ($q = 0$, here we use the Planckian values where $m_{Pl} = 1$). One or two branches occurs and always one of them is asymptotically flat. With the supposition of $\phi_\infty = 0$ (and, as we tested, in this case $1.0 \leq \phi_0 < 2.0$) the **infinum** value of the event horizon is

$$
r_h^{inf} = \sqrt{\lambda}\,\sqrt{4\sqrt{6}}. \tag{9}
$$

The analogous formula but in the other interpretation was studied earlier by Kanti *et. al.*[9]. In the case $q > 0$ such restriction does not exist and regular horizon can take any meaning in the range $[0 \ldots \infty)$.

Consequently, the point r_{hmin} represents the event horizon and the singularity in the same point. One should remember that λ is a combination of the fundamental string constants. That is why formula (9) can be reinterpreted as a restriction to the minimal black hole size (mass) in the given model. This restriction appears in the second order curvature gravity and is absent in the minimal Einstein-Schwarzschild gravity.

3.3. Effects of Moduli Fields

Generalizing the model by taking into account the effective contribution of additional compact dimensions in the most simple form — scalar field, the action must be given as

$$\begin{aligned} S &= \int d^4x\sqrt{-g}\Bigg[-R + 2\partial_\mu\phi\partial^\mu\phi + 2\partial_\mu\psi\partial^\mu\psi \\ &\quad + \left(\lambda_\phi e^{-2\phi} + \lambda_\psi\xi(\psi)\right)S_{GB} \\ &\quad + \text{higher order curvature corrections}\Bigg]. \end{aligned} \tag{10}$$

This model was studied in the details in [18]. For current investigations it should be emphasized that when the contribution of moduli field value is considered, a naked singularity can appear if the size of additional dimensions is greater than the BH size. Therefore, the minimal BH mass must be increased to 10 Planck masses (to avoid being in naked singularity region). It is a key feature because it allows to move away from the Planck region and to use *semiclassical approach*. If these additional dimensions were non-compact [19], the BH minimal mass would be considerably greater.

4. Black Hole Evaporation Law

4.1. Probability of Transition to the Last Stage

According to the analysis given in Ref. [23], the transition from prelast to last stage of BH evaporation is forbidden and evaporating PBHs will never reach the minimal mass state. The shape of the BH mass loss rate law changes and becomes to be similar to the one presented in Fig. 3, analogously to the simplified "toy model" presented in [23]. This Figure represents the moment of the BH transition from "prelast" state to last one that is allowed by the string theory. Prelast state is characterized by the regular horizon with the usual quasi-Schwarzschild configuration. The last state is singular configuration making the transition from prelast to last state forbidden. Different types of similar models for BH evaporation were studied in Lovelock gravity [24], string inspired curvature expansions [25] and in many other theoretical models. The numerical values of Gauss-Bonnet BH (important for experimental search analysis) will be presented later.

4.2. Approximation to Metric Functions

In the WKB approximation of the Hawking evaporation process, everything happens in the neighbourhood of the event horizon. As our metric functions Δ and σ depend upon

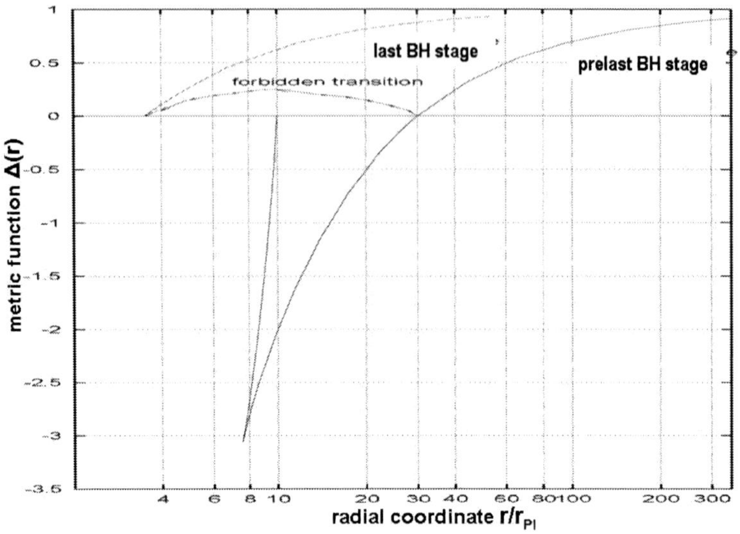

Figure 3. Illustration of the moment of last transition. Prelast state is characterized by the regular horizon with the usual quasi-Schwarzschild configuration. The last state is singular configuration making the transition from prelast to last state forbidden.

radial coordinate r and black hole mass M, i.e. $\Delta = \Delta(M,r)$ and $\sigma = \sigma(M,r)$ (other variables are not important), we can use expansions (7), taking into account only the first terms (partially neglecting the dependence upon radial coordinate r). Using expansions (7), the metric can be written as [26]:

$$\begin{aligned} \Delta(M,r) &= 1-\frac{2M}{r}\epsilon(M) = \frac{1}{2M\epsilon}(r-2M\epsilon(M)), \\ \sigma(M,r) &= \sigma_0(M). \end{aligned} \tag{11}$$

Using the numerically calculated data, fits were performed for $\epsilon(M)$ and $\sigma_0(M)$. As we are interested mostly in the last stages of PBHs evaporation, where the difference from the standard Bekenstein-Hawking picture is considerable, Taylor expansions around M_{min} can be used. Metric function σ and ϵ as a function of the mass M in Planck units for a fixed minimal mass $M_{Min} = 10M_{Pl}$ are shown in Fig. 4. Stars are numerically computed values and solid lines are the fits used to derive the spectrum. This data helps in obtaining good fits of the metric functions which can be considered as polynom expansions (of M or $1/M$) that are valid between $M = M_{min} = 10\ M_{Pl}$ and $M = 1000\ M_{Pl}$ with good accuracy:

$$\begin{aligned} \epsilon &= 1-\frac{\epsilon_1}{M}-\frac{\epsilon_2}{M^2}+\frac{\epsilon_3}{M^3}-\frac{\epsilon_4}{M^4}, \qquad (12) \\ \sigma_0 &= \sigma_2(M-M_{min})^2-\sigma_3(M-M_{min})^3+\sigma_4(M-M_{min})^4 \\ &\quad -\sigma_5(M-M_{min})^5, \end{aligned}$$

where the corresponding coefficients for $M_{min} = 10\ M_{Pl}$ are shown in Table 1.

Using this technique, the PBH evaporation spectra and mass loss rate were derived in an analytical form (valid only near the M_{min} point) [26].

Table 1. The coefficients ϵ_i and σ_i in expansions (12) for $M_{min} = 10\ M_{Pl}$

coefficients ϵ		coefficients σ	
ϵ_1	10.004	σ_2	0.11933×10^{-4}
ϵ_2	13.924	σ_3	0.30873×10^{-7}
ϵ_3	2856.3	σ_4	0.30871×10^{-10}
ϵ_4	25375.0	σ_5	0.11051×10^{-13}

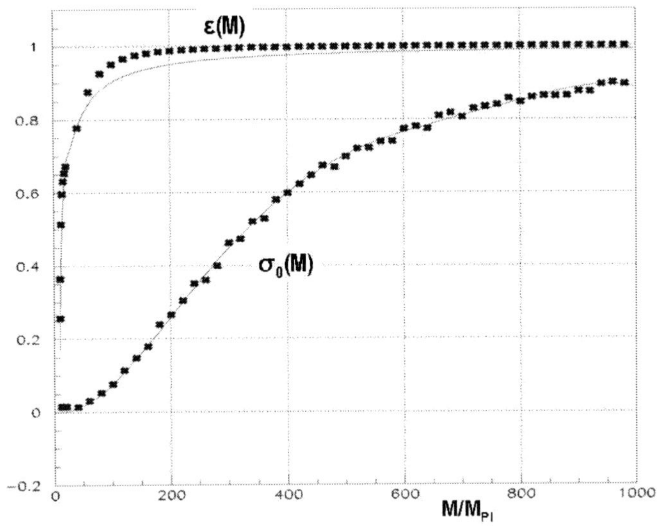

Figure 4. Metric function σ and ϵ as a function of the mass M in Planck units for a fixed minimal mass $M_{Min} = 10 M_{Pl}$. Stars are numerically computed values and solid lines are the fits used to derive the spectrum.

4.3. Black Hole Evaporation Spectra in EDGB Model

In some approaches, black holes are treated as immersed in a thermal bath and the evaporation can be described as a WKB approximation of semiclassical tunnelling in a dynamical geometry. In our investigation, we follow the techniques described in [27, 28] and base our relults on [26]. The same method was also applied in [29]. Some other descriptions of BH evaporation can be found in [30, 31].

The key idea of the method [26, 27, 28] is the following: an energy of a particle changes its sign when crossing the BH horizon. So, a pair created just inside or just outside the horizon can become real with zero total energy after one member of the pair has tunnelled to the opposite side. The energy conservation plays a fundamental role: transitions between states with the same total energy are the only possible ones. Using quantum mechanical rules, it is possible to write the imaginary part of the action for an outgoing positive energy particle which crosses the horizon outwards from r_{in} to r_{out} as:

$$Im(S) = Im \int_{M}^{M-\omega} \int_{r_{in}}^{r_{out}} \frac{dr}{\dot{r}} dH,$$

where ω is the energy of the particle, H is total Hamiltonian (and total energy) and the metric is written so as to avoid the horizon coordinate singularity. Following [28], Painleve's coordinates are used. The transformation to this metric from the Schwarzschild one can be obtained by changing the time variable:

$$t = t_{old} + r\sqrt{\frac{\sigma^2}{\Delta^2} - \frac{1}{\Delta}}.$$

Substituting t_{old} into (2) one obtains

$$ds^2 \quad = \quad -\Delta dt^2 + 2\sqrt{\sigma^2 - \Delta} dr dt + dr^2 + r^2 d\Omega^2. \tag{13}$$

In WKB approximation, the imaginary part of the semiclassical action $Im(S)$, describing the probability of tunnelling through the horizon, is the following:

$$Im(S) = Im \int\limits_{r_{in}}^{r_{out}} p_r \, dr = Im \int\limits_{r_{in}}^{r_{out}} \int\limits_{0}^{p_r} p_r' \, dr, \tag{14}$$

where p_r is canonical momentum.

For Gauss-Bonnet BH the radial geodesics are described by the equation [11]:

$$\dot{r} = \frac{dr}{d\tau} = \frac{\Delta}{\sqrt{\sigma^2 - \Delta} \mp \sigma} = \mp\sigma - \sqrt{\sigma^2 - \Delta}. \tag{15}$$

After substituting the expression (15) to the equation (14) one obtains:

$$Im(S) \quad = \quad Im \int\limits_{M}^{M-\omega} \int\limits_{r_{in}}^{r_{out}} \frac{dr}{\dot{r}} dH = -Im \int\limits_{0}^{\omega} \int\limits_{2M\epsilon}^{2(M-\omega)\epsilon} \frac{dr d\omega'}{\sigma - \sqrt{\sigma^2 - \Delta}}. \tag{16}$$

Substituting the expression (11) extended in (12) to the equation (16), the imaginary part of the action can be written as:

$$Im(S) \quad = \quad -Im \int\limits_{0}^{\omega} d\omega' \Bigg(\int\limits_{2M\epsilon}^{2(M-\omega')\epsilon} \frac{dr}{\sigma - \sqrt{\sigma^2 - \frac{r}{2(M-\omega')\epsilon} + 1}} \Bigg).$$

We introduce a new variable y according to the following rool:

$$y = \sqrt{\sigma^2 - \frac{r}{2(M-\omega)\epsilon} + 1}.$$

After this substitution $Im(S)$ takes the form as follows:

$$\begin{aligned} Im(S) &= -Im\int_0^{\omega} d\omega' \left(\int_{\sqrt{\sigma^2-\frac{\omega'}{M-\omega'}}}^{\sigma} \frac{4(M-\omega')\epsilon y dy}{y-\sigma} \right) \\ &= -Im\int_0^{\omega} d\omega' \left(4(M-\omega')\epsilon\sigma \int_{\sqrt{\sigma^2-\frac{\omega'}{M-\omega'}}}^{\sigma} \frac{dy}{y-\sigma} \right) \\ &= -\int_0^{\omega} d\omega' \left(4(M-\omega')\epsilon\sigma\pi\right). \end{aligned}$$

As a result one can represent the imaginary part of the action (16) in the form:

$$2Im(S) = \frac{840\pi}{M^2(M-\omega)^2}\alpha,$$

where α is the term of a very long size. The exact expression can be found at [26].

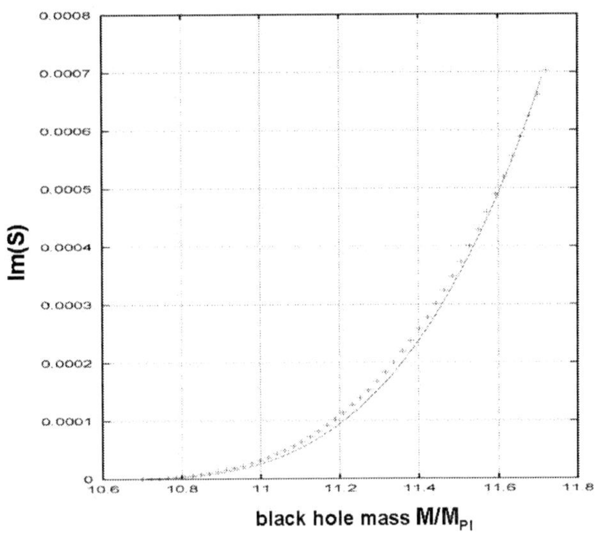

Figure 5. $Im(S)$ (dots) and the fit $\left[(5\cdot 10^{-4})\times(M-M_{min})^3\right]$ (dashed line) versus BH mass M during the last stages of BH evaporation in the Gauss-Bonnet case with $M_{min} = 10.6M_{Pl}$. It is necessary to note that during last stages of evaporation the emitted energy $\omega < M - M_{min} \ll 1$. For fixed values of $\omega = \omega_i^*$ in the vicinity of M_{min} ($O(M_{min}) = 0.01$) the mass $M \in (M_{min}+\omega_i^*, M_{min}+\omega_i^*+O(M_{min}))$. So, for different values of ω^* ($\omega_{i+1}^* = \omega_i^* + O(M_{min}), \omega_1^* = 0.1, i \in N$), M belongs to different (without intersection) intervals. Finally, $Im(S)$ is represented as the connection of such intervals with the most probable values of $\omega_i^* \in (0.1, 1.1)$.

Using the numerical values for a realistic order of M_{min} around 10 Planck masses corresponding ϵ_i and σ_j from Table 1, it is possible to find from (11) the approximate expression for $Im(S)$. Since we are mostly interested in the last stages of BH evaporation where the influence of higher order curvature corrections is important, the limit $M - M_{min} \ll 1$ can be used. This leads to a very different spectrum than the standard Bekenstein-Hawking picture (where $-dM/dt \propto 1/M^2$). Taking into account energy conservation, the value of ω can be bounded: $0 \leq \omega \leq M - M_{min}$. In such a case the approximate expression of $Im(S)(M, \omega)$ for a given M_{min} can be used in the form:

$$Im(S) = k * (M - M_{min})^3, \tag{17}$$

where constant $k = 5 \cdot 10^{-4}$ in Planck unit values with a satisfying accuracy. The plot of $Im(S)$ and its approximation as functions of BH masses in relative units are shown in Fig. 5. One can see from this Figure that the suggested approximation (17) coincides well with the real meanings of $Im(S)$ obtained from numerical integrations.

subsectionEnergy conservation and mass lost rate

Following Ref.[26, 28], the emission spectrum per degree of freedom can simply be written as:

$$\frac{d^2N}{dEdt} = \frac{\Gamma_s}{2\pi\hbar} \cdot \frac{\Theta((M - M_{min})c^2 - E)}{e^{Im(S)} - (-1)^{2s}}, \tag{18}$$

with $\Gamma_s(M, E)$ being the absorption probability for a particle of spin s. The Heavyside function is implemented to take into account energy conservation with a minimal mass M_{min}. In this section and in the following ones, standard units are used instead of Planck ones as numerical results should be obtained for experimental fluxes.

At this point, two questions have to be addressed: which kind of fields are emitted (and which correlative Γ_s have to be used) and which mass range is physically interesting. In order to answer these questions, the mass loss rate is needed to be as following:

$$-\frac{dM}{dt} = \int_0^{(M-M_{min})c^2} \frac{d^2N}{dEdt} \cdot \frac{E}{c^2} dE, \tag{19}$$

where the integration is carried out up to $(M - M_{min})c^2$ in order to ensure that the transition below M_{min} is forbidden. The absorption probabilities can clearly be taken as the limit $GME/\hbar c^3 \ll 1$ as we are considering the endpoint emission, when the cutoff, imposed by M_{min}, prevents the black hole from emitting particles with energies of the order of kT. Using analytical formulae [32] and expanding $\exp(Im(S))$ to the first order with the approximation according to (17), it is easy to show that the emission of spin-1 particles, given by (per degree of freedom)

$$-\frac{dM}{dt} \approx \frac{16}{9\pi} \frac{G^4 M_{Pl}}{\hbar^5 c^2 k} M^4 (M - M_{min})^3, \tag{20}$$

dominates over $s = 1/2$ and $s = 2$ emission whatever the considered energy is in the previously quoted limit. It is interesting to point out that the fermion emission around M_{min} is not strongly modified by the EDGB model as, in the lowest order, $\exp(Im(S)) - (-1)^{2s} \approx 2$. Furthermore, if energy conservation was implemented as a simple cutoff in the Hawking spectrum, the opposite result would be obtained: $s = 1/2$ particles would dominate the mass loss rate as the power of (ME) in the absorption probability is the smallest one. If we restrict ourselves to massless particles, *i.e.* the only ones emitted when M is very close to M_{min}, the metric modification changes the endpoint emission nature from neutrinos to photons. The real mass loss rate is just twice the one given here to account for the electromagnetic helicity states. Shape of BH mass lost rate versus BH mass in Gauss-Bonnet case when the energy conservation is taken into account is presented in Fig 6. Right part of the graph represents the usual Hawking evaporation law when $-dM/dt \sim 1/M^2$. Left part shows the picture at last stages when evaporation decelerates and then stops, distinguishing the minimal possible mass ("ground state").

Knowing the detailed form of the BH evaporation law $-dM/dt = f(M)$, it is possible to compute the mass M at any given time t after formation at the mass M_{init} as:

$$t = \int_{M}^{M_{init}} \frac{dM}{f(M)} \approx \frac{9\pi k\hbar^5 c^2}{32 G^4 M_{Pl}^3} \times \frac{1}{M_{min}^4 (M - M_{min})^2}, \tag{21}$$

where only the dominant term in the limit $t \to \infty$ is taken from the analytical primitive of the function. As it was expected, the result does not depend on M_{init} which is due to the fact that the time needed to go from M_{init} to a few times M_{min} is much less than the time taken to go from a few times M_{min} to M as long as $M_{init} << 10^{15}$ g for $t \approx 10^{17}$ s. At time t after formation, the mass is given by:

$$M \approx M_{min} + \sqrt{\frac{9k\pi\hbar^5 c^2}{8 M_{min}^4 G^4 M_{Pl}^3 t}}. \tag{22}$$

This mass can be implemented in the emission spectrum formula (18):

$$\frac{d^2N}{dEdt} \approx \frac{32}{3\pi}\left(\frac{8}{9\pi}\right)^{\frac{3}{2}} G^{10}\hbar^{-\frac{25}{2}} c^{-15} M_{Pl}^{\frac{15}{2}} M_{min}^{10}$$

$$k^{-\frac{5}{2}} \times t^{\frac{3}{2}} E^4 \Theta\left(\sqrt{\frac{9k\pi\hbar^5 c^6}{8 M_{min}^4 G^4 M_{Pl}^3 t}} - E\right), \tag{23}$$

that leads to a frequency f given by the following expression:

$$f = \int_0^{(M-M_{min})c^2} \frac{d^2N}{dEdt} dE \approx \frac{36}{15} \cdot \frac{1}{t}. \tag{24}$$

If we want to consider the possible relic emission produced now from PBHs formed in the early universe with small masses, this leads to a frequency around $6 \cdot 10^{-18}$ Hz with a typical energy of the order of $1.8 \cdot 10^{-6}$ eV. This emission rate is very small as it corresponds to the evaporation into photons with wavelength significantly longer that the radius of the black hole. Nevertheless, it should be emphasized that the spectrum represents a monophonically increasing function of energy, up to the cutoff, with a E^4-like dependence. Furthermore, the evaporation, beeing even very small in intensity, never stops and it leads to a mass evolution dependence as $1/\sqrt{t}$.

5. Experimental Detection

In this section we study the possibility of measurement of the previously given relic emission. Let R to be a distance from an observer; z to be a redshift corresponding to the

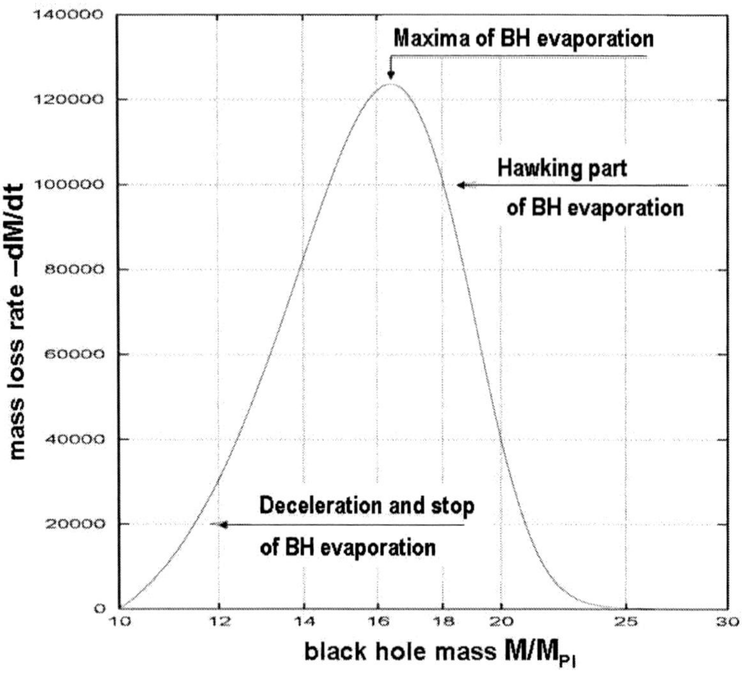

Figure 6. Shape of BH mass lost rate versus BH mass in Gauss-Bonnet case when the energy conservation is taken into account. Right part of the graph represents the usual Hawking evaporation law when $-dM/dt \sim 1/M^2$. Left part shows the picture at last stages when evaporation decelerates and then stops, distinguishing the minimal possible mass ("ground state").

distance R; θ to be an opening angle of a detector (chosen so that the corresponding solid angle is equal to $\Omega = 1\ sr$); $d^2N/dEdt(E,t)$ to be an individual differential spectrum of a black hole relic (BHR) at time t; $\rho(R)$ to be a numerical BHR density taking into account the cosmic scale factor variations; R_{max} to be the horizon in the considered energy range; t_{univ} to be the age of the Universe and H to be the Hubble parameter. The "experimental" spectrum F (J^{-1}·s^{-1}·sr^{-1}) can be written as:

$$F = \int_0^{Rmax} \frac{d^2N}{dEdt}\left(E(1+z), t_{univ} - \frac{R}{c}\right) \times \frac{\rho(R) \cdot \pi R^2 tan^2(\theta)}{4\pi R^2} \mathrm{d}R, \tag{25}$$

which leads to:

$$F = tan^2(\theta)\frac{8}{3\pi}\left(\frac{8}{9\pi}\right)^{\frac{3}{2}} G^{10}\hbar^{-\frac{25}{2}}c^{-15}M_{Pl}^{\frac{15}{2}}M_{min}^{10} \times k^{-\frac{5}{2}}E^4\int_0^{R_{max}} \rho(R)\left(\frac{1+\frac{HR}{c}}{1-\frac{HR}{c}}\right)^2 (t_{univ} - \frac{R}{c})^{\frac{3}{2}}$$

$$\times\Theta\left(\sqrt{\frac{9k\pi\hbar^5c^6}{8M_{min}^4G^4M_{Pl}^3(t-\frac{R}{c})}}-E\sqrt{\frac{1+\frac{HR}{c}}{1-\frac{HR}{c}}}\right)dR. \quad (26)$$

This integral can be analytically computed. Here one has to take into account both the facts that BHRs, beeing far away from the Earth, must be taken at earlier stages of their evolution, and that energies must be redshifted. Even assuming the highest possible density of BHRs ($\Omega_{BHR}=\Omega_{CDM}\approx 0.3$) and R_{max} to be around the Universe radius, the resulting flux is obtained to be extremely small:

$$F\approx 1.1\cdot 10^7\ \mathrm{J}^{-1}\ \mathrm{s}^{-1}\ \mathrm{m}^{-2}\ \mathrm{sr}^{-1}$$

in the particle energy range of around 10^{-6} eV, that is nearly 20 orders of magnitude below the background. This completely closes the question on possible direct detection of BHRs emission.

Another way to investigate differences between EDGB black hole particle emission and a pure Hawking spectrum is to study the mass region where dM/dt is maximal. Taking into account that the mass loss rate becomes considerably higher in the EDGB case than in the usual Hawking picture, it could have been expected that the extremely high energy flux was strongly enhanced. In particularly, it could revive the interest in PBHs as candidates to solve the enigma of measured cosmic rays above the GZK cutoff. Nevertheless, the spectrum modification becomes important only when the mass is quite near to M_{min}. Depending on the real numerical value of M_{min}, it can vary substantially (increasing with increasing M_{min}) but remains a few Planck masses above M_{min}. This is far too small to account for a sizeable increase of the flux.

The number of particles emitted above 10^{20}eV in a pure Hawking model is of the order of 10^{15} [34]. This value should be taken with a caution as it based on the use of leading log QCD computations of fragmentation functions far beyond the energies reached by colliders, but the order of the magnitude is correct. On the other hand, even if all the energy available, when EDGB modifications becomes important, were released in 10^{20} eV particles, BHR would generate only a few times 10^9 particles. The pure Hawking flux would be modified by less than 0.01% by these particles. It would not allow to generate, as it was previously expected, a spectrum harder than E^{-3}.

6. Primordial Black Holes as Dark Matter Candidates

The idea of PBH relics as a serious candidate for cold dark matter was first mentioned in [35]. It was shown that in a Friedman universe without inflation, Planck-mass remnants of evaporating primordial black holes could be expected to have close to the critical density. Nevertheless, the study was based on the assumption that either a stable object forms with a mass around M_{Pl} or a naked space-time singularity is left. Our study provides new arguments favouring massive relic objects, probably to be one order of magnitude above Planck mass, and could revive the interest in such non-baryonic dark matter candidates.

An important problem is still to be addressed to the standard inflationary cosmology: the rather large size of the horizon at the end of inflation. The standard formation mechanism of PBHs requires the mass of the black holes to be of the order of the horizon mass at the formation time. So, only these created after inflation should be taken into account as the huge increase of the scale factor would extremely dilute all the ones possibly formed before. It is easy to show that under such assumptions the density of Planck relics is very small:

$$\Omega_{Pl} = \Omega_{PBH}\frac{\alpha-2}{\alpha-1}M_H^{1-\alpha}M_*^{\alpha-2}M_{min}, \tag{27}$$

where Ω_{PBH} is the density of PBHs not yet evaporated, α is the spectral index of the initial mass spectrum (=5/2 in the standard model for a radiation dominated Universe), M_* is the initial mass of a PBH which evaporating time is the age of the Universe ($\approx 5\cdot 10^{14}$g) and M_H is the horizon mass at the end of inflation. This latter can be expressed as

$$M_H = \gamma^{\frac{1}{2}}\frac{1}{8}\frac{M_{Pl}}{t_{Pl}}t_i \approx \gamma^{\frac{1}{2}}\frac{1}{8}\frac{M_{Pl}}{t_{Pl}}\frac{0.24}{(T_{RH}/1\mathrm{Mev})^2}, \tag{28}$$

where T_i is the formation time and T_{RH} is the reheating temperature.

In order to obtain the estimation of PBH density we take the highest possible value for T_{RH} that is equal to the order of 10^{12} GeV (according to Ref. [33]). If the reheating temperature is more than 10^9 GeV, BHR remnants should be present nowadays. Hence, taking the upper limit on Ω_{PBH} to be of the order of 6×10^{-9} (from gamma-ray astronomy), the resulting density appears to be extremely small: $\Omega_{Pl}\approx 10^{-16}$.

Nevertheless, there are at least two different ways to revive the interest in PBH dark matter. The first one is related to relics that would be produced from an initial mass spectrum decreasing fast enough, so as to overcome the gamma-rays limit. The second one is to have a large amount of big PBHs with masses to be between 10^{15}g and 10^{25}g, where experiments are completely blind: such black holes are too heavy to undergo Hawking evaporation and are too light to be seen by microlensing experiments (mostly because of the finite size effect [36]). The most natural way to produce spectra with such features is inflationary models with a scale, corresponding to a change in the spectral index of the fluctuations power spectrum [37, 38].

7. Black Hole Formation at Colliders

7.1. Generic Remarks

The "large extra dimensions" scenario [39] is a very attractive way to address geometrically the hierarchy problem (among others), allowing only the gravity to propagate in the bulk. The Gauss law relates the Planck scale of the effective 4D low-energy theory M_{Pl} with the fundamental Planck scale M_D through the volume of the compactified dimensions, V_{D-4}, via:

$$M_D = \left(\frac{M_{Pl}^2}{V_{D-4}}\right)^{\frac{1}{D-2}}. \tag{29}$$

Thus, it is possible to set $M_D\sim TeV$ without being in contradiction with any currently available experimental data. This translates into radii values to be between a fraction of a millimeter and a few Fermi ones for the compactification radius of the extra dimensions (assumed to be of the same size and flat, *i.e.* of a toroidal shape). Furthermore, such a small value of the Planck energy can minimize the difference between the weak scale and Planck one, as it is motivated by the construction of this approach. In such a scenario, at sub-weak energies, the Standard Model (SM) fields must be localized to a 4-dimensional manifold of weak scale "thickness" in the extra dimensions. As shown in [39], as an example based on a dynamical assumption with D=6, it is possible to build such a SM field localization. However, this is the non-trivial task of those models.

Another important way for realizing TeV scale gravity arises from properties of warped extra-dimensional geometries used in the Randall-Sundrum scenarios [40]. If the warp factor is small in the vicinity of the standard model brane, particle masses can take TeV values, thereby giving rise to a large hierarchy between the TeV and conventional Planck scales [41, 42]. Strong gravitational effects are therefore also expected in high energy scattering processes on the brane.

In the frameworks of model discussed, black holes could be formed by the Large Hadron Collider (LHC) interactions. Two partons with a center-of-mass energy $\sqrt{s}$, moving in opposite directions with an impact parameter less than the horizon radius r_+, should form a black hole of mass $M \approx \sqrt{s}$ with a cross section expected to be of the order of $\sigma \approx \pi r_+^2$. In reality, these values are approximations as the black hole mass will be equal to only a fraction of the center-of-mass energy whose exact value depends on the dimensionality of the spacetime and on the angular momentum of the produced black hole [43, 44].

Furthermore, suppression effects in the cross section should be considered and be taken into account as it was discussed in [45]. Although the accurate values are not yet known, a semiclassical analysis of quantum black hole formation is now being constructed and the existence of a closed trapped surface in the collision geometry of relativistic particles is demonstrated. To compute the real probability of black hole formations at the LHC, it is necessary to take into account the fact that only a fraction of the total center-of-mass energy is carried out by each parton, and to convolve the previous estimate with the parton luminosity [46]. Many clear experimental signatures are expected at the LHC[41]. In particularly, very high multiplicity events with a large fraction of the beam energy will be converted into transverse energy with a growing cross section. Depending on the value of the Planck scale, up to approximately a billion black holes could be produced at the LHC.

7.2. Schwarzschild-Gauss-Bonnet Black Holes

The classical Einstein theory can be considered as the weak field and the low energy limit of a some quantum gravity model which is not yet built. That is why the curvature expansion of string gravity provides an interesting step in the modelling of a quasiclassical approximation of quantum gravity. As it was pointed out in [47], among higher order curvature corrections to the general relativity action, the quadratic term is especially important as it is the leading one and as it can affect the graviton excitation spectrum near flat space. If, like the string itself, its slope expansion is to be ghost free, the quadratic term *must* be the Gauss - Bonnet combination : $L_{GB} = R_{\mu\nu\alpha\beta}R^{\mu\nu\alpha\beta} - 4R_{\alpha\beta}R^{\alpha\beta} + R^2$. Furthermore, this term is naturally generated in heterotic string theories [48] and makes possible the localization of the graviton zero-mode on the brane [49]. It has been successfully used in cosmology, for instance, in the cosmological constant problem (see *e.g.* [50] and references therein) and in the black hole physics. In the last case it is especially important in the consideration of the endpoint of the Hawking evaporation problem (see *e.g.* [11, 26] and references therein). Here we consider black holes described by such an action:

$$\begin{aligned} S &= \frac{1}{16\pi G} \int d^D x \sqrt{-g} \Big[R \\ &+ \lambda (R_{\mu\nu\alpha\beta} R^{\mu\nu\alpha\beta} - 4 R_{\alpha\beta} R^{\alpha\beta} + R^2) + \ldots \Big], \end{aligned} \tag{30}$$

where λ is the Gauss-Bonnet coupling constant. The measurement of this λ term would allow an important step forward in the understanding of the ultimate gravity theory.

Following [51], we assume the metric to be of the following form:

$$ds^2 = -e^{2\nu} dt^2 + e^{2\alpha} dr^2 + r^2 h_{ij} dx^i dx^j, \tag{31}$$

where ν and α are functions depending only against radial coordinate r, and $h_{ij}dx^i dx^j$ represents the line element of a $(D-2)$-dimensional hypersurface with constant curvature $(D-2)\times(D-3)$. The substitution of this metric into the action (30) [47] leads to the following solutions:

$$\begin{aligned} e^{2\nu} &= e^{-2\alpha} = 1 + \frac{r^2}{2\lambda(D-3)(D-4)} \\ &\times \left(1 \pm \sqrt{1 + \frac{32\pi^{\frac{3-D}{2}} G\lambda (D-3)(D-4) M \Gamma(\frac{D-1}{2})}{(D-2) r^{D-1}}} \right). \end{aligned} \tag{32}$$

The mass of the black hole can then be expressed [47, 51] in terms of the horizon radius r_+ as following:

$$M = \frac{(D-2)\pi^{\frac{D-1}{2}} r_+^{D-3}}{8\pi G \Gamma\left(\frac{D-1}{2}\right)} \left(1 + \frac{\lambda(D-3)(D-4)}{r_+^2} \right),$$

where Γ stands for the Gamma function.

The BH temperature is obtained by the usual requirement that no conical singularity appears at the horizon in the euclidean sector of the hole solution, and looks like as:

$$T_{BH} = \frac{1}{4\pi} (e^{-2\alpha})' \,|_{r=r_+} = \frac{(D-3) r_+^2 + (D-5)(D-4)(D-3)\lambda}{4\pi r_+ \left(r_+^2 + 2\lambda (D-4)(D-3) \right)}. \tag{33}$$

In the case $D = 5$, those black holes have a singular behavior [51] and, depending on the value of λ, can become thermodynamically unstable or form stable relics. For $D > 5$, which is the only relevant hypothesis for this study (as the case of $D = 5$ is excluded by the solar system dynamics if the Planck scale is expected to belong the range of $\sim$TeV), a quantitatively different evaporation scenario is expected.

Figure 7 shows the ratio of the temperatures with and without the Gauss-Bonnet term for different physically motivated values of space-time dimensions D and string coupling constant λ. One can conclude from these plots that the difference between different curves in some cases is rather significant (more than 5%) and, therefore, these different kinds of black holes could be distinguished in real experiments (depending on their uncertanties) in the future collider studies. It should be pointed out that the non-monotonic behavior makes an unambiguous measurement quite difficult and requires to take advantage of the full dynamics of the evaporation. The next sections focus on this issue in order to investigate the λ parameter reconstruction.

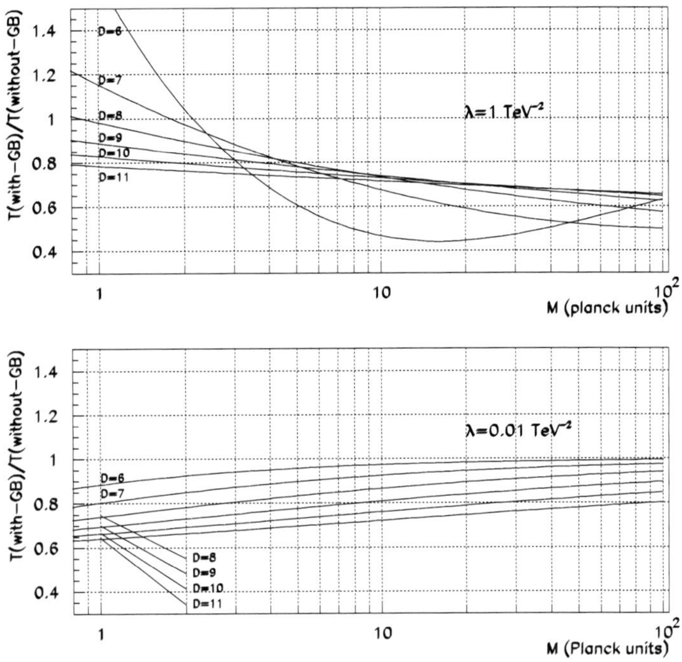

Figure 7. Ratio of the temperatures with and without the Gauss-Bonnet term for $D = 6, 7, 8, 9, 10, 11$ as a function of mass with $\lambda = 1\text{TeV}^{-2}$ (up) and $\lambda = 0.01\text{TeV}^{-2}$ (down).

7.3. Flux Computation

Using the high-energy limit of multi-dimensional grey-body factors [52], the spectrum per unit of time t and of energy Q can be written, for each degree of freedom, for particles of type i and spin s as:

$$\frac{\mathrm{d}^2 N_i}{\mathrm{d}Q\mathrm{d}t} = \frac{4\pi^2 \left(\frac{D-1}{2}\right)^{\frac{2}{D-3}} \left(\frac{D-1}{D-3}\right) r_+^2 Q^2}{e^{\frac{Q}{T_{BH}}} - (-1)^{2s}}. \tag{34}$$

This is an approximation as modifications might arise when the exact values of the grey-body factors are taken into account due to their dependence, in the low energy regime, on both the dimensionality of the spacetime and on the spin of the emitted particle. Fortunately, as it was demonstrated in the 4-dimensional case [53], the *pseudo-oscillating* behaviour induces compensations that makes the differences, probably quantitatively, quite small. As it was shown in the previous section, as long as $D > 5$, the horizon radius r_+ cannot be explicitly given as a function of the mass. So, in order to compute the experimental-like integral spectrum $\mathrm{d}N_i/\mathrm{d}Q$, where the following change of variable is convenient:

$$\frac{\mathrm{d}N_i}{\mathrm{d}Q} = \int_{r_{init+}}^{0} \frac{1}{\frac{\mathrm{d}M}{\mathrm{d}t}} \frac{\mathrm{d}M}{\mathrm{d}r_+} \frac{\mathrm{d}^2 N_i}{\mathrm{d}Q\mathrm{d}t} \mathrm{d}r_+ \tag{35}$$

where

$$\frac{\mathrm{d}M}{\mathrm{d}r_+} = \frac{(D-2)\pi^{\frac{D-1}{2}} r_+^{D-6}}{8\pi G\Gamma(\frac{D-1}{2})} \left[(D-3)r_+^2 + (D-5)(D-4)(D-3)\lambda\right]$$

and

$$\frac{\mathrm{d}M}{\mathrm{d}t} = -\frac{4\pi^6}{15}\left(\frac{D-1}{2}\right)^{\frac{2}{D-3}} \left(\frac{D-1}{D-3}\right) r_+^2 T_{BH}^4 \left[\frac{7}{8}N_f + N_b\right],$$

with N_f and N_b being the total fermionic and bosonic degrees of freedom.

The mean number of emitted particle can then be written as

$$N_{tot} = \frac{15(D-2)\pi^{\frac{D-9}{2}}\zeta(3)}{\Gamma(\frac{D-1}{2})G} \frac{\frac{3}{4}N_f + N_b}{\frac{7}{8}N_f + N_b} \left[\frac{r_{init+}^{D-2}}{D-2} + 2(D-3)\lambda r_{init+}^{D-4}\right], \tag{36}$$

where r_{init+} is the initial horizon radius of a black hole with mass M_{init}. It is necessary to emphasis, that the ratio of a given species i to the total emission is given by:

$$\frac{N_i}{N_{tot}} = \frac{\alpha_s g_i}{\frac{3}{4}N_f + N_{tot}}, \tag{37}$$

where α_s is equal to 1 for bosons and is equal to $3/4$ for fermions. g_i represents the number of internal degrees of freedom for the considered particles. The mean number of particles emitted by a Schwarzschild-Gauss-Bonnet black hole ranges from 25 to 4.7 depending on the values of λ and of D, for $M_D \sim 1$ TeV and $M_{init} \sim 10$ TeV. These values are decreased respectively to 5 and to 1.05 if M_{init} is fixed at the value of 2 TeV. Fig. 8 shows the integrated flux of evaporation particles N_{tot} (36) for different values of λ and D. The presented plots confirm the previous conclusion that the difference between different curves is significant and, therefore, these different kinds of black holes could be distinguished in the future collider experiments. In spite of some combinations seem to be strongly degenerated, future experimental results would allow us to distinguish the different types of black holes and different cases of space-time dimensions.

The next section shows that in any case the values of λ and D can be well reconstructed.

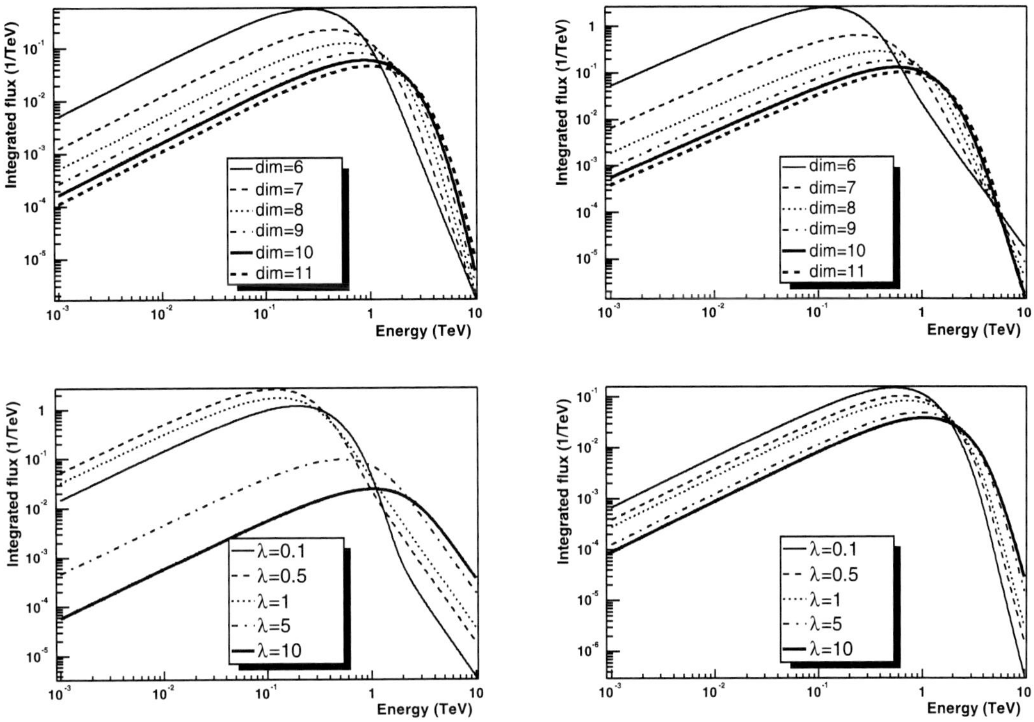

Figure 8. Integrated flux as a function of the total energy of the emitted quanta for an initial black hole mass $M = 10\text{TeV}$. Upper left : $\lambda = 0$, $D = 6, 7, 8, 9, 10, 11$. Upper right : $\lambda = 0, 5\ \text{TeV}^{-2}$, $D = 6, 7, 8, 9, 10, 11$. Lower left : $D = 6$, $\lambda = 0.1, 0.5, 1, 5, 10\text{TeV}^{-2}$. Lower right : $D = 11$, $\lambda = 0.1, 0.5, 1, 5, 10\text{TeV}^{-2}$.

7.4. String Coupling Constant Measurement

In order to investigate the LHC capability to reconstruct the fundamental parameter λ, we have fixed the Planck scale at 1 TeV. A small perturbation range around this value would not change dramatically our conclusions. Nevertheless, one cannot take these values to be much above because of the very fast decrease of the number of formed black holes with increase of M_D.

Following [46], we consider the number of black holes produced in the mass range between 1 TeV and 10 TeV, as a bin width of 500 GeV (much larger than the energy resolution of the detector) is modified with the value of r_+ controlled by the Gauss-Bonnet term.

In the simulations for each signal of black hole borning, the emitted particles are randomly chosen by a Monte-Carlo simulation in accordance with the spectra given by (34), that are weighted by the appropriate number of degrees of freedom. As the Hawking radiation takes place predominantly in the S-wave channel [54], therefore, bulk modes can be neglected. So, the evaporation can be considered as occurring within the brane. Since the intrinsic spectrum $\mathrm{d}N_i/\mathrm{dQ}$ is very strongly modified by a fragmentation process, only the direct emission of electrons and photons with energies above 100 GeV is considered.

Using the Pythia hadronization program [55], we have checked that only a small fraction of directly emitted γ-rays and electrons fall within an hadronic jet, making them impossible to distinguish from the background of decay products. Furthermore, the background from standard model $Z(ee)$+jets and γ+jets remains much lower than the expected signal.

The value of the Planck scale is assumed to be known from an expected clear threshold effect that should appear in the experimental data. Uncertainties of these measurments are expected to be neglible. For each BH signal, the initial mass of the black hole is also assumed to be known as it can be easily determined with the full spectrum of decay products (only 5% of total energy is expected to me missed due to the small number of degrees of freedom of neutrinos and gravitons).

The energy resolution of the detector is taken into account and it is parametrized [56] by $\sigma/E = \sqrt{a^2/E + b^2}$ with $a \approx 10\%\sqrt{\text{GeV}}$ and $b \approx 0.5\%$. In spite of the consideration made in [46], we also take into account the time evolution of the black holes and perform a full fit for each signal.

Once all the particles have been generated, spectra are reconstructed for all the mass bins and are compared with theoretical computations. The supposed values of D and λ, that are compatible with the simulated data, are then analysed. Figure 9 shows the $\chi^2/d.o.f.$ for the reconstructed spectra for two different constant couples (λ $[\,TeV^{-2}]$, D)=(1,10) and (λ $[\,TeV^{-2}]$, D)=(5,8). From these plots it is clear that the difference between specta is serious and, so, different kinds of black holes could be distinguished in the future collider experiments. The statistical significance of this χ^2 should be taken with a caution since a real statistical analysis would require a full Monte-Carlo simulation of the detector.

Nevertheless, the "input" values can clearly be extracted from the simulation data. Furthermore, it is important to notice that for reasonable values of λ (around the order of the quantum gravity scale, *i.e.* around a TeV^{-2} in our case), one can unambiguously distinguish the case *with* and the case *without* a Gauss-Bonnet term. Table 2 summarizes the LHC reconstruction capability requiring the $\chi^2/d.o.f.$ to remain smaller than $2\chi^2_{min}/d.o.f.$ where $\chi^2_{min}/d.o.f.$ corresponds to the "physical" case (*i.e.* $\lambda = \lambda_{input}$ and $D = D_{input}$). This estimation is quite conservative and should be translated into high confidence levels which would require a considerable more detailed simulation of the detector.

For each set of parameters, the cross sections have been taken as πr_+^2, $\pi r_+^2/10$, $\pi r_+^2/100$ and $\pi r_+^2/1000$ in order to account for uncertainties on the production process for $D > 4$ with a non-zero impact parameter. Based on the methods developed by Penrose and D'Eath & Payne [43] and on the hoop conjecture [44], several estimations have been derived. They confirm the formation of an apparent horizon. The wide range of BH formation analysed in this study should cover all physical cases of space-time dimensions and of string coupling constant values.

8. Discussion and Conclusions

In this paper, the BH type solution of $D = 4$ and $D > 4$ effective string gravity action with higher order curvature corrections was applied to the description of BHRs. A corrected version of the evaporation law near the minimal BH mass was established. It was shown that the standard Bekenstein-Hawking evaporation forluma must be modified in the

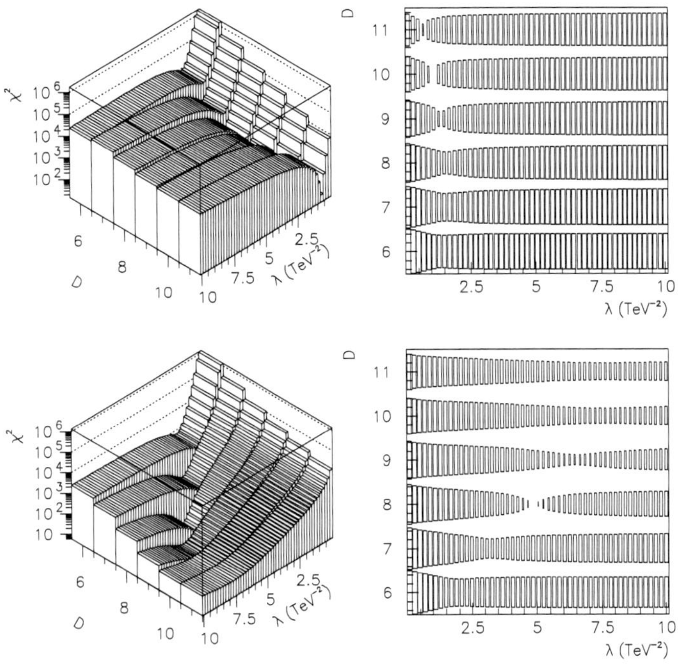

Figure 9. Upper part : values of the $\chi^2/d.o.f.$ for the reconstructed spectra as a function of D and λ for "input" values $\lambda = 1$ TeV^{-2} and $D = 10$; the right side shows rectangles proportional to the logarithm of the $\chi^2/d.o.f.$ Lower part (left and right) : values of the $\chi^2/d.o.f.$ for the reconstructed spectra as a function of D and λ for "input" values $\lambda = 5$ TeV^{-2} and $D = 8$; the right side shows rectangles proportional to the logarithm of the $\chi^2/d.o.f.$

neighbourhood of the last stages. Our main conclusion is to show that contrarily to what is usually thought the evaporation does not end up by the emission of a few quanta with energy around Planck values but goes asymptotically to zero with an infinite characteristic time scale.

The direct experimental registration of the products of evaporation of BHRs is impossible. This gives an opportunity to consider these BHRs as one of the main candidates for cold dark matter in our Universe.

In the case the of the Planck scale to be in the TeV range due to extra dimensions, this study shows that, beyond the dimensionality of space, the next generation of colliders should be able to measure the coefficient of a possible Gauss - Bonnet term in the gravitational action. This would allow an important step forward in the construction of a full quantum theory of gravity. It is also interesting to notice that this would be a nice example of the convergence between astrophysics and particle physics in the final understanding of black holes and gravity in the Planckian region.

Finally, it would be very interesting to extend this study to Kerr-Gauss-Bonnet black

Table 2. Reconstructed values for D and λ (TeV^{-2}) as a function of the "real" input values requiring $\chi^2 < 2\chi^2_{min}$. The first line assumes $\sigma = \pi r_+^2$, the second line $\sigma = \pi r_+^2/10$, the third line $\sigma = \pi r_+^2/100$ and the fourth line $\sigma = \pi r_+^2/1000$.

Allowed values (min/max)	$\lambda = 0.5$ TeV^{-2}	$\lambda = 1$ TeV^{-2}
D=6	$\lambda: 0.39/0.58; D: 6/6$ $\lambda: 0.39/0.58; D: 6/6$ $\lambda: 0.39/0.58; D: 6/6$ $\lambda: 0.39/0.58; D: 6/6$	$\lambda: 0.78/1.18; D: 6/6$ $\lambda: 0.78/1.18; D: 6/6$ $\lambda: 0.78/1.18; D: 6/6$ $\lambda: 0.78/1.32; D: 6/7$
D=7	$\lambda: 0.39/0.58; D: 7/7$ $\lambda: 0.39/0.58; D: 7/7$ $\lambda: 0.39/0.58; D: 7/7$ $\lambda: 0.16/0.58; D: 7/8$	$\lambda: 0.78/1.18; D: 7/7$ $\lambda: 0.78/1.18; D: 7/7$ $\lambda: 0.78/1.18; D: 7/8$ $\lambda: 0.18/1.37; D: 7/11$
D=8	$\lambda: 0.39/0.58; D: 8/8$ $\lambda: 0.39/0.58; D: 8/8$ $\lambda: 0.39/0.58; D: 8/8$ $\lambda: 0.20/0.79; D: 7/9$	$\lambda: 0.99/1.18; D: 8/8$ $\lambda: 0.99/1.18; D: 8/8$ $\lambda: 0.77/1.18; D: 8/9$ $\lambda: 0.22/1.56; D: 7/11$
D=9	$\lambda: 0.39/0.58; D: 9/9$ $\lambda: 0.39/0.58; D: 9/9$ $\lambda: 0.18/0.58; D: 9/10$ $\lambda :< 0.96; D: 8/11$	$\lambda: 0.99/1.18; D: 9/9$ $\lambda: 0.99/1.18; D: 9/9$ $\lambda: 0.37/1.18; D: 9/11$ $\lambda: 0.22/1.58; D: 8/11$
D=10	$\lambda: 0.18/0.58; D: 10/11$ $\lambda: 0.18/0.58; D: 10/11$ $\lambda: 0.18/0.58; D: 10/11$ $\lambda: 0.18/0.97; D: 9/11$	$\lambda: 0.99/1.18; D: 10/10$ $\lambda: 0.58/1.18; D: 10/11$ $\lambda: 0.58/1.58; D: 9/11$ $\lambda: 0.39/1.96; D: 8/11$
D=11	$\lambda: 0.39/0.99; D: 10/11$ $\lambda: 0.39/0.99; D: 10/11$ $\lambda: 0.39/0.99; D: 10/11$ $\lambda: 0.39/1.56; D: 9/11$	$\lambda: 0.99/1.58; D: 10/11$ $\lambda: 0.98/1.58; D: 10/11$ $\lambda: 0.75/1.77; D: 10/11$ $\lambda: 0.75/2.37; D: 9/11$

holes (first steps are already made in [57]) as the holes possibly produced at colliders are expected to be spinning. In the last case the results are expected to be more realistic.

Acknowledgments

S.A. has the pleasant opportunity to thank his co-authors M.Sazhin, A.Barrau, M.Pomazanov, O.Sazhina and all the other colleagues involved to these studies.

This work was supported by RFBR grant No. 07-02-01034-a.

References

[1] R.Penrose, *Quantum Nonlocality and Complex Reality*, In *Trieste 1992 Proceedings, The renaissance of General Relativity and cosmology*, 314 (1992).

[2] S.W.Hawking and R.Penrose, *The Nature of Space and Time*, Princeton, USA, University Press, 141 p. (1996) (The Isaak Newton Institute series of lectures).

[3] E.Poisson, *"Black hole interiors and strong cosmic censorship"*, In *Haifa 1997, Internal structure of black holes and spacetime singularities* 85-100.

[4] R.M.Wald, *"Gravitational Collapse and Cosmic Censorship"*, In *Iyer, B.R. (ed.) et al.: Black holes, gravitational radiation and the universe* 69-85.

[5] L.M.Burko, *Phys. Rev. Lett.* **79**, 4958 (1997).

[6] D.V.Gal'tsov, E.E.Donets and M.Yu.Zotov, *"Internal structure of non-Abelian black holes and nature of singularity"*, In *Haifa 1997, Internal structure of black holes and spacetime singularities* 142-162.

[7] S.W.Hawking and G.F.R.Ellis, *"Large Scale Structure of Space-Time"*, Cambridge University Press, Cambridge, (1973).

[8] I.D.Novikov and V.P.Frolov, *"Physics of black holes"*, Moscow, Nauka, (1986) (in russian).

[9] P. Kanti, N.E. Mavromatos, J. Rizos, K. Tamvakis and E. Winstanley, *Phys. Rev.* **D54**, 5049 (1996); P. Kanti and K. Tamvakis, *Phys. Lett.* **B392**, 30 (1997).

[10] T. Torii, H. Yajima and K. Maeda, *Phys. Rev.* **D 55**, 739 (1997).

[11] S.O. Alexeyev and M.V. Pomazanov, *Phys. Rev.* **D 55**, 2110 (1997); S.O. Alexeyev, M.V. Sazhin, M.V. Pomazanov, *Int. J. Mod. Phys. D* **10**, 225 (2001); S.O. Alexeyev, M.V. Sazhin, *Gen. Relativ. Grav.* **30**, 1187 (1998).

[12] S.O. Alexeyev, *Grav. Cosmol.* **3 (4)**, 161 (1997).

[13] S.O. Alexeyev and M.V.Pomazanov, *Grav. Cosmol.* **3 (11)**, 191 (1997).

[14] S. Mignemi and N.R. Stewart, *Phys. Rev.* **D47**, 5259 (1993).

[15] M.C. Bento and O. Bertolani, *Phys.Lett.B* **368**, 198 (1996).

[16] D. Garfincle, G. Horowitz and A. Strominger, *Phys. Rev.* **D43**, 3140 (1991), *Phys. Rev.* **D45**, 3888 (1992); G.W. Gibbons and K.Maeda, *Nucl. Phys.* **B298**, 741 (1988).

[17] M. Natsuume, *Phys.Rev.* **D50**, 3945 (1994).

[18] S.Alexeyev, S.Mignemi, *Class. Quant. Grav.* **18**, 4165 (2001).

[19] L.Randall, R.Sundrum, *Phys. Rev. Lett.* **83**, 4690 (1999).

[20] R.R. Metsaev, A.A. Tseytlin, *Phys. Lett.* **B 185**, 52 (1987); M.C. Bento, O. Bertolami, *Phys. Lett.* **B 228**, 348 (1989), B 368, 198 (1996).

[21] S.O. Alexeyev, O.S.Khovanskaya, *Grav.& Cosmol,* **6**, 14 (2000).

[22] P.R. Branoff, D.R. Brill, "Instantons for black hole pair production", Published in a festschrift for J.V. Narlikar, Kluwer Academic Pub., 1999, gr-qc/9811079

[23] S.Alexeyev, A.Barrau, G.Boudoul, O.Khovanskaya, M.Sazhin, *Astronomy Letters,* **28**, 428 (2002).

[24] R.C.Myers, J.Z.Simon, *Phys. Rev*. **D38**, 2434 (1988); *Gen. Rel. Grav.* **21**, 761 (1989).

[25] J.D.Barrow, E.J.Copeland, A.R.Liddle, *Phys. Rev.* **D46**, 645 (1992).

[26] S.Alexeyev, A.Barrau, G.Boudoul, O.Khovanskaya, M.Sazhin, *Class.Quant.Grav.* **19**, 4431 (2002).

[27] S. Massar & R. Parentani, *Nucl.Phys.* **B575**, 333-356 (200)

[28] M.K.Parikh, F.Wilczek, *Phys. Rev. Lett.* **85**, 24 (2000).

[29] K.Srinivasan, T. Padamanabhan, *Phys. Rev.* **D60**, 24007 (1999).

[30] T.Damour, R.Ruffini, *Phys. Rev*. **D14**, 332 (1976).

[31] D.N.Page, *Phys.Rev.* **D13**, 198 (1976); D14, 3260 (1976); D16, 2402 (1977).

[32] J.M.Blatt, V.F.Weisskopf, *Theoretical Nuclear Physics* (Wiley, New York, 1952), p. 520; A.A.Starobinsky, S.M.Churilov, *Zh. Eksp. Fiz.* **65**, 3 (1973); J.H.MacGibbon, B.R.Webber, *Phys. Rev.* **D41**, 3052 (1990).

[33] S.Alexeyev, O.Khovansaya, M.Sazhin, Astronomy Letters (*Pisma v Astronomichesky Zhurnal*) **28**, 139 (2002).

[34] A. Barrau, *Astropart.Phys.* **12**, 269 (2000).

[35] A.G.Polnarev, M.Yu.Khlopov, *Astronomicheskii Zhurnal (Astronomical Journal)* **58**, 706 (1981), in Russian; M.Yu.Khlopov, B.A.Malomed, Ya.B.Zeldovich, *MNRAS* **215**, 575 (1985); J.H.MacGibbon, *Nature* **329**, 308 (1987); J.H.MacGibbon, B.Carr, *Ap.J.* **371**, 447 (1991); V.I.Manko, M.A.Markov, *Phys.Lett.* **A172**, 331 (1993).

[36] C. Renault et al., *Astronomy & Astrophysics*, **324**, L69 (1997).

[37] H.I. Kim, *Phys.Rev.* **D62**, 063504 (2000).

[38] T. Bringmann, C. Kiefer, D. Polarski, *Phys. Rev.* **D65**, 024008 (2002).

[39] N. Arkani-Hamed, S. Dimopoulos & G.R. Dvali, *Phys. Lett. B* **429** (1998) 257
I. Antoniadis et al., *Phys. Lett. B* **436** (1998) 257
N. Arkani-Hamed, S. Dimopoulos & G.R. Dvali, *Phys. Rev. D* **59** (1999) 086004

[40] L. Randall & R. Sundrum, *Phys. Rev. Lett.* **83** (1999) 3370

[41] S.B. Giddings & S. Thomas, *Phys. Rev. D* **65** (2002) 056010

[42] S.B. Giddings& E. Katz, *J. Math. Phys.* **42** (2001) 3082

[43] D.M. Eardley & S.B. Giddings, *Phys. Rev. D* **66** (2002) 044011

[44] H. Yoshino & Y. Nambu, *Phys. Rev. D* **66** (2002) 065004

[45] A. Barrau, J. Grain, S. Alexeyev, *Phys. Lett. B,* **584** (2004) 114.

[46] S. Dimopoulos & G. Landsberg, *Phys. Rev. Lett.* **87** (2001) 161602

[47] D.G. Boulware & S. Deser, *Phys. Rev. Lett.* **55** (1985) 2656

[48] B. Zwiebach, *Phys. Lett. B* **156** (1985) 315
N. Deruelle & J. Madore, Mod. *Phys. Lett. A* **1** (1986) 237
N, Deruelle & L. Farina-Busto, *Phys. Rev. D* **41** (1990) 3696

[49] S. Nojiri, S.D. Odintsov & S. Ogushi, *Phys. Rev. D* **65** (2002) 023521
M.E. Mavrotamos & J. Rizos, *Phys. Rev. D* **62** (2000) 124004
Y.M. Cho, I.P. Neupane and P.S. Wesson, *Nucl. Phys. B* **621** (2002) 388

[50] B.C. Paul & S. Mukherjee, *Phys. Rev. D* **42** (1990) 2595
B. Abdesselam & N. Mohammedi, *Phys. Rev. D* **65** (2002) 084018
C. Charmousis & J.-F. Dufaux, *Class. Quantum .Grav.* **19** (2002) 4671
J.E. Lidsey & N.J. Nunes, *Phys. Rev. D* **67** (2003) 103510

[51] R.-G. Cai, *Phys. Rev. D* **65** (2002) 084014
A. Padilla, *Class. Quantum Grav.* **20** (2003) 3129

[52] C. M. Harris & P. Kanti, *JHEP* **010** (2003) 14

[53] A. Barrau et al., *Astronom. Astrophys.* **388** (2002) 676

[54] R. Emparan, G. T. Horowitz and R. C. Myers, *Phys. Rev. Lett* **85** (2000) 499

[55] T. Tjöstrand, *Comput. Phys. Commun.*, **82** (1994) 74

[56] ATLAS TDR 14, vol 1 CERN/LHCC/99-14 (1999)

[57] S. Alexeyev, N. Popov, A. Barrau and J. Grain, *Journal of Phys.: Conf. Series* **33**, 343 (2006); S.Alexeyev, A.Popov, M.Startseva, A.Barrau, J.Grain *JETP* **133**, 710 (2008).

In: Black Holes and Galaxy Formation ISBN 978-1-60741-703-3
Editors: A.D. Wachter and R.J. Propst, pp. 249-294

Chapter 10

X-RAY EMISSION FROM ACCRETION DISKS OF AGN: SIGNATURES OF SUPERMASSIVE BLACK HOLES

Predrag Jovanović and Luka Č. Popović
Astronomical Observatory, Volgina 7, 11160 Belgrade, Serbia

Abstract

In this chapter we discuss the X-ray radiation from relativistic accretion disks around supermassive black holes, supposed to exist in the centers of Active Galactic Nuclei (AGN). Our focus is on the X-ray radiation, especially in the Fe Kα line which originates in the innermost parts of an accretion disk. Moreover, here we discuss some effects which can disturb the Fe Kα profile and cause its rapid and irregular variability, observed in the X-ray spectra of some AGN. We will pay attention to three such effects: perturbations in the disk emissivity, absorbtion by warm absorbers and gravitational microlensing. The X-ray emission from accretion disks around non-rotating (Schwarzschild metric), as well as rotating (Kerr metric) supermassive black holes, is discussed. The X-ray radiation of AGN is probably produced in a compact region near their central supermassive black holes, and can provide us some essential information about the plasma conditions and the space-time geometry in these regions. The goal of this chapter is mainly to present a short overview of some important and recent investigations in this field.

1. Introduction

Active Galactic Nuclei are powerful sources of radiation in a wide spectral range: from γ rays to radio waves. They derive their extraordinary luminosities (sometimes more than 10.000 times higher than luminosities of "ordinary" galaxies) from energy release by matter accreting towards, and falling into, a central supermassive black hole (see e.g. Peterson, 2004).

According to the unification model of AGN (Antonucci, 1993), their central engine consists of a supermassive black hole (SMBH) with mass ranging from 10^5 to 10^9 solar masses $M_\odot$ (Kaspi et al., 2000; Peterson et al., 2004), which is surrounded by an accretion disk that radiates in the X-ray band (Fabian et al., 1989). The vast majority of the X-ray sources in the Universe are AGN. The integral emission of AGN reflects the history of accretion onto SMBH over cosmic time. Emission lines are usually seen in the X-ray spectra of AGN. The broad emission Fe Kα spectral line (6.4–6.9 keV, depending on ionization state) with asymmetric profile (narrow bright blue peak and a wide faint red wing) has been observed in a number of type 1 AGN (see e.g. Nandra et al., 2007). Early results from the ASCA (*Advanced Satellite for Cosmology and Astrophysics*) era suggested that broad relativistic lines might be common in type 1 AGN. However, surprisingly, they have been detected in significant number and their features described in only a small fraction of those sources (Reynolds & Nowak, 2003). The first and the best studied one is MCG-6-30-15 (Tanaka et al., 1995; Fabian & Vaughan, 2003).

In some cases the line width corresponds to one third of the speed of light, indicating that its emitters rotate with relativistic velocities. Therefore, the line is probably produced in a very compact region near the central black hole of AGN and can provide us some essential information about the plasma conditions and the space-time geometry in the vicinity of the black hole. Consequently, if the line is emitted close enough to the SMBH, it shows a broad relativistic profile affected by SMBH spin (Reynolds & Fabian, 2008) and gravitational redshift (as well as other general relativistic effects) (Fabian et al., 1989; Laor, 1991).

Several studies have been performed over samples of local AGN (see e.g. Nandra et al., 1997; Yaqoob et al., 2005; Nandra et al., 2007; Bianchi et al., 2008; Markowitz et al., 2008, etc), as well as from distant quasars (Corral et al., 2008) in order to characterize the Fe Kα emission. Nandra et al. (2007) performed a spectral analysis of a sample of 26 type 1 to 1.9 Seyferts galaxies ($z < 0.05$) observed by *XMM-Newton*. They found that a relativistic line is significantly detected in a half of their sample (54 $\pm$10 percent) with a mean equivalent width (EW) of $\sim$ 80 eV, but around 30% of selected AGN showed a relativistic broad line that can be explained by the emission of an accretion disk.

Accretion disks could have different forms, dimensions, and emission, depending on the type of central black hole, whether it is rotating (Kerr metric) or non-rotating (Schwarzschild metric). They represent an efficient mechanism for extracting gravitational potential energy and converting it into radiation, giving us the most probable explanation for the main characteristics of AGN (high luminosity, compactness, jet formation, rapid time variations in radiation and the profile of the Fe Kα spectral line).

Here we discuss the X-ray radiation from relativistic accretion disks around supermassive black holes, supposed to exist in the centers of AGN. Especially, we discuss the Fe Kα line profile which originates from the accretion disk. Moreover, we also present some effects which can disturb the Fe Kα profile, such as: perturbations in the disk emissivity, absorbtion by warm absorbers and gravitational microlensing.

The aim of this chapter is to present a short overview of results of some recent investigations in this field, and it is divided into the following six sections: *Active Galactic Nuclei as Hosts of Supermassive Black Holes* - where the main features, classification and unified model of active galaxies are briefly presented, *Space-Time Geometry in Vicinity of Supermassive Black Holes* - where the basic definitions of Schwarzschild and Kerr metrics are given, *Accretion Disk Around a Supermassive Black Hole* - where we explain the standard model of an accretion disk, including its emission, accretion rate, luminosity, structure and spectral distribution, *Supermassive Black Holes and X-ray Emission* - where the focus is on the modeling of the observed X-ray radiation from relativistic accretion disk around a supermassive black hole of AGN in both the Fe Kα spectral line and X-ray continuum, *Variability of X-ray Emission Around Supermassive Black Hole* - where we present the main causes of rapid and irregular variability of the X-ray emission which can be due to disk instability, reflecting in perturbations of its emissivity, or it could be caused by some external effects, such as gravitational microlensing and absorption by X-ray absorbers, and finally *Conclusion* - where the most important results from previous sections are pointed out and their brief summary is given.

Finally, we should note that a huge number of papers devoted to investigation of the X-ray emission of AGN was published in the last two decades, so it was not possible to mention all of them here, but more details and references can be found in review papers such as, e.g. Brandt & Hasinger (2005); Miller (2007); Harris & Krawczynski (2006), etc.

2. Active Galactic Nuclei as Hosts of Supermassive Black Holes

Active galaxies differ from the so-called "normal" galaxies in the amount of energy emitted from their nuclei. The term *Active Galactic Nuclei* (AGN) refers to the energetic phenomena in the central region of a galaxy which cannot be solely or directly produced by stars. The nebular-like emission spectra of NGC 1068, NGC 4051 and NGC 4151, with Hα, [O II] λ3727, [Ne III] λ3869 and [O III] $\lambda\lambda$ 4363,4959,5007 were observed by Hubble (1926), but the first classification of these objects was made by Seyfert (1943). He recognized these objects as a class of galaxies with strong and high-excitation optical emission lines localized in the nuclei[1].

Active galaxies have been the subject of an intensive astrophysical investigations for the last 3 – 4 decades. As the most luminous objects with one of the most powerful energy release rates and with the most compact dimensions, AGN are of great interest in modern astrophysics. AGN (i.e. quasars) are the brightest objects in the Universe and thus, these objects are important for the studies of the early Universe and cosmology in general. For instance, studies of their luminosity functions or quasar host galaxies are crucial for understanding the formation and evolution of galaxies in general. The class of active galaxies contains many different objects, such as quasars, gigantic elliptic radio galaxies, luminous spiral Seyfert galaxies, blazars, etc.

2.1. Main characteristics of AGN

AGN are the most luminous objects in the Universe and they have luminosities in the range from $\sim 10^{42}$ to 10^{48} erg s^{-1}, that can be 10^4 times greater than in the case of a typical galaxy. Apart from their great luminosities, AGN are emitting in the broad band of electromagnetic spectrum. The observed emission of AGN is in the continuum and in lines, from the γ and X domains to the far infrared and radio bands (see e.g. Peterson, 2004).

One of the characteristics of AGN is violent and fast variability observed in different parts of electromagnetic spectra. Fast changes in the AGN brightness are observed sometimes just for a couple of days, e.g. in the case of galaxy NGC 4151 (Shapovalova et al., 2008). This leads to the conclusion that the emission regions in the AGN have small

[1] For more detailed historical review see Osterbrock (1989)

dimensions (from only couple of light days up to couple of light months). Also, the brightness of AGN can be so high that the brightness of the host galaxy can be neglected in the total brightness (for instance in the case of quasars). Consequently, they are compact regions which could not be directly observed in the most bands of electromagnetic spectrum. Therefore, AGN and their physical/kinematical properties are indirectly studied through the analysis of their spectra.

In contrast to the spectral energy distribution (SED) of an ordinary galaxy (that represents a sum of stellar spectra, thus the most of its luminosity comes within no more than one decade of frequency), AGN emit across the broad range of frequencies (of order 10^5 Hz). The AGN continuum in a number of high-luminosity sources peaks in the ultraviolet, i.e. in the so-called Big Blue Bump (BBB) (Shilds, 1978; Zhou et al., 1997). The BBB is attributed to the thermal emission from an optically thick region of the accretion disk (Shilds, 1978; Ulrich et al., 1980), or from optically thin regions, i.e. free-free emission (Antonucci & Barvainis, 1988; Ferland et al., 1990). The observed SED in AGN is remarkably different from the thermal (black-body) spectrum of a star or a regular galaxy (Oke & Sargent, 1968). This featureless continuum observed in AGN characterizes the main source of energy input and suggests that the mechanisms that produce it are common to all types of AGN. The SED can be approximated with the power law function $F \approx \nu^{\alpha}$ where α is the spectral index (Krolik, 1999; Peterson, 2004).

The observed continuum of a typical Seyfert 1 galaxy is very strong compared with Seyfert 2 galaxies; as a result Seyfert 1 galaxies appear to be more luminous than Seyfert 2-s. Some AGN have the SED almost flat from the infrared to X-ray part of the spectrum, so that the spectral index is $\alpha \sim 1$, although it is usually steeper. In the radio band, the brightness of active galaxies is usually an order of magnitude higher than of normal galaxies. But still, even if it is so intensive, the radio luminosity is never higher by more than 1% of the bolometric luminosity.

Another property of AGN is strong X-ray flux, with the X-ray continuum produced by lower energy photons which are Compton scattered to higher energies by relativistic electrons (Sunyaev & Titarchuk, 1980). The fraction of the power emitted in the X-ray emission is three or four times larger in AGN than in normal galaxies (Krolik, 1999; Peterson, 2004), therefore the X-emission can indicate the presence of AGN.

Today, it is widely accepted that the mechanism which powers AGN is accretion onto supermassive black hole (SMBH), i.e., the greatest part of the continuum emission comes from an accretion disk and its corona. The mass of the supermassive black hole is estimated to be from 10^5 to 10^9 $M_{\odot}$ (Peterson et al., 2004; Peterson & Wandel, 2000). Thus, in the center of an AGN there is a supermassive black hole with an accretion disk surrounded by gas and dust in the form of a torus.

There are also other emitting regions present in the vicinity of the supermassive black hole, such as ionized gas clouds that produce intensive emission lines or jets of matter that are mostly visible in the radio band, but sometimes they can also be seen in the optical band. The kinematics of these regions is very complex as well as the physical processes that produce such specific spectra of these objects. In the central part, there are mainly three emitting line regions: Fe Kα (in the X-ray range)[2], Broad and Narrow Line Emitting

[2]See §5.2 for more details

Regions (BLR and NLR, respectively; in optical/UV spectra).

In the optical and UV band, the total flux of emission (and occasionally absorbtion) lines contributes from several percent to tens of percent of the continuum flux. The existence of the broad and narrow emission lines comes from the fact that an AGN contains two separated emission line regions with different kinematics, density, ionization, optical thickness and radiation transfer.

The basic characteristics of the BLR and NLR can be described as follows:

- BLR is a compact region, with dimensions from only couple of light days up to couple of light weeks (Kaspi et al., 2000; Peterson et al., 2004), that is located in the vicinity of the black hole ($<$1 pc). The structure of this region is very complex and most likely it consists from at least two separated subregions (see e.g. Popović et al. (2004); Ilić et al. (2006); Bon et al. (2006); Popović et al. (2008), etc). The ionized gas in this region is of relatively high density ($n > 10^9$ cm^{-3}) and temperature $T \sim 10^4$ K. Emitters are moving with high velocities, up to 10000 km/s. In this region the broad emission lines (BELs) from allowed transitions are formed.

- NLR extends even up to 1 kpc far from the central source (Peterson, 2004). The density in this region is significantly smaller ($n \sim 10^3$cm^{-3}) than in the BLR, while the temperature is of the same order of magnitude as in the BLR. From this region the narrow emission lines (NELs) arise, indicating a random motion of the emitting gas smaller than 1000 km s^{-1}. NELs are often coming from the forbidden transitions. The physical properties of this region are much closer to the properties of emission nebulae than to the BLR.

According to the different characteristics mentioned above, there are several classes of AGN.

2.2. Classification

The first group of stellar systems with active nuclei consists of Seyfert galaxies. These are spiral galaxies with very bright nuclei which spectra have strong emission lines of neutral and multi-ionized emitters. The presence of highly ionized emission lines indicates the existence of a non-stellar ionization continuum. The presence or absence of BELs (i.e. BLR) has been historically used to separate Seyfert galaxies into two classes. Seyfert 1 galaxies have broad permitted (H I, He I and He II) and narrow permitted and forbidden lines (such as [O III] λ5007), while Seyfert 2 galaxies have only narrow permitted and forbidden lines (Khachikian & Weedman, 1974). This simple Seyfert classification scheme can be further sub-divided according to specific spectral properties (Osterbrock, 1989), as e.g. Seyfert galaxies with intermediate Balmer profiles are classified as Seyfert 1.5-s with apparent narrow Balmer line components superimposed on broad wings.

Many low luminosity galaxies have a nucleus that resembles the Seyfert 2 nucleus, but with the spectrum that shows the forbidden lines from lower ionization states. Therefore, these galaxies are called Low-Ionization Nuclear Emission-line Regions, i.e LINERs.

Active galaxies which emit intensively in the radio band are so called radio galaxies (RG). Radio galaxies are dominantly elliptic galaxies that are divided into two morphologically different types: FR I and FR II. The classification is done according to the relative position of the surface brightness maximum (called the hot spot) in the radio lobes. For example, in the FR I radio galaxy the distance between the hot spot in the radio lobes is less than half of the maximum diameter of the radio source, while in the FR II type this distance is larger (Binney & Merrifield, 1998). Like Seyfert galaxies, radio galaxies can be divided into subgroups according to the widths of their emission lines, i.e. Narrow Line Radio Galaxies (NLRG) that emit narrow emission lines characteristic for Seyfert 2 galaxies, and Broad Line Radio Galaxies (BLRG) that emit broad lines as Seyfert 1 galaxies.

Being the most powerful sources in the Universe, quasars belong to the group of active galaxies with the absolute magnitude $M_B < -23$ (Osterbrock, 1989). These quasi stellar (quasars) objects emit intensively in the radio domain and their spectrum shows broad emission lines significantly redshifted. In this group of AGN there are also Quasi-Stellar Objects (QSOs) which have the same observed properties as quasars, except they are weak radio sources. Besides, there is a group of QSOs with broad absorption lines, so called Broad Absorption Line Quasars (BAL QSOs).

Finally, the separate class of AGN are so called BL Lacerte (Lacs) objects. BL Lacs are nuclei of elliptical galaxies that emit highly variable and polarized radiation, and have a non thermal optical continuum and a strong radio emission. With this group of galaxies, we often connect another type of AGN, Optically Violently Variable (OVVs) quasars. The OVVs, in contrast to BL Lacs which have featureless continuum, emit broad optical emission lines characteristic for quasars. Both these groups of objects are often referred to as one class called blazars.

2.3. Unified Model

As it was mentioned above, all types of AGN have some common properties. It is widely accepted that central engine of an AGN is accretion onto supermassive black hole. First observational evidence in favor of a unified model of AGN were spectropolarimetric observations of Seyfert 2 galaxy NGC 1068 (Antonucci & Miller, 1985). Antonucci & Miller (1985) found broad Balmer lines and Fe II emission, characteristic of a Seyfert 1 spectrum, in the polarized spectrum of NGC 1068.

In the unified model of AGN (Antonucci, 1993) both types of Seyfert galaxies are intrinsically the same. The difference in spectral characteristics appears due to different angle of the visibility of the central regions (see Fig. 1).

A geometrically and optically thick dusty molecular torus-like structure surrounds the central source, as well as the BLR (see Fig. 1). Therefore the visibility of the nuclear engine depends on viewing angle with respect to the torus. The broad-line polarization in Seyfert 2-s is probably due to electron scattering. This picture is consistent with the differences observed between continuum fluxes of the Seyfert 1 and Seyfert 2 galaxies (Clavel et al., 2000).

Following the unified model, the line-of-sights to Seyfert 2 galaxies (or Narrow Line Objects) are obstructed by optically thick material corresponding to hydrogen column densities of $N_H > 10^{22}\ \mathrm{cm}^{-2}$ (Risaliti et al., 1999). For column densities $N_H < 10^{24}\ \mathrm{cm}^{-2}$, photons above a few keV can penetrate the torus creating an un-obstructed view of the nuclear source. In this case the source is called *Compton thin*. For column densities $10^{25} > N_H > 10^{24}\ \mathrm{cm}^{-2}$, only high energy X-ray emission (tens of keV) can pass through

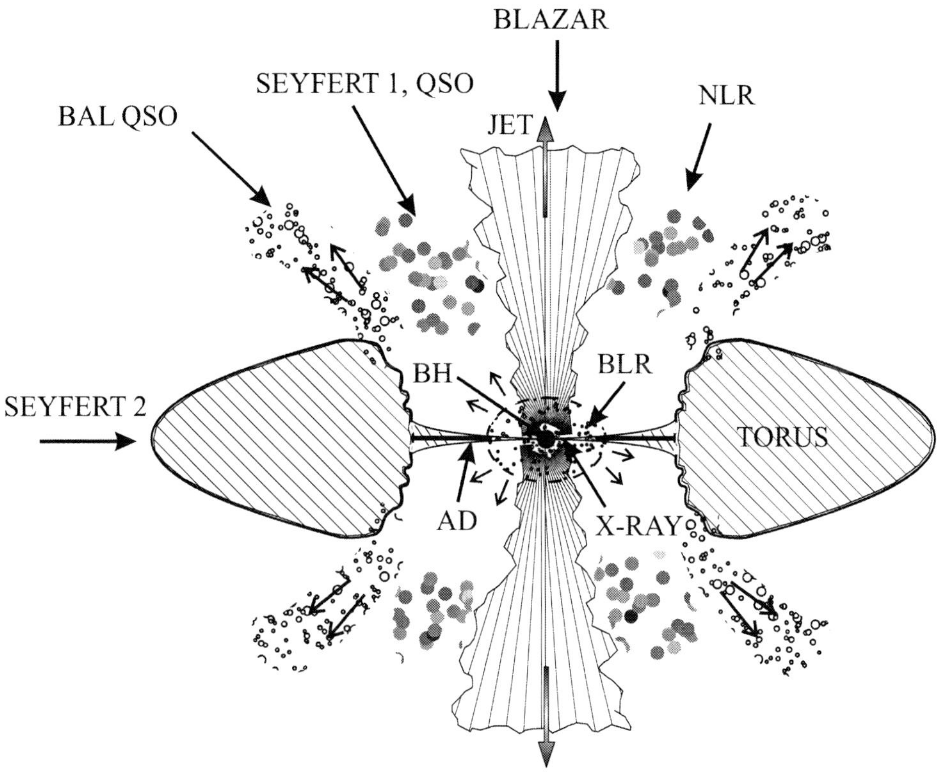

Figure 1. The scheme of unified model of AGN (Figure courtesy: Vesna Borka Jovanović).

the obscuring material (Turner et al., 1997). But in the case $N_H > 10^{25}\ \mathrm{cm}^{-2}$ even high energy X-rays, above a few tens of keV, are Compton scattered and the nuclear source is completely hidden from our direct view. Therefore, by observing near the edge-on torus, one can detect narrow line objects, with weak (or without) X-ray emission; while by observing near the face-on torus, one can detect Blazars or OVVs. The broad line objects can be detected if one observes an inclined torus.

3. Space-Time Geometry in Vicinity of Supermassive Black Holes

As it was shown in the previous section, according to unified model of AGN, in their heart there is a supermassive black hole with mass up to $10^9\ M_\odot$, surrounded by an accretion disk. A black hole is a region of space-time around some collapsed mass which gravitational field became so powerful that nothing (including electromagnetic radiation) could escape from its attraction, after crossing the certain boundary called the event horizon (Hawking, 1988).

All black holes in nature are commonly classified according to their mass as: supermassive black holes (with masses $M_{BH} \sim 10^5 - 10^9\ M_\odot$), intermediate-mass black holes ($M_{BH} \sim 10^2 - 10^5\ M_\odot$), stellar-mass black holes ($M_{BH} < 10^2\ M_\odot$), mini and micro black holes ($M_{BH} \ll M_\odot$). In this chapter our focus will be on supermassive black holes located in the centers of AGN, because they are surrounded by accretion disks which X-ray radiation is the main subject of this discussion. Space-time geometry in vicinity of a supermassive black hole depends on its type (more precisely, on its angular momentum), being either non-rotating, where the local space-time geometry is described by Schwarzschild metric, or rotating, where the Kerr metric determines the geometry of local space-time.

3.1. Historical Background

The term black hole was introduced in 1969 by the American scientist John Wheeler, but the predecessor of the modern idea of black holes emerged in the late eighteenth century. It was independently developed by the British geologist John Michell in 1783 and the French mathematician and astronomer Pierre-Simon Laplace in 1796. This idea was based on Newton's theory of gravity and particle theory of light, and is commonly referred to as a dark star. A dark star is, according to Michell, a sufficiently massive and compact star which have such a strong gravitational field that the corresponding surface escape velocity equals or exceeds the speed of light. Therefore, any light emitted from the surface of such star would be dragged back by the gravitational field and we would not be able to see it, but we could still feel its gravitational attraction (Hawking, 1988). After the wave theory of light was founded, the idea about a dark star was forgotten since, according to the wave theory, it was not clear that light could be affected by gravity at all.

Modern theory of black holes was founded in the twentieth century, more precisely in 1915, shortly after Einstein introduced the theory of general relativity, when the German physicist and astronomer Karl Schwarzschild found a particular exact solution to the Einstein field equations, for the limited case of a single spherical non-rotating mass.

The implications of the general relativity to the stellar evolution were firstly understood by Subrahmanyan Chandrasekhar in 1928 (Hawking, 1988). He calculated a maximum limit for the mass of a star, above which the star would not be able to support itself (by repulsion of electron degeneracy pressure which arises from Pauli exclusion principle) against its own gravity, after it had used up all its fuel. This limit ($\approx 1.5\ M_{\odot}$) is nowadays known as the Chandrasekhar limit. If a star's mass is less than this limit, it can eventually stop contracting at a final state known as a "white dwarf" (Hawking, 1988). A similar discovery was made in 1932 by the Russian scientist Lev Davidovich Landau. He found that there is another limiting mass ($\approx 1 - 2\ M_{\odot}$), for which the corresponding final state is much smaller than a white dwarf and which would be supported against its gravity by the exclusion principle repulsion between neutrons and protons, rather than between electrons. Such stars are therefore called neutron stars (Hawking, 1988). All these results led to a surprising conclusion that a star which mass is above the Chandrasekhar limit, when it comes to the end of its fuel, could experience a catastrophic gravitational collapse to a point with an infinite density. American scientist Robert Oppenheimer was the first who explained in 1939 that, according to general relativity, stars above $\approx 3\ M_{\odot}$ would at the end of their lives collapse into black holes, due to reasons given by Chandrasekhar (Hawking, 1988).

With the discovery of QSOs in the early 60s, the speculations about how their enormous luminosities could be produced within such a small region occurred. Thermonuclear reactions were not enough efficient, and therefore, they were quickly eliminated. Hoyle and Fowler suggested in 1963 that such luminosities could be produced during the collapses and disintegrations of stelar-type objects with masses of $10^5 - 10^8\ M_{\odot}$ (Ferrarese &

Ford, 2005). Soon, it became clear that the energy source was gravitational. For example, Wheeler suggested a mechanism in which gravitational singularity at the center of a galaxy converted the falling matter into energy (Ferrarese & Ford, 2005). Zel'dovich, Novikov and Salpeter proposed in 1964 the release of energy through accretion due the growth of a massive object at the center of a galaxy, while Lynden-Bell in 1969 made an attempt to explain the phenomena observed in QSOs and Seyfert galaxies in terms of a black hole formalism (Ferrarese & Ford, 2005). A crucial event for the acceptance of black holes was the discovery of pulsars by Anthony Hewish and Jocelyn Bell in 1967, because it was the clear evidence of the existence of neutron stars, and therefore, confirmation of Chandrasekhar limit. The first detection of a solar mass black hole came in 1972, when the mass of the rapidly variable X-ray source Cygnus X-1 was proven to be above the maximum allowed for a neutron star (Ferrarese & Ford, 2005). The first convincing evidence for the existence of supermassive black holes is the discovery of the Fe $K\alpha$ spectral line in X-ray spectrum of Seyfert galaxy MGC-6-30-15 by Tanaka et al. (1995).[3]

Nowadays, the black hole theory is very well established, mainly due to the contributions of numerous authors during the second half of the twentieth century, like Roy Kerr, Werner Israel, John Wheeler, Brandon Carter, Roger Penrose, Stephen Hawking and many others.[4]

3.2. Schwarzschild Metric - Non-rotating Black Hole

Schwarzschild metric describes the space-time geometry in spherically symmetric gravitational field around a time-steady non-rotating black hole in vacuum. The solution of the Einstein field equations under these conditions was found by Karl Schwarzschild in 1916 (Schwarzschild, 1916a,b). If we denote the Schwarzschild radius which corresponds to a mass M by $R_S = \dfrac{2GM}{c^2} \approx 3\dfrac{M}{M_\odot}\mathrm{km}$ (where G is the gravitational constant and c is the speed of light), then the squared element of space-time interval ds^2 in the case of Schwarzschild metric is given by:

$$ds^2 = \left(\frac{1}{1-\dfrac{R_S}{r}}\right) dr^2 - \left(1-\frac{R_S}{r}\right) c^2dt^2 + r^2(d\theta^2 + \sin^2\theta d\phi^2), \tag{1}$$

where t is the time coordinate measured by a stationary clock at infinity (which must be distinguished from the proper time τ, measured by a clock moving with the particle), r is the radial coordinate, θ is the colatitude (angle from North) in radians and ϕ is the longitude in radians.

It is obvious from Eq. (1) that Schwarzschild metric depends only on the mass M through the Schwarzschild radius R_S. The Schwarzschild metric has an infinite space-time curvature (i.e. a singularity) at $r = 0$. In the case when $r \to \infty$ this metric is reduced to the Minkowski metric of flat space-time: $ds^2 = dr^2 - c^2dt^2 + r^2(d\theta^2 + \sin^2\theta d\phi^2)$.

If the complete mass M collapses below its Schwarzschild radius R_S, then it will form a spherically symmetric Schwarzschild black hole. A Schwarzschild black hole has no charge

[3]See §5.2 for more details

[4]For more detailed historical overview on this topic, see e.g. Hawking (1988)

or angular momentum and it can be distinguished from any other Schwarzschild black hole only by its mass.

To a first approximation, the matter surrounding a black hole may be assumed to rotate in circular Keplerian orbits with velocity (e.g. Shakura & Sunyaev, 1973): $v^2 = \frac{GM}{R}$. Therefore, taking into account that a maximum allowed rotation velocity is speed of light ($v = c$), one can obtain so called gravitational radius R_g:

$$R_g = \frac{GM}{c^2} = \frac{R_S}{2} \approx 1.5\frac{M}{M_\odot}\text{km}, \tag{2}$$

which is usually used as a unit for distance around a black hole.

3.3. Kerr Metric - Rotating Black Hole

Roy Kerr, a New Zealand mathematician, found an exact solution of the Einstein field equations of general relativity in the case of gravitational field outside an uncharged rotating black hole. The properties of Kerr solution, known as Kerr metric, are given in numerous papers and books (see e.g. Chandrasekhar, 1983; Carter, 1973) and therefore, here we will present only its definition and some basic features, necessary for the further discussion.

Contrary to the Schwarzschild case, the Kerr metric is no longer spherically symmetric. In the case of a rotating black hole with mass M and angular momentum $J = aM$, this metric is given by (see e.g. Fanton et al., 1997):

$$ds^2 = -\left(1 - \frac{2Mr}{\Sigma}\right)dt^2 - \frac{4Mar}{\Sigma}\sin^2\theta dt d\phi + \frac{A}{\Sigma}\sin^2\theta d\phi^2 + \frac{\Sigma}{\Delta}dr^2 + \Sigma d\theta^2, \tag{3}$$

where

$$\Sigma = r^2 + a^2\cos^2\theta, \quad \Delta = r^2 + a^2 - 2Mr, \quad A = (r^2 + a^2)^2 - a^2\Delta\sin^2\theta. \tag{4}$$

Above definition is given in Boyer-Lindquist coordinates for $c = G = 1$. As it can be seen from expressions (3) and (4), in addition to black hole mass M, the Kerr metric also depends on its specific angular momentum a. Here $a \leq M$ because any additional angular momentum would increase the energy of the black hole and, therefore, its mass (Krolik, 1999). In the case of a non-rotating black hole, i.e. for $a \to 0$, it reduces to the Schwarzschild metric.

By solving the equation $\Delta = 0$ and taking only a root with $+$ sign, we obtain the radius of event horizon of a black hole: $r_h = M + \sqrt{M^2 - a^2}$. Thus, when $a \to 0$ then $r_h \to 2M = R_S$ (because we assumed that $c = G = 1$), whereas when $a \to M$ (i.e. for a maximally rotating black hole) then $r_h \to M$.

The minimum allowed radius r_{ms} of a stable circular equatorial orbit, or so called marginally stable orbit, is given by the roots of the equation (Fanton et al., 1997): $r^2 - 6Mr \mp 8a\sqrt{Mr} - 3a^2 = 0$, where the upper sign refers to co-rotating orbits, while the lower one to counter-rotating orbits. In the case of Schwarzschild metric ($a = 0$) $r_{ms} = 6M$, whereas for a maximally rotating black hole ($a = M$) $r_{ms} = M$ (Fanton et al., 1997).

Rotating black holes are surrounded by a region called the ergosphere, in which there can be no static observers and the negative energy orbits are possible (Krolik, 1999). Also, there is a general relativistic effect known as frame-dragging, according to which a rotating black hole drags the space-time around itself, but such effects will not be discussed here in more details.

4. Accretion Disk Around a Supermassive Black Hole

Some observed quasars have luminosities of up to $L_{bol} \sim 10^{47}$ erg/s (Schneider, 2006). The corresponding total energy emitted during the lifetime of such quasars can be estimated to $E \geq 3 \times 10^{61}$ erg, assuming that their minimum age is about 10^7 years and that their luminosities do not change significantly over the lifetime (Schneider, 2006). Only two known mechanisms could produce such enormous amount of energy: nuclear fusion and accretion. The maximum efficiency ϵ (defined as the mass fraction of fuel that is converted into energy, according to $E = \epsilon mc^2$) of thermonuclear reactions is $\epsilon \leq 0.8\%$. Schneider (2006) showed that this value is too low to explain such high energies of $E \geq 3 \times 10^{61}$ erg. Accretion is the only remaining mechanism which can yield larger ϵ. Its maximum efficiency is about $\epsilon \sim 6\%$ for a non-rotating black hole and $\epsilon \sim 29\%$ for a maximally rotating one (Schneider, 2006). Therefore, AGN release their enormous energy by matter accreting towards, and falling into, a central supermassive black hole.

If the cold matter, which was initially at rest and without magnetic field, was subjected to free radial infall it would accrete to the central black hole without any energy release or observational effects (Shakura & Sunyaev, 1973). However, in the case of AGN the accreting matter has a significant angular momentum which does not allow its free infall. At the marginally stable orbit of central black hole with mass of $10^8 M_{\odot}$, the specific angular momentum is $\approx 1 \times 10^{24}$ cm^2/s, which is much less even in comparison to a typical galaxy where the speciffic angular momentum of orbiting material is $\approx 6 \times 10^{28}$ cm^2/s (Krolik, 1999). It means that approach of accreting material toward a black hole requires a loss of the greatest fraction of its initial angular momentum. Mechanisms which can contribute to such loss of angular momentum are viscosity, nonaxisymmetric gravitational forces, magnetic forces, etc. (Krolik, 1999).

Since the orbit of minimum energy for fixed angular momentum in any spherically symmetric potential is a circle, the infall of accreting material due to loss of its angular momentum will be in form of successively smaller and smaller concentric circles. Matter traveling along orbits inclined to each other will eventually collide in the plane of intersection, and as a result, the angular momenta of different gas steams will be mixed and equalized. Consequently, all accreting matter will orbit in a single plane and will have the same specific angular momentum at any given radius, meaning that accretion is most likely performed through an accretion disk (Krolik, 1999).

The accreting material is able to approach the marginally stable orbit of central black hole only if there is an efficient mechanism for transporting angular momentum outward. Magnetic field that exists in matter flowing into the disk, as well as its turbulent motions, enable angular momentum to be transferred outward. Hence, the accretion disk is one of the best candidates for a such mechanism, since the particles in it lose their angular momentum due to friction between its adjacent layers and spiral towards black hole, releasing their gravitational energy. Part of this energy increases the kinetic energy of rotation and the rest is turned into the thermal energy and irradiated from the disk surface (Shakura & Sunyaev, 1973).

The efficiency of angular momentum transport in the disk is characterized by viscosity parameter $\alpha \leq 1$, which can be assumed as constant in most cases, since the spectrum of disk radiation and its surface temperature do not strongly depend on it, except in the case of a supercritical accretion regime (Shakura & Sunyaev, 1973).

4.1. Accretion Rate and Luminosity of AGN

The total energy release and the spectrum of emitted radiation are mainly determined by the rate of matter inflow into the disk on its outer boundary, i.e. by its accretion rate which is usually denoted by $\dot{M}$. If the accretion converts matter to radiation with fixed radiative efficiency η (in rest-mass units), then a characteristic scale for the accretion rate is the Eddington accretion rate (Krolik, 1999; Shakura & Sunyaev, 1973):

$$\dot{M}_E = 3 \times 10^{-8} \, \frac{0.06}{\eta} \frac{M}{M_\odot} \frac{M_\odot}{\text{yr}}, \tag{5}$$

for witch the total release of energy in the disk $L = \eta \dot{M} c^2$ is equal to the Eddington luminosity:

$$L_E = 1.51 \times 10^{38} \frac{M}{M_\odot} \frac{\text{erg}}{\text{s}}. \tag{6}$$

Eddington luminosity defines a critical luminosity for any given mass, beyond which the radiation force overpowers gravity.

Observed AGN have luminosities from $\sim 10^{42}$ to $\sim 10^{48}$ erg/s, which means that their central black holes must have masses from 10^5 to 10^9 $M_\odot$, respectively (Krolik, 1999). If we assume that accretion in AGN occurs with rate of $\dot{M} \sim 0.5\ M_\odot/\text{yr}$ and with radiative efficiency $\eta \sim 0.1$, the resulting luminosity would be $\sim 3 \times 10^{45}$ erg/s, which is the middle of the observed luminosity distribution of low redshifted AGN (Krolik, 1999).

Actual accretion rates $\dot{M}$ of matter inflow into the accretion disks of AGN could be many times less or higher than the critical value of the Eddington accretion rate. At essentially subcritical accretion rates ($\dot{M} \ll \dot{M}_E$), the maximal surface temperatures T_s are on the order of $10^5 - 10^6$ K in the inner regions of the disk, from which the most of energy is released, mainly in UV and soft X-ray bands (Shakura & Sunyaev, 1973). When the accretion rate increases, the luminosity also raises, as well as the effective radiation temperature. At accretion rates comparable to $\dot{M}_E$, the accretion disk becomes a powerful source of X-ray radiation with the effective radiation temperature T_r from 10^7 to 10^8 K. In a strongly supercritical regime of accretion ($\dot{M} \gg \dot{M}_E$), the luminosity becomes fixed at the Eddington critical limit L_E and the most of energy is irradiated from accretion disk in UV and optical spectral bands (Shakura & Sunyaev, 1973).

4.2. Standard Model and Spectral Distribution

By the term "standard Newtonian model of accretion disk", here we will assume the disk model given by Shakura & Sunyaev (1973). This model was originally developed to describe the accretion disks around stellar sized black holes in the binary systems, but with certain modifications it could be also applied on accretion disks around supermassive black holes in AGN.

The standard Newtonian model of accretion disk is based on the supposition that the released gravitational energy is emitted as a multitemperature blackbody radiation, where the surface temperature profile is given by (see Shakura & Sunyaev, 1973):

$$T_s(r) = \left[\frac{3}{8\pi}\frac{GM}{\sigma r^3}\dot{M}\left(1-\sqrt{\frac{r_{in}}{r}}\right)\right]^{1/4}, \tag{7}$$

where σ is the Stefan constant, G is the gravitation constant, M is the mass of the central black hole, $\dot{M}$ is the accretion rate and r_{in} is the inner radius of the accretion disk. In above Equation and throughout this section we will use the dimensionless disk radius r defined as (Popović et al., 2006a):

$$r = \frac{R}{6\,R_g} = \frac{1}{6}\frac{Rc^2}{GM} = \frac{M_\odot}{M}\frac{R}{9\,\mathrm{km}}, \tag{8}$$

where R is disk radius expressed in gravitational radii R_g.

In the standard model of accretion disk, accretion occurs via an optically thick and geometrically thin disk. The effective optical depth in the disk is very high and photons are close to thermal equilibrium with electrons (Popović et al., 2006a). The spectrum of thermal radiation emitted from accretion disk surface depends on its structure and temperature (and therefore on the distance to the black hole), and can have several distributions such as (Shakura & Sunyaev, 1973):

1. Planck distribution in the outer regions of the disk, where free-free and free-bound processes, as well as absorption in the lines of heavy elements broadened by the gas pressure, give the main contribution to the shape of emitted spectrum:

$$F(x) = \frac{2\pi h}{c^2}\left(\frac{kT}{h}\right)^3\frac{x^3}{e^x-1}, \tag{9}$$

   where $x = \dfrac{h\nu}{kT}$, h is Planck and k is Boltzman constant,

2. more complex spectrum which passed through the homogeneous or exponentially varying medium in intermediate regions of the disk where Thomson scattering dominates and

3. Wien distribution in the inner regions where the comptonization processes strongly affect the shape of emitted spectrum:

$$F(x) \sim x^3 e^{-x}. \tag{10}$$

The surface temperature has different radial distributions in different parts of accretion disk and results in the multicolor black body spectrum. In the inner regions of accretion disks of AGN where their X-ray radiation is generated, the radial distribution of surface temperature has the following form (Popović et al., 2006b):

$$T_s(r) = T_0\,r^{-3/2}(1-r^{-1/2})^{4/5}\,\mathrm{K}, \tag{11}$$

where the temperature constant T_0 is chosen so that the corresponding effective radiation temperature is $T_r = 10^7 - 10^8$ K. This modification of standard disk model is made, because the surface temperature given by Eq. (7) could not be successfully applied in the case of accretion disks of AGN. The distribution of the temperature as a function of the accretion disk radius, according to Eq. (11), is presented in Fig. 2 for two different values of angular momentum a of central black hole (Popović et al., 2006a).

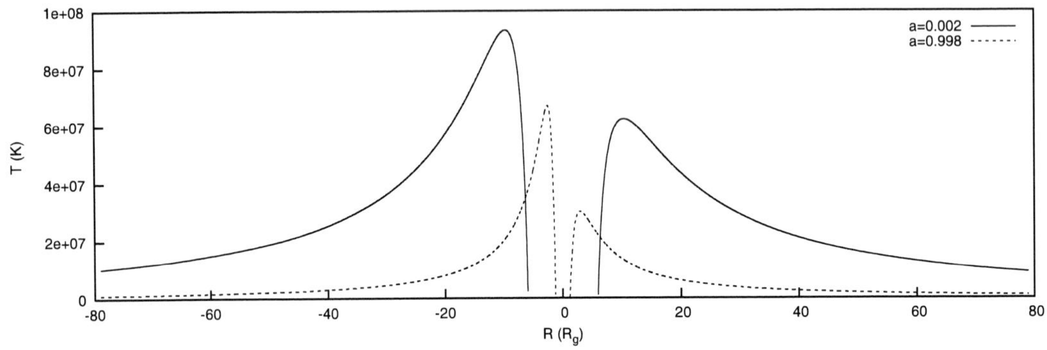

Figure 2. The distribution of the temperature as a function of the disk radius R (in the direction normal to the rotation axis, as seen by distant observer), given for two different values of angular momentum a of central black hole (Popović et al., 2006a). Negative values of R correspond to the approaching and positive to the receding side of the disk.

One can see from Fig. 2 that the temperature depends not only on disk radius, but also on its other parameters, such as e.g. angular momentum a. Emitters located at different radii in accretion disk have different temperatures and therefore make different contributions to the observed flux. Also, it is noticeable that the temperatures at smaller radii of approaching side of the disk are significantly higher than those on its receding side.

Intensity of radiation emitted from the total disk surface can be obtained by integration of all local spectra emitted from the above regions:

$$I_\nu = 4\pi \int_{r_{in}}^{r_{out}} F_\nu \left[T_S\left(r\right)\right] r dr, \tag{12}$$

where r_{in} and r_{out} are inner and outer radii of accretion disk, respectively.

Magnetohydrodynamic simulations of an accretion disk as seen by a distant observer at four different inclinations are presented in Fig. 3 (Armitage and Reynolds, 2003). The most distinctive details in this figure are especially bright regions in form of arcs within the turbulent flow, which trace out the photon trajectories close to the radius of marginally stabile orbit. These arcs are much brighter on approaching side of the disk due to higher surface temperature, as shown in Fig. 2, and also due to Doppler boosting and relativistic beaming which both enhance the flux observed in the direction of motion and diminish the flux in the opposite direction.[5]

[5]More details about the last two relativistic effects can be found in e.g. Krolik (1999)

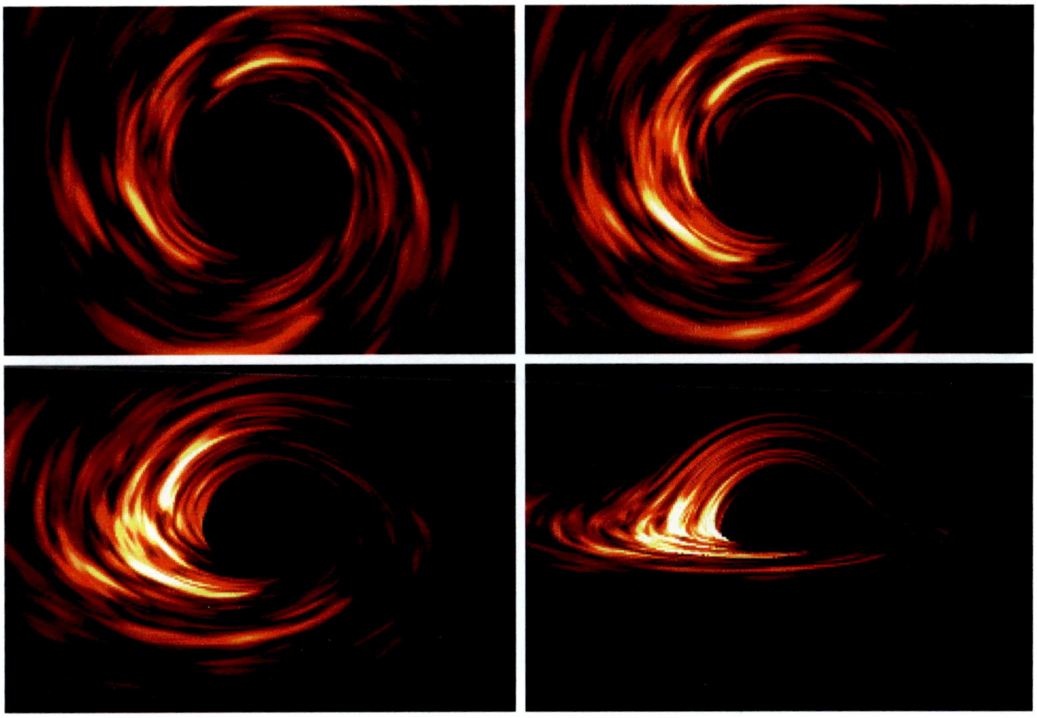

Figure 3. Magnetohydrodynamic simulations of an accretion disk as seen by a distant observer at the following inclination angles (Armitage and Reynolds, 2003): 5 ° (upper left), 30° (upper right), 55° (lower left) and 80° (lower right). The disk rotation is in counter-clockwise direction.

4.3. Structure and Emission

An accretion disk around a supermassive black hole in the center of AGN extends from the radius of marginally stable orbit R_{ms} to the several thousands of gravitational radii. According to its radiation emitted in different spectral bands, it can be stratified in several parts (Jovanović et al., 2008):

1. innermost part close to the central black hole which emits X-rays and which extends from the radius of marginally stable orbit R_{ms} to the several tens of gravitational radii (see e.g. Ballantyne & Fabian, 2005),
2. central part ranging from $\sim 100\ R_g$ to $\sim 1000\ R_g$ which emits UV radiation and
3. outer part extending from several hundreds to several thousands R_g, from which the optical emission is coming (Eracleous & Halpern, 1994, 2003).

Emissivity of the disk has an important role in the line and continuum shapes. In the rest frame of the emitting material (i.e. in disk frame), emissivity $\varepsilon(r)$ is defined as energy emitted per unit proper time per unit proper area (Dabrowski et al., 1997). In the same frame, it is related to the emitted intensity $I(r, \nu_e)$ and emitted flux $F(r, \nu_e)$ by (Dabrowski et al., 1997):

$$F(r, \nu_e) = \pi I(r, \nu_e) = \varepsilon(r)\delta(\nu_e - \nu_0), \tag{13}$$

under assumption that the line emitted at frequency ν_0 can be approximated by a δ function.

Although the standard model of the accretion disk does not predict the power-law for the disk emissivity, such law is usually accepted in the case of the hard X-ray emission from the inner parts of AGN accretion disks (Fabian et al., 1989; Nandra et al., 1997, 1999). In this case the surface emissivity is assumed to vary with radius as power law with emissivity index p (Fabian et al., 1989; Fanton et al., 1997):

$$\varepsilon(r) = \frac{\varepsilon_0}{4\pi} r^{-p}, \tag{14}$$

where ε_0 is the emissivity constant. Fabian et al. (1989) found that emissivity index p varies between 0 and 3 for the Fe Kα line emitting region of Cygnus X-1, where the most probable value is $p \approx 2$. Some authors found that models describing the disk emissivity as a broken power law, rather than a single power law, achieve significantly better statistical fits of the X-ray emission in the Fe Kα line (Brenneman & Reynolds, 2006). One such broken power law between some inner radius r_{in} and outer radius r_{out} with break radius r_{br} is given by (Brenneman & Reynolds, 2006):

$$\varepsilon(r) = \begin{cases} 0, & r < r_{in} \\ (r/r_{br})^{-\alpha_1}, & r_{in} \le r < r_{br} \\ (r/r_{br})^{-\alpha_2}, & r_{br} \le r < r_{out} \\ 0, & r \ge r_{out} \end{cases}, \tag{15}$$

where α_1 and α_2 are some emissivity indices. Using this law Brenneman & Reynolds (2006) fitted the Fe Kα line of MCG–6-30-15, assuming a maximally spinning ($a = 0.998$) black hole, and found that the line emissivity followed such dependence for $\alpha_1 = 4.5 - 6$ and $\alpha_2 \sim 2.5$, where the break radius was $r_{br} \approx 6\ R_g$.

The X-ray continuum can be also modeled e.g. by the following time-independent intrinsic emissivity of the disk as a function of its radius r and photon energy E (Dovčiak et al., 2004; Popović et al., 2006a):

$$I(E, r) \sim E^{-\Gamma} \times r^{-\alpha}, \tag{16}$$

where Γ is photon index and α is radial emissivity power law index of a continuum emitting region. According to some studies of observed X-ray spectra of AGN, the values of Γ and α are estimated to be about 1.5 and 2.5, respectively (Dovčiak et al., 2004; Popović et al., 2006a).

In the case of the disk outer parts (i.e. for optical emitting region), the black-body emissivity law is assumed and, according to the Eq. (9), emitted intensity is given by Planck function (Popović et al., 2006a):

$$I(E, T_s) = \frac{2E^3}{h^2c^2} \frac{1}{e^{E/kT_s} - 1}. \tag{17}$$

However, in the inner parts of the accretion disk Planck function cannot be used properly. In these regions so called "modified" black-body emissivity law is applied. According to the Eq. (10), the corresponding emitted intensity is given by (Popović et al., 2006a):

$$I(E, T_s) \propto x^3 e^{-x}, \qquad x = \frac{E}{kT_s}. \tag{18}$$

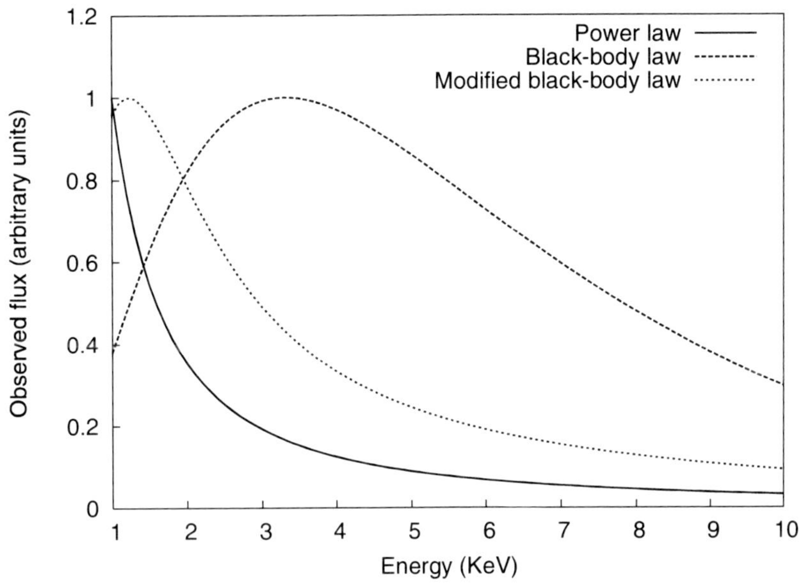

Figure 4. Shapes of the X-ray continuum in 1 – 10 keV energy band, emitted from an accretion disk with the inner and outer radii equal to $R_{in} = R_{ms}$ and $R_{out} = 80\ R_g$, around a supermassive black hole with mass $M = 10^8\ M_\odot$ (Popović et al., 2006a). The continuum shapes are obtained assuming the following three emissivity laws (all normalized to their maximum values): the power law defined by Eq. (14) with emissivity index $p = 2.5$, the black-body and "modified" black-body emissivity laws defined by Eqs. (17) and (18), respectively, assuming the radial distribution of the surface temperature given by Eq. (11).

A comparison between the shapes of the X-ray continuum (1 – 10 keV) emitted from an accretion disk around a supermassive black hole, obtained assuming three different emissivity laws, is presented in Fig. 4 (Popović et al., 2006a). From Fig. 4 it is clear that X-ray continuum shape strongly depends on adopted emissivity law.

5. Supermassive Black Holes and X-ray Emission

A disk geometry for the X-ray emitting regions of AGN could be assumed, since the unified model of AGN includes a supermassive black hole fed by an accretion disk. Fabian et al. (1989) calculated spectral line profiles for radiation emitted from inner parts of accretion disks and later on such features of Fe $K\alpha$ lines were discovered by Tanaka et al. (1995) in Japanese *ASCA* satellite data for Seyfert galaxy MGC-6-30-15. The assumption of a disk geometry for the distribution of the X-ray emitters in the central parts of AGN is also supported by the spectral shape of the Fe Kα line (Nandra et al., 1997, 1999, 2007). Moreover, a bump in the UV spectra of AGN is present very often, indicating that the UV and optical continua also originate from an accretion disk (Jovanović et al., 2008).

We should note here that probably most of the X-ray emission in the 1–10 keV energy range originates from inverse Compton scattering of photons from the disk by electrons in a tenuous hot corona (Jovanović et al., 2008). Proposed geometries of the hot corona of AGN include a spherical corona sandwiching the disk and a patchy corona made of a few compact regions covering a small fraction of the disk (Malzac, 2007). On the other hand, it is known that part of the accretion disk that emits in the 1–10 keV rest-frame band (e.g. the region that emits the continuum Compton reflection component and the fluorescent emission lines) is very compact and may contribute to the X-ray variability in this energy range.

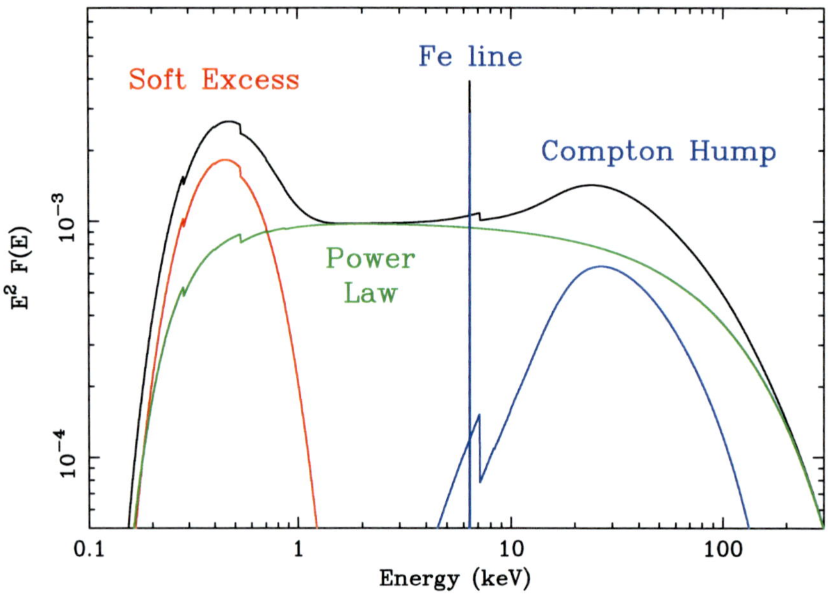

Figure 5. A typical X-ray spectrum of an AGN (Fabian, 2006).

A typical X-ray spectrum of AGN (see Fig. 5) is composed from the following components (Fabian, 2006):

- an underlying power-law continuum due to thermally Comptonized soft photons,
- a soft excess at low energies below 1 keV due to thermal (black-body) emission from an optically-thick accretion disk,
- a fluorescent/recombination Fe Kα line and
- a Compton hump due to X-ray reflection from the disk.

In the following text we will discuss in more details power law component of X-ray continuum and the Fe $K\alpha$ spectral line.

5.1. X-ray Continuum of AGN

AGN are powerful sources of X-ray radiation in the continuum from 0.1 to 100 keV, which contains two components (see Fig. 6): the soft X-rays with a steep spectrum and the hard component with spectrum in form of a power law (Fabian et al., 1989; Fabian, 2006). Therefore, the observed continuum flux is very often fitted with one or two black-body components in the soft X-rays, in addition to a power law component in the hard X-rays. It is believed that both, soft and hard components mostly arise from the inner region of AGN disk, close to the central supermassive black hole ($\sim 10\ R_S$). The first component is probably formed in the accretion disk, which is also a strong source of the soft UV and optical photons, or at least in cold ($T < 10^7$ K) accreting gas clouds. The second one is caused by high-energy (most likely relativistic) electrons in hot corona above the disk, when they multiply inverse-Compton scatter some of the low-energy UV and optical photons from the disk to X-ray energies (Fabian et al., 1989, 2000). The resulting hard X-ray power law component irradiates the accretion disk and produces a reflection component which causes the observed spectrum to flatten above 10 keV (see Fig. 6), as Compton recoil reduces the backscattered flux (Fabian et al., 2000). Apart from its own thermal radiation, the cold gas around a black hole is also irradiated by these hard X-ray photons, producing different spectral features through photoelectric absorption, fluorescence (responsible for occurrence of fluorescent iron lines) and scattering (Fabian et al., 1989). Therefore, the observed X-ray spectra of AGN consist of an iron line spectrum superposed on a power law continuum (Fabian et al., 2000).

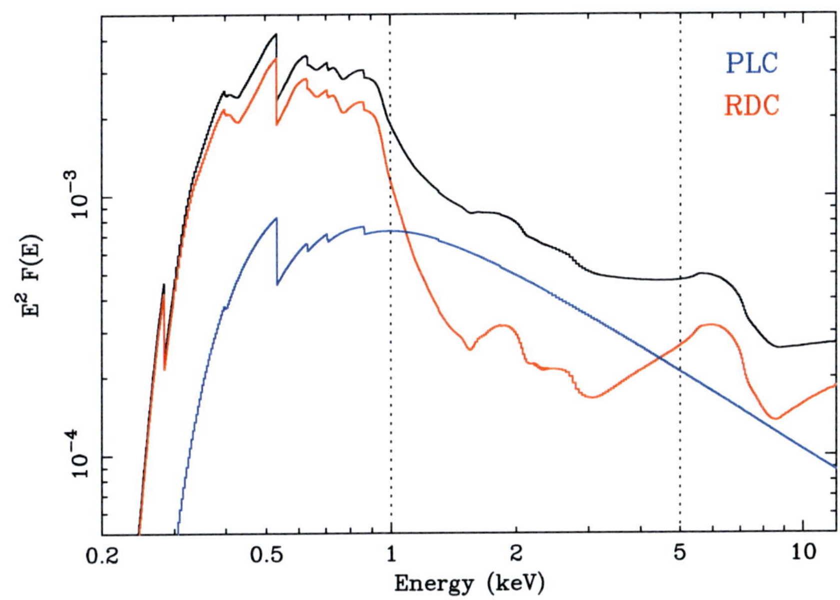

Figure 6. The two-component model of the X-ray continuum in the case of Seyfert 1 galaxy 1H 0707–495, composed from a power law (PLC) and a reflection (RDC) component (Fabian, 2006).

Fluctuations of the X-ray radiation on timescales from several parts of an hour until several days are a common property of all AGN (Krolik, 1999). Such fast variations confirm assumption that X-ray radiation is emitted from a very compact region in the center of AGN.

Fabian (2006) showed that a two-component model, consisting of a power-law and a reflection component, could better describe the observed variability of X-ray spectra in the case of several bright Seyferts than a simple power-law model. This two-component model is presented in Fig. 6.

5.2. Fe Kα Spectral Line

Contrary to the UV and optical spectra of AGN, there are very few strong spectral lines in their X-ray spectra. The most important one is the Fe Kα line. Iron abundance in accretion disks of AGN is sufficient to produce very strong emission in this line. The CCD detectors on Japanese *ASCA* satellite were the first instruments with sufficient spectral resolution and sensitivity in the X-ray band, by which Tanaka et al. (1995) obtained the first convincing proof for the existence of the Fe Kα line in AGN spectra. This discovery was made after four-day observations of Seyfert 1 galaxy MCG-6-30-15 (see Fig. 7).

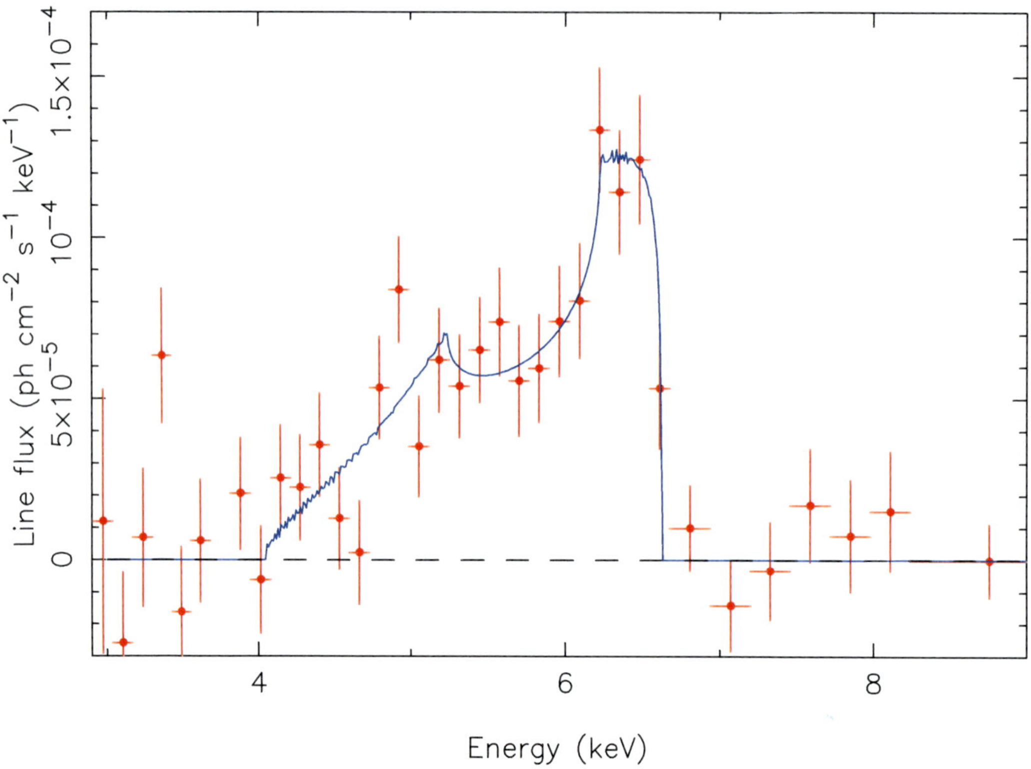

Figure 7. The profile of the Fe Kα line from Seyfert 1 galaxy MCG-6-30-15 observed by *ASCA SIS* detector (Tanaka et al., 1995) and the best fit (blue solid line) obtained by a model of the accretion disk in Schwarzschild metric, extending between 3 and 10 Schwarzschild radii (Fabian et al., 1989). Image Credit: Tanaka et al., Copyright: Nature, 375, 659, 1995. Image generated by Dr. Paul Nandra NASA/GSFC.

The fluorescent/recombination iron Kα line is an important indicator of accreting flows around compact objects, because it is produced in inner parts of their accretion disks. At the same time, it is the strongest line of the X-ray radiation, and it can be found in the spectra of all types of accreting sources: binary black holes and neutron star systems, cataclysmic variable stars and AGN.

The Fe Kα line is produced when plasma is subjected to the influence of the hard X-ray radiation so that one of the two K-shell ($n = 1$, where n is the principal quantum number) electrons of an iron atom (or ion) is ejected following the photoelectric absorption of an

X-ray (Fabian et al., 2000). The threshold for the absorption by neutral iron is 7.1 keV (Fabian et al., 2000). The resulting excited state decays when an L-shell ($n = 2$) electron drops into the K-shell, releasing 6.4 keV of energy. This energy is either emitted as an emission-line photon (34% probability) or internally absorbed by another electron (66% probability) which is consequently ejected from the iron ion (Auger effect).

The fluorescent yield (i.e. the probability that photoelectric absorption is followed by fluorescent line emission rather than the Auger effect) is a weak function of the ionization state from neutral iron (Fe I) up to Fe XXIII (Fabian et al., 2000). For lithium-like iron (Fe XXIV) through to hydrogen-like iron (Fe XXVI), the lack of at least two electrons in the L-shell means that the Auger effect cannot occur. For He and H-like iron ions, the line is produced by the capture of free electrons (recombination) and the equivalent fluorescent yield is high and it depends on the plasma conditions (Fabian et al., 2000).

For the neutral iron, the Fe Kα line energy is 6.4 keV (more precisely, there are two components of the line (Fabian et al., 2000): Fe Kα_1 at 6.404 and Fe Kα_2 at 6.391 keV), while in the case of ionization, the energy of both the photoelectric threshold and the Fe K α line are slightly increased. Even for such high ionization states of He and H-like iron ions, the Fe Kα line energy is increased only to 6.7 and 6.9 keV, respectively (Krolik, 1999).

Fe Kα line is pretty narrow in itself, but in case when it originates from a relativistically rotating accretion disk of AGN it becomes wider due to kinematical effects, and also its shape (or profile) is changed due to Doppler boosting and gravitational redshift. Such broadening of the line is very often observed in spectra of Seyfert galaxies and is one of the main evidences for the existence of a relativistic accretion disk which extends deeply in the gravitational field of the central black hole (Życki, 2004). If the line originated from an arbitrary radius of a nonrelativistic (Keplerian) accretion disk it would have a symmetrical profile (due to Doppler effect) with two peaks: a "blue" one which is produced by emitting material from the approaching side of the disk in respect to an observer, and a "red" one which corresponds to emitting material from the receding side of the disk. The widest parts of the Fe Kα line arise from the innermost regions of the disk, where the rotation of emitting material is the fastest. Using *ASCA* satellite observations, Nandra et al. (1997) found that, in the case of 14 Seyfert 1 galaxies, Full-Widths at Half-Maximum (FWHM) of their Fe Kα lines correspond to velocities of $\approx 50,000$ km/s. In some cases (e.g. for Seyfert 1 galaxy MCG-6-30-15), FWHM velocity reaches 30% of speed of light (see e.g. Nandra et al., 2007). It means that in the vicinity of the central black hole, orbital velocities of the emitting material are relativistic, causing the enhancement of the Fe Kα line "blue" peak in regard to its "red" peak (relativistic beaming). Taking into account the integral emission in the line over all radii of accretion disk, one can obtain the line with asymmetrical and highly broadened profile (Fabian et al., 2000). The "blue" peak is then very narrow and bright, while the "red" one is wider and much fainter (see Fig. 7). Besides, the gravitational redshift causes further deformations of the Fe Kα line profile by smearing the "blue" emission into "red" one. Since the observed Fe Kα line profiles are strongly affected by such relativistic effects, they represent a fundamental tool for investigating the plasma conditions and the space-time geometry in the vicinity of the supermassive black holes of AGN.

One of the important features of the Fe Kα line is variability of both, its shape and intensity. Observed variations of this line are surprisingly less than those of the high energetic continuum, which is assumed to give rise to the line emission (Życki, 2004). Also, it seems that there is a lack of corresponding line response to the continuum variations on time scales from several minutes to several days, or that these line and continuum variations are uncorrelated (Życki, 2004). In §6 of this chapter we will pay attention to some possible causes of such behavior of the Fe Kα line and X-ray continuum.

5.3. Modeling of X-ray Emission Using Ray-tracing in Kerr Metric

The disk emission can be analyzed by numerical simulations, based on so called ray-tracing method in Kerr metric (Bao et al., 1994; Bromley et al., 1997; Fanton et al., 1997; Čadež et al., 1998), taking into account only photon trajectories reaching the observer's sky plane. In this method one divides the image of the disk on the observer's sky into a number of small elements (pixels). For each pixel, the photon trajectory is traced backward from the observer by following the geodesics in a Kerr space-time, until it crosses the plane of the disk (see Fig. 8). Then, the flux density of the radiation emitted by the disk at that point, as well as the redshift factor of the photon are calculated. In that way, one can obtain the color images of the accretion disk which a distant observer would see by a high resolution telescope. The simulated line profiles can be calculated taking into account the intensities and received photon energies of all pixels of the corresponding disk image. All illustrations of the accretion disk and the line shape in this chapter are obtained using such numerical simulations. Here we will briefly describe a pseudo-analytical approach of ray-tracing proposed by Čadež et al. (1998).

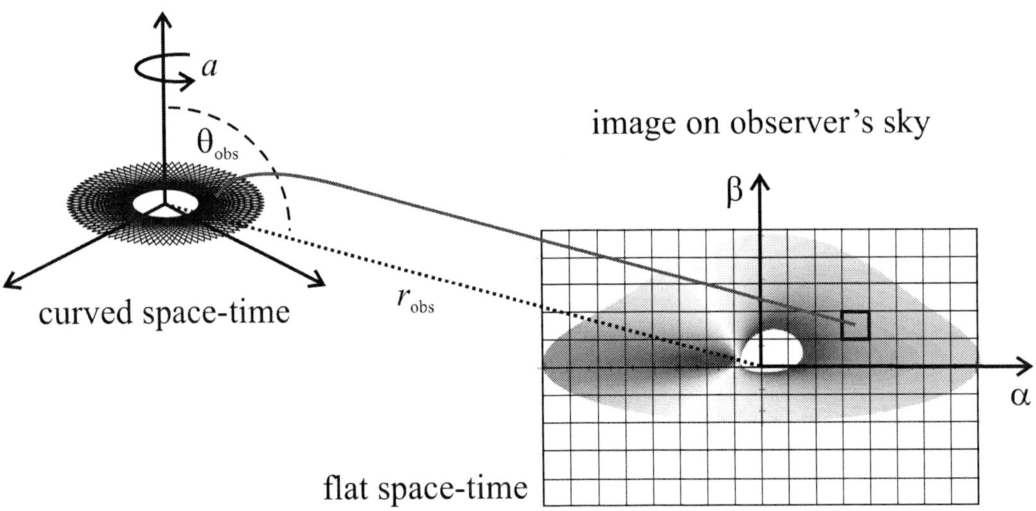

Figure 8. Schematic illustration of the ray-tracing method in the Kerr metric, showing a light ray emitted from some radius of accretion disk in coordinate system defined by a rotating black hole with angular momentum a, and observed at a pixel with coordinates (impact parameters) α, β on the disk image in the observer's reference frame (Figure courtesy: Vesna Borka Jovanović).

This method is based on the pseudo-analytical integration of the geodesic equations which describe the photon trajectories in the general case of a rotating black hole having some angular momentum J, which gravitational field is therefore described by the Kerr metric (see §3.2 for more details). The Kerr metric depends on the angular momentum normalized to the mass M of black hole: $a = J/Mc$, $0 \leq a \leq M$.

A photon trajectory in the Kerr metric can be described by three constants of motion (the energy at infinity and two constants related to the angular momentum, respectively) which have the following forms when natural units $c = G = M = 1$ are assumed (Čadež et al., 1998): $E = -p_t$, $\Lambda = p_\phi$ and $Q = p_\theta^2 - a^2E^2cos^2\theta + \Lambda^2cot^2\theta$. Here, (r, θ, ϕ, t) are the usual Boyer-Lindquist coordinates and p is the 4-momentum. As the trajectory of a photon is independent on its energy, it may be expressed using the two dimensionless parameters $\lambda = \Lambda/E$ and $q = Q^{1/2}/E$ which are very simply related to the two impact parameters α and β describing the apparent position on the observer's celestial sphere: $\alpha = -\dfrac{\lambda}{\sin\theta_{obs}}$ and $\beta = \pm\left(q^2 + a^2\cos^2\theta_{obs} - \lambda^2\cot^2\theta_{obs}\right)^{\frac{1}{2}}$, where the sign of β is determined by $\left(\dfrac{dr}{d\theta}\right)_{obs}$.

In order to find the photon trajectories (null geodesics) which originate in the accretion disk at some emission radius r_{em} and reach the observer at infinity, one must solve the following integral equation (Čadež et al., 1998):

$$\pm\int\limits_{r_{em}}^{\infty}\frac{dr}{\sqrt{R\left(r,\lambda,q\right)}} = \pm\int\limits_{\theta_{em}}^{\theta_{obs}}\frac{d\theta}{\sqrt{\Theta\left(\theta,\lambda,q\right)}}, \tag{19}$$

$$\begin{aligned} R\left(r,\lambda,q\right) &= \left(r^2 + a^2 - a\lambda\right)^2 - \Delta\Big[\left(\lambda - a\right)^2 + q^2\Big], \\ \Theta\left(\theta,\lambda,q\right) &= q^2 + a^2\cos^2\theta - \lambda^2\cot^2\theta. \end{aligned} \tag{20}$$

Above integral Equation (19) can be solved in terms of Jacobian elliptic functions, and therefore it is a pseudo-analytical integration. For the exact expressions of the solutions, see e.g. Čadež et al. (1998), or more recent paper from Li et al. (2005).

Due to relativistic effects, photons emitted at frequency ν_{em} will reach infinity at frequency ν_{obs}, and their ratio determines the shift due to these effects: $g = \dfrac{\nu_{obs}}{\nu_{em}}$. The total observed flux at the observed energy E_{obs} is given by (Fanton et al., 1997):

$$F_{obs}\left(E_{obs}\right) = \int\limits_{image}\varepsilon\left(r\right)g^4\delta\left(E_{obs} - gE_0\right)d\Xi, \tag{21}$$

where $\varepsilon\left(r\right)$ is the disk emissivity, $d\Xi$ is the solid angle subtended by the disk in the observer's sky and E_0 is the rest energy.

A simulated accretion disk image is obtained in the following way:

1. values of the following input parameters are specified: inner and outer radii (R_{in} and R_{out}) of the disk, angular momentum a of the central black hole, observer's viewing angle (disk inclination) θ_{obs} (also, often denoted by i) and parameters defining the disk emissivity
2. constants of motion λ and q are calculated for each pixel on imaginary observer's photographic plate (i.e. for each pair of impact parameters α and β)
3. geodesic Equation (19) is integrated for each pair of λ and q

4. values of shift due to relativistic effects g and observed flux F_{obs} are calculated
5. pixels on imaginary observer's photographic plate are colored according to the value of shift g and a simulated disk image is obtained.

The simulated line profiles can be calculated from the corresponding disk images by binning the observed flux at all pixels over the bins of shift g. The examples of simulated disk images obtained in such way are presented in left panels of Figs. 9 - 11, and the corresponding simulated line profiles are presented in the right panels of the same figures.

5.4. Observational Effects of Strong Gravity in the Vicinity of Supermassive Black Holes

In general, black holes have three measurable parameters (not including the Hawking temperature): charge, mass (and hence gravitational field) and angular momentum (or spin). In the case of supermassive black holes of AGN, only the latter two are of sufficient importance because they are responsible for several effects which can be detected in the observed Fe Kα line shapes (Jovanović and Popović, 2008a).

In order to study the size of the Fe Kα line emitting region, as well as its location in the disk, one can assume that the line is emitted from a region in form of a narrow ring. For example, Jovanović and Popović (2008a) assumed a line emitting region with width equals to 1 R_g, located between: a) R_{in} = 6 R$_g$ and R_{out} = 7 R$_g$ and b) R_{in} = 50 R$_g$ and R_{out} = 51 R$_g$. These two cases are presented in Fig. 9.

From Fig. 9 one can see how the Fe Kα line profile is changing as the function of distance from central black hole. When the line emitters are located at the lower radii of the disk, i.e. closer to the central black hole, they rotate faster and the line is broader and more asymmetric (see Fig. 9 top-right). If the line emission is originating at larger distances from the black hole, its emitting material is rotating slower and therefore the line becomes narrower and more symmetric (see Fig. 9 bottom-right). In majority of AGN, where the broad Fe Kα line is observed, its profile is more similar to the modeled profile as obtained under assumption that the line emitters are located close to the central black hole (Tanaka et al., 1995; Nandra et al., 2007; Jovanović and Popović, 2008a).

Angular momentum or spin of the central supermassive black hole of AGN is a property of the space-time metric. To demonstrate how it affects the observed line profiles we will now assume that the Fe Kα line emitting region extends between the following inner and outer radii: $R_{in} = R_{ms}$ and $R_{out} = 20\ R_g$. We will analyze the following two cases for accretion disk inclination in both Schwarzschild and Kerr metrics: (i) $i = 35°$ and (ii) $i = 75°$. In the case of Schwarzschild metric we have a stationary black hole and hence, angular momentum (normalized to the mass of black hole) is $a = 0$. In Kerr metric (i.e. for a rotating black hole), it can take any value from the $[0, 1]$ range, but in these two examples we will assume an almost maximally rotating black hole with $a = 0.998$.

Illustrations of an accretion disk and the corresponding Fe Kα line shapes in the first case are presented in Fig. 10. As one can see from Fig. 10, the red peak of the Fe Kα line is brighter in case of almost maximally rotating black hole, but at the same time it is also more embedded into the blue peak wing and therefore less separable from it.

Fig. 11 contains illustrations of the line emitting regions and the corresponding line shapes in the case of a highly inclined disk ($i = 75°$). Here, the line profiles are broader

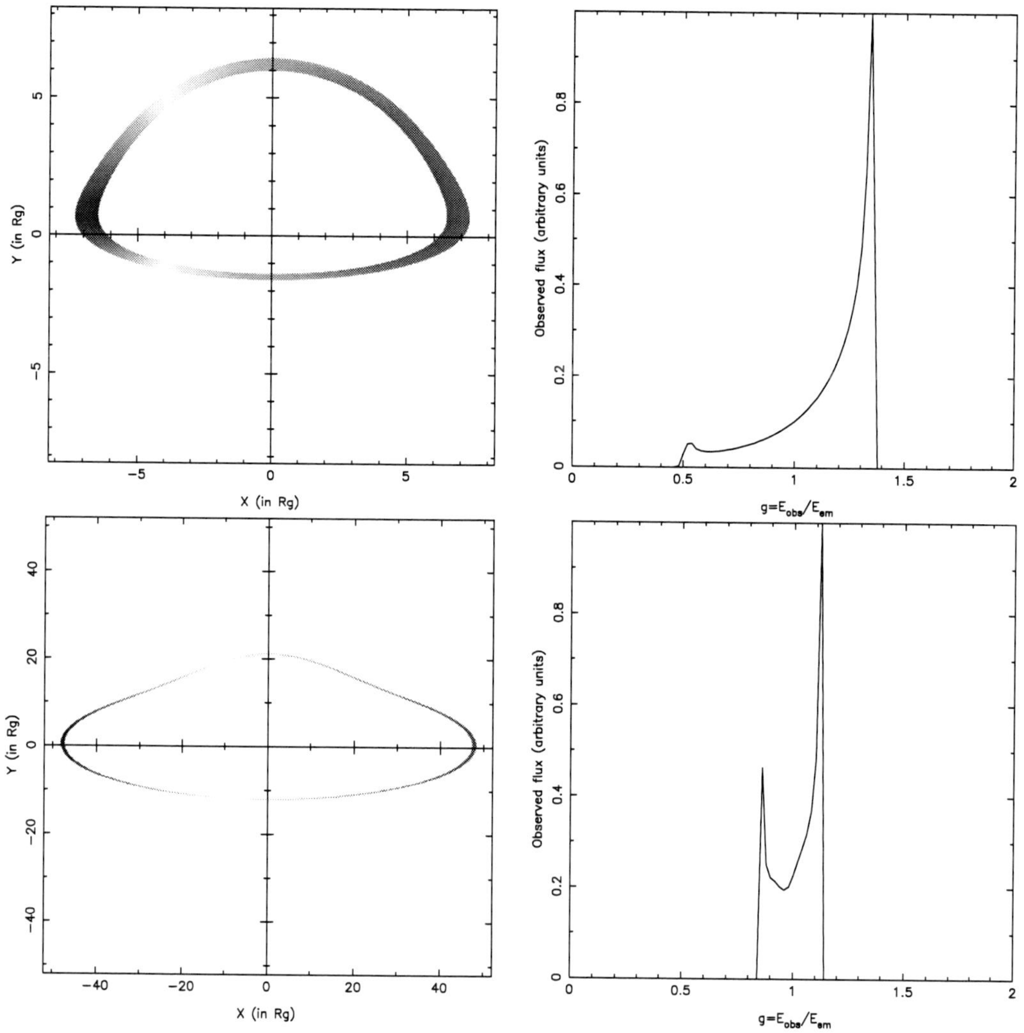

Figure 9. *Left:* illustrations of the Fe Kα line emitting region in form of narrow ring with width $= 1R_g$, extending from: $R_{in} = 6$ R$_g$ to $R_{out} = 7$ R$_g$ (top) and $R_{in} = 50$ R$_g$ to $R_{out} = 51$ R$_g$ (bottom). *Right:* the corresponding Fe Kα line profiles (Jovanović and Popović, 2008a).

than in the first case, mostly due to higher inclination. As it can be seen from Fig. 11, in the case of the Kerr metric, the red peak of the line is again more embedded into its blue peak wing (as in the first case) and it confirms that this effect can be most likely attributed to angular momentum (Jovanović and Popović, 2008a). Consequently, angular momentum of the central black hole has significant influence on the line shape which supports assumption that the line originates from the innermost part of accretion disk, close to the central black hole (see e.g. Ballantyne & Fabian, 2005). This fact can be used for estimation of angular momentum of the central black hole in observed AGN.

Above simulations of the strong gravitational field influence on the Fe K α line show that such effects can be detected in the observed line shapes and therefore, comparisons between the observed and modeled Fe Kα line profiles (see e.g. Fig. 7) can bring us some essential

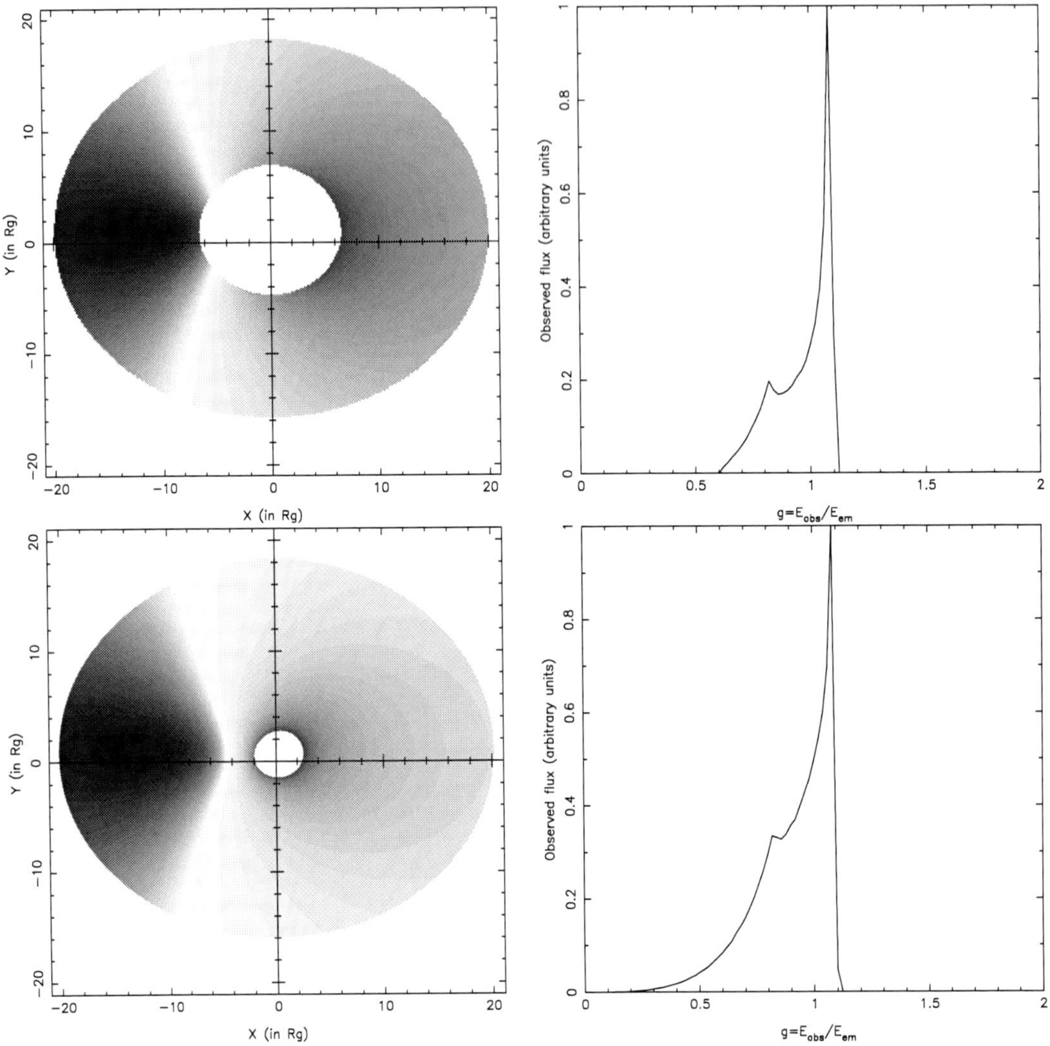

Figure 10. Illustrations of accretion disk (left) and the corresponding Fe K α line profiles (right) in the case of Schwarzschild (top) and Kerr metric with angular momentum $a = 0.998$ (bottom). The disk inclination is $i = 35°$ and its inner and outer radii are $R_{in} = R_{ms}$ and $R_{out} = 20$ R$_g$, respectively (Jovanović and Popović, 2008a).

information about the strong gravitational field in the vicinity of central supermassive black holes of AGN (Nandra et al., 2007).

6. Variability of X-ray Emission Around a Supermassive Black Hole

Rapid and irregular variability of the observed X-ray emission in the line, as well as in the continuum, is a common property of all AGN. This variability could be due to disk instability, reflecting in perturbations of the disk emissivity, or it could be caused by some

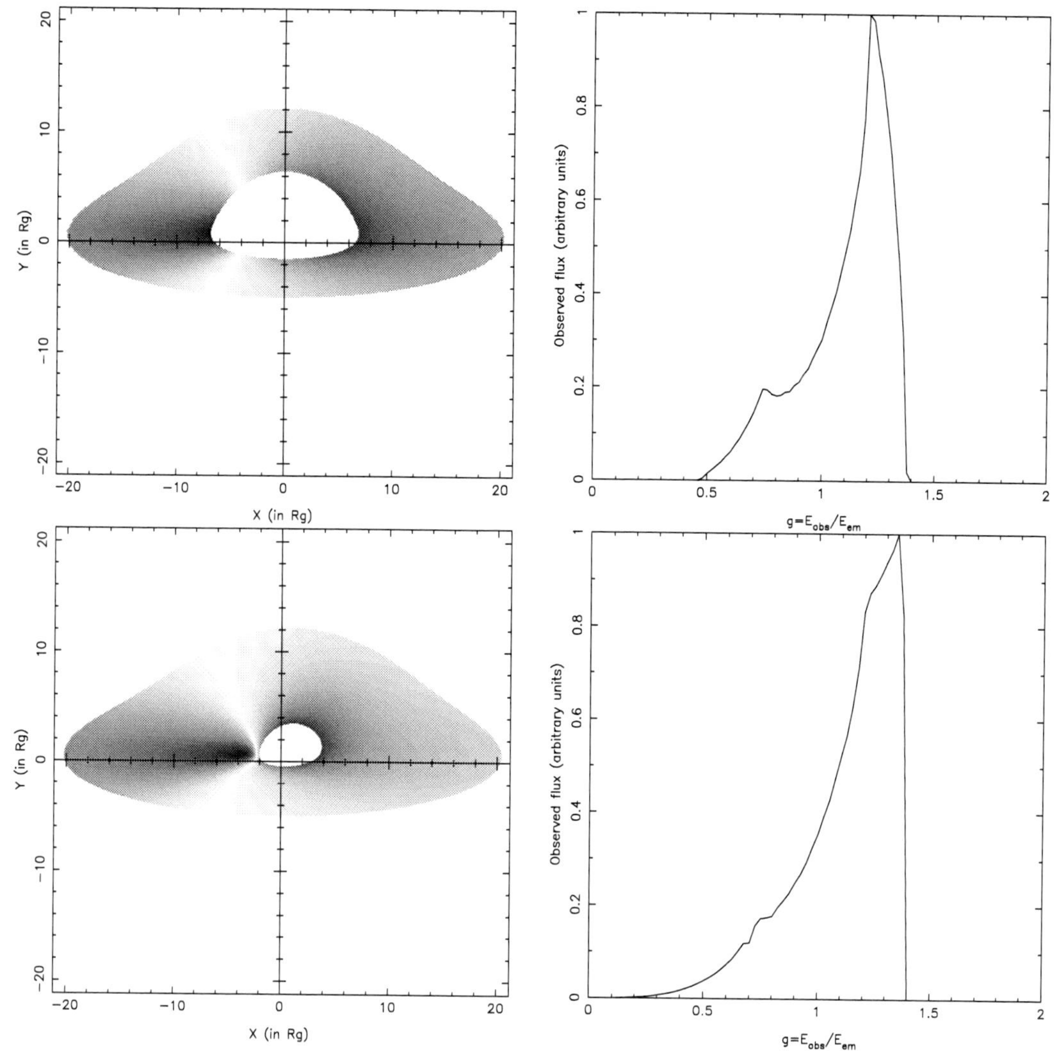

Figure 11. The same as in Fig. 10, but for a highly inclined disk with $i = 75°$ (Jovanović and Popović, 2008a).

external effects, such as gravitational microlensing and absorption by X-ray absorbers. We developed a model of perturbations of the disk emissivity, a model of absorption region and three models of gravitational microlensing. In the following text we will pay more attention to all three mentioned causes of the X-ray variability in AGN.

6.1. Perturbations of Disk Emissivity

In some cases the observed X-ray variability of AGN cannot be explained by the standard model of accretion disk. For example, in addition to the stable 6.4 keV core of the Fe Kα line, a variable "red" feature of the line at 6.1 keV is also detected in X-ray spectrum of Seyfert galaxy NGC 3516 (see Fig. 12), observed by *XMM-Newton* satellite (Iwasawa et al., 2004). This feature varies systematically in the flux at intervals of 25 ks and in energy between 5.7 and 6.5 keV. Iwasawa et al. (2004) found that the spectral evolution of the "red" feature agrees well with hypothesis of an orbiting spot in the accretion disk.

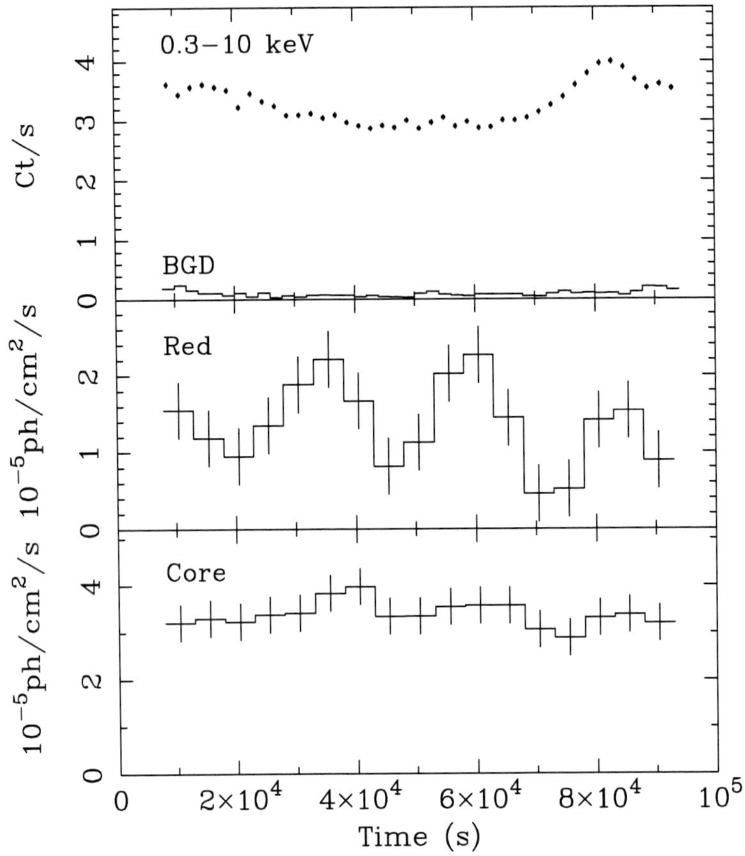

Figure 12. Light curves of Seyfert galaxy NGC 3516 (Iwasawa et al., 2004) for: the 0.3 – 10 keV band (top), the Fe Kα line red feature (middle) and the 6.4 keV line core (bottom).

Many processes in the accretion disk may lead to perturbations in its emissivity, such as self gravity, disk-star collisions and baroclinic vorticity (Flohic & Eracleous, 2008). Different models of emissivity perturbing regions can be used to describe the observed variability in the Fe Kα line profiles and intensities, like stochastically perturbed one given by Flohic & Eracleous (2008), but here we will present a bright spot model given by Jovanović and Popović (2008b). In this model a modification of the power-law disk emissivity is proposed in order to explain the observed profiles. The following emissivity law of the disk is assumed:

$$\varepsilon_1(x_p, y_p) = \varepsilon(r(x_p, y_p)) \cdot \left(1 + \varepsilon_p \cdot e^{-\left(\left(\frac{x - x_p}{w_x}\right)^2 + \left(\frac{y - y_p}{w_y}\right)^2\right)}\right), \qquad (22)$$

where $\varepsilon_1(x_p, y_p)$ is the modified disk emissivity at the given position (x_p, y_p) of perturbing region (in gravitational radii R_g), $\varepsilon(r(x_p, y_p))$ is the power-law disk emissivity at the same position, ε_p is emissivity of perturbing region and (w_x, w_y) are its widths (also in R_g).

An example of the shape of the perturbed emissivity for an accretion disk in Schwarzschild metric, as well as the corresponding perturbed and unperturbed Fe Kα line profiles, are presented in the left and right panels of Fig. 13, respectively.

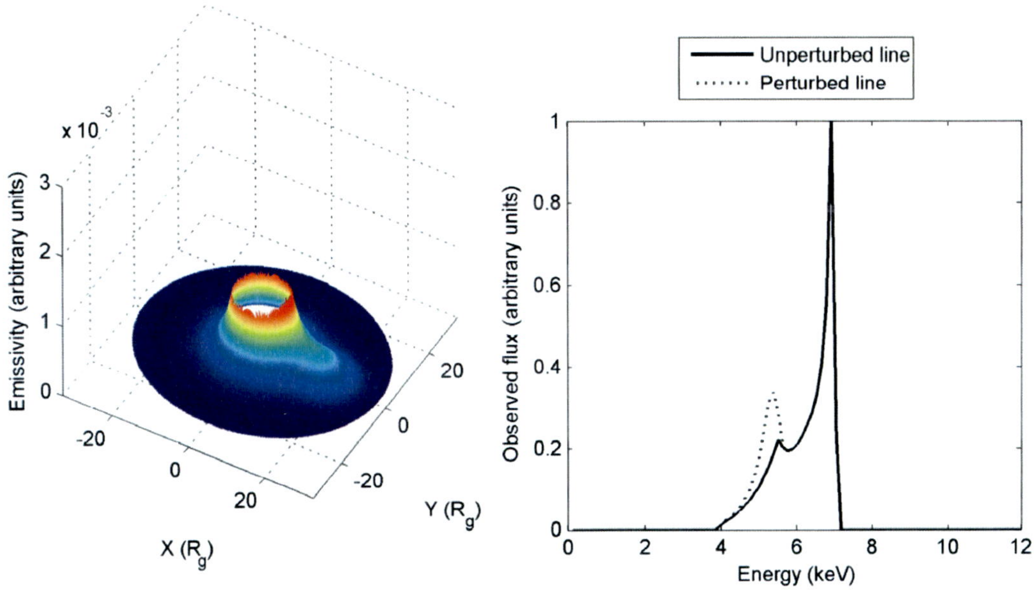

Figure 13. *Left:* shape of perturbed emissivity of an accretion disk in Schwarzschild metric for the following parameters of perturbing region: $x_p = 20\ R_g$, $y_p = 0$ and $w_x = w_y = 7\ R_g$ (Jovanović and Popović, 2008b). *Right:* the corresponding perturbed (dashed line) and unperturbed (solid line) Fe Kα line profiles.

As one can see from Fig. 13, when perturbation moves along receding side of the disk (positive direction of x-axis), the perturbing model affects only "red" part of the line flux, while the "blue" one stays nearly constant, as well as the line core. Therefore, this bright spot model of perturbing region could satisfactorily explain the observed variations of the Fe Kα line flux.

Under assumption that perturbation moves by speed of light c, one can also calculate time $t_p\,[s]$ that corresponds to the current position (x_p, y_p) of perturbation, using the following expression (Jovanović and Popović, 2008b):

$$t_p\,[s] = \frac{r(x_p, y_p)\,[R_g]}{c\,[m \cdot s^{-1}]} = \frac{r_{x,y} \cdot GM_{BH}}{c^3}, \tag{23}$$

where $r_{x,y} = \dfrac{r\,(x_p, y_p)}{R_g}$, G is Newton's gravitational constant and M_{BH} is the mass of central black hole.

Using the time $t_p\,[s]$ we are now able to obtain simulated light curves, produced as perturbation moves along the accretion disk. An example of such light curves, corresponding to displacement of perturbation along the receding side of the disk, is given in Fig. 14. From this figure it can be seen that displacement of perturbing region results in variations

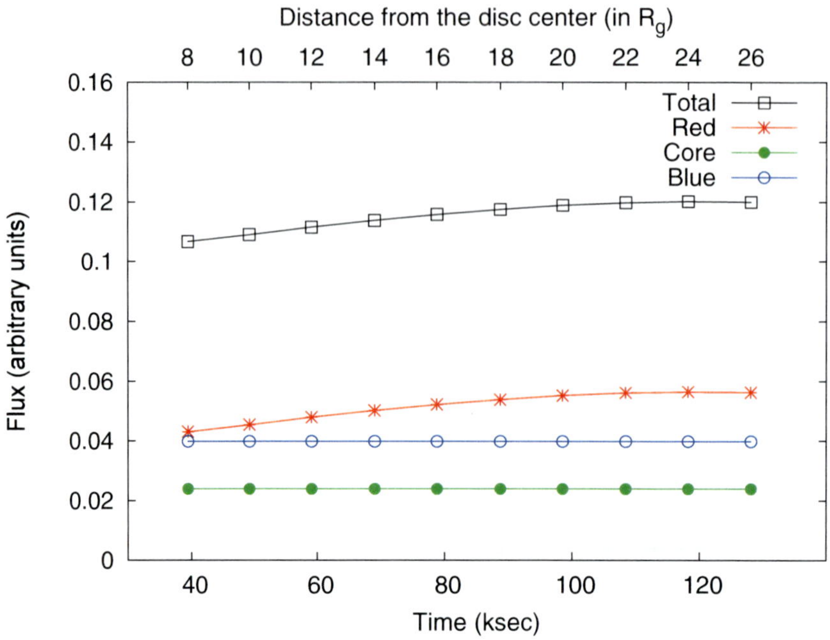

Figure 14. The simulated light curves in case of perturbing region displacement along the receding side of the disk and without rotation (Jovanović and Popović, 2008b). Light curves correspond to the following spectral bands: total flux (black) to 0.1 – 12.8 keV, "red" to 0.1 – 6.1 keV, "core" to 6.1 – 6.7 keV and " blue" to 6.7 – 12.8 keV.

of only "red" light curve (0.1 – 6.1 keV), while the "blue" one (6.7 – 12.8 keV) and the line core (6.1 – 6.7 keV) stay nearly constant.[6] These variations are then reflected in total line flux in 0.1 – 12.8 keV energy band (Jovanović and Popović, 2008b).

Thus, this perturbing model could explain the variable "red" feature of the Fe Kα line observed in NGC 3516. Besides, the realistic durations of disk emissivity perturbations are also obtained if a central supermassive black hole with mass $M_{BH} = 1 \times 10^9 M_\odot$ is assumed (Jovanović and Popović, 2008b).

6.2. Absorption by Warm X-ray Absorbers

The X-ray emission of AGN could be also significantly absorbed by an outflowing wind, especially in case of so-called Low Ionization Broad Absorption Line (LoBAL) quasars. Recent observations of such quasars (e.g. Mrk 231 (Braito et al., 2004) and H 1413+117 (Chartas et al., 2007)) confirmed the presence of X-ray absorbers in these objects. Wang et al. (2001) detected an absorption line at 5.8 keV in nearby ($z = 0.0033$) Seyfert 1.5 galaxy NGC 4151. A variable absorption line at the same energy has been discovered by Nandra et al. (1999) in NGC 3516 (Fig. 15) and was interpreted as a Fe K resonant absorption line, redshifted either by infalling absorbing material or by strong gravity in the vicinity of the black hole.

[6]We should note here that we did not take into account the rotation which can produce periodical peaks in the light curves, as it can be seen from Fig. 12

Done et al. (2007) found an evidence for a P Cygni profile of the Fe Kα line (Fig. 16) in narrow line Seyfert 1 galaxies. According to these authors, complex X-ray spectra of these objects show strong "soft excess" below 2 keV and a sharp drop at $\sim$ 7 keV which can be explained either by reflection or by absorption from relativistic, partially ionized material close to the black hole. They showed that a sharp feature at $\sim$ 7 keV results from absorption/scattering/emission of the iron Kα line in the wind. In the case of 1H 0707-495 (Fig. 16), this absorption feature can be satisfactorily fitted by the P Cygni profile (Done et al., 2007).

There are different models of the X-ray absorbing/obscuring regions, like absorbing medium comprised of cold absorbing cloudlets by Fuerst & Wu (2004), but here we will focus on the model given by Jovanović and Popović (2007), developed in order to study how much warm absorbers can change the Fe Kα spectral line profile, emitted from a relativistic accretion disk. In this model, absorption region is considered to be composed of a number of individual spherical absorbing clouds with the same small radii (see Fig. 17 left), scattered in space so that projections of their centers to the observer's sky plane (X_i, Y_i) have bivariate normal distribution $\mathcal{N}_2(\mu, \Sigma)$. Here, $\mu = [\mu_X, \mu_Y]^T$ and $\Sigma = \begin{bmatrix} \sigma_X^2 & \rho\sigma_X\sigma_Y \\ \rho\sigma_X\sigma_Y & \sigma_Y^2 \end{bmatrix}$, where μ_X and μ_Y are the means of X_i and Y_i, σ_X and σ_Y are their standard deviations and ρ is the correlation between them. Absorbing region presented in Fig. 17 (left) is obtained for the following parameters: $\rho = 0$, $\mu_X = X_A$, $\mu_Y = Y_A$ and $\sigma_X = \sigma_Y = R_A$, where (X_A, Y_A) is the center and R_A is the radius of projection of entire absorbing region.

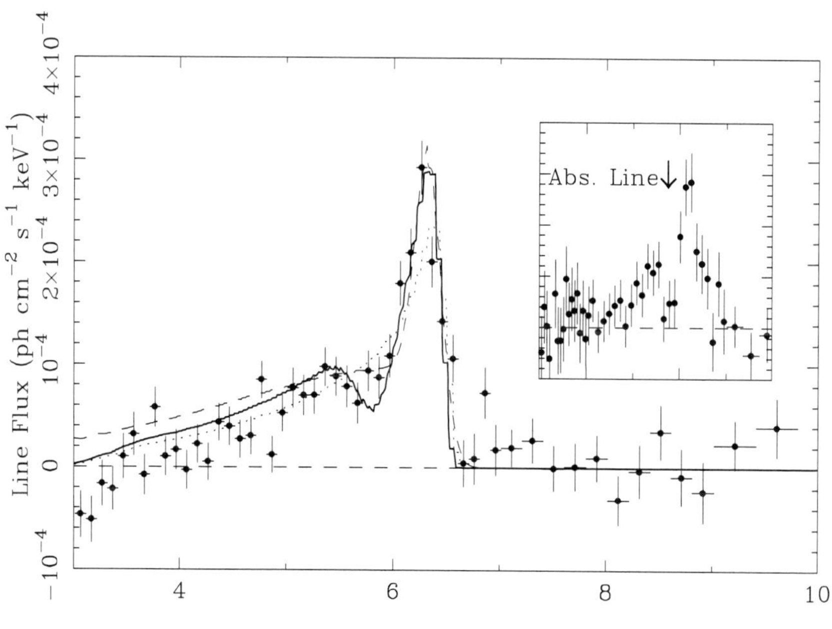

Figure 15. Absorption component at 5.8 keV in the red part of the Fe Kα line of Seyfert 1 galaxy NGC 3516, observed by *ASCA* satellite (Nandra et al., 1999). The dashed line shows best fit with model of an accretion disk around a rotating (Kerr) black hole.

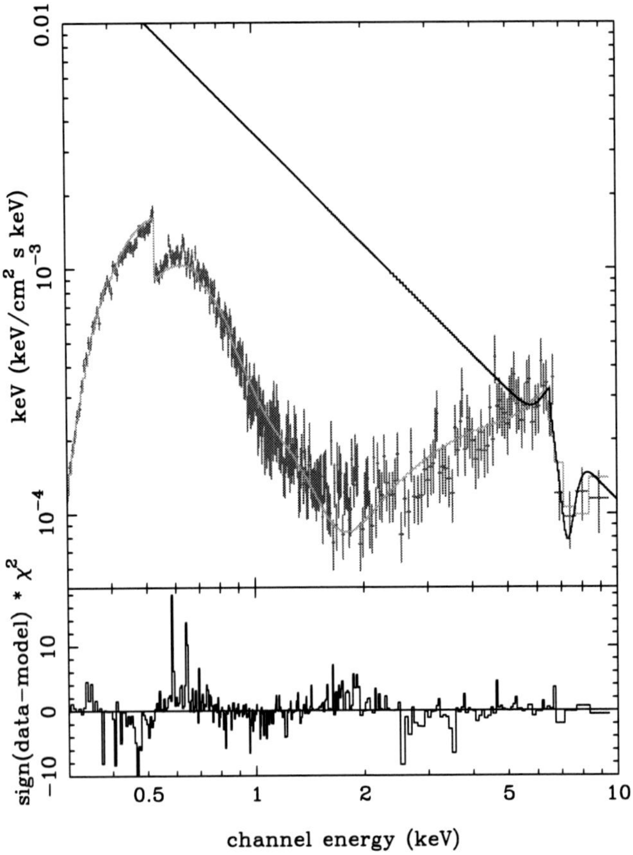

Figure 16. The full *XMM-Newton* spectrum of 1H 0707-495 with the best-fit model which involves a P Cygni profile for the iron features (Done et al., 2007). The lower panel shows residuals to the fit.

The absorption coefficient $A(X, Y)$ for every spherical cloud in the absorbing region is given by (Jovanović and Popović, 2007):

$$A(X,Y) = (1 - I_A(X,Y)) \cdot e^{-\left(\frac{g(X,Y)\, E_0 - E_A}{\sigma_E}\right)^2}, \tag{24}$$

where absorption intensity coefficient $I_A(X, Y)$ describes the distribution of absorption over the whole region, E_A is the central energy of absorption and σ_E is the width of absorption band (velocity dispersion).

A comparison between the unabsorbed Fe Kα spectral line profile and the corresponding absorbed profile obtained by this absorption model is given in Fig. 17 (right). As it can be seen from Fig. 17, when the X-ray radiation from approaching side of the disk is significantly absorbed/obscured by the absorbing region, there is a very strong absorption of the iron line. In such case the emission Fe Kα line looks redshifted at ~ 5 keV and is followed by a strong absorption line at ~ 7 keV (Fig. 17 right), which indicates the P Cygni profile of the iron line. Thus, this model can satisfactorily explain the P Cygni profile of the Fe Kα line in the case when approaching side of the accretion disk is partially blocked from our view by the X-ray absorbing/obscuring material, while the rest of the disk is less absorbed/obscured and therefore is visible (Jovanović and Popović, 2007).

6.3. Gravitational Microlensing

Some recent observational and theoretical studies suggest that gravitational microlensing can also induce variability in the X-ray emission of AGN, especially in the case of gravitationally lensed quasars. The phenomenology of gravitational lensing effects and introduction to this field has been given in several review papers and books. References to some of them can be found in e.g. Popović et al. (2002). Therefore, in this section we will not discuss all aspects of such effects in the universe, but instead, we will briefly present the basic concepts of gravitational microlensing theory, as well as some examples for microlensing influence on the X-ray emission from AGN.

Gravitational lensing is an universal natural phenomenon where the gravitational force of lensing object induce either the amplification of some background source (**microlensing**), or the appearance of its multiple images (**macrolensing**), due to light bending in a gravitational field of the deflector. Separation angle between the images of background source depends on the mass of gravitational lens and therefore, multiple images can be observed only in the case of a massive lensing object, such as galaxy. One of the most famous multiple image lens systems is quasar QSO 2237+0305, also known as Einstein Cross, which is located at redshift $z = 1.695$ (see Fig. 18). Its four images are due to lensing effect of galaxy ZW2237+030, located between us and the quasar at redshift $z = 0.0394$. In the case of a small mass lens (e.g. a star), the separation angle is also small and therefore, different images of background source cannot be resolved. Instead,

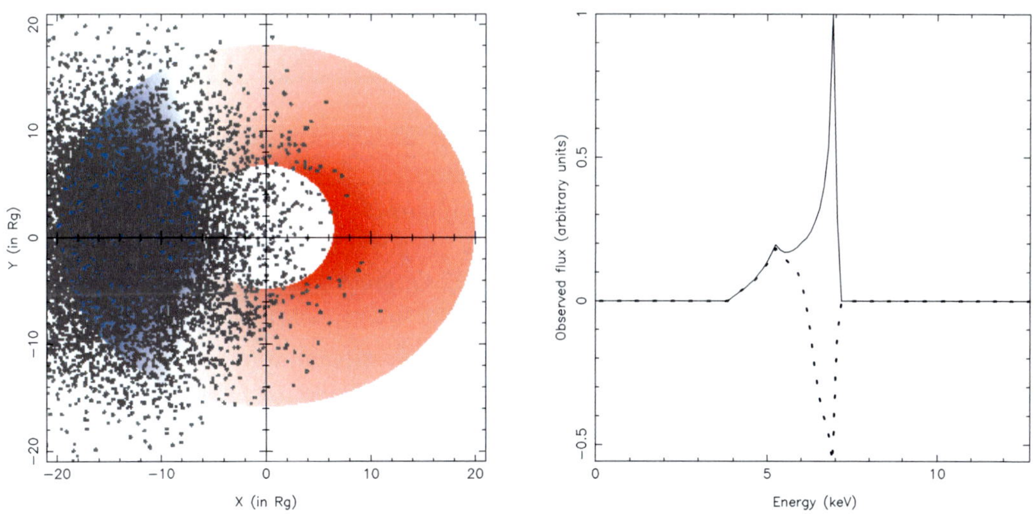

Figure 17. *Left:* Relativistic accretion disk in the Schwarzschild metric partially covered by a cloud of absorbing material (randomly scattered gray dots). *Right:* Comparison between the unabsorbed Fe Kα spectral line profile (solid line) and corresponding absorbed profile (dotted line) caused by the absorbing/obscuring region presented in the left panel (Jovanović and Popović, 2007).

its intensity is amplified, causing the changes in the observed light curve. Common name for both, macrolensing (or simply lensing) and microlensing is **strong lensing**, contrary to **weak lensing**, which causes distortions in observed images of distant objects which can be then used for studying the mass distribution along the line of sight, but such phenomena will not be discussed here.

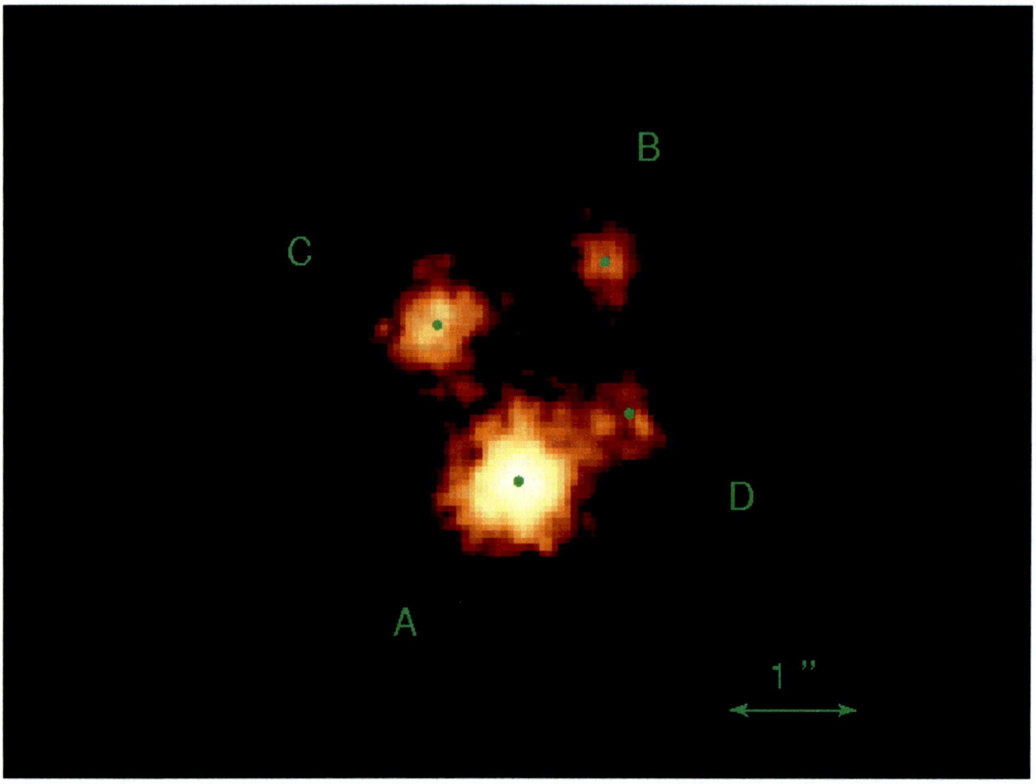

Figure 18. Gravitationally lensed system QSO 2237+0305 (Einstein Cross) observed by the *Advanced CCD Imaging Spectrometer (ACIS)* onboard the *Chandra X-ray Observatory* (Dai et al., 2003). The green circles in each image are the corresponding *Hubble Space Telescope (HST)* image positions provided by *CASTLES* (http://cfawww.harvard.edu/glensdata/Individual/Q2237.html).

Zakharov et al. (2004) found that cosmologically distributed microlenses could significantly contribute to the X-ray variability of high-redshifted ($z > 2$) QSOs. Indeed, microlensing of the Fe Kα line has been reported at least in three macrolensed QSOs: MG J0414+0534 (Chartas et al., 2002), QSO 2237+0305 (Dai et al., 2003), and H 1413+117 (Oshima et al., 2001; Chartas et al., 2004). Chartas et al. (2002) detected an increase of the Fe Kα equivalent width in the image B of MG J0414+0534 which was not followed by the continuum and explained this behavior by assumption that the thermal emission region of the disk and the Compton up-scattered emission region of the hard X-ray source lie within smaller radii than the iron-line reprocessing region. Analyzing the X-ray variability of QSO 2237+0305A, Dai et al. (2003) also measured amplification of the Fe Kα line in component A of QSO 2237+0305 but not in the continuum and suggested that the larger size of the continuum emission region in comparison to the Fe Kα emission region could explain this result. In the case of H 1413+117, Chartas et al. (2004) found that the continuum and the Fe Kα line were enhanced by a different factor. For more detailed discussion about observational and theoretical investigations of the X-ray variability in gravitationally lensed QSOs see e.g. Jovanović et al. (2003); Jovanović (2005); Jovanović & Popović (2005); Jovanović (2006); Jovanović et al. (2008) and Popović et al. (2006a), and also references therein.

The influence of gravitational microlensing on the X-ray emission from a compact accretion disk of AGN can be studied using numerical simulations based on ray-tracing method described in §5.3 of this chapter. In the following text we will briefly review three models of gravitational microlenses (point-like, straight-fold caustic and quadruple microlens), as well as their influence on the X-ray radiation of AGN.

If X and Y are the impact parameters (i.e. the coordinates) which describe the apparent position of each point of the accretion disk image on the celestial sphere as seen by an observer at infinity, then the amplified brightness (due to gravitational microlensing influence) is given by (Jovanović et al., 2003; Popović et al., 2003a,b):

$$I_p(X,Y) = \varepsilon(X,Y)g^4(X,Y)\delta(x - g(X,Y))A(X,Y), \quad (25)$$

where $x = \nu_{obs}/\nu_0$ (ν_0 and ν_{obs} are the transition and observed frequencies, respectively), $g = \nu_{obs}/\nu_{em}$ is the shift due to relativistic effects (ν_{em} is the emitted frequency), ε is the disk emissivity and A is the amplification caused by microlensing. Usually, we do not know the exact form of A, so we are forced to consider the different approximations in order to estimate this quantity.

The first of these approximations is called **point-like microlens**, and it is applied when an isolated compact object (e.g. a star) plays a role of gravitational microlens. Such microlens is characterized by its Einstein Ring Radius in the lens plane (Jovanović et al., 2008):

$$ERR = \sqrt{\frac{4Gm}{c^2}\frac{D_l D_{ls}}{D_s}} \quad (26)$$

or by the corresponding projection to the source plane:

$$\eta_0 = \frac{D_s}{D_l} ERR = \sqrt{\frac{4Gm}{c^2}\frac{D_s D_{ls}}{D_l}}, \quad (27)$$

where G is the gravitational constant, c is the speed of light, m is the microlens mass and D_l, D_s and D_{ls} are the cosmological angular distances between observer-lens, observer-source and lens-source, respectively. In this case, the amplification is given by the following relation (Narayan & Bartelmann, 1999; Popović et al., 2001; Jovanović et al., 2003):

$$A(X,Y) = \frac{u^2(X,Y) + 2}{u(X,Y)\sqrt{u^2(X,Y) + 4}}, \quad (28)$$

where

$$u(X,Y) = \frac{\sqrt{(X - X_0)^2 + (Y - Y_0)^2}}{\eta_0} \quad (29)$$

corresponds to the angular separation between the microlens and a source. X_0, Y_0 are the coordinates of the microlens with respect to the disk center (given in R_g).

The total observed flux is then given by (Jovanović et al., 2003; Popović et al., 2003a,b):

$$F(x) = \int_{image} I_p(x) d\Omega \tag{30}$$

where $d\Omega$ is the solid angle subtended by the disk in the observer's sky and the integral extends over the whole disk image.

An illustration of a point-like gravitational microlens crossing over an accretion disk in the Kerr metric with angular momentum $a = 0.998$ is presented in Fig. 19 (Popović et al., 2002, 2003b). The simulated unamplified and amplified Fe Kα line profiles due to different positions of the microlens are shown in Fig. 20 (Popović et al., 2003a).

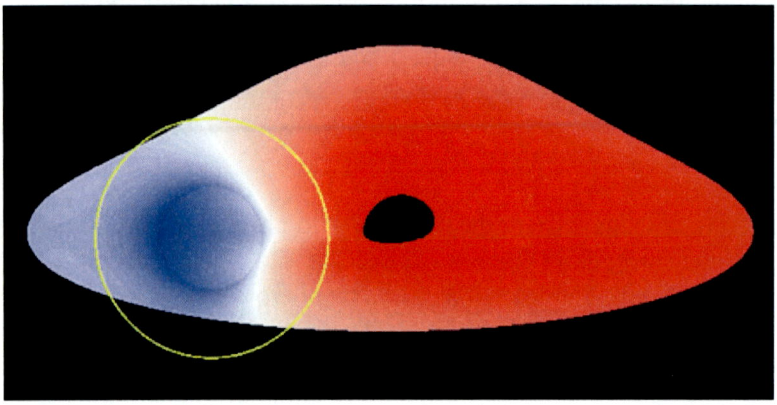

Figure 19. Illustration of a point-like gravitational microlens crossing over an accretion disk in Kerr metric with angular momentum $a = 0.998$ (Popović et al., 2002, 2003b). Einstein Ring of the microlens is schematically presented by yellow Euclidian circle.

From Fig. 20 it can be seen that point-like microlens could induce strong changes of the Fe Kα line shape and intensity, depending on the location of the microlens. In the first place, we have in mind the changes in number of peaks, their relative separation and the peak velocity. Secondly, such transit of the point-like microlens could cause an asymmetrical enhancement of the line profile, and the maximum of the amplification would be obtained for negative values of x-coordinate, i.e. for the approaching side of the rotating accretion disk, and therefore it would affect mainly the blue part of the Fe Kα line.

In most cases we cannot simply consider that microlensing is caused by an isolated compact object, but we should take into account that such micro-deflector is located in an extended object (typically, the lens galaxy). Therefore, when the size of the microlens projected Einstein Ring Radius is larger than the size of the accretion disk and when a number of microlenses form a caustic net, we describe the microlensing effects by a **straight-fold caustic** approximation (Jovanović et al., 2008; Popović et al., 2003a), where the amplification at a point of an extended source (accretion disk) close to the caustic is given by (Chang & Refsdal, 1984):

$$A(X,Y) = A_0 + K\sqrt{\frac{r_{caustic}}{\kappa(\xi - \xi_c)}} \cdot H(\kappa(\xi - \xi_c)). \tag{31}$$

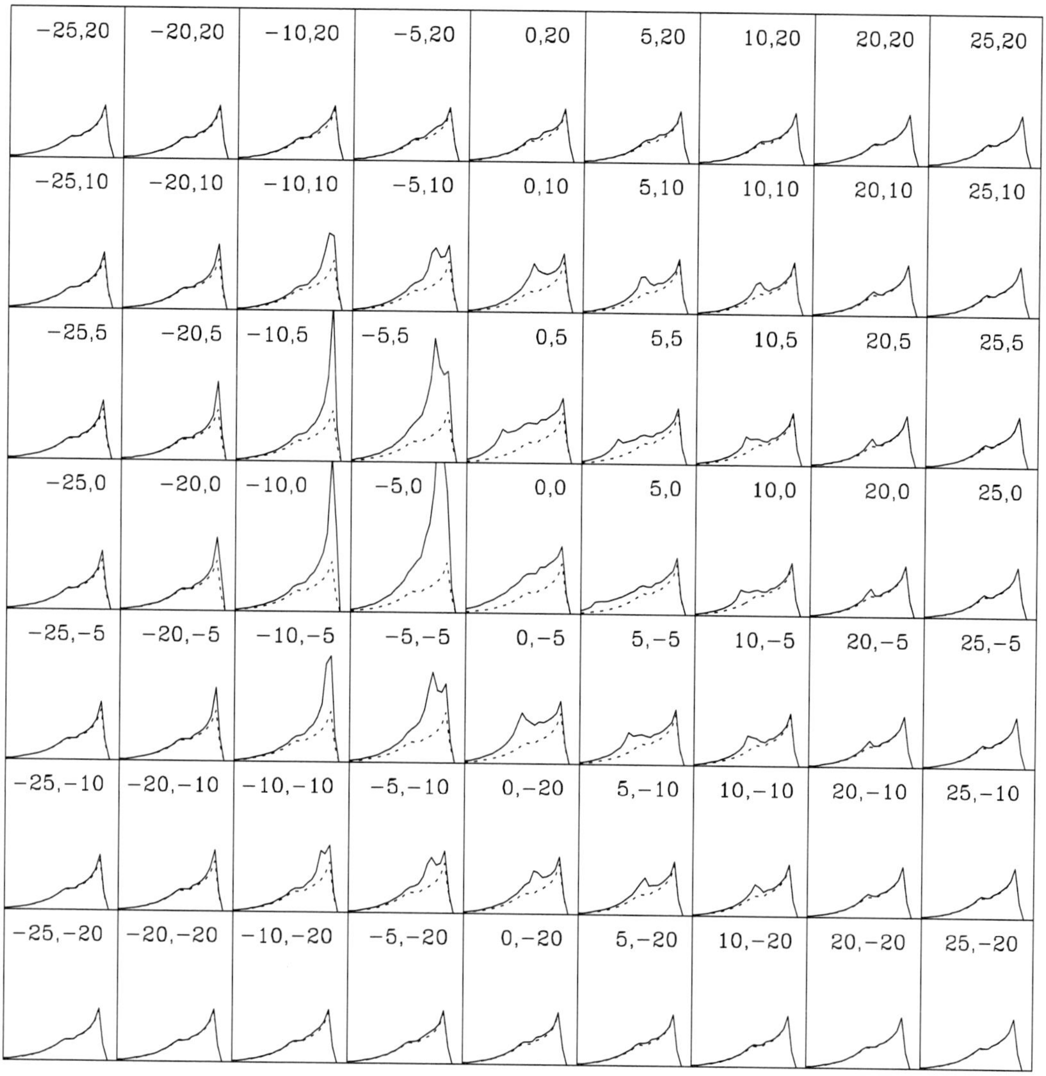

Figure 20. The unamplified (dotted line) and amplified Fe Kα line profiles (solid line) for different positions of the point-like microlens, which coordinates (expressed in R_g) in respect to the accretion disk center are given at the top of each panel (Popović et al., 2003a). The simulations are performed for an accretion disk in Kerr metric with angular momentum $a = 0.998$ and with the following disk parameters: $R_{in} = 6\ R_g$, $R_{out} = 20\ R_g$, $i = 35°$, q=2.5. The Einstein Ring Radius of the point-like microlens is 10 R_g. The relative intensity (y-axis) is in the range from 0 to 3, and energy shift g due to relativistic effects (x-axis) is in the range from 0.4 to 1.2.

In above expression A_0 is the amplification outside the caustic and $K = A_0\beta$ is the caustic amplification factor, where β is constant of order of unity. ξ is the distance perpendicular to the caustic and ξ_c is the minimum distance from the disk center to the caustic (both in R_g). The "caustic size" $r_{caustic}$ is the distance ξ for which the caustic amplification is equal to 1, and therefore this parameter defines a typical linear scale for the caustic. $H(\kappa(\xi - \xi_c))$

is the Heaviside function which equals 1 for $\kappa(\xi - \xi_c) > 0$ and otherwise it is 0. κ is ± 1 depending on the direction of caustic motion: if the direction of the caustic motion is from approaching side of the disk $\kappa = -1$, otherwise it is $+1$. Also, in the special case of caustic crossing perpendicular to the rotating axis $\kappa = +1$ for direction of caustic motion from $-Y$ to $+Y$, otherwise it is -1.

A microlensing event where a caustic crosses over an accretion disk can be described in the following way (Jovanović et al., 2008): before the caustic reaches the disk, the amplification is equal to A_0 because the Heaviside function of Eq. (31) is zero. Just as the caustic begins to cross the disk, the amplification rises rapidly and then decays gradually towards A_0 as the source moves away from the caustic-fold.

An illustration of the straight-fold caustic crossing over an accretion disk in Schwarzschild metric is presented in Fig. 21, and the corresponding effects on the shapes of the X-ray continuum and the Fe Kα line are shown in Fig. 22 (Jovanović, 2006; Popović et al., 2003b). As it can be seen from Fig. 22, caustics could also induce significant amplifications of both, the X-ray continuum and Fe Kα line.

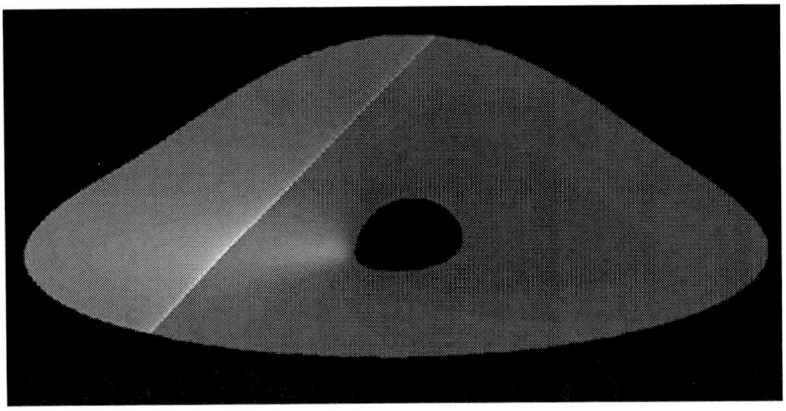

Figure 21. Illustration of a straight-fold caustic gravitational microlens crossing over an accretion disk in Schwarzschild metric (Jovanović, 2006; Popović et al., 2003b).

The most complex approximation for gravitational microlensing amplification is so called **quadruple microlens**. This model is applied to obtain a spatial distribution of magnifications in the source plane (where an accretion disk of AGN is located), produced by a random star field placed in the lens plane. Such spatial distribution of magnifications is called **microlensing map, microlensing pattern or caustic network**. If we consider a set of N compact objects (e.g. stars) which are characterized by their positions x_i and their masses m_i, then the normalized lens equation is given by (Abajas et al., 2007):

$$\vec{y} = \sum_{i=1}^{N} m_i \frac{\vec{x} - \vec{x}_i}{|\vec{x} - \vec{x}_i|^2} + \begin{bmatrix} 1 - \kappa_c + \gamma & 0 \\ 0 & 1 - \kappa_c - \gamma \end{bmatrix} \vec{x}, \tag{32}$$

where $\vec{x}$ and $\vec{y}$ are the normalized image and source positions, respectively. The sum describes the light deflection by the stars and the last term is a quadruple contribution from the galaxy containing the stars, where κ_c is a smooth surface mass density and γ is an external shear. The total surface mass density or convergence can be written as $\kappa = \kappa_* + \kappa_c$,

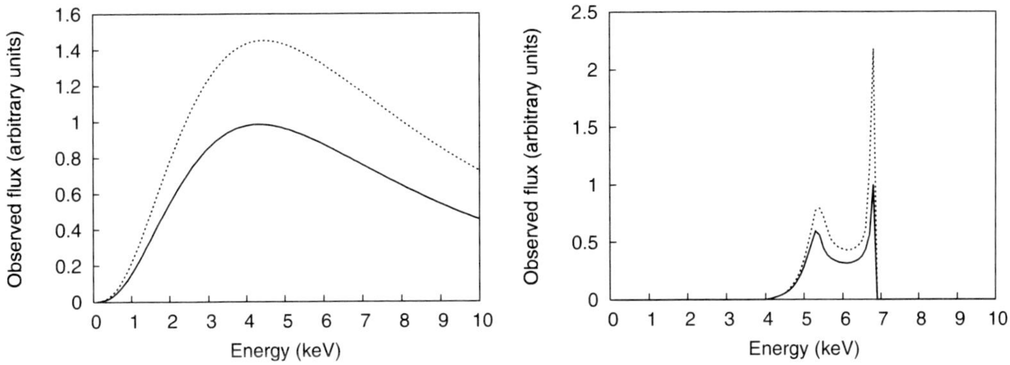

Figure 22. The deformations of the X-ray continuum (left) and the Fe K α line (right) due to influence of straight-fold caustic from Fig. 21. Undeformed profiles are presented by solid and deformed by dotted lines (Jovanović, 2006).

where κ_* represents the contribution from the compact microlenses. The corresponding microlensing map is then defined by two parameters: the convergence - κ, and the shear due to the external mass - γ.

For some specific lensing event one can model the corresponding magnification map using numerical simulations based on ray-shooting techniques (see e.g. Jovanović et al. (2008); Popović et al. (2006a,b) and references therein), in which the rays are shot from the observer to the source, through the randomly generated star field in the lens plane. An example of a magnification map for a "typical" lens system (i.e. for a lens system where the redshifts of microlens and source are $z_l = 0.5$ and $z_s = 2$, respectively) is presented in the left panel of Fig. 23, and the corresponding X-ray, UV and optical continuum variations are given in the right panel of the same figure (Jovanović et al., 2008). The size of this magnification map is 16 $\eta_0 \times 16$ η_0 and it is generated using the following arbitrary values for convergence and shear: $\kappa = 0.45$ and $\gamma = 0.3$. The light curves in the right panel of Fig. 23 are produced when an accretion disk crosses over the magnification pattern in the left panel, along the path denoted by the white solid line.

As one can see from Fig. 23 (right), the variations of the X-ray continuum due to microlensing are much stronger and faster in comparison to the variations in UV and optical spectral bands (Jovanović et al., 2008; Popović et al., 2006a,b). Usually, so called High Amplification Events, i.e. asymmetric peaks in the light curves are also analyzed (Jovanović et al., 2008). The results presented in Fig. 23 (right) show that the rise times of such events are the shortest and their frequency the highest in the X-rays, in comparison to the UV/optical spectral bands.

In order to explain the observed X-ray variability in gravitationally lensed quasars, Popović et al. (2006a) considered a microlensing magnification map for a specific case of the QSO 2237+0305A image (Fig. 24, top). This map with 1 $\eta_0 \times 2$ η_0 on a side was generated using the following parameters: $\kappa = 0.36$, $\gamma = 0.40$ and the mean mass of randomly distributed deflectors $\langle m \rangle = 0.35\ M_\odot$. The corresponding variations of the total Fe Kα line and X-ray continuum fluxes are presented in Fig. 24 (bottom).

As one can see from Fig. 24 (bottom), there is a global correlation between the total

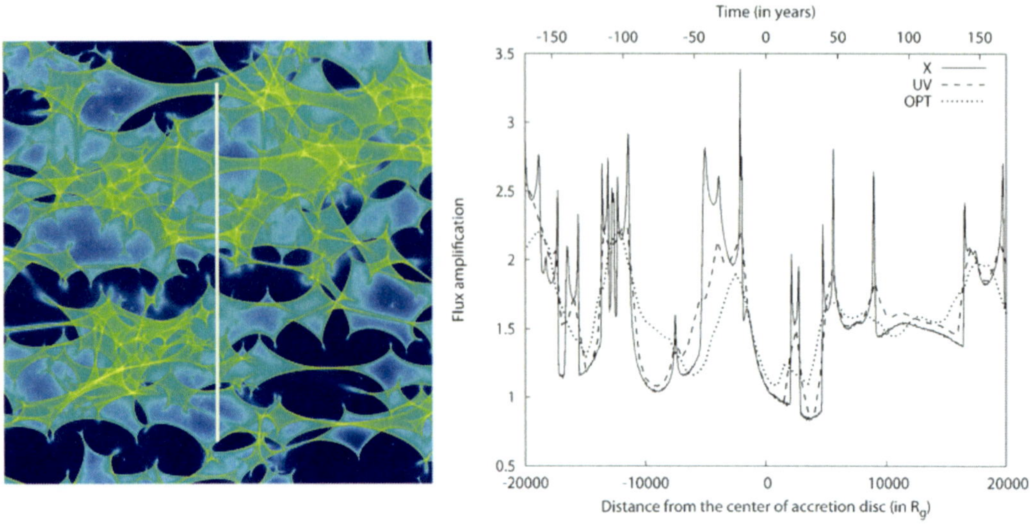

Figure 23. *Left:* Magnification map of a "typical" lens system, where the white solid line represents a path of an accretion disk center. *Right:* Variations in the X-ray (solid), UV (dashed) and optical (dotted) spectral bands corresponding to the path in the magnification map of a "typical" lens system (left).

line and continuum fluxes during the complete simulated microlensing event, but while the continuum flux variations are smooth and slow, the line flux varies much stronger and faster (Jovanović, 2005). Moreover, there is a certain period in the middle of the microlensing event in which the Fe Kα line flux changes rapidly, while the continuum flux remains nearly constant. This indicates that different behavior of the line and X-ray continuum variability, observed in some lensed quasars (Chartas et al., 2002, 2004), may be explained by microlensing hypothesis if the line originates from the innermost part of the accretion disk, while the X-ray continuum is emitted from a larger region in the disk (Jovanović et al., 2003; Jovanović, 2005; Jovanović & Popović, 2005; Popović et al., 2006a).

On the basis of all facts presented in this section, one can conclude that gravitational microlensing can produce significant variations and amplifications of the Fe K α line and X-ray continuum. During a microlensing event, even very small mass objects could produce noticeable changes in the X-ray radiation from accretion disks of AGN. These changes are significantly larger than the corresponding effects in the optical and UV emission. Also, microlensing hypothesis can satisfactorily explain the excess in the iron line emission, observed in some gravitational lens systems.

7. Conclusion

In this chapter we discussed the X-ray radiation from relativistic accretion disks around supermassive black holes, supposed to exist in the centers of all Active Galactic Nuclei. The X-ray radiation is created when the accreting material loses its angular momentum due to friction between the adjacent layers in the disk and spiral towards the central black hole, releasing the gravitational energy. Part of this energy increases the kinetic energy of

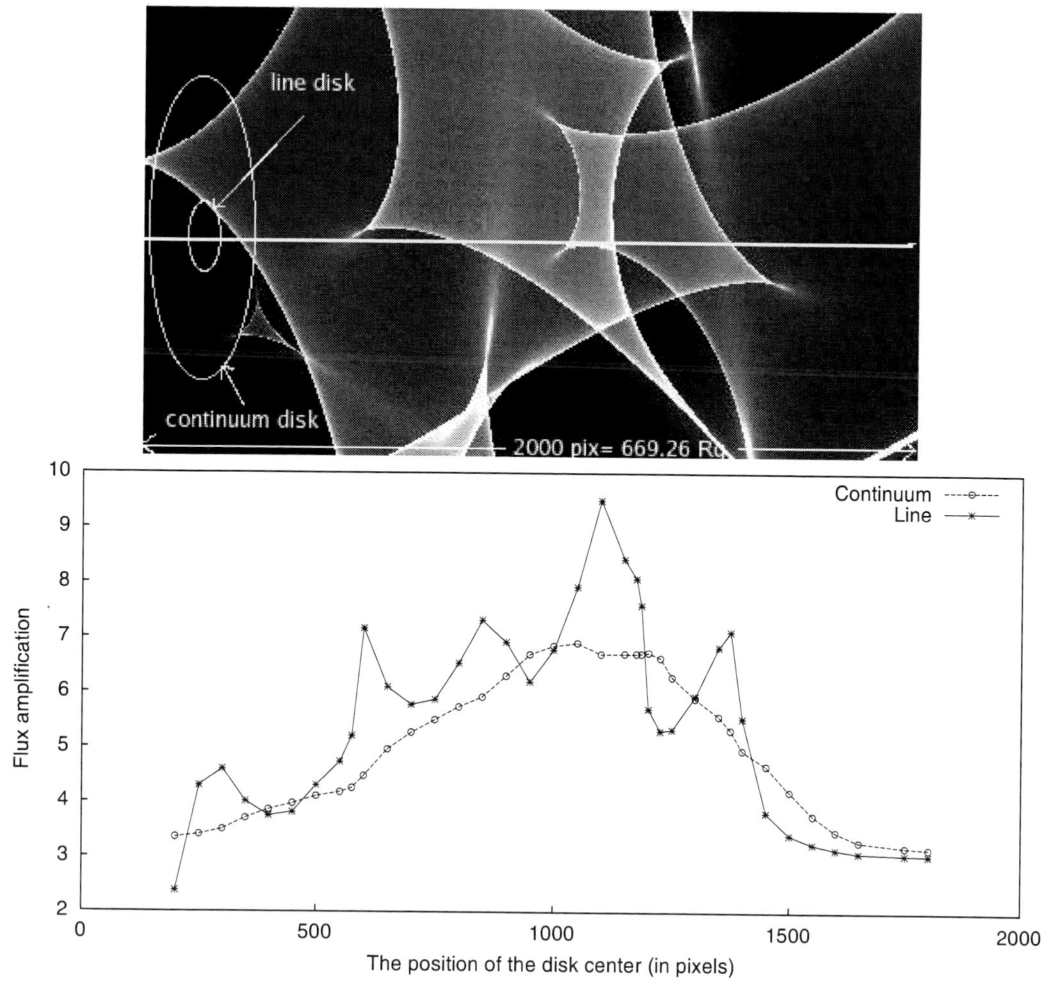

Figure 24. *Top:* Microlensing magnification map of QSO 2237+0305A image with 1 $\eta_0 \times$ 2 η_0 on a side and a scheme of an accretion disk with separated Fe Kα line and X-ray continuum emitting regions which outer radii are 20 and 100 R$_g$, respectively (Popović et al., 2006a). The straight horizontal line represents a path of the disk center. *Bottom:* The corresponding amplifications of the Fe Kα line and X-ray continuum total fluxes for different positions of the disk center along the path on the magnification map in the left panel.

rotation and the rest is turned into the radiation, which is then emitted from the disk surface in the X-ray, UV and optical spectral bands. It is the only known mechanism of converting gravitational potential energy into radiation, which is sufficiently efficient to explain the high luminosities of some observed AGN.

The standard model of an accretion disk, including its emission, accretion rate, luminosity, structure and spectral distribution was described in details. A significant part of this chapter was devoted to modeling of the observed X-ray radiation in both the Fe K α spectral line and X-ray continuum. The modeled X-ray radiation is then used to analyze the several effects of strong gravity in the vicinity of black holes which have an influence on the shapes and intensities of the observed Fe Kα line and X-ray continum. We showed that the X-ray emission strongly depends on the angular momentum and mass of the central black hole, as well as on several parameters describing the X-ray emitting region in the disk. This was demonstrated on several examples, showing various aspects of the X-ray emission from the accretion disk around Schwarzschild and Kerr black holes. Since the X-ray radiation is probably produced in a very compact region near the black holes, such comparison between observed and modeled X-ray radiation can provide us some essential information about the plasma conditions and the space-time geometry in the vicinity of those black holes.

We also studied some possible causes for the rapid and irregular variability of the Fe Kα line and X-ray continuum, observed in the X-ray spectra of some AGN. Our results show that, at least, three mechanisms could explain such variability: perturbations of disk emissivity, absorption by warm X-ray absorbers and gravitational microlensing.

Taking all this into account, we can conclude that the X-ray emission from accretion disks of AGN can be used as a powerful tool for revealing the physics and geometry in the vicinity of their central supermassive black holes. But, to use the Fe Kα line profile for such investigations, one should also take into account the effects which can disturb its profile (such as, e.g., absorption, perturbations in the disk and gravitational lensing).

References

Abajas, C., Mediavilla, E., Muñoz, J. A., Gómez-Álvarez, P., Gil-Merino, R. *ApJ*, 2007, 658, 748

Antonucci, R. *ARA&A*, 1993, 31, 473

Antonucci, R., Barvainis, R. *ApJ*, 1988, 332L, 13

Antonucci, R. R. J., Miller, J. S. *ApJ*, 1985, 297, 621

Armitage, P. J., Reynolds, C. S. *MNRAS*, 2003, 341, 1041

Ballantyne D. R., Fabian, A.C. *ApJ*, 2005, 622, L97

Bao, G., Hadrava, P., Ostgaard, E. *ApJ*, 1994, 435, 55

Bianchi, S., La Franca, F., Matt, G. et al. *MNRAS*, 2008, 389L, 52B

Binney, J., Merrifield, M. *Galactic astronomy*, Princeton University Press: Princeton, New Jersey, 1998

Bon, E., Popović, L. Č., Ilić, D., Mediavlilla, E. G. *New. Ast. Rev*, 2006, 50, 716.

Braito, V., Della Ceca, R., Piconcelli, E., Severgnini, P., Bassani, L., Cappi, M., Franceschini, A., Iwasawa, K., Malaguti, G., Marziani, P., Palumbo, G. G. C., Persic, M., Risaliti, G., Salvati, M. *A&A*, 2004, 420, 79

Brandt, W. N., Hasinger, G. *ARA&A*, 2005, 43, 827

Brenneman, L. W., Reynolds, C. S. *ApJ*, 2006, 652, 1028

Bromley, B.C., Chen, K., Miller, W.A. *ApJ*, 1997, 475, 57

Čadež, A., Fanton, C., Calivani, M. *New Astronomy*, 1998, 3, 647

Carter, B. *Black Holes*, Eds. De Witt, C. and De Witt, B. S., Gordon and Breach: New York, 1973

Chandrasekhar, S. *The Mathematical Theory of Black Holes*, Oxford University Press: New York, 1983

Chang, K., Refsdal, S. *A&A*, 1984, 132, 168

Chartas, G., Agol, E., Eracleous, M., Garmire, G., Bautz, M. W., Morgan, N. D. *ApJ*, 2002, 568, 509

Chartas, G., Eracleous, M., Agol, E., Gallagher, S. C. *ApJ*, 2004, 606, 78

Chartas, G., Eracleous, M., Dai, X., Agol, E., Gallagher, S. *ApJ*, 2007, 661, 678

Clavel, J., Schulz, B., Altieri, B., Barr, P., Claes, P., Heras, A., Leech, K., Metcalfe, L., Salama, A. *A&A*, 2000, 357, 839

Corral, A., Page, M.J., Carrera, F.J., Barcons, X. Mateos, S., Ebrero, J., Krumpe, M., Schwope, A., Tedds, J.A., Watson, M.G. *A&A*, 2008, 492, 71

Dabrowski, Y., Fabian, A. C., Iwasawa, K., Lasenby, A. N., Reynolds, C. S., *MNRAS*, 1997, 288, L11

Dai, X., Chartas, G., Agol, E., Bautz, M. W., Garmire, G.P. *ApJ*, 2003, 589, 100

Done, C., Sobolewska, M. A., Gierliński, M., Schurch, N. J. *MNRAS*, 2007, 374, L15

Dovčiak, M., Karas, V., Yaqoob, T. *ApJ*, 2004, 153, 205

Eracleous, M., Halpern, J. P. *ApJS*, 1994, 90, 30

Eracleous, M., Halpern, J. P. *ApJS*, 2003, 599, 886

Fabian, A. C. *ESASP*, 2006, 604, 463

Fabian A.C., Vaughan S., *MNRAS*, 2003, 340, 28

Fabian, A. C., Rees, M. J., Stella, L. & White, N. E. *MNRAS*, 1989, 238, 729

Fabian, A. C., Iwasawa, K., Reynolds, C. S., Young, A. J. *PASP*, 2000, 112, 1145

Fanton, C., Calivani, M., Felice, F., Čadež, A. *PASJ*, 1997, 49, 159

Ferland, G. J., Korista, K. T., Peterson, B. M. *ApJ*, 1990, 363L, 21

Ferrarese, L., Ford, H. *SSRv*, 2005, 116, 523

Flohic, H. M. L. G., Eracleous, M. *ApJ*, 2008, 686, 138

Fuerst, S. V., Wu, K. *A&A*, 2004, 424, 733

Harris, D. E., Krawczynski, H. *ARA&A*, 2006, 44, 463

Hawking, S. *A Brief History of Time*, Bantam Books: New York, 1988

Hubble, E. *ApJ*, 1926, 54, 368

Ilić, D., Popovć, L. Č., Bon, E. Mediavilla, E. G., Chavushyan, V. H. *MNRAS*, 2006, 371, 1610

Iwasawa, K., Miniutti, G., Fabian, A.C. *MNRAS*, 2004, 355, 1073

Jovanović, P., Popović, L. Č., Dimitrijević, M. S. *PAOB*, 2003, 76, 205

Jovanović, P. *Mem. SAIt*, 2005, 7, 56

Jovanović, P., Popović, L. Č. *PASRB*, 2005, 5, 195

Jovanović, P. *PASP*, 2006, 118, 656

Jovanović, P., Popović, L. Č. *AIP Conf. Proc.*, 2007, 938, 76

Jovanović, P., Zakharov, A.F., Popović, L.Č., Petrović, T. *MNRAS*, 2008, 386, 397

Jovanović, P., Popović, L. Č. *Fortschr. Phys.*, 2008a, 56, No. 4 – 5, 456

Jovanović, P., Popović, L. Č. *PAOB*, 2008b, 84, 467

Kaspi, S., Smith, P. S., Netzer, H., Maoz, D., Jannuzi, B. T., Giveon, U. *ApJ*, 2000, 533, 631

Khachikian, E. Y., Weedman, D. W. *ApJ*, 1974, 192, 581

Krolik, J. H. *Active Galactic Nuclei*, Princeton University Press: Princeton, New Jersey, 1999

Laor A., *ApJ*, 1991, 376, 90

Li, L., Zimmerman, E. R., Narayan, R., McClintock, J. E. *ApJS*, 2005, 157, 335

Malzac, J. *Mem. S.A.It.*, 2007, 78, 382

Markowitz, A., Reeves, J. N., Miniutti, G., et al. *PASJ*, 2008, 60S, 277

Miller, J. M. *ARA&A*, 2007, 45, 441

Nandra, K., George, I.M., Mushotzky, R.F., Turner, T.J. & Yaqoob, T. *ApJ*, 1997, 477, 602

Nandra, K., George, I. M., Mushotzky, R. F., Turner, T. J., Yaqoob, T. *ApJ*, 1999, 523, 17

Nandra K., O'Neill P.M., George I.M., Reeves N., *MNRAS*, 2007, 382, 194

Narayan, R., Bartelmann, M. *in Formation of Structure in the Universe*, Eds. Dekler, A., Ostriker, J.P., Cambridge University Press, 1999, p. 360

Oke, J. B., Sargent, W. L. W. *ApJ*, 1968, 151, 807

Oshima, T., Mitsuda, K., Fujimoto R., Iyomoto N., Futamoto K., et al. *ApJ*, 2001, 563, L103

Osterbrock, D. E. *Astrophysics of Gaseous Nebulae and Active Galactic Nuclei*, University Science Book, 1989

Peterson, B. M. *An Introduction to Active Galactic Nuclei*, Cambridge University Press, 2004

Peterson, B. M., Ferrarese, L., Gilbert, K. M., et al. *ApJ*, 2004, 613, 682

Peterson, B. M., Wandel, A. *ApJ*, 2000, 540L, 13

Popović, L., Č., Mediavilla, E.G., Muñoz J., Dimitrijević, M.S., Jovanović, P. *SerAJ*, 2001, 164, 73

Popović, L.Č., Jovanović, P., Bon, E., Dimitrijević, M.S. *PAOB*, 2002, 73, 49

Popović, L.Č., Mediavilla, E.G., Jovanović, P., Muñoz, J.A. *A&A*, 2003a, 398, 975

Popović, L.Č., Jovanović, P., Mediavilla, E.G., Muñoz, J.A. *Astron. Astrophys. Transactions*, 2003b, 22, 719

Popović, L. Č., Mediavlilla, E. G., Bon, E., Ilić, D., *A&A*, 2004, 423, 909

Popović, L. Č., Jovanović, P., Mediavilla, E.G., Zakharov, A.F., Abajas, C., Muñoz, J.A., Chartas, G. *ApJ*, 2006a, 637, 620

Popović, L.Č., Jovanović, P., Petrović, T., Shalyapin, V. N. *AN*, 2006b, 10, 981

Popović, L. Č., Bon, E., Gavrilović, N. *Rev. Mex. A&A Conf.*, 2008, 32, 99

Reynolds, C. S., Fabian, A. C. *ApJ*, 2008, 675, 1048

Reynolds C. S, Nowak M. A., *Phys.Rep.*, 2003, 377, 389

Risaliti, G., Maiolino, R., Salvati, M. *ApJ*, 1999, 522, 157

Schwarzschild, K. *Sitzungsberichte der Deutschen Akademie der Wissenschaften zu Berlin, Klasse fur Mathematik, Physik, und Technik*, 1916a, 189

Schwarzschild, K. *Sitzungsberichte der Deutschen Akademie der Wissenschaften zu Berlin, Klasse fur Mathematik, Physik, und Technik*, 1916b, 424

Seyfert, C. K. *ApJ*, 1943, 97, 28

Schneider, P. *Extragalactic Astronomy and Cosmology An Introduction*, Springer-Verlag GmbH: Heidelberg, 2006

Shakura, N.I., & Sunyaev, R.A. *A&A*, 1973, 24, 337

Shapovalova, A. I, Popović, L. Č., Collin, S., et al. *A&A*, 2008, 486, 99

Shields, G.A. *Nature*, 1978, 272, 706

Sunyaev, R. A., Titarchuk, L. G. *A&A*, 1980, 86, 121

Tanaka, Y., Nandra, K., Fabian, A. C., Inoue, H., Otani, C., Dotani, T., Hayashida, K., Iwasawa, K., Kii, T., Kunieda, H., Makino, F., Matsuoka, M. *Nature*, 1995, 375, 659

Turner, T. J., George, I. M., Nandra, K., Mushotzky, R. F. *ApJ*, 1997, 488, 164

Ulrich, M. H., Boksenberg, A., Bromage, G., et al. *MNRAS*, 1980, 192, 561

Wang, J., Wang, T., Zhou, Y. *ApJ*, 2001, 549, 891

Yaqoob, T., Reeves, J. N., Markowitz, A., Serlemitsos, P. J., Padmanabhan, U. *ApJ*, 2005, 627, 156

Zakharov, A. F., Popović, L. Č., Jovanović, P. *A&A*, 2004, 420, 881

Zhou, Y. Y., Yu, K. N., Young, E. C. M., Wang, J. M., Ma, E. *ApJ*, 1997, 475L, 9

Życki, P. T. *MNRAS*, 2004, 351, 1180

Reviewed by Dr George Chartas, Sr. Research Associate
Department of Astronomy & Astrophysics
Penn State University, University Park, PA 16802, USA
***e-mail*: chartas@astro.psu.edu**

In: Black Holes and Galaxy Formation ISBN 978-1-60741-703-3
Editors: A.D. Wachter and R.J. Propst, pp. 295-320

Chapter 11

POWERFUL JETS FROM ACCRETING BLACK HOLES: EVIDENCE FROM THE OPTICAL AND INFRARED

David M. Russell*,
University of Amsterdam, Netherlands
Rob P. Fender
University of Southampton, United Kingdom

Abstract

A common consequence of accretion onto black holes is the formation of powerful, relativistic jets that escape the system. In the case of supermassive black holes at the centres of galaxies this has been known for decades, but for stellar-mass black holes residing within galaxies like our own, it has taken recent advances to arrive at this conclusion. Here, a review is given of the evidence that supports the existence of jets from accreting stellar-mass black holes, from observations made at optical and infrared wavelengths. In particular it is found that on occasion, jets can dominate the emission of these systems at these wavelengths. In addition, the interactions between the jets and the surrounding matter produce optical and infrared emission on large scales via thermal and non-thermal processes. The evidence, implications and applications in the context of jet physics are discussed. It is shown that many properties of the jets can be constrained from these studies, including the total kinetic power they contain. The main conclusion is that like the supermassive black holes, the jet kinetic power of accreting stellar-mass black holes is sometimes comparable to their bolometric radiative luminosity. Future studies can test ubiquities in jet properties between objects, and attempt to unify the properties of jets from all observable accreting black holes, i.e. of all masses.

PACS 05.45-a, 52.35.Mw, 96.50.Fm.

Keywords: accretion, accretion disks, outflows, jets, radiation mechanisms, X-ray binaries, optical, infrared.

AMS Subject Classification:

*E-mail address: D.M.Russell@uva.nl

1. Introduction

There is currently just one way in which we can study black holes phenomenologically in the *lab* – investigating the radiation generated by their gravitational influence on matter. Here, the *lab* is outer space, and the detectors are telescopes sensitive to various different wavelengths of radiation (radio, infrared, optical, ultraviolet, X-ray, γ-ray). Black holes (BHs) with masses between a few times the mass of the Sun (solar masses; $M_\odot$) and a billion $M_\odot$ attract local matter, which can convert its gravitational potential energy into heat and light as it approaches the BH. By definition, BHs are dark and do not directly produce radiation we can detect. However it is because they are massive and compact that the gravitational forces close to BHs are the most extreme in the universe; as a result the attracted matter is extremely hot and emits high-energy radiation.

The gravitational force in newtons (ignoring general relativity; this is just for purposes of demonstration) on a particle of mass m kg due to a massive object of mass M kg is

$$F = \frac{GMm}{r^2},$$

where G is the gravitational constant $G = 6.674 \times 10^{-11}$ N m^2 kg^{-2} and r is the distance to the centre of mass of the massive object in metres. The radius of the event horizon (beyond which no radiation or matter can escape) of a (non-spinning) BH is

$$R = \frac{2GM}{c^2},$$

where c is the velocity of light, 3×10^5 km s^{-1}. From the above equations, the maximum gravitational force from a non-spinning BH (i.e. at the event horizon) is

$$F_{\mathrm{max}} \propto \frac{M}{R^2} \propto \frac{1}{M}$$

Close to the event horizon of a $M = 10M_\odot$ black hole, the gravitational force is $\sim 10^{10}$ times that near the surface of the Sun. It is the most compact objects in the universe – the black holes and neutron stars (a neutron star is a collapsed star with a density of the order $\sim 10^{17}$ kg m^{-3}) that exert the largest gravitational force possible on a particle (this holds if general relativistic effects are taken into account). The gravitational potential energy of nearby matter turns into heat and radiation; the extreme gravitational conditions near BHs make them the brightest objects of high energy emission (X-rays and γ-rays) in the universe. We note that in the next decade or so it is likely that the signature of black holes will also be detected via gravitational waves (via e.g. black hole binary mergers; e.g. Flanagan & Hughes, 1998). This will then become the second method in which we can study them observationally.

Black hole candidates in our Galaxy are thought to exist as a result of a massive star collapsing at the end of its life in a supernova explosion. Nuclear fusion in the core of the star ceases because its fuel has been depleted, causing the loss of hydrostatic equilibrium and the core collapses under its own gravitational attraction. Most Galactic BH candidates may be isolated, and accreting matter from the interstellar medium (ISM) at low, undetectable rates. The BHs that we *do* detect are in X-ray binaries (XBs) – binary systems in which the BH candidate (of mass $M \sim 3 - 20M_{\odot}$ or more) and a (relatively normal) star are orbiting a common centre of mass. If the surface of the star lies close to the point between the two objects in which the gravitational force from both are equal (the L1 point), the outer layers leak as a stream towards the BH. The net angular momentum of the gas causes it to form a disc around the BH as the matter spirals inwards. The BH is hence being fed, via this accretion disc, by a semi-regular supply of gas from the companion star. As the matter moves inwards its temperature increases; the highest energy radiation is emitted from the hottest gas close to the BH. A visual illustration of an accreting black hole candidate X-ray binary (BHXB) is given here: `http://www.phys.lsu.edu/~rih/binsim/binarysim_annotate.jpg`

1.1. Powerful Jets from Supermassive Black Holes

It is somewhat counterintuitive that it is in these extreme gravitational fields close to BHs that the highest velocity *outflows* in the universe are formed. One can use the simple argument however that the matter must lose a large amount of its potential energy and angular momentum if it is to reach the event horizon, and outflows are one way of removing this problem by channelling some of the energy and angular momentum away from the BH; some matter then never reaches the event horizon. Jets of matter and energy travelling close to the speed of light were first discovered originating from the vicinity of supermassive ($M \sim 10^5 - 10^9 M_{\odot}$) BHs in active galactic nuclei (AGN) at the centres of galaxies. These are the most massive BHs known in the universe, and their powerful, relativistic jets, and the interactions of those jets with local matter, have been studied for decades (e.g. Baade & Minkowski, 1954; Fanaroff & Riley, 1974). Hotspots indicate where the jets impact with the intracluster gas, and lobes of synchrotron-emitting plasma are inflated which flow back from the hotspot, both of which are often detected at radio frequencies. AGN and their jets are now realised to play an important role in the evolution of matter in the universe, affecting to some extent the rate of galaxy growth in the early universe (e.g. Efstathiou, 1992; Churazov et al., 2002; McNamara & Nulsen, 2007; Schawinski et al., 2009).

There are at least three fundamental questions that remain unanswered regarding AGN jets:

- What is the process of particle acceleration?
- What is their composition?
- How much influence do they have on shaping large-scale structure in the universe?

Difficulties arise in answering these questions because various models predict similar observational properties. For example, the composition of the jets and radio lobes, which are not charged, could be either baryonic (electrons and protons) or leptonic (electrons and positrons), both of which result in synchrotron emission which is observed. Similarly, the energy required to accelerate the jets to relativistic velocities could be tapped either by extraction of the spin energy of the BH, or from the accretion energy in the accretion disc (Blandford & Znajek, 1977; Blandford & Payne, 1982; Meier, 2001); these have not yet been distinguished by observations. The power contained in AGN jets *is* known, at least to some degree. It is possible to use (successfully in some cases) the jet hotspots and lobes to infer the power input into the them, i.e. the energy contained in the jets averaged over time (Burbidge, 1959).

1.2. Powerful Jets from Stellar-Mass Black Holes

It is only in recent years that it has become apparent that accretion onto stellar-mass BH candidates (and neutron stars) in X-ray binaries commonly produce relativistic, synchrotron-emitting jets in analogy with those of AGN (e.g. Mirabel et al., 1992; Falcke et al., 2004; Fender, 2006; Migliari & Fender, 2006). The name 'microquasar' was coined for X-ray binaries with jets due to this analogy. The jets here are found to be associated with specific X-ray regimes (Fender et al., 2004). When the X-ray spectrum is hard, a steady jet exists, the radio luminosity of which correlates with the X-ray luminosity (Gallo et al., 2003, 2006). This correlation extrapolates successfully to the radio and X-ray luminosities of AGN when introducing a mass term which is consistent with that predicted by scalings in black hole physics (Merloni et al., 2003; Falcke et al., 2004), demonstrating that a single jet formation process alone likely governs the physics of all BHs. When BHXBs make transitions to softer X-ray states, bright discrete ejecta are observed and often resolved at large distances from the BH (e.g. Tingay et al., 1995; Gallo et al., 2004). The core jet emission is then quenched, and only returns when the X-ray spectrum again hardens (Fender et al., 2004). XBs vary in luminosity (and hence mass accretion rate) by orders of magnitude on timescales accessible to us – typically over several weeks or months. By comparison, AGN take millions of years to perform these 'outburst' cycles. A recent step forward in our understanding was made when it was proposed that the different X-ray regimes and jet behaviour of BHXBs can be directly linked to the different *classes* of AGN (Körding et al., 2006b). Short term variability probes the inner regions of the accretion flow close to the BH, in the vicinity where the jets are formed. These timescales scale with the mass of the BH; for BHXBs it is milliseconds to minutes, whereas for AGN it can be from hours to years. Minute-to-hour long observations can hence probe these timescales in BHXBs, but dedicated long-term campaigns are required for the same study of AGN.

It is uncertain how powerful XB jets are. Few sites of interaction between the jets of XBs and the ISM have as yet been identified but there are a growing number of examples (see Sections 2.2 and 2.3). These interaction sites can be used to infer the time-averaged power of XB jets (Kaiser et al., 2004; Gallo et al., 2005). Estimates of the jet power based on the core jet luminosity have large errors because both the total spectrum of the jet and its radiative efficiency are poorly constrained (however see Körding et al., 2006a, for a recent advancement of the latter). Estimates of the jet power are dominated by the highest energy electron distributions in the jet, which generate the highest energy photons, but isolation of the jet spectrum can be hampered by other emitting components such as the accretion disc at these higher energies (see Section 2.1). Measuring as accurately as possible the jet power ($P_{\rm J}$) in different sources and at different luminosities is crucial in understanding the process of jet formation and the overall physics of accretion and the matter and energy BHs input into the ISM.

It may be that stellar-mass black holes in BHXBs hold the key to understanding AGN

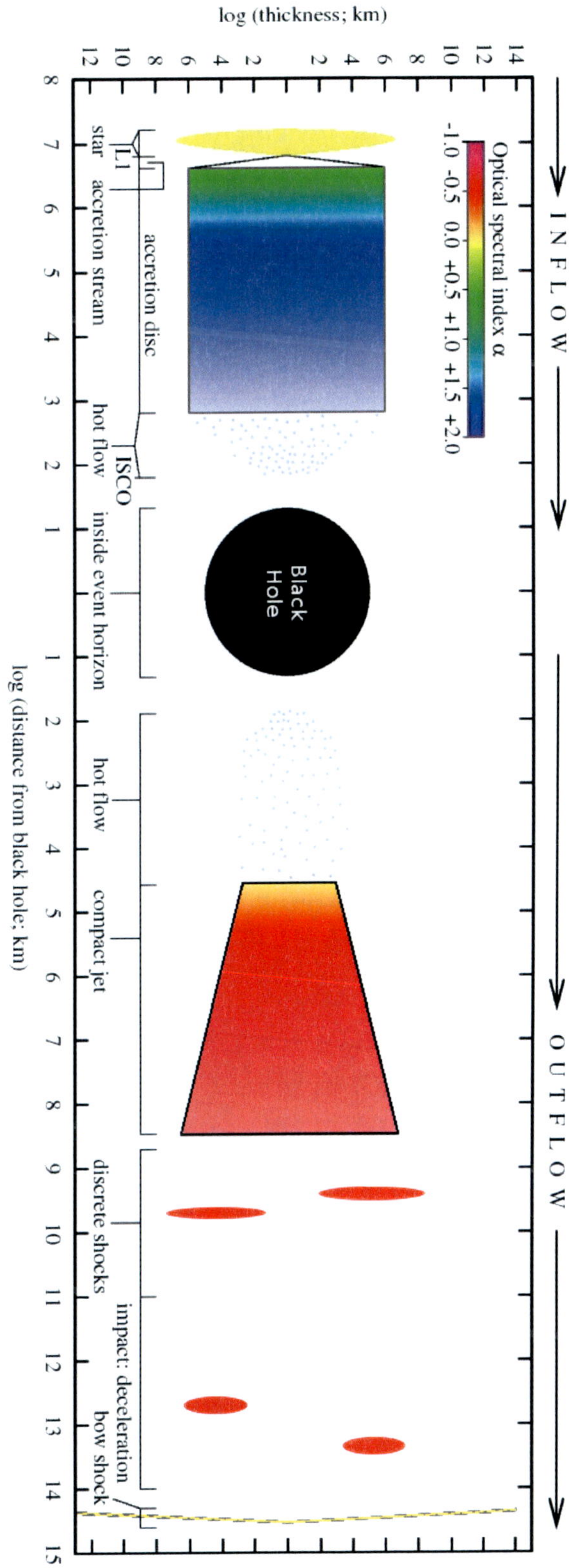

Figure 1. A schematic demonstrating the approximate dimensions and optical colours of the components of a low-mass BHXB (see text for details).

and their jets. If the BHXBs are indeed simply scaled-down versions of AGN, with similar physical structure and behaviour, the three unanswered questions above regarding AGN jets may be answered via studies of BHXBs because they vary on accessible timescales, and their spectra are not complicated by light from the host galaxies or affected (as much) by orientation and obscuration. In addition, it is possible to model the composition of BHXB jets using information from both the compact jet and jet–ISM impact sites (Heinz, 2006).

1.3. Optical and Infrared-Emitting Components of X-ray Binaries

Many physical components and mechanisms are responsible for optical/infrared (OIR) emission from BHXBs and neutron star X-ray binaries. Fig. 1 is a schematic which illustrates the different components of the accretion inflow and outflow (jets), their approximate physical sizes and optical colours, for a BHXB with a low-mass companion star. The colours correspond to approximate intrinsic optical ($\sim$ V-band) spectral indices. The distances from the centre of the black hole are in logarithmic scale. The dimensions of the companion star, accretion stream and accretion disc are based on those quoted in O'Brien et al. (2002) for the BHXB GRO J1655–40. The companion star of GRO J1655–40 has an approximately flat spectral index at V-band ($\alpha = 0$; Migliari et al., 2007) so it appears yellow. The accretion disc, which produces a multi-temperature black body spectrum is hotter at smaller radii, and the optical spectral index steepens in this case from $\alpha \sim +0.5$ at its outer edge to probably $\alpha \sim +2.0$ in the inner regions (based on observed values during outbursts but these change with disc temperature; e.g. Hynes, 2005). Although hotter, the emitting region is smaller in the inner regions of the disc, and the bulk of the optical emission from the disc originates in the outer regions. The inner edge of the accretion disc is at a radius $\sim 10^{-4}$ times that of the L1 radius (O'Brien et al., 2002, L1 is the point at which the gravitational force from the BH and the star is equal; material from the star is no longer gravitationally bound to the star if it crosses this point towards the BH). The hot, optically thin flow (which may also exist above and below the inner part of the disc) ends at the innermost stable circular orbit (ISCO), which is at three times the radius of the event horizon. The radius of the event horizon here is 21 km (we assume a non-spinning black hole of mass $M = 7M_{\odot}$).

The outflow illustrated in Fig. 1 comprises a compact, continuously replenished jet producing synchrotron emission, internal shocks further down the jet, deceleration of jet knots as they impact the ISM, and a radiative bow shock energized by the jet. All components are based on observed, resolved phenomena and are expected to produce OIR emission. We provide a comprehensive review and discussion of the OIR signatures of jets in Section 2.

The compact, conical jet, which is accelerated to relativistic velocities close to the BH, produces synchrotron radiation from electron distributions with energies decreasing with distance from the BH. The optical depth also decreases as the ejected material expands as it flows down the jet. The result is a series of overlapping, self-absorbed synchrotron spectra which form an approximately flat ($\alpha \sim 0$) radio-to-infrared spectrum, where the synchrotron spectrum from each electron energy distribution peaks at a frequency which is approximately inversely proportional to the distance from the jet origin (Blandford & Konigl, 1979; Falcke & Biermann, 1996). Stirling et al. (2001) resolved the compact steady jet of Cygnus X–1 at 8.4 GHz. At this radio frequency its extent was 15 milli-arcseconds (mas) $\approx 3 \times 10^8$ km adopting a distance to the BHXB of 2.1 kpc (Massey et al., 1995). Assuming the highest energy electron distributions in the inner regions of the jet produce a synchrotron spectrum that peaks in the near-infrared (NIR; at $\nu \sim 10^{14.5}$ Hz, as is implied by some observations; see Section 2.1) the summed spectrum will break to an optically thin one, with $\alpha \sim -0.7$ blueward of this 'jet break'. The corresponding radius of the innermost region of the synchrotron-emitting compact jet is then $\sim 4 \times 10^4$ km. This is also consistent with fast timing observations in the optical, in which variability on timescales corresponding to emitting regions as small as $\sim 2 \times 10^4$ km have been shown to most likely originate in the inner regions of the jet (e.g. Kanbach et al., 2001; Hynes et al., 2003b, see Section 2.1). The opening angle of the radio jet of Cyg X–1 was $< 2°$ (Stirling et al., 2001); here we assume 2°. This is consistent with measurements from other BHXBs (Miller-Jones et al., 2006) but much larger than implied by some models (e.g. Kaiser, 2006). The outer region of the jet that produces the 8.4 GHz radio emission would then be wider than the diameter of the accretion disc (assuming the opening angle is constant along the jet). The compact jet cannot currently be resolved at OIR frequencies. The inner regions closest to the BH ($< 10^5$ km away say) dominate the OIR emission from the compact jet; at a distance of 1 kpc (the closest known BHXBs are at this distance), this corresponds to an angular size of the order of micro-arcseconds, which is not resolvable with current detection limits (although sophisticated interferometric methods may achieve this in the future; Markoff, 2008). The radio emission from the compact jet originates in a region much larger than this, but this spectral component is orders of magnitude fainter at OIR frequencies and cannot be detected.

BHXB jets *have* however been resolved at radio and X-ray frequencies, and can be seen to move at relativistic speeds, at distances larger than $\sim 10^9$ km from the BH (see Section 2.2). The emission process is thought to be optically thin synchrotron which peaks at low frequencies, and could either be energized by shocks between discrete ejecta travelling at different velocities, or by impacts with the ISM, which decelerate the ejecta. In Fig. 1 discrete shocks are shown at distances 10^{9-10} km (which have been seen in the radio in Sco X–1, actually a neutron star XB; Fomalont et al., 2001) and impacts with the ISM are shown at $\sim 10^{13}$ km (here observed from the BHXB XTE J1550–564; Corbel et al., 2002). There is just one tentative claim so far of this resolved synchrotron emission detected at OIR frequencies: from GRS 1915+105 (Sams et al., 1996). For the above case of XTE J1550–564 the authors point out that the synchrotron spectrum from radio to X-ray is just below the detection limits of optical data taken with the Very Large Telescope.

It has been shown that the impact of the jets of BHXBs with the ISM inflate lobes of radio plasma, like those seen in FR II radio galaxies powered by the jets of AGN (e.g. Kaiser et al., 2004). Ahead of these lobes there is a radiative shock wave, which produces both thermal continuum and line emission from the recombination of ionized atoms. Optical line emission and thermal radio emission from a bow shock powered by a BHXB jet has recently been detected – at a distance $\sim 3 \times 10^{14}$ km ($\sim$ 10 parsecs) from Cygnus X–1, with a diameter approximately half that distance. This is the furthest and largest structure illustrated in Fig. 1 (see Section 2.3 for discussion of these structures).

2. OIR Identifications of Jets from Black Hole X-ray Binaries

Traditionally, jets have been identified in BHXBs from the radio regime. This is firstly because they are sometimes resolved, which is a very strong argument for a jet, and secondly because no other known emitting components can account for the radio fluxes and spectra (see Fender, 2006, for a review). There is now strong empirical evidence for the spectrum of the compact jet to extend to, and be detected at, OIR frequencies and higher. Although these detections are complicated by other spectral components (which is usually not the case at radio frequencies), many methods can be adopted to successfully disentangle the components and isolate the jet emission. It is in some cases *easier* to detect BHXB jets in the OIR with current facilities. Signatures of OIR jet emission include its spectrum, timing properties and polarimetry properties, all of which differ from other OIR–emitting components such as the accretion disc. In Section 2.1 we review these signatures and discuss the observations of OIR jets in the literature. We stress that although our review focusses on the jet emission, the process that *dominates* the optical (and in some cases the NIR too) emission of most transient BHXBs is X-ray reprocessing on the surface of the accretion disc. For BHXBs with high-mass stars, and for BHXBs at low accretion rates in quiescence, the companion star usually dominates the OIR light.

As was discussed in Section 1.3, large-scale, *resolved* OIR emission can be produced directly and indirectly by the jets of BHXBs. In Sections 2.2 and 2.3 we summarise the known resolved OIR jets and jet–ISM interaction sites. In Section 3 we briefly discuss how these observations (OIR compact jets, resolved jets and jet–ISM interactions) can be used to infer the jet properties, and their implications for jets and accretion onto supermassive BHs in AGN. A summary is provided in Section 4, and we comment on future work that will help to answer the open fundamental questions regarding jets and accretion.

2.1. The Compact Jet

Models of synchrotron emission from compact jets of BHXBs can reproduce the approximately flat ($\alpha \sim 0$), optically thick radio spectrum of AGN and BHXBs, which breaks to one which is optically thin (the aforementioned 'jet break') at some higher frequency (e.g. Blandford & Konigl, 1979; Markoff et al., 2001, 2005; Kaiser, 2006; Jamil et al., 2008, see Fig. 2 for an example spectrum). For BHXBs the jet break probably lies within the OIR regime (see below). The jet break frequency ($\nu_{\rm b}$) depends on the distance of the jet launch region from the BH, which is related to the BH mass, and may also have a weak dependence on mass accretion rate and therefore luminosity (Heinz & Sunyaev, 2003; Nowak et al., 2005). In addition to the spectrum, the OIR-emitting jet has been identified via polarimetry and correlations with other wavebands over short timescales ($\sim$ seconds and less; fast timing) and long timescales ($\sim$ weeks to years; 'slow timing', i.e. over many orders of magnitude of mass accretion rate).

A synchrotron OIR spectrum:

A flat ($\alpha \sim 0$) radio-to-OIR spectrum implied from radio and OIR data alone is consistent with (but not direct evidence for) a significant jet contribution to the OIR, since radio emission in BHXBs originates in the jet. Fender (2001) pointed out that this is the case in

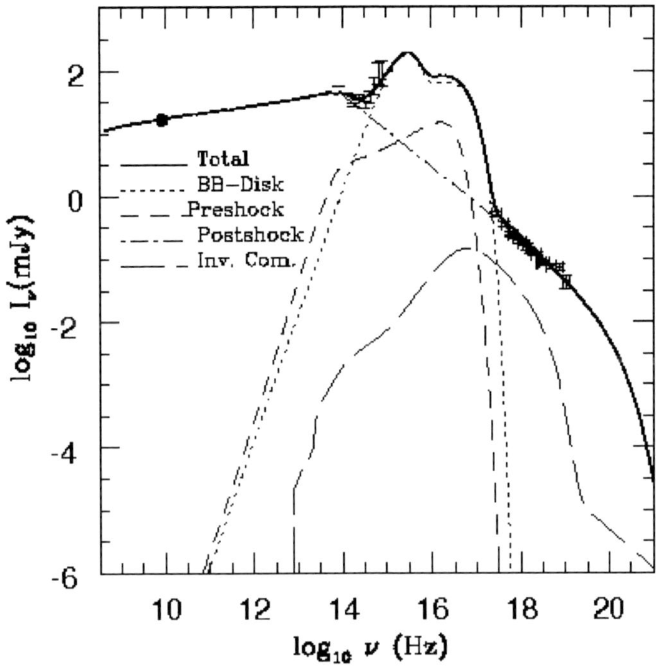

Figure 2. An example of a jet model that successfully reproduces the radio, OIR and X-ray SED of a BHXB (GX 339–4 in this case); from Markoff et al. (2003). The NIR is modelled as optically thin synchrotron emission from the jet.

the spectral energy distributions (SEDs) of the BHXBs V404 Cyg and GRS 1915+105. In fact, Han & Hjellming (1992) showed that the radio and optical light curves of the decay of V404 Cyg follow similar shapes (three separate power-law decay slopes) while remaining proportional in flux ($F_{\nu,\mathrm{radio}} \propto F_{\nu,\mathrm{optical}}$). Similar flat radio–OIR SEDs are seen for GS 1354–64 (BW Cir) and GRO J0422+32 in the hard state (i.e. when the jet is on; Brocksopp et al., 2001, 2004) and for GRS 1915+105 when its flickering radio jet is on (e.g. Fender & Pooley, 2000; Klein-Wolt et al., 2002). A global correlation between the OIR and X-ray luminosities of 15 BHXBs in the hard state (Russell et al., 2006, see below) implies that all BHXBs approximately follow the correlation $L_{\nu,\mathrm{radio}} \approx L_{\nu,\mathrm{optical}}$. Nevertheless, this correlation alone is not a *diagnostic* of a jet origin to the OIR luminosity but is consistent with this interpretation.

More convincing is a measure of the intrinsic OIR spectral index. Corbel & Fender (2002) found two components of OIR emission in an SED of GX 339–4 – one red ($\alpha < 0$) and one blue ($\alpha > 0$). The red component dominates the NIR and is consistent with optically thin synchrotron emission from the jet, for which we expect $\alpha \sim -0.7 \pm 0.2$; this value depends on the electron energy distribution. A $\sim$ flat radio–NIR spectrum was also evident in the SED. The red NIR component was seen additionally in separate outbursts of GX 339–4 (Homan et al., 2005), and Motch et al. (1985) measured the power-law optical—NIR spectral index to be $\alpha \sim -0.58$, fairly typical of optically thin synchrotron emission. The SEDs of XTE J1118+480 (Hynes et al., 2000; Fender et al., 2001; Chaty et al., 2003a; Hynes et al., 2003b) and 4U 1543–47 (Buxton & Bailyn, 2004; Kalemci et al., 2005) have similar red spectral indices in the hard state, and a NIR-excess was apparent above the spectrum of the accretion disc in XTE J1859+226 (Hynes et al., 2002). SEDs of LMC X–3 also reveal an anomalous NIR-excess (Pineault, 1984; Treves et al., 1988), which may be the first evidence for a jet from a BHXB in the Magellanic Clouds (see also Section 2.3 for an other). OIR SEDs of V404 Cyg, GS 1354–64, GS 2000+25, XTE J1550–564 and GRO J1655–40 are sometimes red ($\alpha < 0$; Brocksopp et al., 2004; Russell et al., 2006) which is indicative of either a cold disc (possible for sources at low luminosity) or a synchrotron jet. It is possible to use OIR colour-magnitude diagrams to successfully separate thermal disc from non-thermal jet emission since the irradiated disc produces a predictable relation between OIR colour and magnitude as the temperature of the outer disc (and luminosity) changes, whereas the OIR colour of optically thin synchrotron emission is not expected to change with disc temperature or luminosity (Maitra & Bailyn, 2008; Russell et al., 2008). Colours redder than expected for a disc (deviations from a colour–magnitude relation for a heated black body) were seen in five BHXBs; the redder colours of some were argued to be due to the jets.

One claim of the jet break itself observed in the NIR SED of GX 339–4 (Corbel & Fender, 2002) is based on three data points; follow-up near simultaneous photometry is required to confirm this (the source is known to have large amplitude variability on short timescales). However the jet break may have been detected in GRO J0422+32 (a self-absorbed synchrotron model fits the double-power-law optical–UV spectrum better than an accretion disc model; Hynes & Haswell, 1999; Shrader et al., 1994) and V404 Cyg (Brocksopp et al., 2004; Russell et al., 2006) but these are speculative.

A further source of evidence for OIR jets is successful modelling of broadband spectra, incorporating self-consistent physical constituents and emission processes. This approach has led to further confirmation of OIR jets dominating the SEDs of XTE J1118+480 (Markoff et al., 2001) and GX 339–4 (Markoff et al., 2003, an example model from this paper is shown here in Fig. 2). For GRO J1655–40, and in quiescence for A0620–00, XTE J1118+480 and V404 Cyg, the jet dominates only at lower frequencies – in the mid-infrared – but makes a low-level contribution to the OIR (Migliari et al., 2007; Gallo et al., 2007).

OIR–X-ray correlation in the hard state:

Homan et al. (2005) present a correlation between the quasi-simultaneous NIR and X-ray fluxes of GX 339–4 which span two orders of magnitude in $F_{\rm X}$ during its hard state. The correlation, $F_{\rm NIR} \propto F_{\rm X}^{0.5}$ is synonymous to the hard state radio–X-ray correlation $F_{\rm radio} \propto F_{\rm X}^{0.6}$ (Gallo et al., 2006), suggesting again that the radio and NIR emission origins may be identical. In Russell et al. (2006, 2007b) it was found that the X-ray and OIR hard state data (B-band to K-band) of 15 BHXBs are consistent with a global correlation, $L_{\rm OIR} \propto L_{\rm X}^{0.6}$ which holds over eight orders of magnitude in $L_{\rm X}$. Individual BHXBs do not necessarily display this correlation within their own data sets, but all the data lie close to the global correlation. However, a correlation with a similar ($L_{\rm OIR} \propto L_{\rm X}^{0.5}$) slope is expected (van Paradijs & McClintock, 1994) if the OIR is reprocessed emission from X-rays illuminating the accretion disc. From the observed BHXB correlation alone, the two cannot be confidently separated (Russell et al., 2006).

An optical/infrared flux drop when the radio jet is quenched:

A strong indication of an OIR jet is a dramatic change of flux during an X-ray state transition. The compact radio jet is quenched in the soft state, dropping below detection limits (Gallo et al., 2003). Homan et al. (2005) saw a drop of a factor 18 (3 magnitudes) in the NIR flux of GX 339–4 when the source entered the soft state. In that time the X-ray flux also decreased but only by a factor $\sim$ 2. There are a number of examples of OIR rises/drops seen during transitions in/out of the hard state; in Fig. 3 we plot the light curves and amplitudes of this drop as a function of frequency. It appears that the change in flux is in general stronger in the NIR than in the optical, likely due to the accretion disc dominating moreso at higher frequencies because its spectrum is bluer than that of the jet. In at least one case (A0620–00) there is an X-ray rise of similar amplitude to the OIR rise – for this source X-ray reprocessing cannot be ruled out. Moreover, Russell et al. (2006, 2007b) found that the NIR (but not the optical) luminosity of all six BHXBs in their study (the first three sources in the above Fig. 3, plus GRO J1655–40, XTE J1720–318 and XTE J1859+226) is weaker in the soft state than in the hard state at the same X-ray luminosity, by a factor $\sim$ 10 (this cannot be accounted for by the change in X-ray spectrum). The NIR soft state data of these six BHXBs lie below the global hard state OIR–X-ray correlation. In a related result, a NIR-to-UV jet spectrum in GRS 1915+105 is implied by a drop in the NIR emission line Br-γ when the jet continuum drops, which suggests the line is radiatively pumped by high energy photons from the jet (Eikenberry et al., 1998a). However, this seems not to be the case for the Hα line in BHXBs since no apparent change is seen in the equivalent width of this line after state transitions (Fender et al., 2009).

Fast timing signatures:

Rapid (on timescales down to milliseconds), high-amplitude optical variability has been reported from BHXBs since the 1980s (Motch et al., 1982). Fabian et al. (1982) argued that clouds of plasma with surface brightness temperatures $\sim 5 \times 10^9$ K emitting cyclotron radiation can account for the rapid optical flares of GX 339–4. Motch et al. (1983) constructed a cross-correlation function (CCF) of simultaneous fast optical and X-ray timing data of GX 339–4 during its hard state and discovered a complex behaviour in which (a) a weak positive optical response to X-ray flickering peaked after tens of seconds, and (b) an optical–X-ray anti-correlation exists whereby optical variations precede by a few seconds those in X-ray. This 'precognition dip' was also later found in the CCF of XTE J1118+480 in the hard state, in addition to a strong positive optical lag (Kanbach et al., 2001). The optical variability is less smeared than the X-ray, so the optical origin cannot be X-ray reprocessing on the disc. A further recent study of GX 339–4 in the hard state (Gandhi et al., 2008) showed an optical–X-ray CCF with a weak precognition dip and a large but narrow positive optical response.

Two separate lines of reasoning lead to the conclusion that the origin of the fast optical flickering in XTE J1118+480 (and presumably GX 339–4) is the inner regions of the jet. Hynes et al. (2003b) studied the NIR, optical, UV and X-ray variability properties and by isolating the variable component, found the variability to have a single power-law SED from NIR to X-ray, with a spectral index $\alpha = -0.6$, implying an optically thin synchrotron origin. The variability is also more lagged at lower frequencies, which is consistent with synchrotron bubbles which become optically thin at lower frequencies as they expand, and not

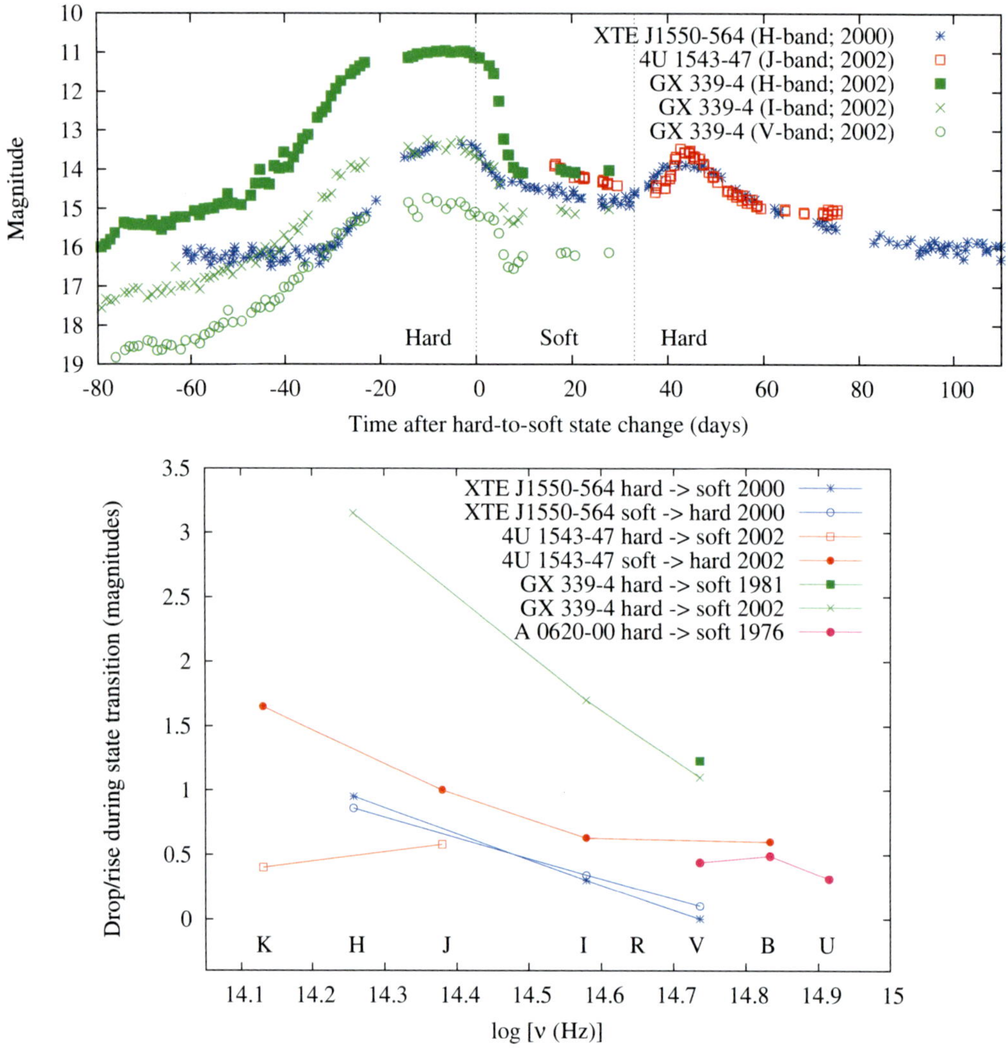

Figure 3. Changes in OIR magnitudes of BHXBs over X-ray state transitions. Top panel: OIR light curves of three BHXBs. Bottom panel: Amplitude of the drop/rise during transition, as a function of frequency. The bandpasses (filters) U (ultraviolet) to K (NIR) are indicated. In all cases, the change of flux is a rise when in transition to the hard state and a drop when the source leaves the hard state (into a softer state). The data are from Motch et al. (1985), Kuulkers (1998), Jain et al. (2001), Buxton & Bailyn (2004) and Homan et al. (2005). Errors are typically less than ~ 0.1 mag.

consistent with either X-ray reprocessing or advection-dominated accretion flow models. Hynes et al. (2006) found a similar spectral index, $\alpha = -0.8$ for the variable component of the NIR emission of XTE J1118+480 during a separate outburst. An alternative approach (Malzac et al., 2004) used a time-dependent model to approximately reproduce the CCF (including the precognition dip) and power spectrum of XTE J1118+480. The model envisages a jet–disc coupling in which accretion (magnetic) energy can be channelled into either the jet power or the X-ray 'corona' in a correlated way, the jet producing the optical emission.

Further clues come from optical variability studies of other BHXBs in outburst and quiescence. Brightness temperature arguments were used to rule out X-ray reprocessing in V4641 Sgr and GRO J0422+32 during hard state outbursts, favouring a non-thermal optical variability origin (Bartolini et al., 1994; Uemura et al., 2002). It was demonstrated that short (< 100 sec) flares from V4641 Sgr are likely synchrotron in nature, and longer (> 1000 sec) flares are thermal (from their SEDs; Uemura et al., 2004a,b). An optical–X-ray CCF of GRO J1655–40 during a *soft* state (Hynes et al., 1998) displays a positive optical lag on light-travel timescales, with no precognition dip. This favours a reprocessing response on the disc surface and further supports the idea that the jet is responsible for the precognition dip. However, the CCF of SWIFT J1753.5–0127 (a BH or neutron star primary is debated in this source) displays a deep, wide precognition dip and weak positive response, at a time in which the contribution of the jet to the OIR SED appears to be minimal (Durant et al., 2008, 2009).

Rather different variability results come from the BHXB GRS 1915+105 in outburst. This source is unique with respect to its X-ray timing and radio jets – it does not remain in the canonical hard or soft X-ray states for weeks to months like other BHXB transients, but instead displays complex states at high accretion rates accompanied by correlated (sometimes resolved) radio jet ejections. Discrete, quasi-periodic NIR, mm and radio flares repeat intermittently on timescales of 30–45 minutes and are correlated with more rapid X-ray variations (e.g. Fender et al., 1997; Ueda et al., 2002; Rothstein et al., 2005). The NIR flares cannot be reprocessed X-rays, and instead are shown to be the precursors of the mm and radio flares – likely single jet ejection events which propagate away from the BHXB, peaking at lower frequencies at larger distances (Mirabel et al., 1998; Fender & Pooley, 1998; Eikenberry et al., 1998b; Fender & Pooley, 2000). A similar process may be evident in SS 433 (this system is most likely a BHXB but may be a neutron star XB). Goranskii et al. (1998) report a variable 'red' component in the optical SED – these optical flares lead radio flares from the SS 433 jets by several hours in time. Chakrabarti et al. (2005) also claim a NIR–radio correlation on longer timescales – a broad dip in the NIR and radio light curves is consistent with the NIR leading the radio by two days.

Numerous fast timing studies of BHXBs have been performed during periods of quiescence, but only one (to our knowledge) makes use of simultaneous X-ray coverage, because most systems are too faint for current X-ray telescopes at these luminosities. Hynes et al. (2004) present a CCF between X-ray flux and Hα optical emission line flux, which shows a positive correlation consistent with zero lag. This result combined with the observation of a double peaked Hα line profile implies line variability in the disc powered by X-ray irradiation. Fast optical continuum variations and flaring were reported in quiescence for at least six BHXBs (A0620–00, GU Mus, MM Vel, V404 Cyg, GRO J0422+32, GS 2000+25; Hynes et al., 2003a; Zurita et al., 2003; Shahbaz et al., 2003, 2004). The rapidity implies (to the authors) localised flares which may occur when magnetic loops reconnect on the disc, although the power density spectrum, when measured, can resemble that of an X-ray power density spectrum of a BHXB in the hard state, in which the X-ray variability comes from the inner regions of the accretion flow. In some cases, the spectral index of the optical flares themselves have been measured and are quite steep: $\alpha = -1.4$ for A0620–00 and

$\alpha = -2.0$ for V404 Cyg, which are consistent with an optically thin gas at a temperature (8–14)$\times 10^3$ K. According to the quiescent broadband SEDs of these two BHXBs (Gallo et al., 2007), the jet likely contributes $< 10\%$ (most likely $\sim 1\%$) of the optical emission at these low luminosities, so the jets are unlikely to be responsible for the flares. However, the flares are short-lived and could be missed in the broadband SEDs, which are based on mean fluxes over longer timescales. If the flares originate in the jets, the optical–radio jet spectrum would have to be inverted; $\alpha \sim> +0.3$ for the jets to be as faint as observed in the radio regime.

Polarization signatures:

Fabian et al. (1982) predicted that if the rapid optical flaring of GX 339–4 (Motch et al., 1982) has a cyclotron origin, the flares should be polarized. Optically thin synchrotron emission can be highly (up to 70%) polarized if the local magnetic field is ordered (e.g. Björnsson & Blumenthal, 1982). This polarization signature is commonly detected (at levels up to tens of per cent) in the radio regime from discrete, optically thin jet ejections (e.g. Hannikainen et al., 2000) and finally confirmed observationally in 2008 at OIR frequencies. One BHXB (GRO J1655–40) and two neutron star XBs (Sco X–1 and Cyg X–2) have linearly polarized NIR emission which is stronger at lower frequencies (Shahbaz et al., 2008; Russell & Fender, 2008). The polarization levels, $\sim$ 1–7% suggest a twisted magnetic field geometry, but the polarization is probably masked by non-polarized light from other spectral components. The polarization of Sco X–1 is variable on timescales of minutes, which could be explained by a variable flux level of the jet or by a changing magnetic field geometry. The polarization angles imply a magnetic field orientation which differs from source to source and may change in time for the same source (Shahbaz et al., 2008; Russell & Fender, 2008). Two further BHXBs, XTE J1118+480 and XTE J1550–564, have low levels of polarization ($\sim 0.5\%$ in the optical and 1–2% in the NIR, respectively) which require follow-up observations to confirm their origin (Schultz et al., 2004; Dubus & Chaty, 2006).

Baryonic jets:

The remarkable precessing jets of SS 433 (of which the compact object may be a BH or a neutron star) have signatures at radio, optical and X-ray frequencies. As well as being resolved in the radio and X-ray regimes (e.g. Schillemat et al., 2004; Migliari et al., 2005), broad optical emission lines with moving velocity shifts up to 40,000 km s^{-1} are shown to be from discrete jets moving ballistically at the same velocity as the radio jets: 26% of the speed of light, at distances $< 10^{10}$ km from their launch region (e.g. Murdin et al., 1980; Margon et al., 1984; Vermeulen et al., 1993). This is the only evidence for baryonic relativistic jets from an X-ray binary. Ionized atoms in the jets are recombining as they cool, and X-ray spectra reveal numerous lines from highly ionized atoms in the jets (Marshall et al., 2002) – their chemical composition therefore includes heavy elements and not pure pair plasma.

SS 433 is surrounded by a large spherical supernova remnant W50. The compact object in SS 433 is thought to be the remnant of the supernova explosion which resulted in W50. The supernova remnant is 10,000 years old. It could be that the recent supernova enriched the photosphere of the companion star with heavy elements, which is now being accreted towards the compact object and swept up into the jets. If this is so, we may not expect to see heavy elements in the jets of other XBs because they are much older, and the photospheres of the companion stars are no longer enriched with heavy elements from the supernova explosion.

A further BHXB that has high-velocity, baryonic outflows is V4641 Sgr. Here, the H α optical and Brγ NIR emission lines are very broad with blue wings and variable, indicating an outflow with velocity $\sim$ 2% of the speed of light (Chaty et al., 2003b). This outflow is however not the same as the highly relativistic (much faster) jet in this source.

2.2. Internal Shocks and Impacts with the ISM

The compact jet, as described in previous Sections, is seen (in the case of Cyg X–1; Stirling et al., 2001) to extend $\sim 10^{8-9}$ km away from the BH. This was measured at one particular radio frequency – at lower frequencies the compact jet may appear longer still, but will also perhaps be fainter. The compact jet makes only a low contribution to the OIR regime at these large distances from the BH. However bright, discrete, moving ejecta are seen at radio and sometimes X-ray frequencies at distances between $\sim 10^9$ and $\sim 10^{14}$ km from the BH, which also emit OIR radiation (Fig. 1). This emission is not part of the overlapping synchrotron spectra that build up the compact jet, but comes from isolated 'knots' downstream. Two processes could be responsible: a fast jet plasma cloud catching up with and colliding with a slower ejection, or a plasma cloud that plows into a denser region of the ISM. Both processes result in synchrotron emission. Although such large-scale moving jets have never been detected at OIR frequencies, we discuss here that OIR jet emission does exist on these scales, and could be detected with future observations.

The large-scale, moving jets of XTE J1550–564 were detected at radio and X-ray frequencies and were shown to be decelerating, likely due to interaction with the ISM (Corbel et al., 2002). The radio spectrum and X-ray flux of the same ejection are consistent with a single power-law of spectral index $\alpha = -0.66$, typical of optically thin synchrotron emission. Deep optical observations were performed with the 8-m Very Large Telescope but no optical counterpart of the ejections were found. The derived optical upper limits were only just greater than the interpolation of the synchrotron power-law, suggesting that with a slightly higher signal-to-noise ratio (a deeper image or a larger telescope) the large-scale ejection would be detected. Deep NIR observations may have been more successful; this source suffers a moderate level of foreground extinction, which affects the optical regime moreso than the NIR.

One claim of a resolved NIR jet from a BHXB does exist in the literature – that of GRS 1915+105 (Sams et al., 1996). The jet here is not resolved itself, but appears as a residual to the south-west of the point-spread-function of GRS 1915+105 in the K-band. The implied distance of the jet from the BH is $\sim 6 \times 10^{11}$ km. This could be the NIR counterpart of a discrete ejection with a broadband radio-to-X-ray SED like the one of XTE J1550–564 above.

Extended radio jets (either moving knots or apparently stationary 'lobes') are now detected from $\sim$ 10 BHXBs. Here, we investigate whether these structures may be observable at OIR frequencies, adopting the assumption that the radio emission is optically thin (observations indicate this is the case), with $\alpha = -0.7$ extrapolated to OIR. We calculate the apparent K-band (2.2μm) magnitude of the knots given the known foreground extinction to each source. The results are given in Table 1. We see that some of the predicted jet magnitudes are bright enough to be detected with current large telescopes. The brightest, at K $\sim$ 18 mag is however only 0.7 arc-seconds from the core, so high-resolution imaging is required, but possible.

Table 1. The predicted NIR K-band magnitudes of known extended radio jets of BHXBs, assuming $\alpha = -0.7$ from radio to NIR. θ is the angular distance in arc-seconds of the radio knot from the core. The 2.2μm values of F_ν are predicted intrinsic flux densities and K mags are the corresponding apparent K-band magnitudes after interstellar extinction is taken into account. For foreground extinction we adopt $A_K = 0.114 A_V$; estimated values of A_V are from Jonker & Nelemans (2004), Chapuis & Corbel (2004) and Eikenberry et al. (2001).

BHXB	Radio jet knot				2.2μm flux		A_V
	F_ν (mJy)	Freq. (GHz)	θ	Reference	F_ν (μJy)	K mag	
				Moving jets:			
XTE J1550–564	3.6	4.8	29"	Corbel et al. (2002)	2.7	21.6	5.0
GRO J1655–40	120	2.3	0.73"	Tingay et al. (1995)	55	18.1	3.7
GX 339–4	0.47	4.8	6.9"	Gallo et al. (2004)	0.36	23.6	3.9
V4641 Sgr	10	4.9	0.6"	Hjellming et al. (2000)	7.7	19.9	1.0
GRS 1915+105	2.1	5.0	0.23"	Miller-Jones et al. (2005)	1.6	23.8	20
				Large-scale lobes:			
1E 1740.7–2942	2	1.5	45"	Mirabel et al. (1992)	0.68	27.1	40
GRS 1758–258	0.52	5.0	23"	Rodríguez et al. (1992)	0.41	23.9	7.9

In addition, the angular size of the resolved X-ray jets of SS 433 are a few arcseconds (Migliari et al., 2005), so it may be possible to detect the OIR counterparts to these jets with deep observations, possibly in line emission such as shifted H α.

SS 433 actually also provides the best evidence yet from the optical regime of a circumbinary disc surrounding an XB – a separate, non-outflowing large-scale structure around the binary (Blundell et al., 2008). This result is implied from two stationary H α emission lines; one redshifted and one blueshifted, that suggest an orbiting velocity of $\sim$ 200 km s^{-1}.

2.3. Bow Shocks, Filaments and Trails

At the 'front' of the jet there should, in analogy with AGN jets, be a hot spot where the jet slams into dense matter and is decelerated, and a bow shock shell of compressed, shocked gas that surrounds synchrotron-emitting radio plasma (e.g. Kaiser et al., 2004, an alternative is a gradual deceleration of the jet over large distances by a low-density ISM). The shell emits bremsstrahlung radiation and recombination lines. The first confirmed bow shock shell associated with the jet of a BHXB was recently discovered as a bremsstrahlung-emitting radio ring and a recombination line-emitting optical ring, $\sim$10 parsecs away from Cyg X–1 in the direction of its known radio jet (Gallo et al., 2005; Russell et al., 2007a).

This (see Fig. 4) is the largest known optically-emitting structure powered by the jet of a BHXB.

Figure 4. Image of the Cyg X–1 ring nebula in optical line emission (Russell et al., 2007a). The bow shock is powered by the jet of Cyg X–1. H α emission is shown in red, [O III] in green and continuum emission in blue. The nebula appears yellow since the [O III]/H α flux ratio is ~ 1, indicating a bow shock velocity > 100 km s^{-1}. The edge of an H II region can be seen red in the lower left of the image. The compact radio jet of Cyg X–1 (lower right) is $\sim 10^6$ times smaller in apparent length (Stirling et al., 2001, see Fig. 1).

Cooke et al. (2007) showed that optical emission line ratios of the nebula close to the BHXB LMC X–1, which was previously thought to be photoionized by the X-rays of the BHXB (Pakull & Angebault, 1986) revealed the nebula is in some regions shock-excited, and argued this shock excitation arises from the jets of LMC X–1. This is then the second BHXB jet-powered bow shock nebula.

Filaments of optical line emission are also seen where the two jets of SS 433 have broken through the W50 supernova remnant that surround the XB (e.g. Boumis et al., 2007). These filaments are shock-excited, compressed gas, energized by the jets of SS 433. Rather than a bow shock morphology, the filaments are instead localised regions of shock waves which may contain reflected or secondary shocks. Similar filaments are observed inside the large bow shock nebula of Cyg X–1. It is interesting to note that the emission line ratios of the Cyg X–1 nebula, LMC X–1 nebula and SS 433 filaments all imply shock waves with velocities around ~ 100 km s^{-1}.

It was argued (Heinz, 2002) that a high density region of the ISM is needed for a jet-powered bow shock to form and appear bright. XBs with high-mass companions typically occur in denser regions of the ISM than those with low-mass companions. The fact that the three above jet–ISM interaction sites are associated with the jets of XBs with high-mass companions supports this hypothesis.

In addition, it was recently demonstrated that most XBs with low-mass companions are probably travelling through the ISM at velocities greater than a required limit for jet-powered bow shock nebulae to form and remain powered by the jet (Heinz et al., 2008). In these cases, synchrotron plasma generated by the jets may form trails behind the XB, if the XB velocity exceeds that of the expanding plasma bubble. These expanding trails will have a shock front where they interact with the ISM. This shocked gas may have been recently detected for a low-mass neutron star XB SAX J1712.6–3739 (Wiersema et al., 2009). Here, two approximately straight stripes of H α emission appear to originate from the location of the XB, analogous to the trails envisaged by Heinz et al. (2008). Follow-up imaging and spectroscopy will constrain the properties and nature of this nebulosity.

Nebulae discovered near the XBs GRO J1655–40, LS 5039 and GRS 1009–45 are candidate jet-powered bow shocks and require follow-up observations to confirm these associations (Russell, 2008; Miller-Jones et al., 2008).

3. The Jet Properties as Revealed from OIR Analyses

In Section 2 it is established that the jets of BHXBs can be detected directly and indirectly via numerous methods from optical or infrared observations. We now discuss how these studies can help to constrain the properties of the jets. Due to simple scaling laws that exist theoretically and empirically between the small BHs in XBs and the massive BHs in AGN (Section 1.2; see Fender, 2009, for a review), it is possible to infer the properties of AGN jets by studying BHXB jets.

From OIR observations of the compact jet, we can essentially uncover its OIR spectrum (which is usually optically thin; $\alpha \sim -0.7$), variability properties and polarization level/orientation. The OIR flux and spectral index of the compact jet, when combined with the spectrum at lower and higher frequencies, reveals the broadband spectrum of the jet which can then be compared to the most advanced jet models to infer its properties; for example its dimensions, Lorentz factor, confinement, and the total jet power. So far, it has been estimated in some cases that the jet power exceeds the accretion power (which can be derived from the bolometric luminosity). A further dimension of information can constrain the properties more tightly: time-dependent jet modelling. In particular, changes in the jet properties as a function of mass accretion rate and X-ray state will probe the jet physics. Scaling relations with accretion rate and black hole mass can be extrapolated to BHs in AGN, providing predictions that can be tested with observations. If consistent, it could be shown that all BH jets are ubiquitous.

The OIR jet emission of BHXBs originates near the jet base where it is launched and accelerated, so fast timing and polarization studies could provide a wealth of information about the physical conditions here (for example the time-dependent magnetic field configuration or the accretion energy extraction mechanism), providing constraints for jet formation theories. A separate piece of information about the compact jets comes from the moving emission lines from precessing jets when they are seen. These lines are signatures of 'heavy' (atomic rather than leptonic) jets and they can also be used to tightly constrain the velocity of the jets.

Extended, moving jet knots detected at OIR frequencies would help to constrain the jet energy losses at large distances downstream. Measurements of deceleration coupled with information about the ISM may tell us how receptive jets are to producing radiation when colliding with dense matter.

Finally, OIR emission lines from ring bow shocks and trailed bow shocks can be used as independent methods of calculating the time-averaged jet power and their composition (Gallo et al., 2005; Heinz, 2006). For BHXBs in which this is known and the broadband spectrum of the compact jet has also been measured, the radiative efficiency of the core jet can be derived. Again, these properties can be scaled up to AGN jets. Essentially, how BHXB jets interact with the ISM has direct implications for how AGN jets energize the intracluster medium, shaping galaxy formation.

4. Conclusion

To summarise, due to the rapid observational and theoretical advances in jets from BHXBs in the last decade or so, including the finding of direct links with AGN jets, this may be the 'golden era' of our understanding of the fastest objects in the universe – relativistic jets – and how they are produced via the process of accretion onto compact objects. Perhaps a crucial and satisfactory future goal would be to unify the properties of jets from *all* accreting objects, including the slowest, weakest jets (for example from young stellar objects) and the fastest, most powerful jets (most likely associated with γ-ray bursts).

From an OIR observer's perspective, there are a number of short term studies in the field of BHXB jets that will help lead us forward in our understanding towards the long term goals:

- OIR polarization and fast timing analyses of the isolated optically thin jet in many BHXBs, and in different X-ray states when the jet is not bright, to constrain the ubiquity of the properties of the inner regions of the jet;
- Track the position of the jet break in the spectrum, with changes in mass accretion rate, which constrains the jet power and jet production theories;
- Resolve the OIR counterparts of discrete, moving radio ejecta;

- Discover more jet-powered bow shock nebulae and infer the time-averaged jet power for a sample of BHXBs, again testing ubiquity;
- Compare the properties of BHXB jet-powered structures to AGN lobes and bow shocks, and also to OIR-emitting shock-excited bubbles associated with ultraluminous X-ray sources (which could be intermediate-mass BHs), of which there are many (e.g. Pakull & Grisé, 2008).
- Eventually, it may be possible to resolve the compact jet at OIR frequencies indirectly using interferometry (Markoff, 2008), which will open up new avenues of studying the inner regions of the jets and accretion flow.

Acknowledgements

DMR would like to thank Piergiorgio Casella for useful discussions regarding the interpretations of multi-wavelength fast-timing analyses. DMR acknowledges support from a Netherlands Organization for Scientific Research (NWO) Veni Fellowship.

References

Baade, W.; Minkowski, R. *ApJ* 1954, 119, 221

Bartolini, C.; Guarnieri, A.; Piccioni, A.; Beskin, G. M.; Neizvestny, S. I. *ApJS* 1994, 92, 455

Björnsson, C.-I.; Blumenthal, G. R. *ApJ* 1982, 259, 805

Blandford, R. D.; Znajek, R. L. *MNRAS* 1977, 179, 433

Blandford, R. D.; Konigl, A. *ApJ* 1979, 232, 34

Blandford, R. D.; Payne, D. G. *MNRAS* 1982, 199, 883

Blundell, K. M.; Bowler, M. G.; Schmidtobreick, L. *ApJ* 2008, 678, L47

Boumis, P.; et al. *MNRAS* 2007, 381, 308

Brocksopp, C.; Jonker, P. G.; Fender, R. P.; Groot, P. J.; van der Klis, M.; Tingay, S. J. *MNRAS* 2001, 323, 517

Brocksopp, C.; Bandyopadhyay, R. M.; Fender, R. P. *NewA* 2004, 9, 249

Burbidge, G. R. *ApJ* 1959, 129, 849

Buxton, M. M.; Bailyn, C. D. *ApJ* 2004, 615, 880

Chakrabarti, S. K.; et al. *MNRAS* 2005, 362, 957

Chapuis, C.; Corbel, S. *A&A* 2004, 414, 659

Chaty, S.; Haswell, C. A.; Malzac, J.; Hynes, R. I.; Shrader, C. R.; Cui W. *MNRAS* 2003a, 346, 689

Chaty, S.; Charles, P. A.; Martí, J.; Mirabel, I. F.; Rodríguez, L. F.; Shahbaz, T. *MNRAS* 2003b, 343, 169

Churazov, E.; Sunyaev, R.; Forman, W.; Böhringer, H. *MNRAS* 2002, 332, 729

Cooke, R.; Kuncic, Z.; Sharp, R.; Bland-Hawthorn, J. *ApJL* 2007, 667, 163

Corbel, S.; Fender, R. P. *ApJ* 2002, 573, L35

Corbel, S.; Fender, R. P.; Tzioumis, A. K.; Tomsick, J. A.; Orosz, J. A.; Miller, J. M.; Wijnands, R.; Kaaret, P. *Science* 2002, 298, 196

Dubus, G.; Chaty, S. *A&A* 2006, 458, 591

Durant, M.; Gandhi, P.; Shahbaz, T.; Fabian, A. P.; Miller, J.; Dhillon, V. S.; Marsh, T. R. *ApJ* 2008, 682, L45

Durant, M.; Gandhi, P.; Shahbaz, T.; Peralta, H. H.; Dhillon, V. S. *MNRAS* 2009, 392, 309

Efstathiou, G. *MNRAS* 1992, 256, P43

Eikenberry, S. S.; Matthews, K.; Murphy, T. W. Jr.; Nelson, R. W.; Morgan, E. H.; Remillard, R. A.; Muno, M. *ApJ* 1998a, 506, L31

Eikenberry, S. S.; Matthews, K.; Morgan, E. H.; Remillard, R. A.; Nelson, R. W. *ApJ* 1998b, 494, L61

Eikenberry, S. S.; Fischer, W. J.; Egami, E.; Djorgovski, S. G. *ApJ* 2001, 556, 1

Fabian, A. C.; Guilbert, P. W.; Motch, C.; Ricketts, M.; Ilovaisky, S. A.; Chevalier, C. *A&A* 1982, 111, L9

Falcke, H.; Biermann, P. L. *A&A* 1996, 308, 321

Falcke, H.; Körding, E.; Markoff, S. *A&A* 2004, 414, 895

Fanaroff, B. L.; Riley, J. M. *MNRAS* 1974, 167, P31

Fender, R. P. *MNRAS* 2001, 322, 31

Fender, R. P. in *Compact Stellar X-Ray Sources*; Cambridge University Press: Cambridge, U.K., 2006; pp 381 (eds. Lewin, W. H. G.; van der Klis, M.)

Fender, R. P. *Disc-jet coupling in black hole X-ray binaries and active galactic nuclei* , in preparation

Fender, R. P.; Pooley, G. G.; Brocksopp, C.; Newell, S. J. *MNRAS* 1997, 290, L65

Fender, R. P.; Pooley, G. G. *MNRAS* 1998, 300, 573

Fender, R. P.; Pooley, G. G. *MNRAS* 2000, 318, L1

Fender, R. P.; Hjellming, R. M.; Tilanus, R. P. J.; Pooley, G. G.; Deane, J. R.; Ogley, R. N.; Spencer, R. E. *MNRAS* 2001, 322, L23

Fender, R. P.; Belloni, T. M.; Gallo, E. *MNRAS* 2004, 355, 1105

Fender, R. P.; Russell, D. M.; Knigge, C.; Soria, R.; Hynes, R. I.; Goad, M. *MNRAS* 2009, in press (arXiv:0811.4260)

Flanagan, É. É.; Hughes, S. A.; *PhRvD* 1998, 57, 4535

Fomalont, E. B.; Geldzahler, B. J.; Bradshaw, C. F. *ApJ* 2001, 558, 283

Gallo, E.; Fender, R. P.; Pooley, G. G. *MNRAS* 2003, 344, 60

Gallo, E.; Corbel, S.; Fender, R. P.; Maccarone, T. J.; Tzioumis, A. K. *MNRAS* 2004, 347, L52

Gallo, E.; Fender, R. P.; Kaiser, C.; Russell, D. M.; Morganti, R.; Oosterloo, T.; Heinz, S. *Nature* 2005, 436, 819

Gallo, E.; et al. *MNRAS* 2006, 370, 1351

Gallo, E.; Migliari, S.; Markoff, S.; Tomsick, J.; Bailyn, C.; Berta, S.; Fender, R.; Miller-Jones, J. *ApJ* 2007, 670, 600

Gandhi, P.; et al. *MNRAS* 2008, 390, L29

Goranskii, V. P.; Esipov, V. F.; Cherepashchuk, A. M. *ARep* 1998, 42, 336

Han, X.; Hjellming, R. M. *ApJ* 1992, 400, 304

Hannikainen, D. C.; Hunstead, R. W.; Campbell-Wilson, D.; Wu, K.; McKay, D. J.; Smits, D. P.; Sault, R. J. *ApJ* 2000, 540, 521

Heinz, S. *A&A* 2002, 388, L40

Heinz, S. *ApJ* 2006, 636, 316

Heinz, S.; Sunyaev, R. A. *MNRAS* 2003, 343, L59

Heinz, S.; Grimm, H. J.; Sunyaev, R. A.; Fender, R. P. *ApJ* 2008, 686, 1145

Hjellming, R. M.; et al. *ApJ* 2000, 544, 977

Homan, J.; Buxton, M.; Markoff, S.; Bailyn, C. D.; Nespoli, E.; Belloni, T. *ApJ* 2005, 624, 295

Hynes, R. I. *ApJ* 2005, 623, 1026

Hynes, R. I.; O'Brien, K.; Horne, K.; Chen, W.; Haswell, C. A. *MNRAS* 1998, 299, L37

Hynes, R. I.; Haswell, C. A. *MNRAS* 1999, 303, 101

Hynes, R. I.; Mauche, C. W.; Haswell, C. A.; Shrader, C. R.; Cui, W.; Chaty, S. *ApJ* 2000, 539, L37

Hynes, R. I.; Haswell, C. A.; Chaty, S.; Shrader, C. R.; Cui, W. *MNRAS* 2002, 331, 169

Hynes, R. I.; Charles, P. A.; Casares, J.; Haswell, C. A.; Zurita, C.; Shahbaz, T. *MNRAS* 2003a, 340, 447

Hynes, R. I.; et al. *MNRAS* 2003b, 345, 292

Hynes, R. I.; et al. *ApJ* 2004, 611, L125

Hynes, R. I.; et al. *ApJ* 2006, 651, 401

Jain, R. K.; Bailyn, C. D.; Orosz, J. A.; McClintock, J. E.; Remillard, R. A. *ApJ* 2001, 554, L181

Jamil, O.; Fender, R.; Kaiser, C. *Proceedings of Science*, 2008, *Proceedings of the VII Microquasar Workshop: Microquasars and Beyond, 1-5 September 2008, Foca, Izmir, Turkey* (arXiv:0811.3320)

Jonker, P. G.; Nelemans, G. *MNRAS* 2004, 354, 355

Kaiser, C. R. *MNRAS* 2006, 367, 1083

Kaiser, C. R.; Gunn, K. F.; Brocksopp, C.; Sokoloski, J. L. *ApJ* 2004, 612, 332

Kalemci, E.; et al. *ApJ* 2005, 622, 508

Kanbach, G.; Straubmeier, C.; Spruit, H. C.; Belloni, T. *Nature* 2001, 414, 180

Klein-Wolt, M.; Fender, R. P.; Pooley, G. G.;p Belloni, T.; Migliari, S.; Morgan, E. H.; van der Klis, M. *MNRAS* 2002, 331, 745

Körding, E.; Fender, R. P.; Migliari, S. *MNRAS* 2006a, 369, 1451

Körding, E. G.; Jester, S.; Fender, R. *MNRAS* 2006b, 372, 1366

Kuulkers, E. *NewAR* 1998, 42, 1

Maitra, D.; Bailyn, C. D. *ApJ* 2008, 688, 537

Malzac, J.; Merloni, A.; Fabian, A. C. *MNRAS* 2004, 351, 253

Margon, B.; Anderson, S. F.; Aller, L. H.; Downes, R. A.; Keyes, C. D. *ApJ* 1984, 281, 313

Markoff, S. *Proceedings of Science*, 2008, *Proceedings of the VII Microquasar Workshop: Microquasars and Beyond, 1-5 September 2008, Foca, Izmir, Turkey* (arXiv:0811.3601)

Markoff, S.; Falcke, H.; Fender, R. *A&A* 2001, 372, L25

Markoff, S.; Nowak, M.; Corbel, S.; Fender, R.; Falcke, H. *A&A* 2003, 397, 645

Markoff, S.; Nowak, M. A.; Wilms, J. *ApJ* 2005, 635, 1203

Marshall, H. L.; Canizares, C. R.; Schulz, N. S. *ApJ* 2002, 564, 941

Massey, P.; Johnson, K. E.; Degioia-Eastwood, K. *ApJ* 1995, 454, 151

McNamara, B. R.; Nulsen, P. E. J. *ARA&A* 2007, 45, 117

Meier, D. L. *ApJ* 2001, 548, L9

Merloni, A.; Heinz, S.; di Matteo, T. *MNRAS* 2003, 345, 1057

Migliari, S.; Fender, R. P.; Blundell, K. M.; Méndez, M.; van der Klis, M. *MNRAS* 2005, 358, 860

Migliari, S.; Fender, R. P. *MNRAS* 2006, 366, 79

Migliari, S.; et al. *ApJ* 2007, 670, 610

Miller-Jones, J. C. A.; McCormick, D. G.; Fender, R. P.; Spencer, R. E.; Muxlow, T. W. B.; Pooley, G. G. *MNRAS* 2005, 363, 867

Miller-Jones, J. C. A.; Fender, R. P.; Nakar, E. *MNRAS* 2006, 367, 1432

Miller-Jones, J.; Russell, D.; Brocksopp, C.; Sokoloski, J.; Stappers, B.; Muxlow, T. *AIPC* 2008, 1010, 50 (arXiv:0802.3446)

Mirabel, I. F.; Rodríguez, L. F.; Cordier, B.; Paul, J.; Lebrun, F. *Nature* 1992, 358, 215

Mirabel, I. F.; Dhawan, V.; Chaty, S.; Rodriguez, L. F.; Marti, J.; Robinson, C. R.; Swank, J.; Geballe, T. *A&A* 1998, 330, L9

Motch, C.; Ilovaisky, S. A.; Chevalier, C. *A&A* 1982, 109, L1

Motch, C.; Ricketts, M. J.; Page, C. G.; Ilovaisky, S. A.; Chevalier, C. *A&A* 1983, 119, 171

Motch, C.; Ilovaisky, S. A.; Chevalier, C.; Angebault, P. *SSRv* 1985, 40, 219

Murdin, P.; Clark, D. H.; Martin, P. G. *MNRAS* 1980, 193, 135

Nowak, M. A.; Wilms, J.; Heinz, S.; Pooley, G.; Pottschmidt, K.; Corbel, S. *ApJ* 2005, 626, 1006

O'Brien, K.; Horne, K.; Hynes, R. I.; Chen, W.; Haswell, C. A.; Still, M. D. *MNRAS* 2002, 334, 426

Pakull, M. W.; Angebault, L. P. *Nature* 1986, 322, 511

Pakull, M. W.; Grisé, F. *AIPC* 2008, 1010, 303 (arXiv:0803.4345)

Pineault, S. *A&A* 1984, 139, 313

Rodríguez, L. F.; Mirabel, I. F.; Martí, J. *ApJ* 1992, 401, L15

Rothstein, D. M.; Eikenberry, S. S.; Matthews, K. *ApJ* 2005, 626, 991

Russell, D. M. *Ph.D. thesis* 2008 (arXiv:0802.0816)

Russell, D. M.; Fender, R. P.; Hynes, R. I.; Brocksopp, C.; Homan, J.; Jonker, P. G.; Buxton, M. M. *MNRAS* 2006, 371, 1334

Russell, D. M.; Fender, R. P.; Gallo, E.; Kaiser, C. R. *MNRAS* 2007a, 376, 1341

Russell, D. M.; Maccarone, T. J.; Körding, E. G.; Homan, J. *MNRAS* 2007b, 379, 1401

Russell, D. M.; Fender, R. P. *MNRAS* 2008, 387, 713

Russell, D. M.; Maitra, D.; Fender, R. P.; Lewis, F. *Proceedings of Science*, 2008, *Proceedings of the VII Microquasar Workshop: Microquasars and Beyond, 1-5 September 2008, Foca, Izmir, Turkey* (arXiv:0811.2919)

Sams, B. J.; Eckart, A.; Sunyaev, R. *Nature* 1996, 382, 47

Schawinski, K.; Virani, S.; Simmons, B.; Urry, C. M.; Treister, E.; Kaviraj, S.; Kushkuley, B. *ApJ* 2009, in press (arXiv:0901.1663)

Schillemat, K.; Mioduszewski, A.; Dhawan, V.; Rupen, M. *AAS* 2004, 20510401

Schultz, J.; Hakala, P.; Huovelin, J. *BaltA* 2004, 13, 581

Shahbaz, T.; et al. *MNRAS* 2003, 346, 1116

Shahbaz, T.; et al. *MNRAS* 2004, 354, 31

Shahbaz, T.; Fender, R. P.; Watson, C. A.; O'Brien, K. *ApJ* 2008, 672, 510

Shrader, C. R.; Wagner, R. M.; Hjellming, R. M.; Han, X. H.; Starrfield, S. G. *ApJ* 1994, 434, 698

Stirling, A. M.; Spencer, R. E.; de la Force, C. J.; Garrett, M. A.; Fender, R. P.; Ogley, R. N. *MNRAS* 2001, 327, 1273

Tingay, S. J.; et al. *Nature* 1995, 374, 141

Treves, A.; Belloni, T.; Bouchet, P.; Chiappetti, L.; Falomo, R.; Maraschi, L.; Tanzi, E. G. *ApJ* 1988, 335, 142

Ueda, Y.; et al. *ApJ* 2002, 571, 918

Uemura, M.; et al. *PASJ* 2002, 54, L79

Uemura, M.; et al. *PASJ* 2004a, 56, 61

Uemura, M.; et al. *PASJ* 2004b, 56, 823

van Paradijs, J.; McClintock, J. E. *A&A* 1994, 290, 133

Vermeulen, R. C.; et al. *A&A* 1993, 270, 204

Wiersema, K.; et al. *MNRAS* 2009, submitted

Zurita, C.; Casares, J.; Shahbaz, T. *ApJ* 2003, 582, 369

In: Black Holes and Galaxy Formation ISBN 978-1-60741-703-3
Editors: A.D. Wachter and R.J. Propst, pp. 321-330 © 2010 Nova Science Publishers, Inc.

Chapter 12

HAWKING RADIATION OF THE ROTATING D3-BRANE FROM GRAVITATIONAL ANOMALY

Zheng Ze Ma[*]
Department of Physics, Southeast University,
Nanjing, 210096, P. R. China

Abstract

In recent years, Hawking radiation of black holes has been reinterpreted by Robinson and Wilczek *et al.* through the method of anomaly cancellation of quantum fields, and such a method has been used in the derivation of the Hawking radiation of many different types of black holes and some other black-hole-like objects with event horizons. In this paper, we study the Hawking radiation of rotating D3-brane which appeared in superstring theories using the method of anomaly cancellation. For the sake of briefness, we study only the rotating D3-brane with one nonzero rotating parameter. We first study the effective field theory of quantum fields near the horizon of the rotating D3-brane. We find that it is equal to a two-dimensional field theory in a curved background. Then we calculate the energy-momentum flux of the rotating D3-brane via the approach of anomaly cancellation of gravitational fields. We find that the energy-momentum flux of the rotating D3-brane derived from the method of anomaly cancellation forms a thermal radiation, and the thermal temperature of the radiation matches the Hawking temperature of the rotating D3-brane obtained from the black brane thermodynamics. Thus, we can derive the Hawking radiation of the rotating D3-brane from the method of anomaly cancellation.

PACS numbers: 04.50.+h, 04.62.+v, 04.70.Dy, 11.30.-j

1. Introduction

Perhaps the most fantastic phenomenon of a black hole is the Hawking radiation [1]. It is a quantum gravitational effect generated at the spacetime near the black hole horizon. After its original discovery by Hawking, people have given many different explanations and

[*]E-mail address: z.z.ma@seu.edu.cn

derivations for such a phenomenon [2–6], and the same result is obtained. Different approaches for the derivation of Hawking radiation have revealed and embodied the different aspects of quantum gravitation.

Recently, a new method for the derivation of Hawking radiation has been proposed by Robinson and Wilczek *et al.* which is named anomaly cancellation [7,8]. They find that the effective field theory of quantum fields near a black hole's horizon is a two-dimensional chiral field theory confronting the fact that a black hole's horizon is a one-way membrane. Hence there exist gauge and gravitational anomalies for the currents near a black hole's horizon. However, the effective action of quantum fields near a black hole's horizon is still gauge invariant and general covariant. Then, to combine the regular conditions for the covariantly anomalous currents on the horizon, gauge and energy-momentum fluxes with the Hawking temperature T_H can be derived outside the horizon. The method of anomaly cancellation for the derivation of Hawking radiation shows the connections between the Hawking radiation and the anomalies of quantum fields. For its precedent mind, one can find that it is related with the study of the quantum radiation of a horizon in a two-dimensional spacetime manifold presented by Christensen and Fulling more than three-decades ago [5].

In [9–11], the method of anomaly cancellation for the derivation of Hawking radiation has been generalized to higher-dimensional rotating black holes. Then, its application to the derivation of Hawking radiation of various types of black holes has been carried out in [12–20]. Some most recent developments and applications about this topic have been carried out in [21–24]. In this paper, we will apply the method of anomaly cancellation for the derivation of Hawking radiation of the rotating D3-brane of superstring theories [25–28]; but for the sake of briefness, we only study the rotating D3-brane with one nonzero rotating parameter in this paper.

The content of this paper is organized as follows. In Sec. 2, we derive the effective field theory of quantum fields near the horizon of the rotating D3-brane. We find that it is equal to a two-dimensional field theory in a curved background. In Sec. 3, we calculate the energy-momentum flux of the rotating D3-brane using the method of anomaly cancellation. We find that the energy-momentum flux of the rotating D3-brane derived from anomaly cancellation matches a thermal radiation, and the thermal temperature of the radiation equals the Hawking temperature of the rotating D3-brane obtained from black brane thermodynamics. Section 4 is devoted to the conclusion.

2. Quantum Fields near the Horizon of the Rotating D3-Brane

The ten-dimensional supergravities are low-energy effective field theories of superstring theories. A lot of rotating black brane solutions have been found in these theories. Due to the internal $SO(6)$ rotational isometry, the complete solution of the rotating D3-brane has three nonzero rotating parameters [27,28]. In this paper, we study the rotating D3-brane with only one nonzero rotating parameter. Its metric is given by [26–28]

$$\begin{aligned} ds^2 &= f^{-1/2}(-hdx_0^2 + dx_1^2 + dx_2^2 + dx_3^2) + f^{1/2}\Bigg[\frac{\Delta dr^2}{1+\frac{l^2}{r^2}-\frac{2m}{r^4}} \\ &+ r^2\Bigg(\Delta d\theta^2 + \Big(1+\frac{l^2}{r^2}+\frac{2ml^2\sin^2\theta}{r^6\Delta f}\Big)\sin^2\theta d\varphi_1^2 + \cos^2\theta d\psi^2 \end{aligned}$$

$$+ \quad \cos^2\theta\Big(\sin^2\psi d\varphi_2^2 + \cos^2\psi d\varphi_3^2\Big)\Big) - \frac{4ml\sin^2\theta\cosh\alpha}{r^4\Delta f}dx_0 d\varphi_1\Bigg] , \qquad (1)$$

where

$$\Delta = 1 + \frac{l^2\cos^2\theta}{r^2} , \qquad h = 1 - \frac{2m}{r^4\Delta} , \qquad f = 1 + \frac{2m\sinh^2\alpha}{r^4\Delta} . \qquad (2)$$

The 4-form gauge field of the rotating D3-brane is represented as

$$C^{(4)} = \frac{1-f^{-1}}{\sinh\alpha}dx_1 \wedge dx_2 \wedge dx_3 \wedge (\cosh\alpha dt - l\sin^2\theta d\varphi_1) . \qquad (3)$$

The brane extends in the x_1, x_2, x_3 directions and rotates in the φ_1 direction. The horizon r_H of the brane is located at

$$r_H^2 = (\sqrt{l^4 + 8m} - l^2)/2 . \qquad (4)$$

The angular velocity of the horizon is obtained as

$$\Omega_H = \frac{l}{\cosh\alpha(r_H^2 + l^2)} . \qquad (5)$$

The Hawking temperature of the brane can be obtained through the method of black hole thermodynamics with the result [27]

$$T_H = \frac{r_H(2r_H^2 + l^2)}{4\pi m\cosh\alpha} . \qquad (6)$$

Robinson and Wilczek *et al.*'s work [7,8] has shown the connections between the Hawking radiation and gravitational anomalies. For quantum field theories near the horizon of a black hole, because of the non-invertibility of the physical processes on the horizon, they are equal to two-dimensional chiral field theories. But to view from the four or higher-dimensional spacetime, field theories near a black hole horizon are not chiral, and there exists the Hawking radiation outside the horizon. This means that when we reduce field theories to two-dimensional spacetime near the horizon of a black hole, there should exist some other currents which have counteracted the anomalies of the two-dimensional chiral field theories. These currents can be determined by certain boundary conditions. To view from the four or higher-dimensional spacetime, it is these currents which have resulted in the Hawking radiation of a black hole. Thus, in order to investigate the Hawking radiation of the rotating D3-brane from the method of anomaly cancellation, we first need to study the effective field theories of quantum fields near the horizon of the rotating D3-brane.

We consider a free scalar field φ in the background of the metric (1). We suppose that the scalar field is zero-mass and it does not couple with the Ramond-Ramond gauge field. Then, the action of this system is given by

$$\begin{aligned} S_{\text{free}}(\varphi) &= \int d^{10}x\sqrt{-g}\, g^{\mu\nu}\partial_\mu\varphi\partial_\nu\varphi \\ &= \int d^{10}x\sqrt{-g}\,[\partial_\mu(\varphi\partial^\mu\varphi) - \varphi\partial_\mu\partial^\mu\varphi] , \end{aligned} \qquad (7)$$

where g is the determinant of the metric (1). The scalar field $\varphi(x)$ can be expanded in terms of the harmonic functions of the D3-brane spacetime. Thus, we can write

$$\varphi(x) = \sum_{l_1,\ldots,l_5,k_1,\ldots,k_3} \varphi_{l_1\ldots l_5 k_1\ldots k_3}(r,t) Y_{l_1\ldots l_5}(\theta_1,\ldots,\theta_5) X(k_1,k_2,k_3) \,. \tag{8}$$

In (8), we use $(\theta_1,\ldots,\theta_5)$ to represent the azimuthal angles $(\theta,\psi,\varphi_1,\varphi_2,\varphi_3)$ of the metric (1), $Y_{l_1\ldots l_5}(\theta_1,\ldots,\theta_5)$ are the normalized spherical harmonics on a five-dimensional unit sphere with the azimuthal angles $(\theta_1,\ldots,\theta_5)$, and

$$X(k_1,k_2,k_3) = e^{i(k_1x^1+k_2x^2+k_3x^3)} \,, \tag{9}$$

they satisfy the condition

$$\begin{aligned}\int d\Omega_5 d^3x Y_{l_1\ldots l_5}(\theta_1,\ldots,\theta_5) X(k_1,k_2,k_3) \times \\ Y_{m_1\ldots m_5}(\theta_1,\ldots,\theta_5) X^*(p_1,p_2,p_3) = \delta_{l_1m_1}\cdots\delta_{l_5m_5}\delta_{k_1p_1}\delta_{k_2p_2}\delta_{k_3p_3} \,,\end{aligned} \tag{10}$$

where the integration of the coordinates x^i takes a unit volume on the three-dimensional transverse space of the rotating D3-brane.

To substitute (8) in (7) and to complexify the first φ inside the integral, then, to take the near horizon limit [8–10,20] and through some tedious calculation, we obtain finally

$$S_{\text{free}}(\varphi) = -\sqrt{-g}|_{r_H} \sum_{l_1,\ldots,l_5,k_1,\ldots,k_3} \int dtdr \varphi_{l_1\ldots l_5k_1\ldots k_3}(r,t)(\partial_t\partial^t + \partial_r\partial^r)\varphi_{l_1\ldots l_5k_1\ldots k_3}(r,t) \,, \tag{11}$$

where in (11), the curved two-dimensional spacetime background is determined by the metric

$$ds^2 = -\frac{A(r)}{B^{1/2}(r)}dt^2 + \frac{B^{1/2}(r)}{A(r)}dr^2 \,, \tag{12}$$

with the functions $A(r)$ and $B(r)$ given by

$$A(r) = 1 - \frac{2m}{r^4+r^2l^2} \,, \qquad B(r) = 1 + \frac{2m\sinh^2\alpha}{r^4+r^2l^2} \,. \tag{13}$$

We obtain that, near the horizon of the rotating D3-brane, the effective field theory of the free scalar field reduces to an infinite set of two-dimensional scalar fields in a curved background with the metric (12) whose horizon is determined by $A(r_H) = 0$. If the scalar field carries the Ramond-Ramond charge, then it will couple with the Ramond-Ramond gauge field of the rotating D3-brane. However, due to the fact that the explicit forms of the covariant couplings of quantum fields with the Ramond-Ramond gauge fields are not clear, we do not consider such a coupling here in this paper; therefore, we will not study the Ramond-Ramond charge flux related with the gauge anomaly in this paper. However, we can expect that if we have involved the coupling of the scalar field with the Ramond-Ramond gauge field in the action, the curved spacetime background of the reduced two-dimensional field theory is still given by the metric (12).

3. Gravitational Anomaly and the Energy-Momentum Flux

We can see that the curved background of the reduced two-dimensional field theory of the rotating D3-brane is just a Schwarzschild type metric, thus, we can use the standard approach of anomaly cancellation to calculate the energy-momentum flux of the metric (12) [7,8]. Following [7,8], to consider the area outside the horizon, we divide it into two parts: $[r_H, r_H + \epsilon]$ and $[r_H + \epsilon, \infty]$. $[r_H, r_H + \epsilon]$ is the area near the horizon, where the physics has certain exotic properties. $[r_H + \epsilon, \infty]$ is the area far away from the horizon, where the physics has the usual properties. The parameter ϵ can be taken arbitrarily small, thus, for the observably physical results, we can take them in the region $[r_H + \epsilon, \infty]$.

In the near horizon area $[r_H, r_H+\epsilon]$, to consider that a black hole's horizon is a one-way membrane, for the $(1 + 1)$-dimensional field theory, the ingoing (left moving) modes will tend to the center singularity, hence they will not affect the physics of the area $[r_H, r_H + \epsilon]$. That is to say in the area $[r_H, r_H + \epsilon]$, only the outgoing (right moving) modes are responsible for the observable physics. This makes the practical field theory be a two-dimensional chiral field theory in the near horizon area. Thus, in the area $[r_H, r_H+\epsilon]$, there exists the gravitational anomaly for the energy-momentum current. In the area $[r_H+\epsilon, \infty]$, the ingoing and outgoing modes are both existing, the field theory is a normal one and there is no gravitational anomaly for the energy-momentum current.

For a black hole or a black brane, it is a thermodynamical equilibrium system, all currents in the spacetime outside the horizon are static. The energy-momentum tensor outside the horizon can be decomposed in the form

$$T^{\mu}_{\nu}(r) = T^{\mu}_{\nu(H)}(r)H(r) + T^{\mu}_{\nu(o)}(r)\Theta_{+}(r) , \tag{14}$$

where $\Theta_+(r) = \Theta(r - r_+ - \epsilon)$ (here we use r_+ to represent the radius of the event horizon), and $H(r) = 1 - \Theta_+(r)$. Thus, $T^{\mu}_{\nu(H)}(r)$ is the energy-momentum tensor in the area $[r_H, r_H + \epsilon]$, $T^{\mu}_{\nu(o)}(r)$ is the energy-momentum tensor in the area $[r_H + \epsilon, \infty]$.

In a two-dimensional spacetime, $T^r_t(r)$ is just the energy-momentum flux in the radial direction. For $T^{\mu}_{\nu(o)}(r)$, as analyzed above, there is no gravitational anomaly, it satisfies the normal conservation equation, therefore we have

$$\partial_r T^r_{t(o)}(r) = 0 . \tag{15}$$

But for $T^{\mu}_{\nu(H)}(r)$ near the horizon, it has the gravitational anomaly. It satisfies the anomalous conservation equation [7,8,29,30]. This results

$$\partial_r T^r_{t(H)}(r) = \partial_r N^r_t(r) , \tag{16}$$

where the right hand side of (16) comes from the gravitational anomaly of the consistent energy-momentum tensor. For the two-dimensional metric (12), to use the result of [7,8], we have

$$N^r_t(r) = \frac{1}{192\pi}\left[\left(\frac{A(r)}{B^{1/2}(r)}\right)'^{2} + \left(\frac{A(r)}{B^{1/2}(r)}\right)''\left(\frac{A(r)}{B^{1/2}(r)}\right)\right] . \tag{17}$$

The integration of (15) and (16) yields

$$T^r_{t(o)}(r) = a_o , \qquad T^r_{t(H)}(r) = a_H + \int_{r_H}^{r} dr \partial_r N^r_t(r) , \tag{18}$$

where a_o, a_H are two integration constants. From (14), we can see that a_o is just the energy-momentum flux for an observer to observe outside the horizon.

On the other hand, the anomaly of the energy-momentum tensor is purely a quantum field effect, it does not affect the general covariance of the effective action. To perform an infinitesimal coordinate transformation for the two-dimensional field theory along the time direction with the parameter ξ^t, we obtain [7,8]

$$\begin{aligned}\delta W &= -\int d^2x\sqrt{-g}\,\xi^t\nabla_\mu T^\mu_t \\ &= -\int d^2x\,\xi^t\left[\partial_r(N^r_t H) + \left(T^r_{t(o)} - T^r_{t(H)} + N^r_t\right)\delta(r - r_+ - \epsilon)\right].\end{aligned} \tag{19}$$

As pointed out in [7,8], the first term in the second line of (19) can be canceled by the quantum effect of the ingoing modes near the horizon. Thus, general covariance of the effective action leads the vanishing of the coefficient of the δ-function. To combine (18), we obtain

$$a_o = a_H - N^r_t(r_H)\,. \tag{20}$$

According to (17), and to consider that on the horizon $A(r_H) = 0$, we obtain

$$N^r_t(r_H) = \frac{1}{192\pi}\left(\frac{A'(r)}{B^{1/2}(r)}\right)^2\Bigg|_{r=r_H}. \tag{21}$$

In order to determine the constant a_H of (20), we need to introduce the covariantly anomalous energy-momentum tensor $\widetilde{T}_{\mu\nu}$ which satisfies the covariant anomaly equation [30]

$$\nabla^\mu\widetilde{T}_{\mu\nu} = \frac{1}{96\pi\sqrt{-g}}\epsilon_{\mu\nu}\partial^\mu R\,, \tag{22}$$

where R is the Ricci scalar. For the component $\widetilde{T}^r_t$ which is necessary for the following calculation, for the metric (12), to use the result of [7,8], we obtain

$$\widetilde{T}^r_t(r) = T^r_t(r) + \frac{1}{192\pi}\left[\left(\frac{A(r)}{B^{1/2}(r)}\right)\left(\frac{A(r)}{B^{1/2}(r)}\right)'' - 2\left(\frac{A(r)}{B^{1/2}(r)}\right)'^2\right]. \tag{23}$$

As pointed out in [7,8], $\widetilde{T}^r_t$ should satisfy the regular boundary condition

$$\widetilde{T}^r_t(r_H) = 0\,, \tag{24}$$

from which the physical quantities are made regular on the future horizon for a free falling observer [7–9]. The combination of (23), (24), and (14) yields

$$T^r_{t(H)}(r_H) = \frac{1}{192\pi}\left[2\left(\frac{A(r)}{B^{1/2}(r)}\right)'^2 - \left(\frac{A(r)}{B^{1/2}(r)}\right)''\left(\frac{A(r)}{B^{1/2}(r)}\right)\right]\Bigg|_{r=r_H}. \tag{25}$$

Because on the horizon $A(r_H) = 0$, we obtain

$$T^r_{t(H)}(r_H) = \frac{1}{96\pi}\left(\frac{A'(r)}{B^{1/2}(r)}\right)^2\Bigg|_{r=r_H}. \tag{26}$$

The combination of (26) and (18) results

$$a_H = \frac{1}{96\pi}\left(\frac{A'(r)}{B^{1/2}(r)}\right)^2\bigg|_{r=r_H} . \tag{27}$$

To substitute (27) and (21) in (20), we obtain

$$a_o = \frac{1}{192\pi}\left(\frac{A'(r)}{B^{1/2}(r)}\right)^2\bigg|_{r=r_H} . \tag{28}$$

We know that for a two-dimensional black body radiation with temperature T, its energy-momentum flux is given by

$$F_E = \frac{\pi}{12}T^2 . \tag{29}$$

As mentioned above, a_o is just the energy-momentum flux for an observer to measure outside the horizon of the metric (12). Thus, to write a_o in the form

$$a_o = \frac{\pi}{12}T_H^2 , \tag{30}$$

where T_H is the proposed Hawking temperature of the metric (12), and to compare (30) with (28), we obtain

$$T_H = \frac{1}{4\pi}\left(\frac{A'(r)}{B^{1/2}(r)}\right)\bigg|_{r=r_H} . \tag{31}$$

Such a temperature is just equal to the Hawking temperature of the metric (12) obtained from the standard method of black hole thermodynamics. Thus, we have derived the existence of the Hawking radiation for the two-dimensional Schwarzschild metric (12) and recovered its Hawking temperature from the method of anomaly cancellation. For the original ten-dimensional metric (1), from the mode decomposition of the scalar field in terms of (8), we can deduce that the distribution of the spectrum is not changed; and, for the original ten-dimensional metric (1), its Hawking temperature is still given by (31). To substitute (13) into (31), we obtain finally

$$T_H = \frac{r_H(2r_H^2 + l^2)}{4\pi m \cosh\alpha} , \tag{32}$$

which is just equal to that of (6) derived from the method of black hole thermodynamics. Thus, the Hawking temperature of the rotating D3-brane has been recovered from the method of anomaly cancellation.

4. Conclusion

There are many different methods for the derivation of the Hawking radiation of black holes [1–6]. Recently, a new method for the derivation of Hawking radiation has been proposed by Robinson and Wilczek *et al.* which is named anomaly cancellation [7,8]. In this paper, we use the method of anomaly cancellation to study the Hawking radiation of the rotating D3-brane that appeared in superstring theories [25–28]. In order to save the length, we have only studied the rotating D3-brane with one nonzero rotating parameter. We first analyze the effective field theory of quantum fields near the horizon of the rotating D3-brane. We find that it is equal to a two-dimensional field theory in a curved background. Then we calculate the energy-momentum flux of the rotating D3-brane via the anomaly cancelation of gravitational fields. We derive the energy-momentum flux of the rotating D3-brane from the method of anomaly cancellation. It is a thermal radiation, and the thermal temperature of the radiation matches the Hawking temperature of the rotating D3-brane obtained from the black brane thermodynamics. The result of this paper can be generalized to the rotating D3-brane with multiple nonzero rotating parameters.

We have not studied the Ramond-Ramond charge fluxes of the rotating D3-brane from the method of anomaly cancellation in this paper. If the scalar field carries the Ramond-Ramond charge, then it will couple with the Ramond-Ramond gauge field of the rotating D3-brane. Due to the fact that the explicit forms of the covariant couplings of quantum fields with the Ramond-Ramond gauge fields are different from that of the $U(1)$ gauge field, and the anomaly equations of the Ramond-Ramond charge currents are also different from that of the electric charge current, there are some difficulties with using the method of anomaly cancellation directly to calculate the Ramond-Ramond charge fluxes of the rotating D3-brane at this time. However, the energy-momentum flux from the gravitational anomaly has composed the main part of the whole radiation. The result of this paper shows that as many other black holes and black-hole-like objects with event horizons [9–24], the Hawking radiation of rotating D3-brane can be interpreted from the method of anomaly cancellation.

References

[1] Hawking, S. W. *Nature* (London) 1974, 248, 30;
Hawking, S. W. *Commun. Math. Phys.* 1975, 43, 199.

[2] Wald, R. M. *Commun. Math. Phys.* 1975, 45, 9.

[3] DeWitt, B. S. *Phys. Rep.* 1975, 19, 295.

[4] Unruh, W. G. *Phys. Rev. D* 1976, 14, 870.

[5] Christensen, S. M.; Fulling, S. A. *Phys. Rev. D* 1977, 15, 2088.

[6] Parikh, M. K.; Wilczek, F. *Phys. Rev. Lett.* 2000, 85, 5042;
Parikh, M. K. *Int. J. Mod. Phys. D* 2004, 13, 2351.

[7] Robinson, S. P.; Wilczek, F. *Phys. Rev. Lett.* 2005, 95, 011303;
Iso, S.; Umetsu, H.; Wilczek, F. *Phys. Rev. Lett.* 2006, 96, 151302.

[8] Iso, S.; Umetsu, H.; Wilczek, F. *Phys. Rev. D* 2006, 74, 044017.

[9] Murata, K.; Soda, J. *Phys. Rev. D* 2006, 74, 044018.

[10] Xu, Z.; Chen, B. *Phys. Rev. D* 2007, 75, 024041.

[11] Iso, S.; Morita, T.; Umetsu, H. *J. High Energy Phys.* 2007, 04, 068.

[12] Setare, M. R. *Eur. Phys. J. C* 2007, 49, 865.

[13] Kui, X.; Liu, W.; Zhang, H. *Phys. Lett. B* 2007, 647, 482.

[14] Jiang, Q. Q.; Wu, S. Q. *Phys. Lett. B* 2007, 647, 200;
Jiang, Q. Q.; Wu, S. Q.; Cai, X. *Phys. Lett. B* 2007, 651, 65;
Jiang, Q. Q.; Wu, S. Q.; Cai, X. *Phys. Rev. D* 2007, 75, 064029.

[15] Vagenas, E. C.; Das, S. *J. High Energy Phys.* 2006, 10, 025;
Das, S.; Robinson, S. P.; Vagenas, E. C. *Int. J. Mod. Phys. D* 2008, 17, 533.

[16] Shin, H.; Kim, W. *J. High Energy Phys.* 2007, 06, 012;
Kim, W.; Shin, H. *J. High Energy Phys.* 2007, 07, 070.

[17] Banerjee, R.; Kulkarni, S. *Phys. Lett. B* 2008, 659, 827;
Banerjee, R.; Kulkarni, S. *Phys. Rev. D* 2008, 77, 024018.

[18] Gangopadhyay, S.; Kulkarni, S. *Phys. Rev. D* 2008, 77, 024038;
Gangopadhyay, S. *Phys. Rev. D* 2008, 78, 044026.

[19] Peng, J. J.; Wu, S. Q. *Phys. Lett. B* 2008, 661, 300;
Wu, S. Q.; Peng, J. J.; Zhao, Z. Y. *Class. Quant. Grav.* 2008, 25, 135001.

[20] Ma, Z. Z. *Int. J. Mod. Phys. A* 2008, 23, 2783.

[21] Murata, K.; Miyamoto, U. *Phys. Rev. D* 2007, 76, 084038;
Miyamoto, U.; Murata, K. *Phys. Rev. D* 2008, 77, 024020.

[22] Iso, S.; Morita, T.; Umetsu, H. *Phys. Rev. D* 2007, 75, 124004;
Iso, S.; Morita, T.; Umetsu, H. *Phys. Rev. D* 2007, 76, 064015;
Iso, S.; Morita, T.; Umetsu, H. *Phys. Rev. D* 2008, 77, 045007.

[23] Kim, W.; Shin, H.; Yoon, M. *J. Korean Phys. Soc.* 2008, 53, 1791;
Bonora, L.; Cvitan, M. *J. High Energy Phys.* 2008, 05, 071;
Bonora, L.; Cvitan, M.; Pallua, S.; Smolic, I. *J. High Energy Phys.* 2008, 12, 021;
Kulkarni, S. *Class. Quant. Grav.* 2008, 25, 225023;
Iso, S. *Int. J. Mod. Phys. A* 2008, 23, 2082;
Umetsu, K. *Prog. Theor. Phys.* 2008, 119, 849.

[24] Shirasaka, A.; Hirata, T. (2008). Higher derivative correction to the Hawking flux via trace anomaly [hep-th]. arXiv: 0804.1910

Ghosh, S. (2008). Unifying the tunneling and anomaly phenomena in Hawking radiation [hep-th]. arXiv: 0804.2999

Banerjee, R.; Gangopadhyay, S.; Kulkarni, S. (2008). Hawking radiation and near horizon universality of chiral Virasoro algebra [hep-th]. arXiv: 0804.3492

Becar, R.; Gonzalez, P.; Pulgar, G.; Saavedra, J. (2008). Anomaly and Hawking radiation from Unruh's and canonical acoustic black hole [hep-th]. arXiv: 0808.1735

Akhmedova, V.; Pilling, T.; de Gill, A.; Singleton, D. (2008). Comments on anomaly versus WKB/tunneling methods for calculating Unruh radiation [hep-th]. arXiv: 0808.3413

Banerjee, R.; Kulkarni, S. (2008). Hawking radiation, covariant boundary conditions and vacuum states [hep-th]. arXiv: 0810.5683

Morita, T. (2008). Hawking radiation and quantum anomaly in AdS2/CFT1 correspondence [hep-th]. arXiv: 0811.1741

Porfyriadis, A. P. (2008). Hawking radiation via anomaly cancelation for the black holes of five-dimensional minimal gauged supergravity [hep-th]. arXiv: 0811.2822

Banerjee, R.; Majhi, B. R. (2008). Connecting anomaly and tunneling methods for Hawking effect through chirality [hep-th]. arXiv: 0812.0497

Papantonopoulos, E.; Skamagoulis, P. (2008). Hawking radiation via gravitational anomalies in non-spherical topologies [hep-th]. arXiv: 0812.1759

[25] Cvetič, M.; Youm, D. *Nucl. Phys. B* 1996, 477, 449.

[26] Russo, J. G. *Nucl. Phys. B* 1999, 543, 183.

[27] Russo, J. G.; Sfetsos, K. *Adv. Theor. Math. Phys.* 1999, 3, 131.

[28] Kraus, P.; Larsen, F.; Trivedi, S. P. *J. High Energy Phys.* 1999, 03, 003.

[29] Alvarez-Gaumé L.; Witten, E. *Nucl. Phys. B* 1984, 234, 269.

[30] Bertlmann, R. A.; Kohlprath, E. *Ann. Phys.* (N.Y.) 2001, 288, 137.

In: Black Holes and Galaxy Formation ISBN 978-1-60741-703-3
Editors: A.D. Wachter and R.J. Propst, pp. 331-353

Chapter 13

SCHWARZSCHILD-LIKE EXTERIORS FOR STARS IN KALUZA-KLEIN GRAVITY

J. Ponce de Leon[*]
Laboratory of Theoretical Physics, Department of Physics
University of Puerto Rico, P.O. Box 23343, San Juan,
PR 00931, USA

Abstract

In this work we examine the effective four-dimensional world that emanates from a general class of static spherical Ricci-flat solutions in Kaluza-Klein gravity in D-dimensions. By means of dimensional reduction we obtain a family of asymptotically flat Schwarzschild-like metrics for which all the components of the Ricci tensor, except for R_{11}, are zero. Although the reduced spacetime is not empty, it is similar to a vacuum in the sense that the effective matter satisfies an equation of state which is the generalization of $(\rho + 3p) = 0$ for "nongravitating matter" in $4D$. In Kaluza-Klein gravity these Schwarzschild-like metrics describe the exterior of a spherical star without rotation. In this framework, we generalize the well-known Buchdahl's theorem for perfect fluid spheres whose mass density does not increase outward. Without any additional assumptions, we develop the most general expression for the compactness limit of a star. We provide some numerical values for it, which in principle are observationally testable and allow us to compare and contrast different theories and exteriors. We find that in Kaluza-Klein gravity the compactness limit of a star can be larger than $1/2$, without being a black hole: the general-relativistic upper limit $M/R < 4/9$ is increased as we go away from the Schwarzschild vacuum exterior. We show how this limit depends on the number of dimensions of spacetime, and demonstrate that the effects of gravity are stronger in $4D$ than in any other number of dimensions.

PACS: 04.50.+h; 04.20.Cv
Keywords: Kaluza-Klein Theory; General Relativity; Stellar Models.

[*]E-mail address: jpdel@ltp.upr.clu.edu; jpdel1@hotmail.com

1. Introduction

The concept of black holes in the context of general relativity theory arises with the well-known Schwarzschild solution, which according to Birkhoff's theorem is the most general spherically symmetric, asymptotically flat, vacuum solution of the Einstein field equations. It describes the gravitational field outside an isolated spherical, non-rotating mass or black hole. For a perfect fluid sphere with a mass density which does not increase outwards, Buchdahl [1] showed that the ratio of its gravitational mass M to the coordinate radius R satisfies the inequality $GM/R \leq 4/9$, i.e. under these conditions no static spherically symmetric star is possible in general relativity with a radius less than 9/8 of the Schwarzschild radius $R_{Schw} = 2GM$. By definition, any non-rotating and non-charged mass that is smaller than the Schwarzschild radius (horizon) constitutes a black hole.

Nowadays there are a number of non-equivalent theories that suggest the existence of extra dimensions. Kaluza-Klein gravity [2]-[4] and braneworld theory [5]-[8] are well-known examples. The study of stellar structure and stellar evolution might constitute an important approach to predict observable effects from extra dimensions. However, there is a fundamental limitation. Namely, that Birkhoff's theorem is no longer valid in more than four dimensions, i.e. there is no an unique asymptotically flat vacuum solution with spatial spherical symmetry. As a consequence, the effective picture in four dimensions allows the existence of different possible non-Schwarzschild scenarios for the description of the spacetime outside of a spherical star [9]-[21].

Therefore, the study of stars in these theories leads to two main questions [22]:

1. How does the existence of stars restrict the class of admissible non-Schwarzschild exteriors?

2. How does a possible deviation from the Schwarzschild vacuum exterior affect the star parameters?

In a recent series of papers we have investigated different aspects of these questions. In the framework of five-dimensional Kaluza-Klein gravity, without making any assumption about the stellar structure, we have shown that the condition that in the weak-field limit we recover the usual Newtonian physics singles out an *unique* effective exterior for a spherically symmetric star [22], [23]. On the other hand, in the framework of Randall-Sundrum II braneworld scenario, without making any assumption about the bulk, or the material medium inside the star, we have proved that for *any* nonstatic spherical star, without rotation, there are only two possible static exteriors; these are the Schwarzschild and the 'Reissner-Nordström-like' exteriors [24]. This is quite distinct from the case of stars in hydrostatic equilibrium which admit a much larger family of non-Schwarzschild static exteriors.

We have also studied the second question mentioned above in the context of static, perfect fluid, spherical stars of *uniform density* [22], [25]. We have shown that, in principle, the compactness limit of such a star can be larger than $1/2$, without being a black hole . In this work we generalize this result to *any* static spherical perfect fluid star in Kaluza-Klein gravity.

Thus, here we get rid of the unrealistic assumption of uniform density, but keep the isotropy condition[1]. The generalization is attained by using an extension, to more than four dimensions, of an inequality originally discovered by Buchdahl [1]. This inequality and the matching conditions allow us to obtain the compactness limit of a star for *any* given exterior spacetime. This limit is important because it indicates how much perfect fluid matter can be packed in a given volume, without provoking gravitational collapse. Here we develop the most general expression for this limit and provide some numerical values for it, which in principle are observationally testable and allow us to compare and contrast different theories and exteriors.

This paper is organized as follows. In section 2, we derive and analyze a general class of static, spherical vacuum solutions in Kaluza-Klein gravity in D-dimensions, where the metric functions are independent of the extra coordinates. They generalize some well-known solutions in the literature as the Schwarzschild-Tangherlini spacetimes and other ones with spherical symmetry in the three usual space dimensions. In section 3, we carry out the dimensional reduction of the solutions. We obtain a family of asymptotically flat Schwarzschild-like metrics for which all the components of the Ricci tensor, except for R_{11}, are zero. In section 4, we use them to construct the spacetime outside of a spherical star. In this framework, we generalize the well-known Buchdahl's theorem for perfect fluid spheres whose mass density does not increase outward. Without any additional assumptions, we show that the compactness limit of a star in Kaluza-Klein gravity can be larger than $1/2$, without being a black hole: the general-relativistic upper limit $M/R < 4/9$ increases as we go away from the Schwarzschild vacuum exterior. Our results are consistent with our previous findings and show that, as in general relativity, all these inequalities can be saturated in the case of uniform proper density from the condition that the isotropic pressure does not become infinity at the center. In section 5 we summarize our results. Finally, in the Appendix following our previous work [28] we present some general inequalities for the metric functions in the interior of a static star, which extend to D-dimensions those obtained by Buchdahl.

2. Static Spherical Vacuum Solutions in Kaluza-Klein Gravity

In five-dimensional general relativity the most general time-independent vacuum solutions, with spherical symmetry in the three usual space dimensions and a Killing vector in the fifth direction, have been found and classified by Chodos and Detweiler [29]. The physical properties of these solutions with zero electric charge have been discussed by a number of authors [4], [30]-[39]. Although they are frequently described as black holes, many researchers have noticed that instead they represent naked singularities because the event horizon is reduced to a singular point at the center of ordinary space [40]-[43]. Truly higher dimensional extensions of the Schwarzschild black hole metric have been obtained by Tangherlini [44] and generalized by Myers and Perry [45], who considered spacetimes with spherical symmetry in $(D-1)$, rather than three spatial dimensions.

In this work we consider static spacetimes with topology $R^1 \times S^{n+3} \times K^m$, where R^1 corresponds to the time dimension and K^m is a m-dimensional manifold. The $D =$

[1]Is important to emphasize the role of isotropic pressures, because for anisotropic pressures there is no upper bound on the gravitational potential of a star. See [26], [27] and references therein.

$(4+n+m)$ metric is assumed to be independent of the $(n+m)$ extra coordinates and in isotropic coordinates can be written as

$$dS^2 = A^2(r)dt^2 - B^2(r)\left[dr^2 + r^2 d\Omega^2_{(n+2)}\right] - C^2(r)\delta_{ab}dy^a dy^b. \tag{1}$$

Here n and m represent the number of "internal" and "external" dimensions, respectively; $d\Omega_{(2+n)}$ is the metric on a unit $(n+2)$-sphere $(n = 0, 1, 2..)$; the coordinates along the external dimensions are denoted as $y^{a,b}$ with $a, b = 1, 2...m$. These spacetimes generalize the ones discussed by Kramer [46], Davidson and Owen [47], Chatterjee [42], Millward [48], Dereli [40], Tangherlini [44], as well as the non-rotating black holes of and Myers and Perry [45].

Notation: Before going on, it is useful to establish our notations and conventions: Indices labeled by Latin capital letters A, B run over the full $D = (4+n+m)$ space; indices labeled by $\tilde{\mu}, \tilde{\nu}$ run over the $(4+n)$ subspace and the metric is called $\gamma_{\tilde{\mu}\tilde{\nu}}$; the indices in the conventional spacetime, with signature is $(+,-,-,-)$, are labeled $\mu, \nu = (0, 1, 2, 3)$; primes denote differentiation with respect to r; we follow the definitions of Landau and Lifshitz [49]; and the speed of light, as well as the gravitational constant are taken to be unity.

The metric functions in (1) are solutions of the vacuum Einstein field equations $R_{AB} = 0$. The non-vanishing equations are

$$\frac{A''}{A'} + (n+1)\frac{B'}{B} + \frac{n+2}{r} + m\frac{C'}{C} = 0, \tag{2}$$

$$\frac{A''}{A} + (n+2)\frac{B''}{B} + m\frac{C''}{C} + \frac{B'}{B}\left[\frac{n+2}{r} - \frac{A'}{A} - (n+2)\frac{B'}{B} - m\frac{C'}{C}\right] = 0, \tag{3}$$

$$\frac{B''}{B} + \frac{B'}{B}\left[\frac{2n+3}{r} + \frac{A'}{A} + n\frac{B'}{B} + m\frac{C'}{C}\right] + \frac{1}{r}\left[\frac{A'}{A} + m\frac{C'}{C}\right] = 0, \tag{4}$$

$$\frac{A'}{A} + (n+1)\frac{B'}{B} + \frac{n+2}{r} + (m-1)\frac{C'}{C} + \frac{C''}{C'} = 0, \tag{5}$$

which correspond to $R^0_0 = 0$, $R^1_1 = 0$, $R^2_2 = ... = R^{n+2}_{n+2} = 0$ and $R^a_a = 0$, respectively. Now combining (2) and (5) we get an equation that can be easily integrated, viz., $A'/A = -k\ C'/C$, where k is a constant of integration. Thus, $A \propto C^{-k}$ and $B^{n+1} \propto C^{k+1-m}/r^{n+2}C'$. Substituting into (4) we obtain an equation for $C(r)$, whose solution is

$$C(r) = C_0\left(\frac{ar^{n+1}+1}{ar^{n+1}-1}\right)^{\sigma}, \tag{6}$$

where a, C_0 are constants of integration and σ is subjected to the condition

$$\sigma^2\left[(n+2)k^2 - 2km + m(m+n+1)\right] = n+2, \tag{7}$$

which is imposed by (3). The final form of the remaining metric functions is given by

$$A(r) = \left(\frac{ar^{n+1} - 1}{ar^{n+1} + 1}\right)^{\sigma k}, \quad B^{n+1}(r) = \frac{1}{a^2 r^{2(n+1)}} \frac{(ar^{n+1} + 1)^{\sigma(k-m)+1}}{(ar^{n+1} - 1)^{\sigma(k-m)-1}}. \tag{8}$$

We will see bellow that the constant a is related to the total gravitational mass. Therefore, in what follows we take $a > 0$. Also, without loss of generality we set $C_0 = 1$. In the case where $D = 5$ with $n = 0$ and $m = 1$, (7) reduces to $\sigma^2(k^2 - k + 1) = 1$ which is the consistency condition in Davidson-Owen solution [47].

In the above equations k is an arbitrary real number, i.e.

$$-\infty < k < \infty, \tag{9}$$

and, as a consequence of (7), σ is bounded from bellow and above. Namely,

$$|\sigma| \leq \sigma_{max} = \frac{n + 2}{\sqrt{m(n + 1)(n + m + 2)}}, \tag{10}$$

where the maximum value is attained at $k = m/(n + 2)$. Thus, the range of σ decreases, moving closer to zero, with the increase of m. Therefore, for any fixed n the effects of the external extra dimensions (measured by the deviation of σ from zero) become weaker with the increase of m. Also, for $m = 1$, σ_{max} steadily decreases with the increase of n. However, for $m > 1$ this is not so. Indeed, in this case σ_{max} first decreases and subsequently increases with the increase of n. As an illustration, let us take

$$m = (1, 2, 5, 7, 11),$$

then

$$n = 0, \quad |\sigma| \leq (1.15, 0.71, 0.34, 0.25, 0.17),$$

$$n = 1, \quad |\sigma| \leq (1.06, 0.67, 0.34, 0.25, 0.17),$$

$$n = 5, \quad |\sigma| \leq (1.01, 0.67, 0.37, 0.29, 0.20).$$

The behavior of the metric functions near $ar^{n+1} \sim 1$ depends on the choice of σ and k. For example, when $\sigma > 0$ we find that $B \to 0$ as $ar^{n+1} \to 1^+$, in the whole range of k. However, when $\sigma < 0$ the same limit gives: (i) $B \to \infty$ for $k < (m-1)/2$, (ii) $B \to 0$ for $k > (m-1)/2$, and (iii) $B \to 2^{2/(n+1)}$ for $k = (m-1)/2$. In addition, for $\sigma > 0$, $k \geq 0$ we find that $A \to 0$, $B \to 0$, $C \to \infty$ as $ar^{n+1} \to 1^+$ for any number of dimensions n and m. Consequently, in the allowed range of σ and k, there are several families of solutions with different geometrical and physical properties. For $n = 0$ and $m = 1$, these have been thoroughly discussed in the literature [38], [39].

A simple classification of the solutions can be obtained from the analysis of the physical radius $R(r)$ of a $(n + 2)$-sphere. In the present case it is given by

$$R(r) = rB(r) = \frac{1}{a^{2/(n+1)} r} \left[\frac{(ar^{n+1} + 1)^{\sigma(k-m)+1}}{(ar^{n+1} - 1)^{\sigma(k-m)-1}}\right]^{1/(n+1)}. \tag{11}$$

For $ar^{n+1} \gg 1$, $R \approx r$, regardless of the number of dimensions and the choice of σ and k. However, near $ar^{n+1} \sim 1$ we find

$$\lim_{ar^{n+1}\to 1^+} R(r) = \begin{cases} 0, & \text{for } \sigma > 0, -\infty < k < \infty, \text{ and } \sigma < 0,\ k > (m-1)/2, \\ \infty, & \text{for } \sigma < 0, -\infty < k < (m-1)/2, \\ \left(\frac{4}{a}\right)^{1/(n+1)}, & \text{for } \sigma = 0, \text{ and } \sigma < 0,\ k = (m-1)/2. \end{cases} \tag{12}$$

For the parameters σ, k in the first expression of (12) we cannot interpret (6)-(8) as a black hole solution because the "event horizon"[2] $ar^{n+1} = 1$ occurs at $R = 0$, which by itself is a singular point, line, plane...for $m = 0, 1, 2..$ etc. The second expression in (12) is a consequence of the fact that dR/dr vanishes at some finite value of r, say $\bar{r}$. Since $R_{min} = R(\bar{r}) > 0$, the solutions with $\sigma < 0$, $k < (m-1)/2$ can be used to generate the higher dimensional counterpart of the wormholes solutions discussed by Agnese *et al* [34]. The solutions with $\sigma = 0$ and $\sigma < 0$, $k = (m-1)/2$ are specially simple because now k and σ are fixed. Bellow we discuss them separately.

Solutions with $\sigma = 0$: In the limit $\sigma \to 0$, from (7) it follows that $\sigma^2 k^2 \to 1$. If we take $\sigma k = 1$, then

$$A(r) = \frac{ar^{n+1} - 1}{ar^{n+1} + 1}, \quad B^{n+1}(r) = \frac{(ar^{n+1} + 1)^2}{a^2 r^{2(n+1)}}. \tag{13}$$

Far away from a stationary source $g_{00} \sim (1 + 2\phi)$, where ϕ is the Newtonian gravitational potential which goes as $-M/r^{n+1}$. Therefore, in the present case the total gravitational mass M is given by

$$M = \frac{2}{a}. \tag{14}$$

(If we had taken $\sigma k = -1$, we would have obtained a negative mass, viz., $M = -2/a$). With the transformation

$$R = \frac{\left[ar^{n+1} + 1\right]^{2/(n+1)}}{a^{2/(n+1)} r}, \tag{15}$$

the solution becomes

$$dS^2 = \left(1 - \frac{2M}{R^{n+1}}\right) dt^2 - \left(1 - \frac{2M}{R^{n+1}}\right)^{-1} dR^2 - R^2 d\Omega^2_{(n+2)} - \delta_{ab} dy^a dy^b, \tag{16}$$

which, up to the innocuous m flat extra dimensions, describes the so-called Schwarzschild-Tangherlini black holes with spherical symmetry in $(n+3)$ rather than three spatial dimensions. The radius R_h of the horizon of the black hole is given by $R_h = (4/a)^{1/(n+1)}$, which in isotropic coordinates corresponds to $ar_h^{n+1} = 1$, as expected. For $n = 0$ they reduce to the conventional Schwarzschild solution of general relativity.

[2]For black holes in general relativity, the event horizon is defined as the surface where the norm of the timelike Killing vector vanishes. In our case the Killing vector is just $(1, 0, 0, ...0)$ so its norm vanishes where g_{00} does.

Solutions with $\sigma < 0$, $k = (m-1)/2$: Substituting $k = (m-1)/2$ into (7) we find $\sigma = -2/(m+1)$. To illustrate the properties of this class of solutions let us momentarily take $n = 0$, $m = 2$. For this choice, using (6)-(8) we obtain

$$dS^2 = \left(\frac{ar+1}{ar-1}\right)^{2/3} dt^2 - \frac{(ar+1)^4}{a^4 r^4}\left[dr^2 + r^2 d\Omega^2_{(2)}\right] - \left(\frac{ar-1}{ar+1}\right)^{4/3}\left[dy_1^2 + dy_2^2\right]. \tag{17}$$

The interpretation of this metric is difficult because far from the origin $g_{00} \sim [1 + (4/3ar)]$, which implies a negative mass parameter, viz., $M = -(2/3a)$. However, using a double Wick rotation $t \rightarrow iy_1$, $y_1 \rightarrow it$ we generate a solution of the field equations with positive mass. Namely,

$$\begin{aligned} dS^2 &= \left(\frac{ar-1}{ar+1}\right)^{4/3} dt^2 - \frac{(ar+1)^4}{a^4 r^4}\left[dr^2 + r^2 d\Omega^2_{(2)}\right] \\ &- \left(\frac{ar+1}{ar-1}\right)^{2/3} dy_1^2 - \left(\frac{ar-1}{ar+1}\right)^{4/3} dy_2^2, \end{aligned} \tag{18}$$

for which $M = (4/3a)$. Using the transformation of coordinates $aR = [(ar+1)^2/ar]$ this metric becomes

$$\begin{aligned} dS^2 &= \left(1 - \frac{3M}{R}\right)^{2/3} dt^2 - \frac{dR^2}{(1 - 3M/R)} - R^2 d\Omega^2_{(2)} \\ &- \frac{dy_1^2}{(1 - 3M/R)^{1/3}} - \left(1 - \frac{3M}{R}\right)^{2/3} dy_2^2. \end{aligned} \tag{19}$$

We see that $g_{TT} = 0$ and $g_{RR} = -\infty$ at $R = 3M$. However, this is *not* a horizon but a singularity. This may be verified by evaluating the invariant geometric scalars. For example, the Kretschmann curvature scalar $I = R_{ABCD}R^{ABCD}$ is

$$I = \frac{12M^2(9R^2 - 52RM + 76M^2)}{(R - 3M)^2 R^6}, \tag{20}$$

which is manifestly divergent at $R = 3M$. The above analysis can be extended to any n and m. The result is that the line element (19) is a member of the family of solutions

$$\begin{aligned} dS^2 &= F^{2/(m+1)} dt^2 - F^{-1} dR^2 - R^2 d\Omega^2_{(n+2)} \\ &- F^{(1-m)/(1+m)} dy_1^2 - F^{2/(m+1)}\left[dy_2^2 + dy_3^2 + \ldots.. + dy_m^2\right], \end{aligned} \tag{21}$$

with[3]

$$F = 1 - \frac{(m+1)M}{R^{n+1}}. \tag{22}$$

[3] A generalization of Schwarzschild-Tangherlini's spacetimes to a chain of several Ricci-flat internal spaces has been obtained in [50]. In turn, those generalized solutions are extended to include a massless scalar field in [51].

These are singular at $R = [(m+1)M]^{1/(n+1)}$ for any $m \neq 1$. However, for $m = 1$ they yield Schwarzschild-Tangherlini spacetimes with one flat extra dimension. It is important to note that (16) and (21) are the *only* solutions for which $g_{TT} = 0$ and $g_{RR} = -\infty$ in the same region of the spacetime. In fact, from (8) and (11) it follows that

$$B^2(r)dr^2 = \frac{\left[a^2 r^{2(n+1)} - 1\right]^2 dR^2}{\left[a^2 r^{2(n+1)} + 2\sigma(m-k)ar^{a(n+1)} + 1\right]^2} = -g_{RR}dR^2. \tag{23}$$

Thus, in general $g_{RR} \to 0$ as $ar^{(n+1)} \to 1^+$, except in the case where the denominator in (23) also vanishes in this limit, i.e. for

$$1 + \sigma(m-k) = 0. \tag{24}$$

This occurs in two cases only: (i) $\sigma = 0$, ($\sigma k = 1$) which corresponds to Schwarzschild-Tangherlini's spacetimes, and (ii) $\sigma(k-m) = 1$, which by virtue of (7) implies $k = (m-1)/2$ and thus generates the family (21)-(22).

To conclude this section, we notice that for $k = 0$ the solution (6)-(8) generalizes the well-known zero-dipole moment soliton of Gross and Perry [52] to any number of internal dimensions[4]. As an illustration, let us take $m = 1$,

$$dS^2 = dt^2 - \left(1 - \frac{1}{ar^{n+1}}\right)^{4/(n+1)} \left[dr^2 + r^2 d\Omega^2_{(n+2)}\right] - \left(\frac{ar^{n+1}+1}{ar^{n+1}-1}\right)^2 dy_1^2. \tag{25}$$

We note that here a does not have to be positive. If we assume $a = -\alpha^2 < 0$ and perform a double Wick rotation $t \to iy_1$, $y_1 \to it$, then we obtain the metric

$$dS^2 = \left(\frac{1-\alpha^2 r^{n+1}}{1+\alpha^2 r^{n+1}}\right)^2 dt^2 - \left(1 + \frac{1}{\alpha^2 r^{n+1}}\right)^{4/(n+1)} \left[dr^2 + r^2 d\Omega^2_{(n+2)}\right] - dy_1^2, \tag{26}$$

which after a simple coordinate transformation reduces to (16) with mass parameter $M = 2/\alpha^2$. This suggests that, as in $5D$, the above $(4+n)$ metrics represent limiting configurations that are unstable to metric perturbations [39].

3. Dimensional Reduction

The class of Kaluza-Klein solutions discussed in the previous section clearly demonstrates the wealth of possible physical scenarios in higher dimensions. In order to study their observational implications, and test possible deviations from general relativity, we have to examine the effective four-dimensional world that emanates from them. Thus, we consider the case where $n = 0$, which physically corresponds to spherical symmetry in the three usual spatial dimensions. For the sake of generality, we also discuss the dimensional reduction for $n \neq 0$. Also, at the end of this section, we discuss some similarities and differences between spacetimes with different n.

[4]To avoid a misunderstanding we note that the Kaluza-Klein monopole of Gross, Perry and Sorkin [53] does not belong to the class of spacetimes investigated here.

To this end, we recall that the dimensional reduction of $R_{(D)}$, the curvature scalar associated with the metric

$$dS^2 = \gamma_{\mu\nu}(x)dx^\mu dx^\nu - \sum_{i=1}^{m} N_i^2(x)dy_i^2, \tag{27}$$

can be expressed as

$$\sqrt{|g_{(D)}|}\, R_{(D)} \propto \sqrt{|g_{(4)}^{\text{eff}}|}\, R_{(4)} + \text{other terms}, \tag{28}$$

where $R_{(4)}$ is the four dimensional curvature scalar calculated from the effective $4D$ metric tensor[5]

$$g_{\mu\nu}^{\text{eff}} = \gamma_{\mu\nu} \prod_{i=1}^{m} N_i; \tag{29}$$

$g_{(D)}$ and $g_{(4)}^{\text{eff}}$ denote the determinants of the D-dimensional metric (27) and effective $4D$ metric (29), respectively, and the "other terms" are proportional to the Lagrangian of an effective energy-momentum tensor (EMT) $T_{\mu\nu}$ in $4D$. The exact form for these terms is, in general, quite cumbersome. However, if we use the fact that the metric functions satisfy the field equations $R_{AB} = 0$, then we obtain a nice expression for (28), viz.,

$$\sqrt{|g_D|}R_{(D)} = \sqrt{|g_{(4)}^{\text{eff}}|}\left[R_{(4)} - \sum_{a=1}^{m}\frac{\partial_\mu N_a \partial^\mu N_a}{N_a^2} - \frac{1}{2}\sum_{a=1}^{m}\sum_{b=1}^{m}\left(\frac{\partial_\mu N_a}{N_a}\right)\left(\frac{\partial^\mu N_b}{N_b}\right)\right]. \tag{30}$$

Here $D = 4 + m$ and $m \geq 1$. It shows that the choice of the factor $\prod_{i=1}^{m} N_i$ in (29) assures that the first term in (30) yields the conventional general relativity action. In the case under consideration $N_i = C(x)$. Therefore, the physics in $4D$ can be extracted from the four-dimensional effective action

$$S_{(4)} = -\frac{1}{k_4}\int d^4x\sqrt{|g_{(4)}^{\text{eff}}|}\left[R_{(4)} - m\left(1 + \frac{m}{2}\right)\frac{\partial_\mu C\, \partial^\mu C}{C^2}\right], \tag{31}$$

where k_4 is a positive constant. From the variational principle, $\delta S_{(4)} = 0$, we get the effective equations in $4D$, viz.,

$$R_{\mu\nu} - \frac{1}{2}g_{\mu\nu}R = \frac{m(m+2)}{2C^2}\left[\partial_\mu C\, \partial_\nu C - \frac{1}{2}g_{\mu\nu}\partial_\alpha C\, \partial^\alpha C\right] \equiv 8\pi T_{\mu\nu}. \tag{32}$$

Here $g_{\mu\nu} \equiv g_{\mu\nu}^{\text{eff}}$, and $R_{\mu\nu}$ as well as R are calculated with $g_{\mu\nu}$.

Thus, for the Kaluza-Klein metric (1), (6), (8), with $n = 0$, the physics in $4D$ is concentrated in the effective line element

$$ds^2 = \left(\frac{ar-1}{ar+1}\right)^{2\varepsilon} dt^2 - \frac{1}{a^4r^4}\frac{(ar+1)^{2(\varepsilon+1)}}{(ar-1)^{2(\varepsilon-1)}}\left[dr^2 + r^2 d\Omega^2\right], \tag{33}$$

[5]For the effective action in $4D$ to contain the exact Einstein Lagrangian, i.e. to deal with a constant effective gravitational constant, $g_{\mu\nu}^{\text{eff}}$ should be identified with the physical metric in ordinary $4D$ spacetime [2], [47], [54].

with

$$\varepsilon \equiv \frac{\sigma(2k - m)}{2}. \tag{34}$$

The effective $4D$ energy-momentum tensor $T_{\mu\nu}$ can be calculated by substituting (33) into the l.h.s of (32). However, it is easier to use the r.h.s., viz.,

$$8\pi T_0^0 = \frac{m(m+2)}{4B^2 C^{m+2}} \left(\frac{dC}{dr}\right)^2, \quad T_1^1 = -T_0^0, \quad T_3^3 = T_2^2 = T_0^0. \tag{35}$$

Taking B and C from (6), (8) and using the condition (7), we find

$$8\pi T_0^0 = \frac{4a^6 r^4 (1 - \varepsilon^2)(ar - 1)^{2(\varepsilon - 2)}}{(ar + 1)^{2(\varepsilon + 2)}}. \tag{36}$$

It should be noted that for an observer in $4D$, who is not aware of the extra dimensions, ε is a free parameter in the solution and measures the deviation from Schwarzschild.

The line element (33) acquires a more familiar form in terms of the Schwarzschild-like coordinate R defined by

$$R = r\left(1 + \frac{1}{ar}\right)^2, \tag{37}$$

and setting $a = (2/\varepsilon M)$. Indeed, (33) becomes

$$ds^2 = \left(1 - \frac{2M/\varepsilon}{R}\right)^{\varepsilon} dt^2 - \frac{dR^2}{\left(1 - \frac{2M/\varepsilon}{R}\right)^{\varepsilon}} - R^2 \left(1 - \frac{2M/\varepsilon}{R}\right)^{1-\varepsilon} \left(d\theta^2 + \sin^2\theta d\phi^2\right). \tag{38}$$

This $4D$ metric has been obtained previously by the present author in the context of Davidson-Owen solutions [47] (these correspond to $m = 1$). This means that the effective metric in $4D$ is independent of the number of external extra coordinates. It "generalizes" the Schwarzschild vacuum metric ($\varepsilon = 1$), and has been widely discussed as a possible non-Schwarzschild exterior for stellar models within the context of Kaluza-Klein gravity [22]-[23]. We have shown that (38) is compatible with (i) Newtonian physics, in the weak-field limit; (ii) the general-relativistic Schwarzschild limit for $\varepsilon = 1$; (iii) the dominant energy condition, and (iv) the (weak) equivalence principle, even in the non-Schwarzschild case where $\varepsilon \neq 1$.

In terms of R the EMT becomes

$$8\pi T_0^0 = \frac{(1 - \varepsilon^2) M^2}{\varepsilon^2 R^4} \left(1 - \frac{2M/\varepsilon}{R}\right)^{(\varepsilon - 2)}, \quad T_1^1 = -T_0^0, \quad T_3^3 = T_2^2 = T_0^0. \tag{39}$$

Since $(T_0^0 - T_1^1 - T_2^2 - T_3^3) = 0$, this spacetime is 'similar' to Schwarzschild vacuum in the sense that it has no effect on gravitational interactions. However, it is more realistic because instead of being absolutely empty $T_0^0 = T_1^1 = T_2^2 = T_3^3 = 0$, it is consistent with the existence of quantum zero-point fields [55]. It is not difficult to show that (38) and (39) are equivalent to the static, spherically symmetric solution of the coupled Einstein-massless scalar field equations originally discovered by Fisher [56] and rediscovered by Janis, Newman and Winicour [57].

3.0.1. Dimensional Reduction for $n > 0$

Although from an observational point of view the models with $n > 0$ might not be of prime concern, it is of theoretical interest to study how the physics depends on the number of internal dimensions.

It is not difficult to verify that $g^{\text{eff}}_{\tilde{\mu}\tilde{\nu}}$, the effective metric in $D = (4+n)$, is obtained from $\gamma_{\tilde{\mu}\tilde{\nu}}$ as

$$g^{\text{eff}}_{\tilde{\mu}\tilde{\nu}} = \gamma_{\tilde{\mu}\tilde{\nu}} \prod_{i=1}^{m} N_i^{q(n)}, \quad q(n) = \frac{2}{n+2}. \tag{40}$$

Similar to the discussion leading to (30), with this choice the dimensional reduction of the m external dimensions yields

$$\sqrt{|g_{(D)}|}\, R_{(D)} \propto \sqrt{|g^{\text{eff}}_{(4+n)}|}\left[R_{(4+n)} + \text{constant} \times \frac{\partial_\mu C\, \partial^\mu C}{C^2}\right], \tag{41}$$

where the constants depend on the choice of n and m. In this way, the gravitational action has the standard form

$$S_{(D)} = -\frac{1}{k_D}\int d^D x \sqrt{|g^{\text{eff}}_{(D)}|}\, R_{(D)}, \tag{42}$$

in any number of dimensions.

In the case under consideration $g^{\text{eff}}_{00} = C^{2m/(n+2)}A^2$, $g^{\text{eff}}_{11} = -C^{2m/(n+2)}B^2$, $g^{\text{eff}}_{22} = -C^{2m/(n+2)}B^2 r^2$, etc. Thus the effective gravity in $4+n$ is governed by the line element

$$ds^2 = \left(\frac{ar^{n+1}-1}{ar^{n+1}+1}\right)^{2\varepsilon} dt^2 - \frac{1}{(ar^{n+1})^{4/(n+1)}}\frac{(ar^{n+1}+1)^{2(\varepsilon+1)/(n+1)}}{(ar^{n+1}-1)^{2(\varepsilon-1)/(n+1)}}\left[dr^2 + r^2 d\Omega^2_{(n+2)}\right], \tag{43}$$

where

$$\varepsilon = \frac{\sigma\left[(n+2)k - m\right]}{n+2}. \tag{44}$$

From the asymptotic behavior of the metric we get the total mass

$$M = \frac{2\varepsilon}{a}. \tag{45}$$

Now, introducing the Schwarzschild-like coordinate

$$R = r\left(1 + \frac{1}{ar^{n+1}}\right)^{2/(n+1)} \tag{46}$$

the metric becomes

$$ds^2 = \left(1 - \frac{2M/\varepsilon}{R^{n+1}}\right)^{\varepsilon} dt^2 - \left(1 - \frac{2M/\varepsilon}{R^{n+1}}\right)^{-(n+\varepsilon)/(n+1)} dR^2 - R^2\left(1 - \frac{2M/\varepsilon}{R^{n+1}}\right)^{(1-\varepsilon)/(n+1)} d\Omega^2_{(n+2)} \tag{47}$$

For $n = 0$ we recover (38). Also, for $n \neq 0$ and $\varepsilon = 1$ it reduces to the Schwarzschild-Tangherlini black hole solutions (16). For any other ϵ the effective $(4+n)$ spacetime is not empty. In fact, the EMT is given by

$$8\pi T_0^0 = \frac{(n+1)(n+2)(1-\varepsilon^2)M^2}{2\,\varepsilon^2 R^{2(n+2)}}\left(1 - \frac{2M/\varepsilon}{R^{n+1}}\right)^{(\varepsilon-n-2)/(n+1)},$$
$$T_1^1 = -T_0^0, \qquad T_2^2 = T_3^3 = \cdots = T_{n+3}^{n+3} = T_0^0, \tag{48}$$

which can be interpreted as a massless scalar field in $(4+n)$ dimensions. For $n = 0$ it reduces to (39), as expected.

3.1. Properties of the Effective Spacetimes

It should be noted that the dimensional reduction eradicates the geometrical and physical differences between the three families of higher-dimensional solutions discussed in (12). In particular, the effective $(4+n)$ spacetime shows no evidence of the different nature of the singularity of g_{RR} near $ar^{n+1} = 1$, which is revealed by (23). Thus, regardless of their specific properties, all the solutions discussed in section 2 yield the same effective spacetime in $(4+n)$.

From (48) we find that the components of the EMT satisfy the equation

$$T = (n+2)T_0^0. \tag{49}$$

Substituting this into (A-1) we find that the effective spacetimes satisfy the equations

$$R_0^0 = R_2^2 = R_3^3 = \cdots = R_{n+3}^{n+3} = 0, \quad R_1^1 = 16\pi T_1^1. \tag{50}$$

We now recall that in the case of a constant, asymptotically flat, gravitational field there is an expression for the total energy of matter plus field, which is an integral of R_0^0 over the volume V occupied by the matter[6] [49], viz., $M = \alpha\int\sqrt{|g|}R_0^0 dV$, where the constant of proportionality α depends on the number of dimensions, e.g., $\alpha = 1/4\pi$ in $4D$. In conventional general relativity this expression is known as the Tolman-Wittaker formula.

Since $R_0^0 = 0$, it follows that the gravitational mass of any spherical shell is just zero. This conclusion holds for any n and ε, which include the Schwarzschild-Tangherlini black holes, as well as the familiar Schwarzschild solution of general relativity. Clearly, this is a consequence of the fact that the scalar field is *massless*.

We note that

$$R_0^0 = \rho + p_r + (n+2)p_\perp = 0, \quad p_r = -T_1^1, \quad p_\perp = -T_2^2, \tag{51}$$

generalizes to n dimensions the well-known equation of state $(\rho + p_r + 2p_\perp) = 0$ for nongravitating matter in $4D$, which in turn generalizes to anisotropic matter the equation of state $(\rho + 3p) = 0$ for a perfect fluid that has no effect on gravitational interactions [58]-[60].

[6] In Landau and Lifshitz [49] the discussion is in $4D$, however it can be extended to any number of dimensions

Finally, we note that at large distances from the origin, i.e. for $R \gg (2M)^{1/(n+1)}$, the line element (47) becomes

$$ds^2 = ds_0^2 - \frac{2M}{R^{n+1}} \left[dt^2 + \frac{(\varepsilon + n)}{\varepsilon(n+1)} dR^2 + \frac{(1-\varepsilon)}{\varepsilon(n+1)} R^2 \, d\Omega_{(n+2)} \right] + O\left(\left(\frac{M}{R^{n+1}} \right)^2 \right). \tag{52}$$

The second term represents a small correction to the Minkowski metric ds_0^2 in $(4+n)$. Since at large distances from the sources every field appears centrally symmetric, it follows that (52) determines the metric at large distances from any system of bodies [49]. For $\varepsilon = 1$ the correction is identical to the one in general relativity, for any n. However, this is not so for $\varepsilon \neq 1$. This could serve in astrophysical observations to detect possible deviations from general relativity.

4. Physical Interpretation

The metric of the effective spacetime (43) contains a singularity at $ar^{n+1} = 1$, which, in principle, can be visible to an external observer. Only in the Schwarzschild-Tangherlini limit ($\varepsilon \to 1$) it is covered by an event horizon; for any other value of ε the horizon is reduced to a singular point.

However, the presence of naked singularities makes everybody uncomfortable and according to the cosmic censorship hypothesis they should not be realized in nature. In order to avoid them, we have to exclude the central region and require $ar^{n+1} > 1$. What this suggests is that the asymptotically flat metric (43) should be used to describe the gravitational field outside of the core of a spherical matter distribution. The "interior" region has to be described by some solution of the field equations, which must be regular at the origin and not necessarily asymptotically flat.

In this interpretation, the effective exterior is not Ricci-flat (50), except for $\varepsilon = 1$. However, we have just seen that the exterior scalar field is gravitationally innocuous, in the sense that, as in conventional general relativity, any shell outside of the source carries zero gravitational mass.

We now proceed to investigate this interpretation. We use the standard boundary conditions to study the question of how a possible deviation from the Schwarzschild vacuum exterior can affect the star parameters. Instead of restricting our discussion to a particular equation of state for the stellar interior, here we only assume that (1) the matter inside the star is a perfect fluid, and that (2) the energy density is positive and does not increase outward. Under these conditions we will be able to extend the well-known Buchdahl's theorem of general relativity to stellar models in Kaluza-Klein gravity.

Thus, we assume that the spacetime outside of a spherical body is described by the line element (47), which we now denote us[7]

$$ds^2 = e^{\alpha(r)} dt^2 - e^{\beta(r)} dr^2 - r^2 e^{\mu(r)} \, d\Omega_{(n+2)}, \tag{53}$$

[7]Note that the r in (53), as well as in the rest of this section, represents what we have previously called 'Schwarzschild-like coordinate', e.g., in (47) and (48). Please, do not confuse it with the radial coordinate in isotropic coordinates (1).

with

$$e^{\alpha(r)} = \left(1 - \frac{2M/\varepsilon}{r^{n+1}}\right)^{\varepsilon}, \quad e^{\beta(r)} = \left(1 - \frac{2M/\varepsilon}{r^{n+1}}\right)^{-(n+\varepsilon)/(n+1)} \quad e^{\mu(r)} = \left(1 - \frac{2M/\varepsilon}{r^{n+1}}\right)^{(1-\varepsilon)/(n+} \tag{54}$$

The interior of a spherical star is assumed to be described by

$$ds^2 = e^{\nu(R)}dt^2 - e^{\lambda(R)}dR^2 - R^2 d\Omega_{(2+n)}, \tag{55}$$

which is the most general line element describing the interior of a spherical non-rotating star in $(4+n)$ dimensions. The functions $e^{\nu(R)}$ and $e^{\lambda(R)}$ are solutions of the Einstein field equations (A-1).

The exterior boundary of the star is a hypersurface Σ defined as $R = R_b$, and $r = r_b$ from inside and outside respectively. Standard matching conditions require continuity of the first and second fundamental forms at Σ [61]. For the metrics under consideration they demand

$$e^{\nu(R_b)} = e^{\alpha(r_b)}, \quad R_b = r_b e^{\mu(r_b)/2}, \tag{56}$$

and

$$\begin{aligned} e^{-\lambda(R_b)/2}\left(\frac{d\nu}{dR}\right)_{|R=R_b} &= e^{-\beta(r_b)/2}\left(\frac{d\alpha}{dr}\right)_{|r=r_b}, \\ e^{-\lambda(R_b)/2} &= e^{-\beta(r_b)/2}\left[\frac{1}{r} + \frac{1}{2}\left(\frac{d\mu}{dr}\right)\right]_{|r=r_b} r_b\, e^{\mu(r_b)/2}. \end{aligned} \tag{57}$$

Setting

$$e^{-\lambda(R)} = 1 - \frac{2m(R)}{R^{n+1}}, \tag{58}$$

from the second equation in (57) we get

$$\frac{m(R_b)}{R_b} = \frac{1}{2}\left\{1 - \left[1 - \frac{(\varepsilon+1)\phi_g}{\varepsilon}\left(\frac{R_b}{r_b}\right)^{n+1}\right]^2 \left[1 - \frac{2\phi_g}{\varepsilon}\left(\frac{R_b}{r_b}\right)^{n+1}\right]^{-1}\right\}, \tag{59}$$

where ϕ_g is the surface gravitational potential, viz.,

$$\phi_g \equiv \frac{M}{R_b^{n+1}}. \tag{60}$$

The above conditions require continuity of T_1^1, i.e. the "radial" pressure, across Σ. We find,

$$8\pi R_b^2 p(R_b) = \frac{(n+1)(n+2)(1-\varepsilon^2)\phi_g^2}{2\varepsilon^2}\left(\frac{R_b}{r_b}\right)^{2(n+2)}\left[1 - \frac{2\phi_g}{\varepsilon}\left(\frac{R_b}{r_b}\right)^{n+1}\right]^{(\varepsilon-n-2)/(n+1)} \tag{61}$$

Buchdahl's limit for $\varepsilon = 1$: We note that for $\varepsilon = 1$, we recover the Schwarzschild case, i.e., $R_b = r_b$ and

$$m(R_b) = M, \quad p(R_b) = 0. \tag{62}$$

Substituting this into (A-21), after a simple algebra we obtain

$$\frac{M}{R_b^{n+1}} \leq \frac{2(n+2)}{(n+3)^2}. \tag{63}$$

Thus,

$$g_{00} \geq \left(\frac{n+1}{n+3}\right)^2. \tag{64}$$

For $n = (0, 1, 2, 3, 4, 5, 6, 7)$, i.e. $D = (4, 5, 6, 7, 8, 9, 10, 11)$ we find

$$\begin{aligned} \frac{M}{R_b^{n+1}} &\leq (4/9,\ 0.375,\ 0.320,\ 0.278,\ 0.245,\ 0.219,\ 0.198,\ 0.180), \\ g_{00} &\geq (1/9,\ 0.250,\ 0.360,\ 0.444,\ 0.510,\ 0.563,\ 0.605,\ 0.640), \end{aligned} \tag{65}$$

respectively. We note that for $n = 0$, which corresponds to a matter distribution that has spherical symmetry in three spatial dimensions (rather than in $n + 3$), we recover the usual Schwarzschild values as expected.

An important observational parameter is the redshift $Z(R) = 1/\sqrt{g_{00}(R)} - 1$ of the light emitted from a point R inside the sphere to infinity. For Schwarzschild-Tangherlini's's exteriors Z_b, the redshift of the light emitted from the boundary surface, is given by

$$Z_b^{max} \leq \frac{2}{n+1} \tag{66}$$

Buchdahl's limit for $\varepsilon \neq 1$: In this case $m(R_b) \neq M$ and $p(R_b) \neq 0$. From the second equation in (56) we get

$$\phi_g = \frac{\varepsilon}{2\left(R_b/r_b\right)^{n+1}}\left[1 - \left(\frac{R_b}{r_b}\right)^{2(n+1)/(1-\varepsilon)}\right]. \tag{67}$$

Consequently, at the surface of a body

$$g_{00}(R_b) = e^{\nu(R_b)} = \left[1 - \frac{2\phi_g}{\epsilon}\left(\frac{R_b}{r_b}\right)^{n+1}\right]^{\varepsilon} = \left(\frac{R_b}{r_b}\right)^{2\varepsilon(n+1)/(1-\varepsilon)}. \tag{68}$$

Substituting (59), (61) and (67) into (A-21) we obtain an inequality for (R_b/r_b). Unfortunately, it is very cumbersome, so we omit it here. However, it can be solved numerically for any given value of n and $0 < \varepsilon < 1$. Then, using the allowed values of (R_b/r_b) in (67) we obtain the range of ϕ_g. Here we present the solution for $n = 0, 1, 2$ and some selected values[8] of ε, viz.,

[8] We emphasize that for $\varepsilon > 1$ the boundary conditions do not admit solutions in the realm of positive real numbers.

$$\varepsilon = (0.90,\ 0.80,\ 0.75,\ 0.60)\,. \tag{69}$$

We also present the redshift of the light emitted from the boundary surface $Z_b = Z(R_b) = 1/\sqrt{g_{00}(R_b)} - 1$, which in the present case is given by

$$Z_b = \left(\frac{R_b}{r_b}\right)^{\varepsilon(n+1)/(\varepsilon-1)} - 1. \tag{70}$$

• For $n = 0$, which corresponds to spherical symmetry in ordinary three space, the boundary conditions are satisfied for

$$(0.882,\ 0.746,\ 0.669,\ 0.383) \leq \frac{R_b}{r_b} < 1 \quad \Longrightarrow \quad 0 < \phi_g \leq (0.469,\ 0.508,\ 0.538,\ 0.777). \tag{71}$$

It is important to note that the upper values of ϕ_g give the Buchdahl's limit for a star of uniform density (See equation (83) in [22]). Using (68) and (70) we find

$$g_{00}(R_b) \geq (0.104,\ 0.096,\ 0.090,\ 0.056), \quad \Longrightarrow Z_b \leq (2.096, 2.229, 2.340, 3.219).$$

• For $n = 1$, and the same values selected in (69), we get

$$(0.962,\ 0.915,\ 0.887,\ 0.778) \leq \frac{R_b}{r_b} < 1 \quad \Longrightarrow \quad 0 < \phi_g \leq (0.383,\ 0.397,\ 0.407,\ 0.455), \tag{72}$$

and

$$g_{00}(R_b) \geq (0.248,\ 0.241,\ 0.237,\ 0.222) \quad \Longrightarrow \quad Z_b \leq (1.008, 1.035, 1.053, 1.124).$$

• For $n = 2$, under the same conditions, we find

$$(0.981,\ 0.958,\ 0.944,\ 0.888) \leq \frac{R_b}{r_b} < 1 \quad \Longrightarrow \quad 0 < \phi_g \leq (0.326,\ 0.329,\ 0.334,\ 0.356), \tag{73}$$

and

$$g_{00}(R_b) \geq (0.355,\ 0.357,\ 0.354,\ 0.343) \quad \Longrightarrow \quad Z_b \leq (0.679, 0.673, 0.680, 0.707).$$

The above calculations show how Buchdahl's limit depends on n and ε: (i) For a fixed n, the upper limit of ϕ_g increases as we go away from the Schwarzschild vacuum exterior, i.e., with the increase of $(1 - \varepsilon)$; (ii) For a fixed ε, the upper limit of ϕ_g decreases with the increase of n, which implies that the effects of gravity are stronger in $4D$ than in any other number of dimensions.

5. Summary and Concluding Remarks

The main question under investigation here has been how a possible deviation from the Schwarzschild vacuum exterior can affect the compactness of spherical stars in equilibrium. We have discussed this question within the context of Kaluza-Klein gravity, by using a general class of Ricci-flat metrics in $D = (4 + n + m)$-dimensions, namely (6)-(8), which generalize a number of solutions in the literature .

Following a standard technique, based on the assumption that the gravitational action has the standard form (42) in any number of dimensions, we have reduced the m external dimensions. We have seen that the reduction procedure flattens out the rich diversity of higher-dimensional solutions. The effective metrics in $(4 + n)$ constitute a one-parameter family of asymptotically-flat metrics given by (47), which contain Schwarzschild-Tangherlini's spacetimes in $(4 + n)$ dimensions for $\varepsilon = 1$ $(m = 0, \sigma = 0)$. For any other value of ε the effective spacetime is not Ricci-flat because $R_{11} \neq 0$, while all the other components of the Ricci tensor vanish identically (50). The fact that $R_0^0 = 0$ implies that the effective (or geometrical) matter in $(4 + n)$ satisfies the equation of state $\rho + p_r + (n + 2)p_\perp = 0$, which generalizes to n dimensions the well-known equation of state $(\rho + 3p) = 0$ for nongravitating matter in $4D$ (gravitational or Tolman-Wittaker mass is proportional to R_0^0). Thus, for any value of ε, the effective spacetime is similar to Schwarzschild vacuum in the sense that it has no effect on gravitational interactions. Consequently, we can interpret it as describing the gravitational field outside of a spherical star.

To put the discussion in perspective, let us notice that in the Randall & Sundrum braneworld scenario the concept of empty space requires only the vanishing of the Ricci scalar, while the components of the Ricci tensor are unknown without specifying the metric in the bulk (see, e.g., [24]). Here the situation is much more restricted because only R_1^1 is allowed to be different from zero. The crucial point is that Kaluza-Klein and Braneworld theories are alike in one important aspect: the effective vacuum spacetime outside of an isolated star does not have to be Ricci-flat, as in conventional $4D$ general relativity.

The line element (47) provides the Kaluza-Klein corrections to the Minkowski metric at large distances from any system of bodies (52), which could serve in astrophysical observations to detect possible deviations from general relativity. Also, it allowed us to study the question under consideration. Namely, using the standard matching conditions, i.e. the continuity of the first and second fundamental forms across the boundary, as well as the generalized Buchdahl's inequality (A-21), we have obtained the compactness limit for various values of ε and n, for any perfect fluid star with a mass density which does not increase outward.

Our analysis shows that in Kaluza-Klein gravity the compactness limit of a star can be larger than $1/2$, without being a black hole: the general-relativistic upper limit $M/R < 4/9$ is increased as we go away from the Schwarzschild vacuum exterior. Our results are consistent with our previous findings in [22], [28]. They show that, as in general relativity, the compactness limit can be saturated in the case of stars uniform proper density from the condition that the isotropic pressure does not become infinity at the center.

It should be noted that, for any n, the boundary conditions require $0 < \varepsilon \leq 1$, otherwise they have no real solutions. From (44) it follows that $\varepsilon \to 0$ as $k \to m/(n + 2)$, which according to (10) corresponds to the maximum value of σ. In the other extreme, for $\varepsilon = 1$ we recover the Schwarzschild-Tangherlini spacetimes. From a physical point of view $0 < \varepsilon \leq 1$ ensures the positivity of the effective energy density (48).

Our approach allowed us to determine some similarities and differences between spacetimes with a different number of internal dimensions: (i) Regardless of ε, they satisfy similar equations, viz., (49), (50); (ii) The corrections to Minkowski metric (52) manifestly depend on n; (iii) the effects of gravity decrease with the increase of n.

Appendix A: Buchdahl's Inequalities in D-dimensions

In this appendix, following our previous work [28] we show how Buchdahl's inequalities can be extended to any number of internal dimensions. We start with the Einstein field equations in D dimensions,

$$R_{AB} = 8\pi G\left[T_{AB} - \frac{1}{D-2}g_{AB}T\right], \tag{A-1}$$

where G, T_{AB}, and T represent: the gravitational constant; the energy momentum tensor in D-dimensions; and its trace respectively. In what follows we set $G = 1$.

We will consider the D-dimensional spherically symmetric metric, given by

$$ds^2 = e^{\nu(R)}dt^2 - e^{\lambda(R)}dR^2 - R^2 d\Omega_{(2+n)}, \tag{A-2}$$

where $d\Omega_{(2+n)}$ is the line element on a unit $(n+2)$ sphere; $n = D - 4$.

Now, let us assume that the D-dimensional energy-momentum tensor has the form

$$T_A^B = \text{diag}(\rho, -p, -p, ..., -p), \tag{A-3}$$

where ρ is the energy density and p is the isotropic pressure. With this choice the field equations (A-1) reduce to

$$e^{-\lambda(R)}\left[\frac{1}{R}\left(\frac{d\lambda}{dR}\right) - \frac{n+1}{R^2}\right] + \frac{n+1}{R^2} = \frac{16\pi\rho}{n+2}, \tag{A-4}$$

$$e^{-\lambda(R)}\left[\frac{1}{R}\left(\frac{d\nu}{dR}\right) + \frac{n+1}{R^2}\right] - \frac{n+1}{R^2} = \frac{16\pi p}{n+2}, \tag{A-5}$$

$$\frac{1}{2}\frac{d^2\nu}{dR^2} + \frac{1}{4}\left(\frac{d\nu}{dR}\right)^2 - \frac{1}{4}\left(\frac{d\lambda}{dR}\right)\left(\frac{d\nu}{dR}\right) - \frac{1}{2R}\left[(n+1)\left(\frac{d\lambda}{dR}\right) + \left(\frac{d\nu}{dR}\right)\right] + \frac{n+1}{R^2}\left[e^{\lambda(R)} - 1\right] = 0. \tag{A-6}$$

We note that the density and pressure satisfy the relation

$$\frac{dp}{dR} = -\frac{(\rho + p)}{2}\frac{d\nu}{dR}, \tag{A-7}$$

which is equivalent to the conservation equation ${T^B}_{A;B} = 0$.

The isotropy condition given by (A-6) takes a remarkable simple form with the introduction of the following notation [62]

$$e^{-\lambda} = 1 - \frac{2m(R)}{R^{n+1}} = Z, \;\; e^{\nu(R)} = Y^2, \;\; R^2 = x. \tag{A-8}$$

and

$$u = \int_0^x \frac{dx'}{\sqrt{Z(x')}}. \tag{A-9}$$

Indeed, (A-6) reduces to

$$2\frac{d^2Y}{du^2} = (n+1)\,Y\frac{d}{dx}\left(\frac{m}{R^{n+3}}\right). \tag{A-10}$$

The function $m(R)$ can be obtained from the integration of (A-4), viz.,

$$m(R) = \frac{8\pi}{n+2}\int_0^R \rho(\bar{R})\bar{R}^{n+2}d\bar{R}, \tag{A-11}$$

where the constant of integration has been set equal to zero to remove singularities at the origin. Thus,

$$\frac{d}{dR}\left(\frac{m}{R^{n+3}}\right) = \frac{8\pi\rho(R)}{(n+2)R} - \frac{(n+3)m(R)}{R^{n+4}} \tag{A-12}$$

Now, following Buchdahl we assume that the energy density is positive and does not increase outward, i.e.,

$$\frac{d\rho}{dR} \leq 0. \tag{A-13}$$

This implies that $\rho(\bar{R}) \geq \rho(R)$ in (A-11). Consequently,

$$m(R) \geq \frac{8\pi R^{n+3}\rho(R)}{(n+2)(n+3)}.$$

Substituting into (A-12) we find

$$\frac{d}{dR}\left(\frac{m}{R^{n+3}}\right) \leq 0. \tag{A-14}$$

We note that the quantity m/R^{n+3} can be identified with the mean density of the fluid sphere in D dimensions.

Now Eq. (A-10) gives

$$\frac{d^2Y}{du^2} \leq 0, \tag{A-15}$$

which means that dY/du decreases monotonically. This in turn implies

$$\frac{dY}{du} \leq \frac{Y(u) - Y(0)}{u}. \tag{A-16}$$

Since both $Y(0)$ and u are non-negative, it follows that

$$Y^{-1}\frac{dY}{du} \leq \frac{1}{u}. \tag{A-17}$$

In terms of the original variables this equation reads

$$\left(1-\frac{2m(R)}{R^{n+1}}\right)^{1/2}\frac{d\nu}{dR}\leq 2R\left[\int_0^R \bar{R}\left(1-\frac{2m(\bar{R})}{\bar{R}^{n+1}}\right)^{-1/2}d\bar{R}\right]^{-1}. \tag{A-18}$$

Using the fact that the average density decreases outward (A-14), we can evaluate the integral in (A-18) as follows

$$\begin{aligned}\int_0^R \bar{R}\left(1-\frac{2m(\bar{R})}{\bar{R}^{n+1}}\right)^{-1/2}d\bar{R} &\geq \int_0^R \bar{R}\left(1-\frac{2m(R)}{R^{n+3}}\bar{R}^2\right)^{-1/2}d\bar{R}\\ &= \frac{R^{n+3}}{2m(R)}\left[1-\left(1-\frac{2m(R)}{R^{n+1}}\right)^{1/2}\right].\end{aligned} \tag{A-19}$$

On the other hand, using (A-5) and (A-8), $(d\nu/dR)$ can be expressed as

$$\frac{d\nu}{dR}=\frac{2\left[(n+2)(n+1)m(R)/R^{n+1}+8\pi pR^2\right]}{(n+2)\left(1-2m(R)/R^{n+1}\right)\,R}. \tag{A-20}$$

Now, substituting (A-19) and (A-20) into (A-18) we obtain

$$\frac{(n+2)(n+1)m(R)/R^{n+1}+8\pi pR^2}{(n+2)\left(1-\frac{2m(R)}{R^{n+1}}\right)^{1/2}}\leq\frac{2m(R)}{R^{n+1}}\left[1-\left(1-\frac{2m(R)}{R^{n+1}}\right)^{1/2}\right]^{-1}. \tag{A-21}$$

Evaluating this expression at the outer surface of a static spherical star, and using the standard matching conditions, i.e. the continuity of the first and second fundamental forms across the boundary, we obtain the compactness limit of a such star for *any* given exterior spacetime. In particular, for the Schwarzschild exterior, (A-21) leads to the well-known upper mass limit $M/R_b \leq 4/9$.

References

[1] H. A. Buchdahl, *Phys. Rev.* **116** (1959) 1027.

[2] G.C. Segré, "Physics in more than four-dimensions, another look at the Kaluza-Klein theory", in Cosmology and Elementary particles. Proceedings of the first winter school of physics. World Scientific Publishing Co. Pte. Ltd. 1989.

[3] J. M. Overduin and P. S. Wesson, *Phys.Rept.* **283** (1997) 303, arXiv:gr-qc/9805018

[4] P.S. Wesson, *Space-Time-Matter* (World Scientific Publishing Co. Pte. Ltd. 1999).

[5] L. Randall and R. Sundrum, *Phys. Rev. Lett.* **83** (1999) 4690; arXiv:hep-th/9906064.

[6] N. Arkani-Hamed, S. Dimopoulos, G. Dvali and N. Kaloper, *Phys.Rev.Lett.* **84** (2000) 586, arXiv:hep-th/9907209.

[7] N. Arkani-Hamed, S. Dimopoulos and G. Dvali, *Phys.Lett.* **B429** (1998) 263, arXiv:hep-ph/9803315.

[8] N. Arkani-Hamed, S. Dimopoulos and G. Dvali, *Phys.Rev.* **D59** (1999) 086004, arXiv:hep-ph/9807344.

[9] I. Antoniadis, *Phys. Lett.* **B246** (1990) 3171.

[10] R. Maartens, *Phys. Rev.* **D62** (2000) 084023, arXiv:hep-th/0004166.

[11] Roy Maartens, Frames and Gravitomagnetism, ed. J Pascual-Sanchez et al. (World Sci., 2001) p93-119, arXiv:gr-qc/0101059.

[12] N. Dadhich and S.G. Gosh, *Phys. Lett.* **B518** (2001) 1, arXiv:hep-th/0101019.

[13] M. Govender and N. Dadhich, *Phys.Lett.* **B538** (2002) 233, arXiv:hep-th/0109086.

[14] C. Germani and Roy Maartens, *Phys. Rev.* **D64** (2001) 124010, arXiv:hep-th/0107011.

[15] M. Bruni, C. Germani and R. Maartens, *Phys. Rev. Lett.* **87** (2001) 231302, arXiv:gr-qc/0108013.

[16] G. Kofinas and E. Papantonopoulos, *J. Cosmol. Astropart. Phys.* **12** (2004) 11, arXiv:gr-qc/0401047.

[17] M. Visser and D. L. Wiltshire, *Phys.Rev.* **D67** (2003) 104004, arXiv:hep-th/0212333.

[18] N. Dadhich, R. Maartens, P. Papadopoulos and V. Rezania, *Phys.Lett.* **B487** (2000) 1, arXiv:hep-th/0003061.

[19] K.A. Bronnikov, H. Dehnen and V.N. Melnikov, *Phys.Rev.* **D68** (2003) 024025, arXiv:gr-qc/0304068.

[20] K.A. Bronnikov and S-W Kim, *Phys.Rev.* **D67** (2003) 064027, arXiv:gr-qc/0212112.

[21] R. Casadio, A. Fabbri and L. Mazzacurati, *Phys.Rev.* **D65** (2002) 084040, arXiv:gr-qc/0111072.

[22] J. Ponce de Leon, *Class.Quant.Grav.* **24** (2007)1755, arXiv:gr-qc/0701129

[23] J. Ponce de Leon, "The equivalence principle in Kaluza-Klein gravity", arXiv:gr-qc/0703094. Accepted for publication in *Int. J. Mod. Phys.* **D**.

[24] J. Ponce de Leon, *Class.Quant.Grav.* **25** (2008) 075012, arXiv: gr-qc/0711.4415.

[25] J. Ponce de Leon, *Grav.Cosmol.* **14** (2008) 65, arXiv:0711.0998.

[26] L. Bowers and E.P.T. Liang, *Astrophys. J.* **188** (1974) 657.

[27] J. Ponce de Leon, *Phys. Rev.* **D37** (1988) 309.

[28] J. Ponce de Leon and N. Cruz, *Gen.Rel.Gravit.* **32** (2000) 1207, arXiv:gr-qc/0207050.

[29] A. Chodos and S. Detweiler, *Gen. Rel. Gravit.* **14** (1992) 879.

[30] J.A. Casas, C.P. Martin and A.H. Vozmediano, *Phys. Lett.* **B 186** (1987) 29.

[31] V.S. Gurin and A.P. Trofimenko, *Phys. Lett.* **B 241** (1990) 328.

[32] P.S. Wesson and J. Ponce de Leon, *Class. Quantum Grav.* **11** (1994) 1341.

[33] P. Lim, J. Overduin and P.S. Wesson, *J. Math. Phys.* **36** (1995) 6907.

[34] A.G.Agnese, A.P. Billyard, H. Liu and P.S. Wesson, *Gen. Rel. Gravit.* **31** (1999) 527.

[35] W.N. Sajko and P.S. Wesson, *Gen. Rel. Gravit.* **32** (2000) 1381.

[36] H. Liu and J. Overduin, *Astrophys. J.* **538** (2000) 386, arXiv:gr-qc/0003034.

[37] J. Overduin, *Phys. Rev.* **D62** (2000) 102001, arXiv:gr-qc/0007047.

[38] K. Lake, *Class. Quant. Grav.* **23** (2006) 5876, arXiv:gr-qc/0606005.

[39] J. Ponce de Leon, *Int.J.Mod.Phys.* **D17** (2008) 237, arXiv:gr-qc/0611082.

[40] T. Dereli, *Phys. Lett.* **B161** (1985) 307.

[41] L. Sokolowski and B. Carr, *Phys. Lett.* **B 176** (1986) 334.

[42] S. Chatterjee, *Astron. Astrophys.* **230** (1990) 1.

[43] H. Liu, *Gen. Rel. Gravit.* **23** (1991) 759.

[44] F. R. Tangherlini, *Nuovo Cimento* **27** (1963) 636.

[45] R.C. Myers and M.J. Perry, *Annals of Physics* **172** (1986) 304.

[46] D. Kramer, *Acta Phys. Polon.* **B2** (1970) 807.

[47] A. Davidson and D. Owen, *Phys. Lett.* **B 155** (1985) 247.

[48] R.S. Millward, "A five-dimensional Schwarzschild-like solution" arXiv:gr-qc/0603132.

[49] L.D. Landau and E.M. Lifshitz, *The Classical Theory of Fields*, Fourth Edition (Butterworth-Heinemann, 2002).

[50] S.B. Fadeev, V.D. Ivashchuk and V.N. Melnikov, *Phys. Lett.* **A 161** (1991) 98.

[51] V.D. Ivashchuk and V.N. Melnikov, *Grav. Cosmol.* **1** (1995) 133, arXiv:hep-th/9503223.

[52] D.J. Gross and M.J. Perry, *Nucl. Phys.* **B226** (1983) 29.

[53] D.J. Gross, M.J. Perry and R.D. Sorkin, *Phys. Rev. Lett.* **51** (1983)) 87.

[54] L. Dolan and M.J. Duff, *Phys. Rev. Lett.* **52** (1984) 14.

[55] P.S. Wesson, *Phys. Essays, Orion* **5** (1992) 591.

[56] Z. Fisher, *Zh. Eksp. Teor. Fiz.* **18** (1948) 636 (in Russian), arXiv:gr-qc/9911008.

[57] A.I. Janis, E.T. Newman and J. Winicour, *Phys. Rev. Lett.* **20** (1968) 878.

[58] J.R. Gott and M.J. Rees, *Mon. Not. R. Astron. Soc.* **227** (1987) 453.

[59] E. Kolb, *Astrophys J.* **344** (1989) 543.

[60] J. Ponce de Leon, *Gen. Rel. and Gravit.* **11** (1993) 1123.

[61] W. Israel, *Nuovo Cimento* **B44**, (1966) 1;[Erratum-ibid. **B48** (1967) 463].

[62] J. Ponce de Leon, *J. Math. Phys.* **29** (1988) 197.

In: Black Holes and Galaxy Formation ISBN 978-1-60741-703-3
Editors: A.D. Wachter and R.J. Propst, pp. 355-390

Chapter 14

Scalar Potential Model of Galaxies: Review and New Speculations

J.C. Hodge*
Blue Ridge Community College, Flat Rock, NC, USA

Abstract

Models of observations from our solar system scale to the galaxy cluster scale are only moderately successful. Such observations are the pioneer anomaly; mid-range, flat rotation curves (RCs); asymmetric RCs; the spiral galaxy central mass and velocity dispersion correlation with outer galaxy parameters; and galaxy structure. Each model describes only one type of observation. The Scalar Potential Model (SPM) posits the existence of a plenum that influences matter and that is influenced by matter. The plenum obeys the heat equation, is sourced at the center of spiral galaxies, and is sunk in elliptical galaxies. The SPM is consistent with all the types of observation listed above; with rising and falling RCs in addition to flat RCs; and with inner-range and outer-range RCs. The SPM suggests a shift in fundamental principles and suggests a new model for the structure, formation and evolution of galaxies. Qualitative speculations about models of QSOs; of black holes; of clouds of shocked, outflowing hydrogen near the center of spiral galaxies; of the relationship of galaxy luminosity to many galaxy parameters; of the cause of the X-rays emanating from the center of spiral galaxies; of why the mid-range H I and Hα RCs differ; of why inner-range RCs of different trace materials differ; of the cause of disks in spiral galaxies; of the cause of warps in the disks; of cooling flows; and of why radio galaxies are preferentially near elliptical galaxies follow from the SPM.

Keywords: galaxies:fundamental parameters – galaxies:kinematics and dynamics – galaxies:nuclei – cosmology:theory

1. Introduction

Observations of galaxies and galaxy clusters have been found inconsistent with cosmological models (the very big) (Sellwood & Kosowsky, 2001) and with sub–atomic models (the

*E-mail address: jch9496@blueridge.edu; scjh@citcom.net.
Visiting from XZD Corp., 16 Hosta Ln., Brevard, NC, 28712.

very small). The understanding of the galaxy scale will need elements of both the very big and the very small and may hold the key to uniting the big and the small.

A large amount of data on galaxies have been collected. Although the current models yield calculations consistent with much of the data, many irregularities exist. Examples of these model challenges are (1) the need for and the ad hoc introduction of nonbaryonic, dark matter to describe rotation curves of spiral galaxy disks (RCs) that are flat, (2) the rising and declining RCs that have also been observed in spiral disks, (3) the asymmetric RCs, (4) the Pioneer Anomaly (PA), (5) the galaxy central mass and central velocity dispersion correlation with outer galaxy parameters, (6) the clouds of shocked, outflowing hydrogen near the center of spiral galaxies, (7) the relationship of galaxy luminosity L to many galaxy parameters, (8) the cause of the periodic X-ray bursts emanating from the center of the Galaxy, (9) the differing disk H I and Hα RCs, (10) the differing inner RCs of different trace materials, (11) the cause of disks in spiral galaxies, (12) the cause of warping of disks, (13) the cooling flows between spiral and elliptical galaxies, (14) the radio galaxies that are preferentially near elliptical galaxies, (15) the galaxy structure, (16) the need for fine – tuning of significant parameters, (17) the lack of a galaxy and galaxy cluster evolution model consistent with observations, (18) the lack of confirmation of the Strong Equivalence Principle (SEP), and (19) the lack of application of Mach's Principle. Other examples (Arp, 1998; Pecker & Narklikar, 2006) that proponents of the standard model dispute the interpretation of the observational data include evidence for QSO association with nearby galaxies, evidence that spiral galaxies are forming by ejection from other galaxies, and evidence for discrete redshift.

Because considerable time and effort has been expended attempting to develop a model for galaxies consistent with currently popular cosmological models, considering a new model has merit.

The scalar potential model (SPM) was created to be consistent with galaxy scale observations (Hodge, 2006a). The focus has been to posit the necessary forces to describe the problematic galaxy observations. By positing a scalar potential for all forces, the possibility of consistency with observations of the big and the small is improved.

The Scalar Potential Model (SPM) posits the existence of a massless plenum that influences matter and that is influenced by matter. The plenum obeys the heat equation and is ubiquitous. The plenum density causes a scalar potential ρ field. The plenum acts on the cross–section of matter m_s. Matter m_g is posited to cause a static[1] warp in the ρ field in accordance with the Newtonian spherical property. Because the ρ field near matter must attract other matter, matter decreases the ρ field. For matter created ρ field variation, the $\vec{\nabla}\rho$ then causes matter attraction according to Newtonian gravitation. "Static" because matter is neither a Source nor a Sink of energy. Matter merely modifies the energy flowing from Sources to Sinks.

Several differences among galaxy types suggest that Sources of ρ are located in spiral galaxies and that Sinks of ρ are located in elliptical, lenticular, and irregular galaxies. The scalar potential $\rho \propto R^{-1}$ derived from a heat differential equation, where R is the radius from the Source, Sink or matter being evaluated. The SPM suggests the $L_{\text{source}} \propto \epsilon$, where L_{source} is the luminosity of a Source galaxy and $\epsilon > 0$ is the Source strength. "Strength"

[1]"Static" such as caused by a stationary electron in a stationary field. Sources induce a flow such as a source of heat or fluid.

is a measure of the rate energy flows through the Source or Sink. The force $\vec{F}_s \propto \vec{\nabla}\rho$ at a given R on m_s. Therefore, the m_s/m_i, where m_i is the inertial mass, of particles at a given radius varies with L_s. Similarly, $L_{sink} \propto \eta$, where L_{sink} is the luminosity of a Sink galaxy and $\eta < 0$ is the Sink strength.

The SPM was found to be consistent with observations of redshift; of discrete redshift; of the pioneer anomaly; of spiral galaxy disk, flat RCs; of asymmetric RCs; of the galaxy central mass and central velocity dispersion correlation with outer galaxy parameters; of rising and falling RCs; and of inner and outer RCs. Therefore, the total mass of a galaxy is related to the luminosity L of a galaxy. A cell structure of galaxy groups and clusters was proposed with Sinks at the center and Sources in the outer shell of the cells. The cell model is supported by the data and analysis of Aaronson et al. (1982); Ceccarelli et al. (2005); Hudson et al. (2004); Lilje et al. (1986); and Rejkuba (2004). Because the distance between galaxies is larger than the diameter of a galaxy, the Sources and Sinks were considered as point (monopole) Sources and Sinks, respectively.

This chapter suggests a shift in fundamental principles; reviews the confrontation of data with the SPM; and suggests the SPM of the structure, formation and evolution of galaxies. Qualitative speculations about models of QSOs; of black holes; of clouds of shocked, out flowing hydrogen near the center of spiral galaxies; of the relationship of galaxy luminosity to many galaxy parameters; of the cause of the periodic X-ray bursts emanating from the center of spiral galaxies; of the cause of the differing disk H I and Hα RCs; of the cause of the differing inner RCs of different trace materials; of the cause of disks in spiral galaxies; of the cause of warps in the disks; of cooling flows; and of the cause of radio galaxies being preferentially near elliptical galaxies follow from the SPM.

Section 2. suggests SPM's new principles. Section 3. reviews the data confrontation with the SPM to date. Section 4. suggests a SPM of the structure, formation, and evolution of galaxies that may be helpful in forming testable hypotheses. The conclusion is in Section 5..

2. Principles

2.1. Mach's Principle

Ernst Mach suggested the precept that the inertia of objects results not from their relationship to Newtonian absolute space, but to the rest of the mass and energy distributed throughout the Universe. In the SPM, Mach's Principle is manifest by the dependence of the ρ field on the strength and position of Sources, Sinks, and matter. Because the plenum flows, the plenum is continually changing rather than being a universal rest frame. Because most of the matter is around galaxies, the relative position of galaxies determine the ρ field throughout the universe. In the SPM the energy and momentum are defined by the ρ field. Therefore, the SPM supports Mach's Principle.

2.2. Principle of Negative Feedback

The Principle of Negative Feedback states that any system that has relatively narrow parameter relationships must evolve from a broader system and must have a negative feedback loop to maintain the narrow parameters and achieve balance between processes. The Principle of Negative Feedback may also be called the Principle of Balance. Engineering applications have long used negative feedback to maintain tight control of parameters. A problem for scalar models in theoretical cosmology is that scalar models tend to lead to predictions that are unstable or transitory and to some equation results that are unbounded. This leads to the need for more complex models. If negative feedback relationships are included in the models, the effects of the unstable system of scalar models may be removed without consequential fallout or permanent change. If a time delay exists in the feedback loop, a cycling of the parameters results.

We observe objects that have limited size. So, there is a limiting force or negative feedback condition controlling the size of each object. So too must black holes have a size limitation and a negative feedback condition.

The R^{-1} of Newtonian gravitational potential implies the potential is unbounded at $R = 0$ and the universe model is unstable as $R \to \infty$. The Principle of Negative Feedback suggests that processes other than gravitation exist to better model the observations at these limits.

According to the Principle of Negative Feedback, the philosophical need for an origin (Source) of the universe causes a need for a Sink to maintain energy balance.

2.3. Principle of Minimum Action

The Principle of Minimum Potential Energy can be stated as a Principle of Minimum Action. That is, the path of least energy expenditure will be followed during the change from one state to another. Because a higher power feedback equation would result in the same controlled output and consume more energy, the minimum equations are sufficient. Combining this with the Principle of Negative Feedback yields the conclusion that "linear" or scalar formulations of negative feedback energy equations is sufficient.

2.4. Action by Contact

Matter acts on the plenum. The plenum acts on matter. The plenum is real and ubiquitous. That is, the Newtonian action at a distance and the general relativistic curved space concepts are replaced by the ancient notion of action by contact.

All models of photons have considered the photon structure to be similar to other matter particles. The conclusion drawn from the well-known Michelson-Morley experiment's null result was that the aether for light to "wave" in is nonexistent and that a Fitzgerald-Lorentz type of contraction was a work. The SPM suggests the null result is because of the structure of the photon rather than because of the lack of an interacting medium between particles.

The matter - plenum interaction may have a negative component in all matter-centered coordinate systems. If the sum of all Source strengths is greater than zero, a universe model with negative feedback processes may be "flat", finite, and stable. "Stable" in the sense that the distance equations yield a finite and non-zero prediction for arbitrarily large time and that there exists a central value for each parameter around which the universe's feedback processes cycle. The SPM allows the Source strengths to vary. However, in the cases examined to date, the rate of varying must be small relative to the cycle times of those cases.

2.5. Equivalence Principle

The concept of two forms of energy was born in Galileo's free fall experiments. The interaction of the plenum and matter produces two forms of energy. One depends on the action of the ρ field on matter that is kinetic energy. The other is the effect of Sources, Sinks, and matter on the ρ field that produces the potential energy field.

If the Sink strength varies with the amount of matter flowing into it, a feedback mechanism is established that balances the relative Source and Sink strengths (Hodge, 2006b). Therefore, the potential and kinetic energy balance is maintained. Thus, the resolution of the flatness problem is a consequence of the SPM.

The GR WEP implies that a kinetic property measured by acceleration equals a static property measured by gravitational potential (Unzicker, 2007). The SPM postulates matter experiences a force $\vec{F_s} = G_s m_s \vec{\nabla} \rho$, where G_s is the proportionality constant of the m_s property of lower case particle m of matter. In the case of a ρ field variation produced by matter and the effect of Sources and Sinks are relatively small, the force F_m is

$$F_m = (G_s m_s) \left(- \frac{G_g M_g}{R^2} \right), \tag{1}$$

where M_g is bulk property of matter of the upper case particle M called mass, R is the distance between the center of M_g and the test particle, and G_g is the proportionality constant of the M_g action on the ρ field. The $G_g M_g / R^2$ factor is interpreted as characterizing the acceleration field of ρ variation caused by M_g.

For an assembly of gravitationally bound particles, the total m_s is the sum of the total number of the particles upon which the ρ field acts as individual particles.

The semi empirical mass for the internal energy E^A generated for nuclei of atomic weight Z and atomic number A indicates a gravitational effect for differing nuclei (Will, 1993, Equations 2.5 to 2.12) and is thought to place an upper limit on the WEP violation, if any. Further, isotopes with the same Z and different A have differing E^A values. In the experiments performed to examine such factors, the bulk property of the masses (m_g and M_g) is used in the calculations. Equation (1) implies the WEP test should be done with the same attractor and with the same number of atoms in the pendulum or free-fall body for the differing isotopes. Maintaining equal m_g considers only the bulk property of matter that reduces the SPM effect.

Because F is symmetric, $(G_s m_s)(G_g M_g) = (G_g m_g)(G_s M_s) = G m_g M_s$. If the largest particle upon which the ρ field acts as a single particle (LP) is quarks, the (G_g / G_s) ratio indicates the relative volume to cross-section that may differ for quarks and larger structures. For example, if the LP are quarks, differing A/Z ratios with equal Z will determine differing G values. For elements the number of protons and the number of neutrons are approximately equal. Therefore, the G varies little among atoms, but may vary greatly among sub-atomic particles.

However, the WEP deals with m_i rather than m_s. Therefore, the SPM suggests a test of the WEP different from those done previously is required to differentiate the three mass interaction parameters m_i, m_g, and m_s. Therefore, the SPM rejects both the Weak and Strong Equivalence Principles. Instead, the SPM posits a fundamental particle that influences the plenum and is influenced by the plenum. However, the Weak Equivalence Principles may apply in the limited case of considering only atoms and larger particle assemblies in a non-turbulent ρ field.

3. Galaxy Scale Data Confrontation Review

3.1. Pioneer Anomaly (PA)

That an unexplained blueshift exists in the radio signal from the Pioneer 10 (P10) and Pioneer 11 (P11) spacecraft is well established (Anderson et al., 2002; Toth & Turyshev, 2006). Several models involving conventional mechanisms or involving "new physics" have been proposed to account for the PA (Anderson et al., 2002; Turyshev et al., 2006). A currently popular interpretation of the PA is that the blueshift is due to the Doppler effect. If so, the Pioneer spacecraft are being subjected to a force that causes a sunward acceleration $a_P \approx (8.74 \pm 1.33) \times 10^{-8}$ cm s^{-2}. That $a_P \approx cH_o$, where c (cm s^{-1}) is the speed of light and H_o (s^{-1}) is the Hubble constant, suggest a cosmological connection to the PA. However, the PA exceeds by at least two orders of magnitude the general relativity (GR) corrections to Newtonian motion.

The PA could be an effect other than a real acceleration such as a time acceleration (Anderson et al., 2002; Nieto & Anderson, 2005a) or an effect of an unmodeled effect on the radio signals. Although unlikely, a currently unknown systematics effect is not entirely ruled out.

Data for the existence of a_P from the Galileo, Ulysses, Voyager, and Cassini spacecrafts are inconclusive (Anderson et al., 2002; Nieto et al., 2005b).

In addition to the blueshift, there are other characteristics of the PA (Anderson et al., 2002). The PA has an apparent annual periodicity with an amplitude of approximately 1.6×10^{-8} cm s^{-2}. Although within uncertainty limits, the P11 anomaly $a_{P11} \approx 8.55 \times 10^{-8}$ cm s^{-2} may be slightly larger than the P10 anomaly $a_{P10} \approx 8.47 \times 10^{-8}$ cm s^{-2} (Anderson et al., 2002, Section V.A., Section VI.A., Section VI.D., and Section VIII.G.). The a_P calculation by the *Sigma* and CHASMP program methods for P10 (I) and P10 (II) show a discrepancy while showing consistency for P10 (III) (Anderson et al., 2002, Table I). The a_P of both spacecraft may be declining with distance (Turyshev et al., 1999). The blueshift of the PA is significantly smaller (Nieto & Anderson, 2005a) immediately before and after P11's Saturn encounter. The value of a_P averaged over a period during and after the Saturn encounter had a relatively high uncertainty (Nieto & Anderson, 2005a).

An obstacle to a new gravitational physics explanation of the PA is that a modification of gravity large enough to explain the PA is inconsistent with planetary ephemeredes unless the Equivalence Principle is violated (Anderson et al., 2002). The common opinion is that cosmic dynamics according to general relativity (GR) has far too little influence in galaxies to be measurable and that the expansion of the universe is essentially negligible for scales up to galactic clusters (Cooperstock et al., 1998; Sellwood & Kosowsky, 2001). Further, the expansion of the universe indicated by redshift z has a sign opposite to the PA. Several new physics models have been proposed (Anderson et al., 2002; Bertolami & Páramos, 2004) but fail is some respect. Bertolami & Páramos (2004) concluded a scalar field is able to explain the PA.

In GR a type of redshift dependent on the gravitational potential field derives from the

Weak Equivalence Principle (WEP). If the emitter and observer are in a uniform gravitational field $z \propto R^{-1}$, the signal acquires or looses energy because of the change in gravitational field potential. The Pound-Rebka experiment of 1959-1960 measured the WEP redshift of an signal traveling through a gravitational field (Pound & Rebka, 1960)[2]. The WEP suggests the integral of the signal frequency change experienced at each small increment along the path could be used to calculate a_{P}.

The SPM proposed the ρ field causes a non-Doppler redshift or blueshift of photons traveling through it. The resulting equation was applied to z and discrete redshift of galaxies (Hodge, 2006a). The SPM uses the gravitational potential from all matter in the solar system and from the Source at the center of the Galaxy at each point along the signal's path to calculate a_{P}. The difference between GR and the SPM is that the SPM also considers the turbulence of the gravitational field in the $K_{\mathrm{i}}I$ term and the amount of plenum the signal passes through in the $K_{\mathrm{dp}}D_{\mathrm{l}}P$ term (Hodge, 2006a, see the equations in). Without the $K_{\mathrm{i}}I$ term or the $K_{\mathrm{dp}}D_{\mathrm{l}}P$ term in the calculation, the correlation of measured and calculated a_{P} would be much poorer to the same degree that the WEP method fails.

The amount of warp ρ_{m} at a point in space due to the ith mass M_i was posited to be

$$\rho_{\mathrm{m}} = -\sum_{i=1}^{N} GM_i/R_i, \tag{2}$$

where N is the number of bodies used in the calculation, G is the Newtonian gravitational constant, and R_i is the distance from the center of M_i to the point where ρ is calculated.

The K_{min} term in the equation derived by Hodge (2006a) resulted from the flow from Sources. The K_{vp} term results from the relative movement of galaxies. Therefore, $K_{\mathrm{min}} = 0$ and $K_{\mathrm{vp}} = 0$ for the static warp field of matter in the Solar System. Because the K factors were calculated in a flowing ρ field, the static warp requires other values of the K factors. The resulting equation for the calculated redshift z_{c} is

$$z_{\mathrm{c}} = \mathrm{e}^{-X} - 1, \tag{3}$$

where

$$X = K_{\mathrm{dp}}D_{\mathrm{l}}P_{\rho} + K_{\mathrm{p}}P_{\rho} + K_{\mathrm{f}}F_{\rho}, \tag{4}$$

the terms are defined in Hodge (2006a), $D_{\mathrm{l}} = 2D$ (AU) is the distance the radio signal travels, and D (AU) is the geocentric distance to the spacecraft. The P_{ρ} is a measure of the amount of ρ the signal travels through. The F_{ρ} is a measure of the turbulence of ρ the signal travels through. The constant values are listed in Table 1. Table 2 lists the data dates, D, and the values of the terms of Eq. (4).

The calculated acceleration a_{Pc} is calculated from z_{c} as defined in Anderson et al. (2002). Figure 1 plots the a_{P} and a_{Pc} values versus D for each spacecraft. The correlation coefficient between the a_{P} and a_{Pc} is 0.85 for all data points. Without the P11 80/66 data point, which is the most uncertain measurement (Nieto & Anderson, 2005a), the a_{P} and a_{Pc} correlation coefficient is 0.93.

[2]This should not be confused with the theorized GR time dilation effect $(1 + z)$ caused by the differing gravitation fields of emitter and observer suggested by the Strong EP.

Table 1. The values of the constants of Eq. (4).

Parameter	value	units
$K_{\rm dp}$	1.30×10^{-28}	erg^{-1} AU^{-2}
$K_{\rm p}$	2.10×10^{-26}	erg^{-1} AU^{-1}
$K_{\rm f}$	5.38×10^{-22}	erg^{-1}
$K_{\rm co}$[a]	-5.98×10^{6}	erg AU^{-1}

[a]Part of the F_ρ calculation.

Table 2. The components of Eq. (4) and $A_{\rm sun}$ for each data point.

Craft Date	D (AU)	$K_{\rm dp}DP_\rho$ $\times 10^{-14}$	$K_{\rm p}P_\rho$ $\times 10^{-14}$	$K_{\rm f}F_\rho$ $\times 10^{-14}$	$A_{\rm sun}$ (°)
P10 82/19	25.80	-1.83	-5.74	15.40	124
P10 82/347	27.93	-2.06	-5.97	16.77	165
P10 83/338	30.68	-2.43	-6.42	18.54	167
P10 84/338	33.36	-2.82	-6.85	20.27	175
P10 85/138	36.57	-4.43	-9.81	21.17	12
P10 86/6	36.53	-3.34	-7.42	21.12	144
P10 87/80	40.92	-4.38	-8.66	23.96	70
P10 88/68	43.34	-4.76	-8.90	26.69	82
P10 89/42	45.37	-5.04	-8.99	26.82	109
P11 77/270	6.49	-0.24	-3.05	2.99	43
P11 80/66	8.42	-0.26	-2.49	4.23	166
P11 82/190	11.80	-0.49	-3.34	6.41	109
P11 83/159	13.13	-0.55	-3.42	7.25	148
P11 84/254	17.13	-0.99	-4.68	9.83	71
P11 85/207	18.36	-1.01	-4.47	10.62	121
P11 86/344	23.19	-2.09	-7.31	13.74	16
P11 87/135	22.41	-1.40	-5.06	13.23	151
P11 88/256	26.62	-2.01	-6.12	15.93	88
P11 89/316	30.29	-2.85	-7.62	18.29	36

The error bars in Fig. 1 reflect the uncertainty from Table II of Anderson et al. (2002) except for P11 80/66 where the uncertainty is from Nieto & Anderson (2005a). The stochastic variable of the unmodeled acceleration was sampled in ten-day or longer batches of data (Anderson et al., 2002). Starting at the P11 80/66 data point, the average extended over many months (Nieto & Anderson, 2005a). The "X's" in Fig. 1 plot the calculated data point for a ten-day change from the date of the $a_{\rm P}$. Some data points showed little change between the two dates of calculation. Others showed moderate change. Because the value of $a_{\rm Pc}$ depends on the closeness of matter to the line of sight (LOS), a change over ten days is due to a body close to the LOS and to long integration times. Therefore, the closeness

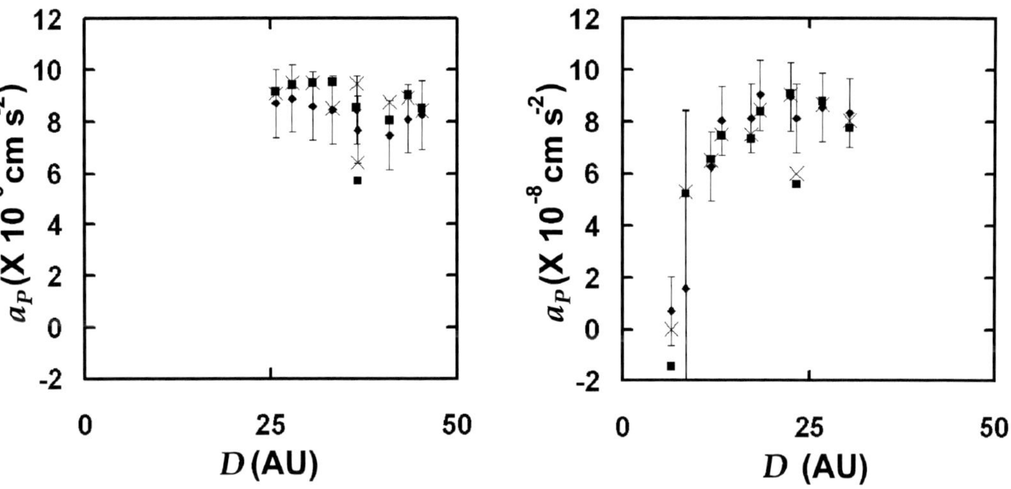

Figure 1. Plots of the $a_{\rm P}$ data versus geocentric distance D for P10 (left figure) and P11 (right figure). The solid diamonds reflect the $a_{\rm P}$ from Nieto & Anderson (2005a), the solid squares are the calculated points for $a_{\rm Pc}$, and the "X's" are calculated points for dates ten days from the date of the $a_{\rm P}$.

of matter to the LOS introduces an uncertainty for even ten-day integration times that was unaccounted in the error budget.

As seen in Fig. 1 the $a_{\rm Pc}$ calculation reproduces the subtler effects of the PA noted by Anderson et al. (2002).

3.1.1. Annual Periodicity

Figure 2 show the ρ value along the LOS versus D on dates when the angle $A_{\rm sun}$ (arcdegrees) between the LOS and a line from the Earth to the Sun was $< 60°$ and when $A_{\rm sun} > 120°$.

Table 2 lists the $A_{\rm sun}$ for each data point. On the dates that $A_{\rm sun} < 60°$, the $a_{\rm P}$ and $a_{\rm Pc}$ were considerably smaller. The correlation coefficient between the $a_{\rm P}$ and $a_{\rm Pc}$ without P10 85/138 ($A_{\rm sun} \approx 12°$), P11 86/344 ($A_{\rm sun} \approx 16°$), and P11 80/66 (Saturn encounter) is 0.97. The small $A_{\rm sun}$ value combined with long integration time causes larger uncertainty. The effect of the Sun is to cause the annual variation of the $a_{\rm Pc}$. Therefore, the cause of the PA is also the cause of the annual periodicity.

3.1.2. Difference of $a_{\rm p}$ between the Spacecraft

The P10 data included one of nine dates (≈ 0.11) with $A_{\rm sun} < 60°$ and five of nine dates (≈ 0.56) with $A_{\rm sun} > 120°$. The P11 data included three of ten dates (≈ 0.30) with $A_{\rm sun} < 60°$ and four of ten dates (≈ 0.40) with $A_{\rm sun} > 120°$. Figure 3 shows the trend of the data points with the Sun excluded $a_{\rm Psun}$ versus D between P10 and P11 data points. At $D > 10$ AU, the $a_{\rm Psun}$ appears to be a linear function of D. At $D < 10$ AU, the Sun's influence is to decrease $a_{\rm P11}$ more than $a_{\rm P10}$.

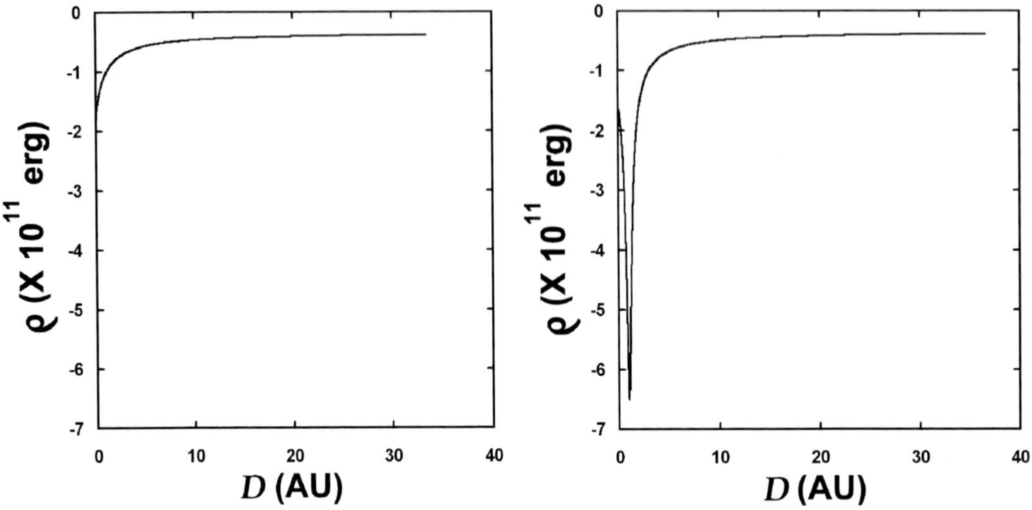

Figure 2. Plots of the ρ versus geocentric distance D along the LOS. Data is for P10 on 84/338 (left figure) $a_{\mathrm{P}} = 8.43 \times 10^{-8}$ cm s^{-2} and for P10 on 85/138 (right figure) $a_{\mathrm{P}} = 7.67 \times 10^{-8}$ cm s^{-2}. The $A_{\mathrm{sun}} \approx 175°$ on P10 84/338 and $A_{\mathrm{sun}} \approx 12°$ on P10 85/138.

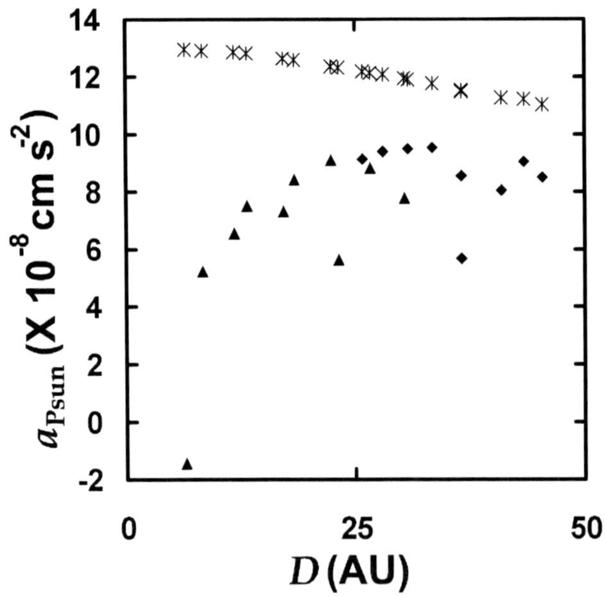

Figure 3. Plots of the a_{Pc} data versus geocentric distance D for P10 (solid diamonds) and P11 (solid triangles). The stars plot the data points for a_{Pc} with the Sun excluded (a_{Psun}).

The SPM also suggests the mass of the planets and the Source of the Galaxy have an influence on the ρ field. Because the outer planets are opposite the Sun for P10, the effect of the planets on a_{Pc} of P10 is less than P11.

From the galaxy scale perspective, the spacecraft in the solar system appears near the large mass of the Sun and inner planets. The effect of the Galaxy Source appears to decrease the a_{Pc} nearly uniformly for P11. The outer P10 data points show a trend of increasing relative effect of the Galaxy Source. The orbit of P10 is closer to the Sun-Sgr.A* axis than P11 and the D of P10 is greater than the D of P11. However, this effect is within the uncertainty level.

The difference in a_{P10} and a_{P11} noted byAnderson et al. (2002) results primarily from the effect of the Sun. A secondary effect is the effect caused by increasing D and a small effect of the planets on P11 which declines as D increases. Because the Sun and planets dominate the $\vec{\nabla}\rho$ field, the SPM expects the Galaxy Source to have a small difference between a_{P10} and a_{P11} caused by their different travel directions.

The a_{P} is not constant as the CHASMP software assumed (Anderson et al., 2002). This variation is especially present when the spacecraft are nearer the inner planets $D < 10$ AU. Therefore, the varying a_{P} may explain the difference between the *Sigma* and CHASMP program methods for P10 (I) and their closer agreement at P10 (III) (Anderson et al., 2002, Table I).

3.1.3. Slow Decline in a_{P}

The plot in Fig. 3 suggests $\mathrm{d}a_{\mathrm{Psun}}/\mathrm{d}D$ at $D < 10$ AU is nearly zero, followed by a decline and then a flattening.

The radio signal measurements are from and to Earth. At small D, the relative effect of the Sun-Earth distance is larger than at farther D. As D increases the Solar System effect appears to approach a single mass located at the barycenter of the Solar System. Therefore, a_{P} declines and approaches a constant value dictated by $\rho \propto -R^{-1}$. However, the SPM expects that at much greater D (near the orbit of Pluto in both directions), the effect of the Galaxy mass and Galaxy Source will increase to cause a difference in the a_{P} values between the spacecraft.

3.1.4. Saturn Encounter

Figure 4 shows a plot of a_{Pc} versus the hours from the closest approach of P11 to Saturn on P11 79/244 ($A_{\mathrm{sun}} \approx 8°$). The plot shows the a_{Pc} varies widely over a period of hours. The negative a_{Pc} is a redshift. As seen in Fig. 1, the SPM is consistent with the P11 77//270 ($A_{\mathrm{sun}} \approx 43°$) data point at the beginning of the Saturn encounter of a near zero blueshift.

3.1.5. Large Uncertainty of P11 80/66

Because the P11 80/66 ($A_{\mathrm{sun}} \approx 166°$) data point extends over a relatively long time, the rapidly varying a_{Pc} seen in Fig. 4 is consistent with the uncertainty in the P11 80/66 data point.

The a_{Psun} data points for P11 77/270 and P11 80/66 in Fig. 3 have only a slightly smaller slope than the later data points. The planets gravity well is in a larger gravity well of the Sun. The change from the Sun ρ versus D curve to a planet gravity well causes a smaller $K_{\mathrm{f}}F_{\rho}$ term relative to $K_{\mathrm{p}}P_{\rho}$. Table 2 lists $|K_{\mathrm{f}}F_{\rho}| < |K_{\mathrm{p}}P_{\rho}|$ for the P11 77/270, where "| |" means "absolute value", and $|K_{\mathrm{f}}F_{\rho}| > |K_{\mathrm{p}}P_{\rho}|$ for other data points. Without

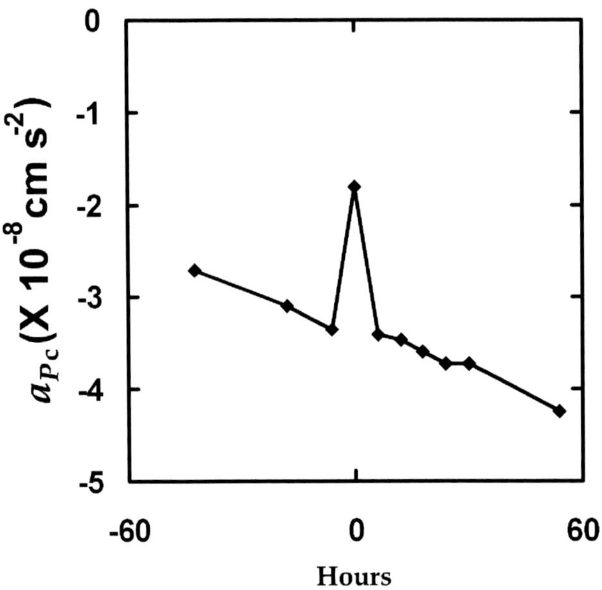

Figure 4. Plot of a_{Pc} versus the hours from the closest approach of P11 to Saturn.

the Sun gravity well in the calculations, $|K_\mathrm{f}F_\rho| > |K_\mathrm{p}P_\rho|$ for all data points. Therefore, the a_{Psun} for the P11 77/270 data point is consistent with the other data points.

3.1.6. Cosmological Connection

The SPM obtains the H_o value by $z \longrightarrow \exp(-X) - 1 \approx -X$, where $X < 0$ in Hodge (2006a) and $X > 0$ in this chapter. Figure 5 is a plot of D_l with the units changed to Mpc versus X. The straight line is a plot of the least squares fit of the data. The line is:

$$\begin{aligned} D_\mathrm{l} &= (2800 \pm 200\mathrm{Mpc})X + (5 \pm 2) \times 10^{-11}\mathrm{Mpc} \\ &\approx -\frac{c}{H_\mathrm{o}}z \end{aligned} \quad (5)$$

at 1σ and with a correlation coefficient of 0.95, where $H_\mathrm{o} = 106 \pm 8$ km s^{-1} Mpc^{-1}.

The PA and the z of cosmology are the result of the same ρ effect on light. In the cosmological z calculation, the z follows the Hubble law if $\rho \propto R^{-1}$. In the gravity well, z follows the negative Hubble law if $\rho \propto -R^{-1}$. The presence of other galaxies near the path of the light causes P_ρ and F_ρ variation of z. This is also the effect of matter close to the LOS in the PA. In the SPM, the Hubble law and $a_\mathrm{P} \approx cH_\mathrm{o}$ are manifestations of the Newtonian spherical property.

The a_{Pc} calculation reproduces the subtler effects of the PA noted by Anderson et al. (2002) including the annual periodicity, the slight difference between P10 and P11, the difference between the programs, the slow decline in a_P, the large uncertainty of a_P during the Saturn encounter, and suggests a cosmological connection.

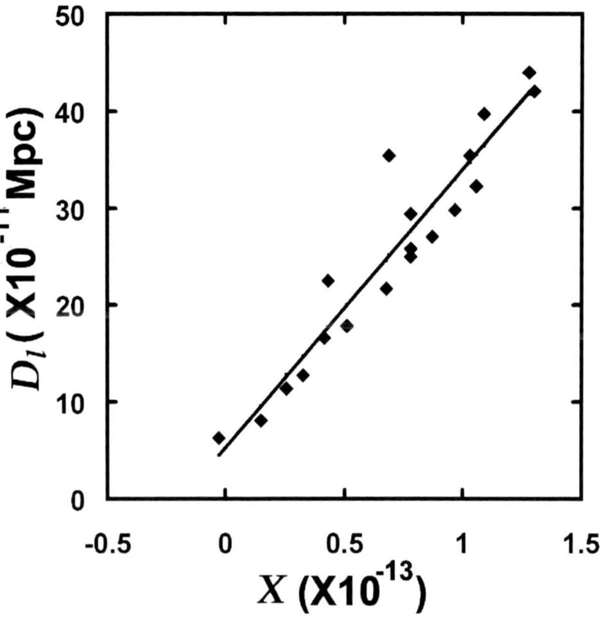

Figure 5. Plot of the light travel distance $D_\mathrm{l}(\times 10^{-11}$ Mpc) versus $X(\times 10^{-13})$ for both spacecraft. The line is a plot of $D_\mathrm{l} = 2800X + 5 \times 10^{-11}$ Mpc.

The variation of a_p through the sidereal day remains to be investigated. The specific location on Earth relative to the Earth's center and the timing of the transmitting and receiving signals are required. The observation that the period is a sidereal day rather than a solar day suggests the SPM will be consistent with this observation, also.

The SPM found the PA is an effect on only the radio signal. The SPM remains the only model consistent with the planetary ephemeris and the data of the PA including the several other noted anomalies and the cosmological connection.

3.2. Rotation Curves

The discrepancy between the Newtonian estimated mass in spiral galaxies and the observation of star and gas kinematics is well established. The rotation curve (RC) relates the square of the tangential rotation velocity v (km s^{-1}) of particles in orbit around spiral galaxies to their galactocentric radius R (kpc). The RC is measured along the galaxy's major axis. Traditionally, the focus has been on accounting for H I RCs that are flat in the disk region immediately beyond the knee (OR). However, observations also include rising RCs, declining RCs, an abrupt change in slope at the extreme outer region (EOR) in many galaxies, and rotational asymmetry with non-relativistic velocities. Battaner & Florido (2000); Ferrarese & Merritt (2002); Ghez et al. (2000); Sofue et al. (1999); Sofue & Rubin (2001); and Takamiya & Sofue (2000) provide a summary of the knowledge of RCs. The RC is an excellent tool to evaluate galaxy models.

The stars near the center of galaxies have rotation velocities more than an order of magnitude larger than v at the knee of the H I RC. The decline of the stellar RCs from the center is Keplerian (Ghez et al., 2000; Ferrarese & Merritt, 2002). The v continues to decline to the radius R_Δ (kpc) at which the rotation velocity v_Δ is at a minimum. The v_Δ is approximately an order of magnitude smaller than v at the knee of the HI RC.

The HI RC rarely extends inward to near R_Δ. When drawing the HI RC, a straight line is often drawn from the innermost measured point through the center of the galaxy. This practice is inconsistent with stellar observations (Schödel, 2002) and is incorrect (Sofue & Rubin, 2001).

From R_Δ to the knee of the RC, the HI RC rises rapidly, the rapidly rising region (RR).

As larger samples of RCs have been measured, researchers attempted to produce a quantitative description of an empirically derived, average RC. Persic et al. (1996) claimed the existence of a universal RC (URC) for a large sample revealed by the feature that only luminosity of the host galaxy dictated the RC. Independent work has confirmed much of the URC (Catinella, 2006, and references therein). Other researchers have discussed the inadequacy of the average RC (e.g. Courteau 1997; Garrido et al. 2004; Willick 1999).

Roscoe (2002) used a dynamical partitioning process and found that the dynamics in the outer part of optical RCs are constrained to occupy one of four *discrete dynamical classes*. The classes are determined by the absolute magnitude, surface brightness, and a parameter for each optical RC that is an exponent of the radius at which v is measured.

If most of the matter of a galaxy is in the bulge region, classical Newtonian mechanics predicts a Keplerian declining RC in the disk. The observation of rising RCs in the RR and of rising, flat, or declining RC's in the OR poses a perplexing problem. Postulates to model the RCs have included the existence of a large mass $M_{\rm DM}$ of non-baryonic, dark matter (DM) and dark energy; the variation of the gravitational acceleration with R; the variation of the gravitational constant G with R; and the existence of an added term to the Newtonian gravitation equation opposing gravitational attraction.

Repulsive unknown matter has been proposed to provide a repulsive force that has the form of an ideal, relativistic gas (Goodman, 2000). In DM models the rising RCs of low surface brightness galaxies (LSBs) require larger relative amounts of DM than flat or declining RCs. This contradiction to the successful Tully-Fisher relation (TF) (Tully & Fisher, 1977) is well established (Sanders & McGaugh, 2002). Another mystery of the DM paradigm is the apparent strong luminous-to-dark matter coupling that is inferred from the study of RCs (Persic et al., 1996).

Currently, the most popular modified gravity model is the Modified Newtonian Dynamics (MOND) model (Bottema et al., 2002; Sanders & McGaugh, 2002, and references therein). MOND suggests gravitational acceleration changes with galactic distance scales by a fixed parameter, which is related to the mass-to-luminosity ratio, with units of acceleration. MOND appears limited to the disk region of spiral galaxies. MOND requires the distance to NGC 2841 and NGC 3198 to be considerably larger than the Cepheid-based distance $D_{\rm a}$ (Mpc) (Bottema et al., 2002). Appling MOND to cosmological scales is being investigated (Bekenstein, 2004; Sanders, 2006).

The possible variation of G or $|\dot{G}/G| \sim H_{\rm o}$ were first proposed by Dirac. This was followed by the Brans-Dicke theory of gravity that postulated G behaves as the reciprocal of a scalar field whose source is all matter in the universe (Narlikar, 2002, pp. 283-5). More recently Brownstein & Moffat (2006, and references therein) have proposed a variation of general relativity in which G varies with R. This model has been applied to RCs (Brownstein & Moffat, 2006; Moffat, 2005) and the postulated gravitational radiation of binary pulsars. This model has results similar to the MOND model. However, it uses a distance to NGC 2841 of 9.5 Mpc rather than $D_{\rm a}$ =14.07 Mpc (Macri et al., 2001).

Alternate models of the RC posit the existence of another term in the Newtonian equation resulting from a scalar field of a "fifth force", a scalar field coupled to matter, or a static scalar field coupled to DM. Padmanabhan & Choudhury (2002, and references therein) explored the possibility of a scalar field acting as DM. The usual "fifth force" models posit a Yukawa like potential (Bertolami & Páramos, 2004). Fay (2004) postulated minimally coupled and massive scalar fields are responsible for flat RCs, only. Mbelek (2004) suggested a real scalar field, minimally coupled to gravity sourced by baryonic and dark matter may also reproduce flat and rising RCs with the assumption that the dark halo mass density dominates the baryonic mass density. The free parameters are a turnover radius $r_{\rm o}$, the maximum rotation velocity, which is at $r_{\rm o}$, and an *integer*. If a natural, theoretical explanation could be found for the relation between $r_{\rm o}$ and the positive integer, the integer may be the only free parameter. The postulate of an unknown "fifth force" is as likely as the postulate of unknown and elusive DM. Note the SPM is not a "fifth force" model because the plenum is responsible for gravity and the repulsive force from the Source. That is, the force of $\vec{\nabla}\rho$ is the "fourth force".

The force $\vec{F}_{\rm s}$ (dyne) of the ρ field from Sources and Sinks that acts on matter is

$$\vec{F}_{\rm s} = G_{\rm s} m_{\rm s} \vec{\nabla}\rho, \tag{6}$$

where $G_{\rm s}$ is a proportionality constant, ρ is the sum of the effects of all galaxies,

$$\rho = K_\epsilon \sum_{i=1}^{N_{\rm source}} \frac{\epsilon_i}{r_i} + K_\eta \sum_{l=1}^{N_{\rm sink}} \frac{\eta_l}{r_l}, \tag{7}$$

where ϵ_i and η_l are numbers representing the strength of the ith Source and l^{th} Sink, respectively, K_ϵ and K_η are proportionality constants, and r_i (Mpc) and r_l (Mpc) are the distance from a Source and a Sink, respectively, to the point at which ρ is calculated, $\epsilon > 0$, $\eta < 0$, and $N_{\rm source}$ and $N_{\rm sink}$ are the number of Sources and Sinks, respectively, used in the calculation.

The force from other matter is treated in this subsection in the traditional Newtonian gravitational force manner[3].

The $K_\epsilon \epsilon = L \propto M_{{\rm t}\epsilon}$, where $M_{{\rm t}\epsilon}$ is the total mass of the Source galaxy; and $K_\eta \eta = L \propto M_{{\rm t}\eta}$, where $M_{{\rm t}\eta}$ is the total mass of the Sink galaxy.

By analogy to heat flow that maintains the temperature of a volume, if the amount of matter flowing into the universe is equal to the amount of matter flowing out, then the total of the Source strengths equals the total of the Sink strength in the long term [4] (Hodge, 2006a). However, if the spiral galaxy mechanisms differ from the elliptical galaxy mechanisms as their morphology difference implies, $K_\epsilon \neq K_\eta$. Define the luminosity ratio $L_{\rm r}$ as the ratio of the sum of the $L_{\rm B}$ of all Sources within a limiting distance $D_{\rm l}$ divided by the sum of the

[3] This paper was presented before the analysis of the PA in which the matters influence on the plenum was explored.

[4] see the second paragraph in Subsection 2.5.

L_B of all Sinks within D_l,

$$L_\mathrm{r} = \frac{\sum_{i=1}^{N_\mathrm{sources}(D<D_\mathrm{l})} \epsilon_i}{\sum_{k=1}^{N_\mathrm{sink}(D<D_\mathrm{l})} \eta_k}. \tag{8}$$

Figure 6 shows a plot of L_r versus D_l for 29,899 sample galaxies from the NED database with $z < 0.03$.

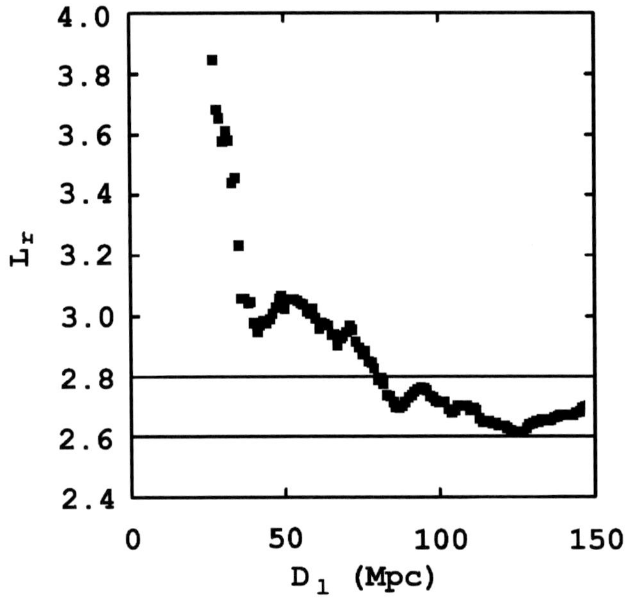

Figure 6. Plot of the luminosity ratio L_r for galaxies with a distance D less than the limiting distance D_l versus D_l.

Therefore, $L_\mathrm{B} = K_\epsilon \epsilon = 10^{-0.4 M_\mathrm{B}}$ erg s^{-1} for Source galaxies or $L_\mathrm{B} = K_\eta \eta = -2.7 \times 10^{-0.4 M_\mathrm{B}}$ erg s^{-1} for Sink galaxies.

The SPM posits the simplified Newtonian calculation of v of a test particle is the first approximation of v in an isolated galaxy, that observed asymmetries and deviations from the simplified Newtonian calculation in spiral galaxies are the result of the influence of neighbor galaxies, and that the v^2 of a particle in orbit around a spiral galaxy is the sum of the effects of F_s, the centrifugal force, and the gravitational force F_g. Because v is measured only along the major axis in the region under consideration (Binney & Merrifield, 1998, p. 725) and if the $\vec{\nabla}\rho$ field from neighbor galaxies is approximately uniform across a galaxy,

$$v^2 = G\frac{M}{R_\mathrm{major}} - \frac{G_\mathrm{s} m_\mathrm{s}}{m_\iota}\frac{L_\mathrm{B}}{R_\mathrm{major}} + |\vec{K} \bullet \vec{a}_\mathrm{o}| R_\mathrm{major}, \tag{9}$$

where (1) the Newtonian simplifications of circular motion ($\ddot{R} = 0$) (Binney & Merrifield, 1998, p. 725) and a spherical mass density field are assumed, (2) the mass of the test particle is assumed to be constant over time, (3) M ($\mathrm{M}_\odot$) is the mass inside the sphere of radius R from Newton's spherical property, (4) the inertial mass m_ι equals gravitational mass m_g of the test particle (Will, 1993), (5) the L_B term is due to the F_s of the host galaxy, which also derives from a spherically symmetric ρ field; (6) $M_{\rm B}$ is the B band magnitude of the galaxy; (7) $|\,|$ indicates absolute value; (8) $\vec{K}$ (km kpc^{-1}) is a constant vector; (9) $R_{\rm major}$ (kpc) is the galactocentric radius along the major axis; and (10) $\vec{a}_{\rm o}$ (km s^{-2}) is the acceleration of the test particle caused by neighboring galaxies,

$$\vec{a}_{\rm o} = \frac{G_{\rm s} m_{\rm s}}{m_\iota} \vec{\nabla} \rho, \tag{10}$$

where the number of galaxies excludes the host galaxy. Note that no assumption about the significance of $\vec{a}_{\rm o}$ has been made. Because only the HI RC is considered in the calculations herein, the units used were $G_{\rm s} m_{\rm s}/m_\iota = 1$ kpc km^2 s^{-1} erg^{-1}.

The sample consisted of 105 galaxies with varying degrees of RC completeness. Of these, 16 (called "select galaxies") had a HI RC available in the literature and $D_{\rm a}$ from Freedman et al. (2001) or Macri et al. (2001). Data for the 16 select galaxies are shown in Table 3 and their HI RCs are plotted in Fig. 7. The $L_{\rm B}$ for the select galaxies was calculated using $D_{\rm a}$, apparent B band magnitude $m_{\rm b}$ from the NED database, and extinction $E_{\rm xt}$ from the NED database. The selection criteria for the other sample galaxies were that a HI RC was available in the literature and that the $L_{\rm B}$ could be calculated using the TF method with the constants developed in Hodge (2006a).

Table 3. Data for the select galaxies.

Galaxy	morphology[a]	lc[b]	RC[c]	$D_{\rm a}$[d]
NGC 0224	SA(s)b	2	F	0.79
NGC 0300	SA(s)d .9	6	R	2.00
NGC 0598	SA(s)cd	4	R	0.84
NGC 0925	SAB(s)d HII	4	D	9.16
NGC 2403	SAB(s)cd	5	F	3.22
NGC 2841	SA(r)b;LINER Sy	1	F	14.07[e]
NGC 3031	SA(s)ab;LINER Sy1.8	2	D	3.63
NGC 3198	SB(rs)c	3	F	13.80
NGC 3319	SB(rs)cd;HII	3.3	R	13.30
NGC 4258	SAB(s)bc;LINER Sy1.9	3.9	D	7.98
NGC 4321	SAB(s)bc;LINER HII	1	D	15.21
NGC 4414	SA(rs)c? LINER	3.6	D	17.70
NGC 4535	SAB(s)c	1.9	D	15.78
NGC 4548	SBb(rs);LINER Sy	2	D	16.22
NGC 5457	SAB(rs)cd	1	D	6.70
NGC 7331	SA(s)b;LINER	2	F	14.72

[a]Galaxy morphological from the NED database.
[b]Galaxy luminosity class code from the LEDA database.
[c]Galaxy's HI RC type according to slope in the OR. R is rising, F is flat, and D is declining.
[d]The distance $D_{\rm a}$ (Mpc) to the galaxy from Freedman et al. (2001) unless otherwise noted.
[e]The $D_{\rm a}$ is from Macri et al. (2001).

This select galaxy sample has LSB, medium surface brightness (MSB), and high surface brightness (HSB) galaxies; has LINER, Sy, HII, and less active galaxies; has galaxies that

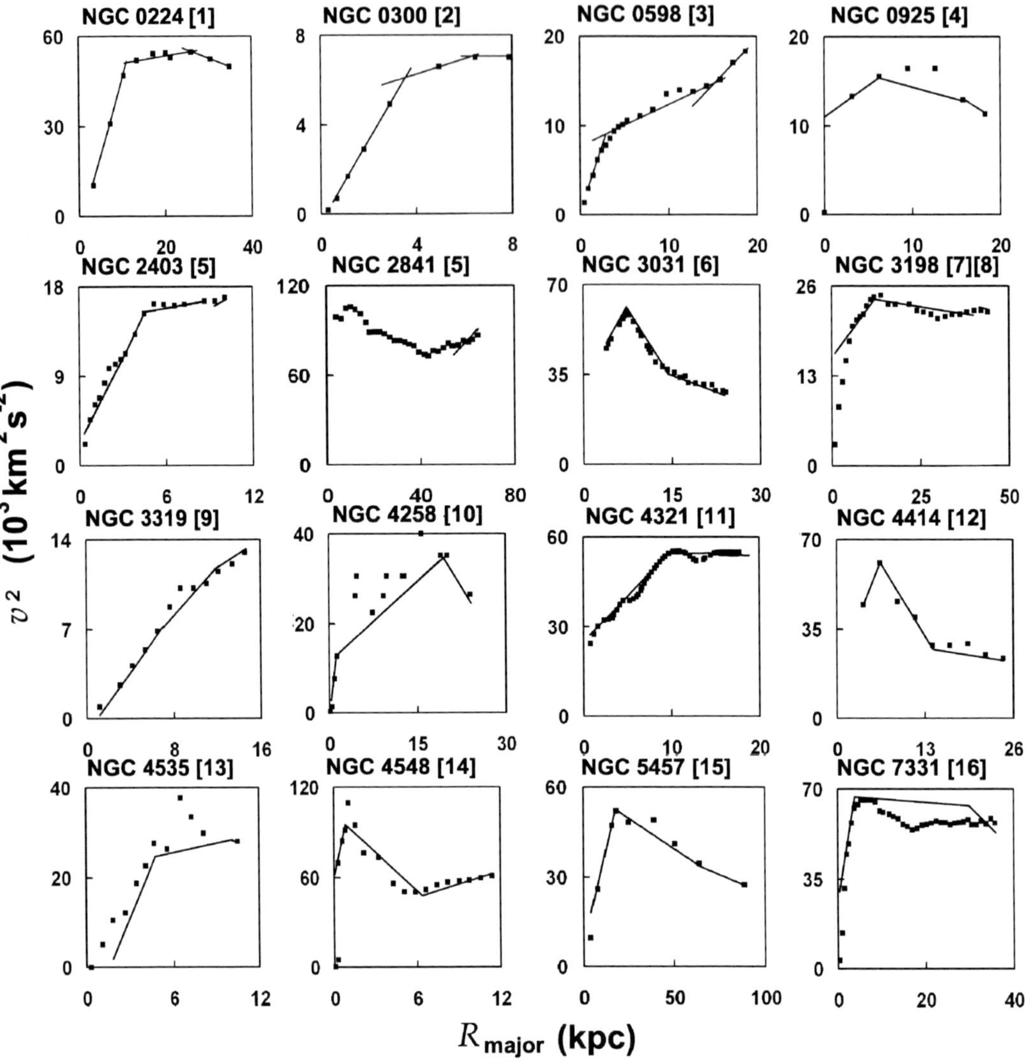

Figure 7. Plots of the square of the HI rotation velocity v^2 (10^3 km^2 s^{-2}) versus galactocentric radius R_{major} (kpc) along the major axis. The piecewise straight lines mark the application of the derived equations to the RCs of the select galaxies. The application of the derived equations to NGC 0224, NGC 0300, and NGC 0598 lacked a $|\vec{K} \bullet \vec{a}_{\mathrm{o}}|$ value. The references for the RCs are noted in brackets and are as follows: [1]Gottesman et al. (1966), [2]Carignan & Freeman (1985), [3]Corbelli & Salucci (2000), [4]Krumm & Salpeter (1979), [5]Begeman et al. (1991), [6]Rots (1975), [7]van Albada et al. (1985), [8]Bosma (1981), [9]Moore & Gottesman (1998), [10]van Albada & Shane (1975), [11]Sempere et al. (1995), [12]Braine et al. (1993), [13]Chincarini & de Souza (1985), [14]Vollmer et al. (1999), [15]Huchtmeier (1975), and [16]Mannheim & Kmetko (1996).

have excellent and poor agreement between the distance D_{tf} (Mpc) calculated using the TF relation and D_{a}; has a D_{a} range of from 0.79 Mpc to 17.70 Mpc; has field and cluster galaxies; and has galaxies with rising, flat, and declining RCs.

When modeling an RC, the knee or turnover of an H I RC is often drawn smooth and relatively rounded. As depicted in Fig. 7 most of RCs of the selected galaxies have an abrupt change of slope at the turnover. Also note another abrupt change of slope at the end of the OR and beginning of the EOR.

Indeed, the RCs are better represented by a piecewise linear curve than rounded curves. The parameters P of the piecewise linear curve are the square of the rotation velocity, the radius, the effective mass (Rv^2/G), and the effective acceleration (v^2/R) at discontinuities in the RC. The slopes of the lines between the parameter points are also parameters that may be calculated in the same manner. The slope in the RR (S_{rr}), the EOR (S_{eor}), and OR (S_{or}) were also calculated. Plots of the parameters of the piecewise linear curves yield discrete variation of the parameters among the sample galaxies. For example, Fig. 8 shows a plot of the maximum rotation velocity v_{rrmax} at the end of the RR versus L_{B}.

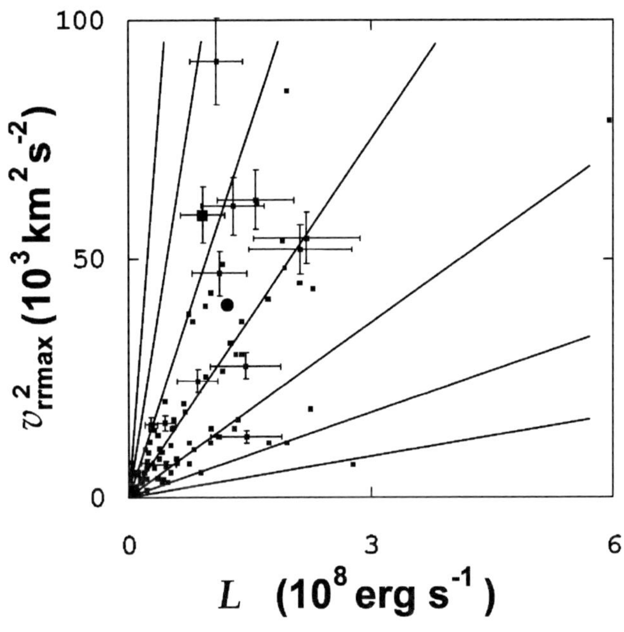

Figure 8. Plots of square of the rotation velocity v^2_{rrmax} ($10^3 \text{km}^2\,\text{s}^{-2}$) at the maximum extent of the RR versus B band luminosity L_{B} ($10^8\,\text{erg}\,\text{s}^{-1}$) for 95 sample galaxies. The other galaxies in the sample lacked a clear definition of v^2_{rrmax}. The 15 select galaxies have error bars that show the uncertainty range in each section of the plot. The data for NGC 2841 are insufficient to determine v^2_{rrmax}. The error bars for the remaining galaxies are omitted for clarity. The straight lines mark the lines whose characteristics are listed in Table 4. The large, filled circle denotes the data point for NGC 5448. The large, filled square denotes the data point for NGC 3031.

Table 4. Data for the integer values of a_1 shown as the plotted lines in Fig. 8.

a_1[a]	No.[b]	Corr.[c]	F Test	S_a[de]	I_a[f]
1	2	1.00	1.00	2.482±0.004	0.006±0.006
2	7	0.97	0.93	6.6 ±0.6	0.1 ±0.7
3	20	0.99	0.96	12.8 ±0.5	-0.8 ±0.8
4	39	0.98	0.92	23.2 ±0.7	0.7 ±0.7
5	24	0.97	0.90	42 ±2	1 ±2
6	2	1.00	1.00	84.3 ±0.3	-0.2 ±0.2
7	1	1.00	1.00	261	0

[a]The integer denoting the exponential value in Eq. (14) for v^2_{rrmax}.
[b]The number of data points for each a_1 value.
[c]The correlation coefficient rounded to two decimal places.
[d]The least squares fit slope of the v^2_{rrmax} - L lines.
[e]The lines were calculated using $(L, v^2_{\mathrm{rrmax}}) = (0,0)$ as a data point.
[f]The least squares fit intercept of the v^2_{rrmax} - L lines.

3.3. Asymmetric RCs

The evidence that the cluster environment affects RC shape appears contradictory (Whitmore et al., 1988; Dale et al., 2001). These studies concentrated on Hα RCs. The HI RC usually extends farther outward than Hα RCs. Also, the SPM suggests the higher $m_{\mathrm{s}}/m_{\mathrm{i}}$ ratio of hydrogen compared to other elements implies HI is more influenced by variation in the ρ field. Therefore, the HI RCs are more sensitive to neighboring galaxy's effects.

Galaxies with a small asymmetry in the RC of a few km s^{-1} are well known (Shane & Bieger-Smith, 1966). Large RC asymmetries are less appreciated (Jog, 2002). Because the observational data from each side of a galaxy RC are generally averaged, only highly asymmetric cases are recognized. RC asymmetry appears to be the norm rather than the exception (Jog, 2002). Weinberg (1995) and Jog (1997) proposed the implied mass asymmetry is due to an imposed lopsided potential caused by galaxy interaction. Dale et al. (2001) found RC asymmetry of early type galaxies falls by a factor of two between the inner and outer regions of clusters. The formation, evolution, and long term maintenance of galactic, kinematical asymmetry remains a mystery.

At equal $R_{\mathrm{major}} > R_{\mathrm{rrmax}}$ on opposite sides of the galaxy, the L_{B} terms of Eq. (9) are equal and the M terms are nearly equal. Define asymmetry A_{sym},

$$A_{\mathrm{sym}} \equiv (v_{\mathrm{h}}^2 - v_{\mathrm{l}}^2)|_R, \tag{11}$$

where v_{h} and v_{l} are the larger v and smaller v at the same R_{major} ($|_R$) on opposite sides of the galaxy, respectively.

The A_{sym} is a function of R_{major}. The A_{sym} and $|\vec{K} \bullet \vec{a}_{\mathrm{o}}|$ are sensitive to both radial (non-circular) motion of particles and to variation of v due to torque. The first approximation of the maximum asymmetry A_{symax} in the OR depends predominately on $|\vec{K} \bullet \vec{a}_{\mathrm{o}}|$. Because A_{symax} is minimally dependant on R, A_{symax} is comparable among galaxies. The $\vec{a}_{\mathrm{o}}$ term depends on the strength and relative position of neighbor galaxies [see Hodge (2006c) for the calculation method].

The calculation of $|\vec{K} \bullet \vec{a}_o|$ also requires knowledge of the orientation of the sample galaxy. The direction of the sample galaxy's polar unit vector $\vec{e}_{polar}$ was defined as along the polar axis of the galaxy. Northward from the center of the sample galaxy was defined as positive. Hu et al. (2006); Pereira & Kuhn (2005); and Trujillo et al. (2006) found an alignment of the polar axis of neighboring spiral galaxies. Therefore, the orientation of $\vec{e}_{polar}$ with the greater $|\vec{e}_{polar} \bullet \vec{a}_o|$ was chosen for the calculations.

The major axis unit vector $\vec{e}_{major}$ was defined as eastward from the center of the sample galaxy along the major axis of the galaxy. The minor axis unit vector $\vec{e}_{minor} \equiv \vec{e}_{major} \times \vec{e}_{polar}$.

The $\vec{K}$ term was calculated by evaluating the $A_{symax} - |\vec{K} \bullet \vec{a}_o|$ relation. The result is

$$\frac{A_{symax}}{10^3\,\mathrm{km}^2\,\mathrm{s}^{-2}} = K_{j1} B_{j1}^{j1} \frac{|\vec{K} \bullet \vec{a}_o|}{10^3\,\mathrm{kpc}^{-1}\,\mathrm{km}^2\,\mathrm{s}^{-2}} \pm 28\%, \tag{12}$$

where the smallest correlation coefficient of the lines $C_{cmin} = 0.98$, $B_{j1} = 2.94 \pm 0.09$, the value of $\vec{K}$ was chosen such that $K_{j1} \approx 1.00$, and

$$\begin{aligned} \vec{K} \;=\; & (3.60 \times 10^{-6})\,\vec{e}_{polar} \\ & -(6.81 \times 10^{-6})\,\vec{e}_{major} \\ & -(1.18 \times 10^{-5})\,\vec{e}_{minor}\,\mathrm{km\,kpc}^{-1} \end{aligned} \tag{13}$$

at 1σ.

The first approximation parameter value selected either L_B or $|\vec{K} \bullet \vec{a}_o|$ depending upon which is predominant. The second approximation considered the less predominant term to be a systematic correction term. Posit the residual between the measured parameter value and the calculated first approximation parameter value is proportional to the less predominant term.

The P include A_{symax} in addition to those listed previously. The equation is

$$\begin{aligned} \frac{P}{unit} = \; & K_1\, B_1^{I_1} \frac{L_B}{10^8\,\mathrm{erg\,s}^{-1}} + \\ & (-1)^s K_2 B_2^{I_2} \frac{|\vec{K} \bullet \vec{a}_o|}{10^3\,\mathrm{kpc}^{-1}\,\mathrm{km}^2\,\mathrm{s}^{-2}} \pm \sigma_e, \end{aligned} \tag{14}$$

where $unit$ is the units of P; K_1, K_2, B_1, and B_2 are constants that are unique for each P; I_1 and I_2 are integers that are unique for each galaxy; $|\vec{K} \bullet \vec{a}_o|$ is the influence of nearby galaxies; s determines the sign of the $|\vec{K} \bullet \vec{a}_o|$ term; and σ_e is the standard deviation of the relative differences ($\delta P/P$) of the sample galaxies. The $\sigma_e < 3\%$ for all except $\sigma_e(v^2_{rrmax}/10^3\,\mathrm{km}^2\,\mathrm{s}^{-2}) = 10\%$ and $\sigma_e(A_{symax}/10^3\,\mathrm{km}^2\,\mathrm{s}^{-2}) = 10\%$

The calculated RCs of the above equations are plotted as solid lines in Fig. 7 for the select galaxies.

The equations show little effect of neighboring galaxies on the H I RC except when orientation and closeness produce a greater $|\vec{K} \bullet \vec{a}|$. Although the neighboring galaxy's effect is less significant on more dense particles, it does induce small perturbations in the orbits of particles that cause the observed asymmetry.

The SPM's $(Gm_gM_g - G_sm_sL_B)/m_\iota R^2$ term has the form of a subtractive change in Newtonian acceleration. The SPM suggests the m_s/m_i change with radius causes the species of matter to change with radius that causes the appearance of a gravitational acceleration change with radius that is related to $L_{\rm B}$. If the $|\vec{K} \bullet \vec{a}|$ term is small, the SPM effective force model reduces to the MOND model as was suggested in Hodge (2006a) for elliptical galaxies.

The use of the $M_{\rm B}$ derived from the TF and the degree of correlation achieved herein suggests the mass term is totally determined by baryonic matter. The $\vec{\nabla}\rho$ field serves the same role in the equations as DM. Equation (9) suggests a much larger spiral galaxy baryonic mass than traditional Newtonian mechanics calculates. Therefore, the $\vec{\nabla}\rho$ field causes Newtonian kinematic measurements to considerably underestimate mass in a galaxy ($M \approx Rv^2/G + G_{\rm s}m_{\rm s}L_{\rm B}/Gm_\iota$). The difference between the SPM and other RC models of the "missing mass problem" is that the added mass follows the luminous mass as suggested by the $m_{\rm s}/m_\iota$ in the equations. Therefore, the added mass is baryonic matter rather than DM, a changing gravitation constant, a changing acceleration, or other form of "fifth force". The "missing mass" is non-luminous, baryonic matter.

The deviation of the data of NGC 5448 from Eq. (14) for $v^2_{\rm rrmax}$ in Fig. 8 suggest a physical mechanism behind the quantized galaxy parameters. The clear departure from circular motion and the significant mass transfer inward (the $\ddot{R} = 0$ condition of Eq. (9) is violated) found by Fathi et al. (2005) suggests this galaxy is in transition from one virial state to another. Further, the noted stellar and gas velocity difference decreases at larger radii. NGC 3031 shows strong, non-circular motion in the disk (Gottesman et al., 1966). This suggests the integer variation is caused by the accumulation of mass at potential barriers such as at R_Δ and $R_{\rm rrmax}$. Continued nucleosynthesis and changing $|\vec{K} \bullet \vec{a}_{\rm o}|$ causes an occasional, catastrophic rupture of one or more of the potential barriers, $\ddot{R} \neq 0$, and, therefore, the transition of the galaxy from one integer classification to another. A smoothly varying transition from the RR to the OR for flat or declining RCs such as NGC 4321 suggests mass is accumulating at a potential barrier at the end of the RR and is being depleted from the outer parts of the OR.

Approximately 66% of the sample galaxies have $a_1 = 4$ or $a_1 = 5$ as seen in Fig. 8 and Table 4. If $a_1 = 4.5$ is used in Eq. (14), $v^2_{\rm rrmax} \propto L_{\rm B}$ for a majority of sample galaxies. Only NGC 4258 of the select galaxies would appear to be an outlier, which may suggest the $v_{\rm rrmax}$ is larger than the measured point chosen herein. Further, the neighboring galaxy effect would fail to improve the $v^2_{\rm rrmax} \propto L_{\rm B}$ relation. The effect of the integer values is to broaden the applicability of the $P - L_{\rm B}$ relations and to establish relations wherein the neighboring galaxy effect improves the calculation. Because the correction to an integer line may be increasing or decreasing, the difficulty in discovering neighboring galaxy effects on galaxies is accounted.

In a similar manner, the URC is consistent with approximately 80% of the galaxies in the present sample. That is, by restricting the integer values for each parameter, the URC curves may be derived. The penalty is that other galaxies with differing integer values are inconsistent with the derived URC. This may be the basis of the disagreement about the URC.

The SPM does not use the mass-to-light ratio to fit RCs. Indeed, Eq. (14) for the effective mass at the end of the RR and EOR imply the effective mass-to-light ratio varies among galaxies as observed.

That neighbor galaxies must have some effect on galaxies is intuitive and is supported by galaxy alignment observations. The SPM incorporates neighbor galaxies in a quantative manner.

The SPM is consistent with the rising, flat, or declining RCs in the OR. The Source of the scalar field acts as a monopole at distances of a few kpc from the center of spiral galaxies. In addition to fitting the RR and OR parameters, the SPM is consistent with EOR and asymmetry observations.

3.4. Central Velocity Dispersion

Because the amplitude and shape of galaxy RCs correlate with galaxy luminosity (Burstein & Rubin, 1985; Catinella, 2006; Hodge, 2006c; Persic et al., 1996), relationships between galaxy central parameters and large scale galaxy parameters are unexpected by Newtonian dynamics.

Whitmore et al. (1979) and Whitmore & Kirshner (1981) found the ratio of the rotation velocity v_c (km s^{-1}) in the flat region of the RC and the central velocity dispersion σ_c (km s^{-1}) $(v_c/\sigma_c) \approx 1.7$ for a sample of S0 and spiral galaxies. Gerhard et al. (2001) found the maximum circular velocity of giant, round, and nearly non-rotating elliptical galaxies is correlated to the central velocity dispersion σ_c. Ferrarese (2002) discovered a power law relationship between circular velocity v_{c25} (km s^{-1}) beyond the radius R_{25} of the 25th isophote and σ_c for a sample that also included elliptical galaxies (see her Fig. 1). Baes et al. (2003) expanded on the data for spiral galaxies with flat and smooth RCs. NGC 0598 was a clear outlier. The v_{c25} for NGC 0598 used in Ferrarese (2002) was 135 km s^{-1} that is the highest data point of a rising RC (Corbelli & Salucci, 2000) Figure 7 shows this data point is in the EOR and subject to strong neighbor galaxy effects. Galaxies with $\sigma_c < 70$ km s^{-1} ($v_{c25} < 150$ km s^{-1}) also deviate from the linear relation. NGC 4565 was excluded because of warps in the HI disk. Also, galaxies with significant non-circular motion of the HI gas such as NGC 3031, NGC 3079, and NGC 4736 were omitted in Ferrarese (2002). NGC 3200 and NGC 7171 were also excluded from Ferrarese (2002) because of discrepant σ_c values in the literature.

Pizzella et al. (2005) found results similar to Ferrarese (2002) for high surface brightness (HSB) galaxies. Pizzella et al. (2005) also found the data consistent with a linear v_c–σ_c relation. HSB galaxies with flat RCs were chosen for the sample. Also, galaxies with highly asymmetric RCs and galaxies with RCs not characterized by an outer flat portion were excluded. Also, Pizzella et al. (2005) found the v_c–σ_c linear relation for low surface brightness (LSB) galaxies is offset with a larger v_c intercept relative to HSB galaxies. Buyle et al. (2004) confirmed this distinction between HSB and LSB galaxies for $\sigma_c > 80$ km s^{-1} and found that the distinction is less pronounced for galaxies with $\sigma_c < 80$ km s^{-1}. They concluded that the scatter of the v_c–σ_c relation is a function of galaxy mass or that the v_c–σ_c relation changes at $\sigma_c \approx 80$ km s^{-1}. Hodge (2006c) suggested the offset of the measurement of v_c between HSB and LSB galaxies is because the measurement is in an area of the rotation curve (RC) more influenced by neighboring galaxies and in an area of the generally rising RC of LSB galaxies. Further, the criteria for excluding galaxies describe characteristics of a large influence from neighbor galaxies. Using the v_{rrmax} as defined by Hodge (2006c) and Subsection 3.2. may be more appropriate.

The sample host galaxies consisted of 82 galaxies. Of the host galaxies, 60 are Source galaxies and 22 are Sink galaxies. This host galaxy sample has LSB, medium surface brightness (MSB), and HSB galaxies; includes LINER, Sy, HII, and less active galaxies; field and cluster galaxies; galaxies with rising, flat, and declining RCs; and galaxies with varying degrees of asymmetry. The host sample includes NGC 0598, NGC 3031, NGC 3079, NGC 3200, NGC 4565, NGC 4736, and NGC 7171 that were excluded from Ferrarese (2002), six galaxies with $\sigma_c < 70$ km s^{-1}, and galaxies that Pizzella et al. (2005) would exclude.

Appling the same procedure used by Hodge (2006c) and Subsection 3.2. for finding the parametric equations [Eq. (14)] yields:

$$\frac{\sigma_c^2}{10^3\,\mathrm{km}^2\,\mathrm{s}^{-2}} = K_{\sigma 1}\,B_{\sigma 1}^{m_1}\,\frac{L}{10^8\,\mathrm{erg\,s}^{-1}} + (-1)^s K_{\sigma 2} B_{\sigma 2}^{m_2}\,\frac{|\vec{K}\bullet\vec{a}_o|}{10^3\,\mathrm{kpc}^{-1}\,\mathrm{km}^2\,\mathrm{s}^{-2}} \pm 6\%, \quad (15)$$

where $K_{\sigma 1} = 1.6\pm 0.3$, $K_{\sigma 2} = (2.5\pm 0.7)\times 10^{-3}$, $B_{\sigma 1} = 1.88\pm 0.06$, $B_{\sigma 2} = 2.52\pm 0.09$, and $\sigma_e = 18\%$ for the L term, only.

3.5. Central Mass

The masses of compact stellar clusters at the center of low- and intermediate-luminosity galaxies also correlate with the mass of the host galaxy (Ferrarese et al., 2006; Wehner & Harris, 2006). Ferrarese et al. (2006) suggested the compact stellar clusters and the supermassive black hole (SBH) modeled as being at the center of high-luminosity galaxies should be grouped together under the terminology of "Central Massive Objects" (CMOs) with mass M_{cmo}. The finding of the correlation between M_{cmo} and the total mass in a galaxy M_{gal} suggests a similar galaxy formation process (Ferrarese et al., 2006; Wehner & Harris, 2006).

Ghez et al. (2000) and Ferrarese & Merritt (2002) have observed Keplerian motion to within one part in 100 in elliptical orbits of stars that are from less than a pc to a few 1000 pc from the center of galaxies. The stars within nine light hours of the Galaxy center have velocities of 1300 km s^{-1} to 9000 km s^{-1}(Schödel, 2002) and high accelerations (Ghez et al., 2000). A huge amount of mass M_c ($M_\odot$) such as millions of black holes, dense quark stars (Prasad & Bhalerao, 2003, and references therein), and ionized iron (Wang et al., 2002) must be inside the innermost orbit of luminous matter (Dunning-Davies, 2004; Ghez et al., 2000, 2003, 2005; Schödel, 2002).

The central mass M_c varies among galaxies from 10^6 $M_\odot$ to 10^{10} $M_\odot$ (Ferrarese & Merritt, 2000; Gebhardt et al., 2000a). Ferrarese (2002) found the ratio of the M_c to the mass M_{DM} of the dark matter halo thought to be around spiral galaxies is a positive value that decreased with M_{DM}. The M_c can be distributed over the central volume with a density of at least 10^{12} $M_\odot$ pc^{-3} (Dunning-Davies, 2004; Ghez et al., 1998, 2000, 2005). The orbits of stars closest to the center of the Galaxy are following elliptical paths (Ghez et al., 2000) that suggests a net, attractive central force consistent with the Newtonian spherical property (Ghez et al., 2003; Schödel, 2003).

That M_c is crowded into a spherical volume with a radius of less than 45 AU is proven (Ghez et al., 2005). That the structure of M_c is a SBH or CMO is widely accepted, but unproven (see Kormendy & Richstone 1995 for a discussion). Kormendy & Richstone (1995); Magorrian et al. (1998) suggested the Newtonian model implies the M_c must either quickly dissipate or must quickly collapse into a SBH . The long-term maintenance of M_c rules out the first possibility. Mouawad et al. (2004) suggested there is some extended mass around Sgr. A*. Observations have ruled out many models of the nature of M_c of galaxies (Ghez et al., 2003; Schödel, 2003). Shu et al. (2005) and Silk & Rees (1998) suggested a third alternative that a repulsive force, called a "wind" (a gas), acted on the cross sectional area of particles. Therefore, denser particles such as black holes move inward relative to less dense particles. Less dense particles such as hydrogen gas move outward. The SPM suggests that there is a plenum Source that induces a large repulsive force on matter from the center of spiral galaxies.

Observations inconsistent with the SBH model include shells of outward flowing, shocked gas around galactic nuclei (Binney & Merrifield, 1998, page 595)(Königl, 2003); the apparent inactivity of the central SBH (Baganoff et al., 2001; Baganoff, 2003a; Nayakshin & Sunyaev, 2003; Zhao et al., 2003); and the multitude of X-ray point sources, highly ionized iron, and radio flares without accompanying large variation at longer wavelengths reported near the center of the Milky Way (Baganoff et al., 2001; Baganoff, 2003a,b; Binney & Merrifield, 1998; Genzel et al., 2003; Zhao et al., 2003; Wang et al., 2002).

McLure & Dunlop (2002) found that M_c is 0.0012 the mass of the bulge. Ferrarese & Merritt (2002) reported that about 0.1% of a galaxy's luminous mass is at the center of galaxies and that the density of SMB's in the universe agrees with the density inferred from observation of quasars. Merritt & Ferrarese (2001c); Graham et al. (2008) found similar results in their study of the $M_c - \sigma$ relation. Ferrarese (2002) found a "tight" relation between v_c in the outer disk region and bulge velocity dispersion σ_c ($v_c - \sigma_c$), which is strongly supporting a relationship of the center force with total gravitational mass of a galaxy. Wandel (2003) showed the M_c of AGN galaxies and their bulge luminosity follow the same relationships as their ordinary (inactive) galaxies, with the exception of narrow line AGN. Graham et al. (2003,a); Graham et al. (2003b) found correlations between M_c and structural parameters of elliptical galaxies and bulges. Either the dynamics of many galaxies are producing the same increase of mass in the center at the same rate or a feedback controlled mechanism exists to evaporate the mass increase that changes as the rate of inflow changes as suggested by Merritt & Ferrarese (2001a). The RCs imply the dynamics of galaxies differ so the former is unlikely.

The M_c correlation with Blue band luminosity L_{bulge} of the host galaxy's bulge (Kormendy & Richstone, 1995) has a large scatter. The $M_c \propto \sigma_c^{\alpha}$, where α varies between 5.27±0.40 (Ferrarese & Merritt, 2000) and 3.75±0.3 (Gebhardt et al., 2000a). The $M_c - \sigma_c$ relation appears to hold for galaxies of differing Hubble types, for galaxies in varying environments, and for galaxies with smooth or disturbed morphologies. Tremaine et al. (2002, and references therein) suggested the range of α is caused by systematic differences in the velocity dispersions used by different research groups. Merritt &and Ferrarese (2001b) found the range of α is partly due to the type of regression algorithm used and partly due to the velocity dispersion of the Galaxy sample selected. Also, discrepancies have been noted among the methods used to measure M_c (Gebhardt et al., 2000b; Merritt & Ferrarese, 2001a, and references therein). Bernardi et al. (2006) found a selection bias or large scatter in the M_c–σ and M_c–L_{bulge} correlations that may be the result of more fundamental relations among M_c, σ, and $L_{\rm bulge}$.

The sample of 29 host galaxies of Subsection 3.4. were used in the M_c calculation.

Appling the same procedure used by Hodge (2006c) and Subsection 3.2. for finding the parametric equations yields:

$$\begin{aligned}\frac{M_c}{10^8\,M_\odot} = &\quad K_{\rm M1}\,B_{\rm M1}^{n_1}\,\frac{L}{10^8\,{\rm erg\,s^{-1}}} + \\ &(-1)^{s_{\rm M}}K_{\rm M2}B_{\rm M2}^{n_2}\frac{|\vec{K}\bullet\vec{a}_{\rm o}|}{10^3\,{\rm kpc^{-1}\,km^2\,s^{-2}}} \pm 3\%, \qquad (16)\end{aligned}$$

where $K_{M1} = 0.7 \pm 0.1$, $K_{M2} = (5.1 \pm 0.8) \times 10^{-3}$, $B_{\rm M1} = 1.73 \pm 0.04$, and $B_{\rm M2} = 1.70 \pm 0.03$, and $\sigma_{\rm e} = 18\%$ for the L term, only.

Equation (16) was derived assuming the Newtonian spherical property applied. That is, the Source was internal to the orbits of the mass under consideration. The applicability of Eq. (14) to σ_c and M_c suggests the Source or Sink acts as a monopole internal to the particles of M_c.

4. SPM Speculations of Galaxy Structure, Formation, and Evolution

Spiral galaxies form around Sources. The stuff of the plenum and fundamental matter particles come through the Source into our universe. The plenum flows outward. The fundamental particles form photons and elementary particles that eventually form hydrogen.

The $\vec{\nabla}\rho$ of the plenum forces hydrogen outward. The orbits of stars closest to the center are approximately 1,169 times (Schödel, 2002) the Schwartschild radius of the theorized SBH (Ghez et al., 2000). The SPM suggests a strong repulsive force at the center of spiral galaxies would explain the shells of shocked gas around galactic nuclei (Binney & Merrifield, 1998, page 595), (Königl, 2003), the apparent inactivity of the central object (Baganoff et al., 2001; Baganoff, 2003a; Nayakshin & Sunyaev, 2003; Zhao et al., 2003), and the surprising accuracy of reverberation mapping as a technique to determine M_c (Merritt & Ferrarese, 2001a).

The hydrogen may leave the galaxy, combine with existing stars, or combine in a weak ρ field to form hydrogen stars. The hydrogen stars produce radiation and heavier elements through nucleosynthesis.

Matter in a spiral galaxy is subjected to the force of ρ from the Source, from other Sources and Sinks, and from other matter (gravity). The ρ field from other Sources and Sinks cause the stars and gases to gain angular momentum in latitudinal orbits. Because the stars in small circle orbits have a force component of $|\vec{K} \bullet \vec{a}_{\rm o}|$ moving the star to a great circle orbit, disks are formed. The alignment of disk axis with close galaxies has been noted (Hu et al., 2006; Pereira & Kuhn, 2005; Trujillo et al., 2006). An indication of this disk forming proposition was seen in Subsection 3.3. wherein the polar vector component of K was the smallest. That is, the $\vec{\nabla}\rho$ from neighbor galaxies in the plane of the disk had more influence on asymmetry and the RCs than the $\vec{e}_{\rm polar}$ component.

Further, the changing relative position and orientation of neighbor galaxies causes the $|\vec{K} \bullet \vec{a}_{\rm o}|$ to change. The outer disk is the most sensitive to this change. Thus, warped disks result.

Spectra coming from HII regions are correlated systematically with R and little else (Binney & Merrifield, 1998, pp. 516-522). The interstellar abundances of metals in a disk galaxy decrease with increasing radius (Tadross, 2003). Also, M_{B} of a galaxy is correlated with the metallicity obtained by extrapolating [O/H] within the disk to the galactic center. Low luminosity galaxies tend to be metal poorer than high luminosity galaxies.

Steinmetz & Navarro (2002) found in a series of N-body/gas dynamical simulations that included feedback: that feedback is a necessary component for morphology determination; that the main morphological component is regulated by the mode of gas accretion and intimately linked to *discrete* accretion events; that morphology is a transient phenomenon; and that the Hubble sequence reflects the varied accretion histories of galaxies. If luminosity is proportional to the ϵ, which directly causes the parameters of the RC, then there must exist a feedback mechanism controlling the parameters of the RC. The SPM suggests this posited mechanism controls the number and type of stars in a galaxy that yield a galaxy's luminosity.

Subsections 3.2. and 3.3. suggests the accumulation of mass at potential barriers such as at R_{Δ} and R_{rrmax} and suggests the occasional, catastrophic rupture of one or more of the potential barriers.

The RC differs for different particles. For example, the HI RC and the RCs of stars as shown by the Hα line for NGC 4321 (Sempere et al., 1995) differ in the RR and approach each other in the OR as shown in Fig. 9.

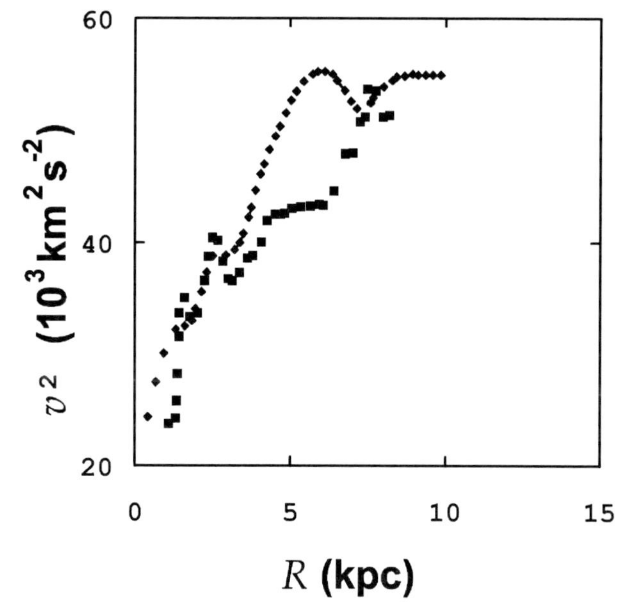

Figure 9. Plots of the square of the rotation velocity v^2 (10^3 km^2 s^{-2}) versus galactocentric radius R (kpc) of the HI RC (filled diamonds) and Hα RC (filled squares) for NGC4321 (Sempere et al., 1995).

In Fig. 9 the HI RC at smaller radius in the RR has two scalloped shapes that suggests spherically symmetric shells of matter. Also, the Hα RC rapidly increases, peaks, and then declines at the beginning of each shell. The Hα lines are generally formed in excited interstellar gas. In the disk region of a galaxy, the gas is usually excited by hot stars (Binney & Merrifield, 1998, pp.576-7). Because the m_s/m_ι factor must be different for different matter types, each shell has a different metallicity star type. Because the Hα RC approaches the HI RC in the disk region such as plotted in Fig. 9 with hot, hydrogen burning stars, the m_s/m_ι factor must be the same for HI and hydrogen stars. This suggests the m_s/m_ι factor varies by element type and acts on atoms at the largest. The metallicity - radius relation follows.

The m_s/m_ι ratio of stars is changing through changing elemental composition by nucleosynthesis in addition to accretion and emission of matter. Therefore, the HI RC is preferred to trace the forces influencing a galaxy outside the bulge.

Inward from the knee of the RC, stars loose angular momentum through velocity reduction until their R decreases. The elemental change of nucleosynthesis is slowed or even stopped if the core is iron. The scalloped shapes of the Hα RC marks the borders of m_s/m_g families. If the LP is a nucleon, an added nucleon may make the nucleus oblong that leaves the m_s unchanged and increases m_g. For example, the addition of a neutron to a proton forming deuterium increases the m_g while leaving the minimum cross-section of one nucleon. The tendency of nuclei to be spherical limits the length of the oblong axis and forms the m_s/m_g family. Therefore, the nuclear cross-section (m_s) changes in steps rather than continuously.

The SPM includes the nucleosynthesis of atoms as they progresses toward the center of galaxies. That is, matter becomes denser and ρ increases as R decreases. Eventually, the matter becomes iron atoms with electrons. At the end of the nucleosynthesis, the SPM suggests the increasing ρ helps cause the posited fundamental particles of matter to compress to neutron stars, quark stars, and black holes. Thus, black holes are the most compressed form of matter.

Black holes become denser (the m_s/m_i decreases) as they grow. The SPM suggests this density growth results in faster travel to the Source. Thus, a limit on black hole size results.

If gravitational collapse of a mass density were sufficient, the atoms and nucleons on Earth would form neutron stars and black holes.

The suggested CMO structure is a central Source of a matter-repulsive $\rho \propto R^{-1}$ surrounded by a spherical shell of dense matter such as black holes, neutron stars, and iron nuclei. Star cores may accumulate hydrogen, which is traveling outward, and form an especially bright star because of the ρ compression in addition to hydrogen fusion.

The SPM suggests the $L \propto \epsilon$ and, therefore, $F_s \propto \nabla\rho$ at a given R. Therefore, the m_s/m_i of particles at a given radius varies with L. Therefore, the galaxies with larger L will have more mass in the center shell to balance the higher F_s with the gravitational force F_g. Therefore, the SPM naturally leads to the smoothness of the $M_{cmo} - M_{gal}$ relation for the full range of CMO spiral galaxies.

Baganoff et al. (2001); Baganoff (2003a,b); Binney & Merrifield (1998); Genzel et al. (2003); Zhao et al. (2003); Wang et al. (2002) reported a multitude of X-ray point sources and radio flares without accompanying large variation at longer wavelengths near the center of the Milky Way. The existence of a mechanism other than matter losing energy as it descends to the center of galaxies is suggested. Wang et al. (2002) also reported highly ionized iron in the central region of our galaxy.

Subsection 3.5. suggest the central volume of spiral galaxies has millions of black holes, dense quark stars (Prasad & Bhalerao, 2003, and references therein), and ionized iron (Wang et al., 2002) distributed inside the innermost orbit of luminous matter. Combining these observations with the SPM suggestion of the formation of black holes yields the proposition that multitudes of black holes within 60 AU of the Source revolve around the Source.

If this speculation is essentially correct, then the correlation of central parameters with spiral galaxy global and RC parameters suggests not only a similar galaxy formation process but also a self - regulatory, negative feedback process continually occurring. Feedback processes have been suggested in several recent studies of galaxies with CMOs (e.g. Li et al. 2006; Merritt & Ferrarese 2001a; Robertson et al. 2006).

The ϵ is the control of the negative feedback process. If the mass of the CMO increases, then F_{g} increases and mass migrates inward. At very high ρ, the high repulsive F_{s} compresses matter, the mass (black hole) cracks like complex molecules in the high heat and pressure of a fractional distillation process, and matter is reclaimed as radiation (X-rays) and other elementary particles that form hydrogen. This accounts for the large amount of hydrogen outflowing from the Galaxy center and shocked gas near the Galaxy center. A single black hole reclamation event is consistent with the periodic X-ray pulses from the Galaxy center. Further, the feedback loop controlled by ϵ is the connection among the central parameters, outer RC parameters, and the global parameters of spiral galaxies. However, the ϵ of a galaxy acts only radially. Therefore, the $|\vec{K} \bullet \vec{a}_{\mathrm{o}}|$ terms effects are the asymmetry and the formation, evolution, and maintenance of the rotation of particles. Thus, a feedback situation is established with the amount of matter in the galaxy and luminosity of the galaxy caused by the Source strength.

The transition of spiral galaxies from one integer value to another appears to be more than a process of initial formation to a stable morphology. Rather, as a large amount of matter is transported inward, the bulge region increases that makes more black holes. The increased central mass increases the amount of evaporation. Thus, a cycling, feedback mechanism is formed controlling the amount of mass in the outer regions of spiral galaxies. Further, the process of moving matter from the outer regions to the inner regions requires time. Therefore, mature spiral galaxies are in a cycle of changing their outer parameters and their morphology. Because the L_{B} is dependent on the type and amount of stars in a galaxy, L_{B} is also cycling.

Supernovae may provide enough outward velocity to cause denser matter to leave the galaxy.

Far from Sources, the $\nabla \rho$ field due to the Sources is very small. The value of the ρ field from Sources is not the governing factor of the kinematics of matter. For example, the volume half way between two equal Sources has $\vec{\nabla} \rho = 0$ due to Sources even if $\rho > 0$. Matter is the predominate influence on $\nabla \rho$ in the intergalactic fields. Matter then condenses in volumes between spiral galaxies. Further, more matter will be between close spiral galaxies. Therefore, this greater depression in the ρ field causes all the matter from the close galaxies to flow to the volume between the galaxies. This forms the structure of galaxy clusters.

As matter flows down the potential well, the $\vec{\nabla} \rho$ increases matter's kinetic energy that it must loose to increase density. It does this through radiation such as radio emission.

Occasionally, enough cooling and density allows a "radio galaxy" to form. Increased cooling and condensation forms elliptical galaxies. The difference between a radio galaxy and an elliptical galaxy is the amount of matter, temperature, and density in the volume. The greater density of the elliptical galaxy allows the hydrogen and other elements to form stars. Unlike spiral galaxies with the high ρ field to aid compression, the elliptical galaxies with a low ρ must allow iron stars to become larger and to migrate to the center of the elliptical galaxy without forming black holes. This forms a deep well in the ρ field.

Radio galaxies are formed by the same mechanism that forms elliptical galaxies. Therefore, radio galaxies are preferentially found in the same region as elliptical galaxies (Binney & Merrifield, 1998, pp. 245-6).

What is this mass in the center of elliptical galaxies? The SPM suggests black holes are formed by gravitational collapse and ρ pressure. The ρ pressure in an elliptical galaxy is very small. An iron ball a few AU in diameter would be non-radiating and as massive as the theorized SBH masses in spiral galaxies.

In previous calculations the η number associated with matter and elliptical galaxies was negative for calculating convenience. However, if the plenum exists, some plenum must exist in the center of elliptical galaxies as part of the structure of the iron. The Feedback Principle implies a mechanism that limits the amount of energy in the universe must exist. Continuing the analogy of the plenum obeying the heat equation, the η represents the rate of flow of energy out of the universe. Energy has two parts, matter and the plenum. The near agreement of the measured temperature of the universe with this model lends support for this speculation (Hodge, 2006b).

The similarity of the central masses of spiral and elliptical galaxies suggests Sources and Sinks are caused by similar mechanisms.

There may be galaxies with greater and smaller values of ϵ than in spiral galaxies. For instance, QSOs may have a higher value of ϵ that ejects matter from a spiral configuration (e.g. see the images of Sulentic & Arp 1987). A smaller value of ϵ would be insufficient to form a disk.

Alternately, Arp (1998) suggests QSOs are associated with and are being generated by nearby spiral galaxies. The SPM would consider QSOs as early forms of spiral galaxies that have not yet formed the mass of stars around them. Without the mass the ϵ would be misrepresented relative to mature Source galaxies. Therefore, their luminosity would be low for their Source strength. Hence, the controversy concerning QSOs distance[5] and luminosity. Also, this accounts for the dimensions of galaxy clusters relative to the distance between clusters. The splitting of Sources would be a feedback situation by limiting Source size.

5. Conclusion

The special focus of the present investigation included galaxies that are problematic in other models. Considering the range of observations and range of galaxy characteristics with which the SPM is consistent, the SPM is a relatively simple model.

[5]Hodge (2006a) suggests z of galaxies is not a Doppler shift. Without the mass around QSOs, the ρ field is modified to increase QSOs z.

This chapter has suggested a shift in fundamental principles; has reviewed the confrontation of data with the SPM; and has suggested the SPM of the structure, formation and evolution of galaxies. Qualitative speculations have been advanced about models of QSOs; of black holes; of clouds of shocked, out flowing hydrogen near the center of spiral galaxies; of the relationship of galaxy luminosity to many galaxy parameters; of the cause of the periodic X-ray bursts emanating from the center of spiral galaxies; of the cause of the differing disk H I and Hα RCs; of the cause of the differing inner RCs of different trace materials; of the cause of disks in spiral galaxies; of the cause of warps in the disks; of cooling flows; and of the cause of radio galaxies being preferentially near elliptical galaxies.

Acknowledgments

The anonymous reviewer of Hodge (2006a) inspired the investigation of the Pioneer Anomaly within the context of the SPM.

This research has made use of the NASA/IPAC Extragalactic Database (NED) which is operated by the Jet Propulsion Laboratory, California Institute of Technology, under contract with the National Aeronautics and Space Administration.

This research has made use of the LEDA database (http://leda.univ-lyon1.fr).

I acknowledge and appreciate the financial support of Maynard Clark, Apollo Beach, Florida, while I was working on this project.

References

Aaronson, M. et al. *ApJ* 1982, 258, 64.

Anderson, J. D. et al. *Phys. Rev. D* 2002, 65, 082004.

Arp, H. *Seeing Red: Redshifts, Cosmology and Academic Science*, Apeiron: Montreal, Canada, 1998.

Baes, M. et al. *MNRAS* 2003, 341L, 44.

Baganoff, F.K. et al. *Nature (London)* 2001, 413, 45.

Baganoff, F.K. *American Astronomical Society* 2003a, HEAD meeting #35, session #03.02.

Baganoff, F.K. *ApJ* 2003b, 591, 891.

Battaner, E.; Florido, E. *Fundamentals of Cosmic Physics* 2000, 21, 1.

Begeman, K.G., Broeils, A. H., Sanders, R. H. *MNRAS* 1991 249, 523.

Bekenstein, J. D. *Phys. Rev D* 2004, 70, 083509.

Bernardi, M.; Sheth, R. K.; Tundo, E.; Hyde, J. B. (2004). preprint http://www.arxiv.org/abs/astro-ph?0609300.

Bertolami, O.; Páramos, J. P. *Clas. Quantum Gravity* 2004, 21, 3309.

Binney, J.; Merrifield, M. *Galactic Astronomy.*; Princeton University Press: Princeton, NJ, 1998.

Bosma, A. *AJ* 1981, 86 1791.

Bottema, R. et al. *A&A* 2002, 393, 453.

Braine J. et al. *A&A* 1993, 280, 451.

Brownstein, J. R.; Moffat, J. W. *ApJ* 2006, 636, 721.

Burnstein, D.; Rubin, V. C. *ApJ* 1985, 297, 423.

Buyle, P. et al. (2006). preprint http://www.arxiv.org/abs/astro-ph?0609286.

Carignan, C.; Freeman, K. C. *ApJ* 1985, 294, 494.

Catinella, B.; Giovanelli, R.; Haynes, M. P. *ApJ* 2006, 640, 751.

Ceccarelli, M. L. et al. *ApJ* 2005, 622, 853.

Chincarini, G.; de Souza, R. *A&A* 1985, 153, 218.

Cooperstock, F. I.; Faraoni, V.; Vollick, D. N. *ApJ* 1998, 503, 61.

Corbelli, E.; Salucci, P. *MNRAS* 2000, 311, 441.

Courteau, S. *AJ* 1997, 114, 2402.

Dale, D. A. et al. *AJ* 2001, 121, 1886.

Dunning-Davies, J. (2004) preprint http://www.arxiv.org/abs/astro-ph?0402290.

Fathi, K. et al. *MNRAS* 2005, 364, 773.

Fay, S. *A&A* 2004, 413, 799.

Ferrarese, L.; Merritt, D. *ApJL* 2000, 539L, 9.

Ferrarese, L.; Merritt, D. (2002). preprint http://www.arxiv.org/abs/astro-ph?0206222.

Ferrarese, L. et al. *ApJL* 2006, 644L, 21.

Ferrarese, L. *ApJ* 2002, 578, 90.

Freedman, W. A. et al. *ApJ* 2001, 533, 47.

Garrido, O.; Marcelin, M.; Amram, P. *MNRAS* 2004, 349, 225.

Gebhardt, K. et al. *ApJL* 2000a, 539L, 13.

Gebhardt, K. et al. *ApJL* 2000b, 543L, 5.

Genzel, R. et al. *Nature (London)* 2003, 425, 934.

Gerhard, O. et al. *AJ* 2001, 121, 1936.

Ghez, A. M. et al. *ApJ* 1998, 509, 678.

Ghez, A. M. et al. *Nature* 2000, 407, 349.

Ghez, A. M. et al. *Astro. Nachr.* 2003, 324, 527.

Ghez, A. M. et al. *ApJ* 2005, 620, 744.

Goodman, J. *NewA* 2000, 5, 103.

Gottesman, S. T.; Davies, R. D.; Reddish, V. C. *MNRAS* 1966, 133, 359.

Graham, A. W. et al. *AJ* 2003b 125, 2951.

Graham, A. W. et al. (2003). preprint http://arxiv.org/PS_cache/astro-ph/pdf/0306/0306023v1.pdf

Graham, A. W. et al. *RMxAC* 2003a. 17, 196.

Graham, A. W. et al. (2008). preprint http://arxiv.org/PS_cache/arxiv/pdf/0807/0807.2549v1.pdf.

Hodge, J. C. *NewA* 2006a 11, 344. preprint http://www.arxiv.org/abs/astro-ph?0602344.

Hodge, J. C. (2006b). preprint http://www.arxiv.org/abs/astro-ph?0603140.

Hodge, J. C. (2006c). preprint http://www.arxiv.org/abs/astro-ph?0611029.

Hodge, J. C. (2006d). preprint http://www.arxiv.org/abs/astro-ph?0611699.

Hodge, J. C. (2006e). preprint http://www.arxiv.org/abs/astro-ph?0612567.

Hu, F. X. et al. *Ap&SS* 2006, 302, 43.

Huchtmeier, W. K. *A&A* 1975, 45, 259.

Hudson, M. J. et al. *MNRAS* 2004, 352, 61.

Jog, C. J. *ApJ* 1997, 488, 642.

Jog, C. J. *A&A* 2002, 391, 471.

Königl, A. K. (2003). preprint http://www.arxiv.org/abs/astro-ph?0302109.

Knapen, J. H. *ApJ* 1993, 416, 563.

Kormendy, J.; Richstone, D. *Annual Review of Astronomy and Astrophysics* 1995, 33, 581.

Krumm, N.; Salpeter, E. E. *AJ* 1979, 84, 1138.

Li, Y. et al. (2006). preprint http://www.arxiv.org/abs/astro-ph?0607444.

Lilje, P. B.; Yahil, A.; Jones, B. J. T. *ApJ* 1986 307, 91.

McLure, R. J.; Dunlop, J. S. (2002). preprint http://www.arxiv.org/abs/astro-ph?0201081.

Macri, L. M. et al. *ApJ* 2001, 559, 243.

Magorrian, J. et al. *AJ* 1998, 115, 2285.

Mannheim, P. D.; Kmetko, J. (1996). preprint http://www.arxiv.org/abs/astro-ph?9602094.

Mbelek, J. P. *A&A* 2002, 424, 761.

Merritt, D.; Ferrarese, L. (2001a). preprint http://www.arxiv.org/abs/astro-ph?0107134.

Merritt, D.; Ferrarese, L. *ApJ* 2001b, 547, 140.

Merritt, D.; Ferrarese, L. *MNRAS* 2001c, 320, L30.

Moffat, J. W. *Journal of Cosmology and Astro-Particle Physics* 2005, 5, 3.

Moore, E. M.; Gottesman, S. T. *MNRAS* 1998, 294, 353.

Mouawad, N. et al. *Astron. Nach.* 2004, 325, 102.

Narlikar, J. V. *An Introduction to Cosmology*. Cambridge University Press: Cambridge, UK, 2002.

Nayakshin, S.; Sunyaev, R. (2003). preprint http://www.arxiv.org/abs/astro-ph?astro-ph/0302084.

Nieto, M. M.; Anderson, J. D. *Class. Quant. Grav.* 2005, 22, 5343. preprint http://www.arxiv.org/abs/gr-qc?0507052 .

Nieto, M. M.; Turyshev, S. G.; Anderson, J. D. *AIPC* 2005 758, 113. preprint http://www.arxiv.org/abs/gr-qc?0411077 .

Padmanabhan, T.; Choudhury, T. R. *Phys. Rev D* 2002, 66, 081301.

Pecker, J.; Narliker, J. V. *Current Issues in Cosmology*, Cambridge University Press: Cambridge, UK, 2006.

Pereira, M. J.; Kuhn, J. R. *ApJL* 2005, 627L, 21.

Persic, M.; Salucci, P.; Stel, F. *MNRAS* 1996, 281, 27.

Pizzella, A. et al. *ApJ* 2005, 631, 785.

Pound, R. V.; Rebka Jr., G. A. *Phys. Rev. Letters* 1960, 4, 337.

Prasad, N.; Bhalerao, R. S. (2003). preprint http://www.arxiv.org/abs/astro-ph?0309472.

Rejkuba, M. *A&A* 2004, 413, 903.

Robertson, B. et al. *ApJ* 2006, 641, 90.

Roscoe, D. F. *A&A* 2002, 385, 431.

Rots, A. H. *A&A* 1975, 45, 43.

Sanders, R. H.; McGaugh, S. S. *ARA&A* 2002, 40, 263.

Sanders, R. H. 2006, In *MOND and Cosmology*; Mamon, G. A.; Combes, F.; Deffayet, C.; Fort, B.;Eds.; EAS Publications Series.pp 231-238.

Schödel, R. *Nature* 2002, 419, 694.

Schödel, R. *ApJ* 2003, 596, 1015.

Sellwood, J. A.; Kosowsky, A. *ASPC* 2001, 240, 311.

Sempere, M. J. et al. *A&A* 1995, 296, 45.

Shane, W. W.; Bieger-Smith, G. P. *Bull. Astron. Inst. Neatherlands* 1966, 18, 263.

Shu, C.; Mo, H. J.; Mao, S. *ChJAA* 2005, 5, 327.

Silk, J.; Rees, M. J. *A&A* 1998, 331, 1.

Sofue, Y. *ApJ* 1999, 523, 136.

Sofue, Y.; Rubin, V. *ARA&A* 2001, 39, 137.

Steinmetz, M.; Navarro, J. F. *NewA* 2002, 7, 155.

Sulentic, J. W.; Arp, H.C. *ApJ* 1987, 319, 687.

Tadross, A. L. *NewA* 2003, 8, 737

Takamiya, T.; Sofue, Y. *ApJ* 2000, 534, 670

Toth, V. T.; Turyshev, S. G. (2006). preprint http://www.arxiv.org/abs/gr-qc?0603016 .

Tremaine, S. et al. *ApJ* 2002, 574, 740.

Trujillo, I. et al. *ApJL* 2006, 640L, 111.

Tully, R. B.; Fisher, J. R. *A&A*, 1977, 54, 661

Turyshev, S. G. et al. (1999). preprint http://www.arxiv.org/abs/gr-qc?9903024.

Turyshev, S. G. et al. (2006). preprint http://www.arxiv.org/abs/gr-qc?0512121v2.

Unzicker, A. (2007) preprint http://www.arxiv.org/abs/gr-qc?0702009.

van Albada, G. D.; Shane, W. W. *A&A* 1975, 42 433

van Albada, T. S. et al. *ApJ* 1985, 295 305

Vollmer, B. et al. *A&A* 1999, 349 411

Wandel, A. (2003) preprint http://www.arxiv.org/abs/astro-ph?0303562.

Wang, Q. D. et al. *Nature (London)* 2002, 145 148.

Wehner, E. H.; Harris, W. E. *ApJL* 2006, 644L, 17.

Weinberg, M. D. *ApJL* 1995, 455L 31.

Whitmore, B. C.; Kirshner, R. P. *ApJ* 1981, 250, 43.

Whitmore B. C. et al. *ApJ* 1979, 234, 68.

Whitmore, B. C.; Forbes, D. A.; Rubin, V. C. *ApJ* 1988, 333, 542.

Will, C. *Theory and Experiment in Gravitational Physics*, Cambridge University Press: Cambridge, United Kingdom 1993.

Willick, J. A. *ApJ* 1999, 516, 47.

Zhao, J. et al. (2003). preprint http://www.arxiv.org/abs/astro-ph?0302062.

INDEX

A

B

C

F

G

H

L

M

N

Q

R

S

T

U

V

W

X

Y